INSTRUCTOR'S M

J. RICHARD CHRISTMAN
U.S. Coast Guard Academy
New London, CT

to accompany

FUNDAMENTALS OF PHYSICS

FOURTH EDITION
INCLUDING EXTENDED CHAPTERS

DAVID HALLIDAY
University of Pittsburgh

ROBERT RESNICK
Rensselaer Polytechnic Institute

JEARL WALKER
Cleveland State University

with the assistance of

Stanley A. Williams
Iowa State University

Walter Eppenstein
Rensselaer Polytechnic Institute

Edward Derringh
Wentworth Institute of Technology

Van E. Nie
Purdue University

JOHN WILEY & SONS, INC.
New York Chichester Brisbane Toronto Singapore

ISBN 0-471-58877-6

Printed in the United States of America

10 9 8 7 6 5 4 3 2 1

Printed and bound by Malloy Lithographing, Inc.

PREFACE

This manual contains material we hope will be useful in the design of an introductory physics course based on the text *FUNDAMENTALS OF PHYSICS* by David Halliday, Robert Resnick, and Jearl Walker. It may be used with either the extended or regular versions. We include material to help instructors choose topics. We also provide lecture notes, outlining the important topics of each chapter, and suggest demonstration experiments, laboratory and computer exercises, films, and video cassettes. Separate sections contain answers to selected end-of-chapter questions, answers to end-of-chapter problems, and a bibliography of related articles that have appeared recently in the American Journal of Physics, The Physics Teacher, and Scientific American.

Some instructors include in their courses computational exercises to be carried out by students on computers, using either spreadsheet software or their own programs. To aid these instructors a selection of computer projects has been included.

The principal author is grateful to Stanley Williams, who co-authored the first edition of the instructor manual for *Fundamentals of Physics*. Much of his material has been retained in this manual. He is also grateful to Edward Derringh, who helped with the answers to some of the questions; to Walter Eppenstein, who helped with demonstration and laboratory experiments; and to Van Nie, who wrote many of the film and video cassette reviews.

The principal author is also grateful to the many good people at Wiley who helped in a host of ways. Special thanks go to Cliff Mills, Joan Kalkut, and Catherine Donovan. The unfailing support of Mary Ellen Christman is joyfully acknowledged.

J. Richard Christman
U.S. Coast Guard Academy
New London, Connecticut 06320

TABLE OF CONTENTS

Section One
Pathways 1

Section Two
Lecture Notes 7

Section Three
Bibliography 157

Section Four
Answers to Selected Questions 185

Section Five
Comparison of Problems with Third Edition . . 225

Section Six
Answers to Problems 251

Section Seven
Computer Projects 293

SECTION ONE
PATHWAYS

Fundamentals of Physics follows the sequence of topics found in most introductory courses. In fact, this text was instrumental in establishing the sequence. It is, however, extremely flexible in regard to both the range of topics and the depth of coverage. As a result, it can be used for a two, three, or four term course along traditional lines. It can also be used with many of the innovative courses that are presently being designed and taught. In many instances sections that discuss fundamental principles and give applications are followed by other sections that go deeper into the physics. Some instructors prefer to cover fewer topics than others but treat the topics they do cover in great depth. Others prefer to cover more topics with less depth. Courses of both types can easily be accommodated by selecting appropriate sections of the text.

By carefully choosing sections of the text to be included, your course might be a two-term, in-depth study of the fundamentals of classical mechanics and electromagnetism. With the addition of another term you might include more applications and the thermodynamics and optics chapters.

On the other hand, you might decide to forgo some depth (or perhaps the thermodynamics section) and include some modern topics. This can be accomplished in two ways. In a two-term course, you can include some or all of the modern topics that appear in special sections of the classical physics chapters. In a three-term course, you might include Chapter 42 (special relativity) and some of the quantum mechanics chapters added in the extended version. Here the ideas are presented again, but this time in the context of a more complete description of relativity or quantum theories. In a four-term course you can cover essentially the entire text in considerable depth.

When designing the course great care must be taken in the selection of topics because many discussions in later chapters presume coverage of prior material. Here are some comments you might find useful in designing your course. Also refer to the *Lecture Notes* section of this manual.

Mechanics. The central concepts of classical mechanics are covered in Chapters 1 through 12. Some minor changes that are possible, chiefly in the nature of postponements, are mentioned in the Lecture Notes. For example, the vector product can be postponed until the discussion of torque in Chapter 11. In addition, some topics are labeled optional or supplementary. The material that follows does not depend of them and, if desired, they may safely be omitted from the course in the interest of time. The modern topics sections mentioned above fall into this category and may be omitted or retained to suit the length and nature of the course.

Coverage of Chapter 5 can be shortened to 2 lectures or elongated to over 4, depending on the time spent on applications. Other sections in the first 12 chapters that can be used to adjust the length of the course are 2–9, 4–10, 6–3, 6–5, 7–7, 7–8, 8–8, 8–9, 9–7, 9–8, 10–5, 10–6, 12–10, and 12–11. Coverage of Chapter 10 (Collisions) can be shortened significantly with safety. Section 11–7, which deals with the calculation of rotational inertias of extended bodies, can be covered in detail or can be shortened by simply stating results once the definition as a sum over particles has been discussed.

The order of the chapters should be retained. For example, difficulties arise if you precede dynamics with statics as is sometimes done in other texts. To do so, you would need to discuss torque, introduced in Chapter 11, and explain its relation to angular acceleration. This involves considerable effort and is of questionable value.

Chapters 13 through 18 apply the fundamental principles of the first 12 chapters to special systems and, in many cases, lay the groundwork for what is to come. Many courses omit one or more of Chapters 13 (Equilibrium and Elasticity), 15 (Gravitation), 16 (Fluids), and 18 (Waves — II). There is some peril in these omissions, however. Chapter 15, for example, is pedagogically important. The central idea of the chapter is a force law and the discussions of many of its ramifications show by example how physics works. Since the chapter brings together many previously discussed ideas it can be used as a review. In addition, Newton's law of gravity is used later to introduce Coulomb's law and the proof that the electrostatic force is conservative relies on the analogy. The basis of Gauss' law is laid in Chapter 15 and inclusion of this chapter makes teaching of the law easier.

The idea of a velocity field is first discussed in Chapter 16 and is used to introduce electric flux in Chapter 25 (Gauss' Law). The concepts of pressure and density are explained in Chapter 16 and are used again in the thermodynamics chapters. If any of these chapters are omitted, you should be prepared to make up for the loss of material by presenting definitions and discussions of velocity field, pressure, and density when they are first used in your course.

Chapter 13 (Equilibrium of Rigid Bodies) can be safely omitted. If it is, a brief description of the equilibrium conditions might be included in the discussion of Chapter 11 or 12. The few problems in later chapters that depend on material in this chapter can be passed over. If Chapter 13 is included be sure you have covered torque and have explained its relation to angular acceleration.

Chapters 14 (Oscillations) and 17 (Waves — I) are important parts of an introductory course and should be covered except when time constraints are severe. Chapter 14 is required for Chapter 17 and both are required for Chapter 18 (Waves — II). Chapter 14 is also required for Chapters 35 (Electromagnetic Oscillations) and 36 (Alternating Currents) and the parts of Chapter 17 not labeled optional are required for Chapters 38 (Electromagnetic Waves), 40 (Interference), 41 (Diffraction), and 44 (Quantum Physics — II). Chapter 17 may be covered immediately following Chapter 14 and both may be covered just prior to their first use in the course

Sections of Chapters 12 through 18 that can be used to adjust the length of the course are 13–5, 13–6, 14–8, 14–9, 15–9, 15–10, 16–12, 17–8, 17–10, and 18–8.

Thermodynamics. Chapters 19 through 22 cover the ideas of thermodynamics. Most two term courses and some three term courses omit these chapters entirely. If they are covered, they can be placed as a unit almost anywhere after the mechanics chapters. The idea of temperature is used in Chapter 28 (Current and Resistance) and in some of the modern physics chapters, as well as in the other thermodynamics chapters. If Chapter 19 is not covered prior to Chapter 28, you should plan to discuss the idea of temperature in connection with that chapter or else omit the section that deals with the temperature dependence of the resistivity.

Electromagnetism. The fundamentals of electricity and magnetism are covered in Chapters 23 through 37. Chapter 38 (Electromagnetic Waves) may be considered a capstone to the electromagnetism chapters or as an introduction to the optics chapters. Sections that can safely be omitted to adjust the length of the course are 23–5, 23–7, 26–12, 27–6, 27–7, 27–8, 28–8, 28–9, 29–7, 29–8, 30–6, 32–7, 33–7, 34–5, 34–6, 34–7, 34–8, 34–9, 35–7, and 36–6.

Sections 27–6, 27–7, and 27–8, on dielectrics, should be included in an in-depth course but may be omitted in other courses to make room for other topics. Similarly, coverage of Chapters 29 (Circuits) and 36 (Alternating Currents) may be adjusted considerably, depending on the extent to which the course emphasizes practical applications. Some courses omit Chapters 35 and 36. However, Chapter 35 must be covered if Chapter 36 is included.

Section 34–4 contains a discussion of Gauss' law for magnetism, one of Maxwell's equations, and should be included in every course. For continuity the first three sections of the chapter should also be included. The rest of the chapter deals with magnetic properties of materials and some of ramifications of those properties. It nicely complements the previous sections on dielectrics. These parts of the chapter might be omitted or passed over swiftly to gain time for other sections. On the other hand, they should be included if you intend to emphasize properties of materials.

Optics. Chapters 39 through 41 are the optics chapters. You might wish to precede them with Chapter 38 (Electromagnetic Waves) or you might wish to replace Chapter 38 with a short qualitative discussion. You can be somewhat selective in your coverage of Chapter 39. The first nine sections are important for an introduction to optics, but sections 39–10 and 39–11 can be covered as lightly or as deeply as desired. Much of the material in this chapter can be covered as laboratory exercises.

Chapters 40 (Interference) and 41 (Diffraction) are important in their own right and are quite useful for the discussion of matter waves in Chapter 44. Chapter 41 cannot be included without Chapter 40 but coverage of both chapters can be reduced somewhat to make room for other topics. The fundamentals of interference and diffraction are contained in Sections 40–1 through 40–6, and 41–1 through 41–4. Other sections of these chapters can be included or excluded, as desired. Section 41–8 is a nice prelude to the discussion of atomic spectra in Section 43–8 but it not required. If you do include it be sure to cover Section 41–7 first.

Modern Physics. As mentioned previously, modern physics topics are scattered throughout the text. They are 2–9 (The Particles of Physics), 4–10 (Relative Motion at High Speeds), 6–5 (The Forces of Nature), 7–7 (Kinetic Energy at High Speeds), 8–8 (Mass and Energy), 8–9 (Energy is Quantized), 10–6 (Reactions and Decay Processes), 12–11 (Angular Momentum is Quantized), 15-10 (A Closer Look at Gravitation), 18–8 (The Doppler Effect for Light), 21–10 (A Hint of Quantum Theory), 23–5 (Charge is Quantized), 30–6 (Cyclotrons and Synchrotrons), 32–7 (The Betatron), 34–9 (Nuclear Magnetism), 40–8 (Michelson's Interferometer), and 41–9 (X-ray diffraction). These sections compare modern ideas to corresponding classical ideas and show students similarities and differences.

Although most of these topics have been categorized as supplementary in the Lecture Notes, they constitute important extensions of the usual course in classical physics. Instructors can include many important modern ideas simply by covering these sections. By studying modern topics when they occur in the text, students are not only introduced to them but also can begin to see their relationship to classical ideas.

In addition, Chapters 42 through 49, all but 42 included in the extended version of the text only, provide a more formal development of some of the important ideas and applications of modern physics. All of these chapters may omitted, of course. Many instructors, however, use some or all of this material to round out their courses.

Strictly speaking, Chapter 42 (The Special Theory of Relativity) is not required for the chapters that follow. The results of relativity theory that are needed, chiefly the definitions of energy and momentum, the mass-energy relationship, and velocity transformations, appear as optional topics in earlier chapters. If you do not desire more than minimum coverage, Chapter 42 can be omitted. However, relativity is treated as a coherent theory in this chapter and flesh is added to the bare bones discussions of earlier chapters. The chapter may be used as a capstone to the mechanics section of the course, a capstone to the entire course, or as an introduction to the modern physics included in the extended version of the text.

The fundamentals of the quantum theory are presented in Chapters 43 (Quantum Physics — I) and 44 (Quantum Physics — II). This material should be treated as a unit and must follow in the order written. If you include these chapters be sure earlier parts of the course include discussions of uniform circular motion, angular momentum, Coulomb's law, electrostatic potential energy, electromagnetic waves, and diffraction. $E = mc^2$ and $E^2 = (pc)^2 + (mc^2)^2$, from relativity theory, are used in discussions of the Compton effect.

The introductory chapters are followed by application chapters: Chapters 45 (All About Atoms), 46 (The Conduction of Electricity in Solids), 47 (Nuclear Physics), 48 (Energy from the Nucleus), and 49 (Quarks, Leptons, and the Big bang). You may choose to end the course with Chapter 44 or you may choose to include one or more of the application chapters.

The ideas of temperature and the Kelvin scale are used in several places in the modern physics chapters: Chapter 43 (cavity radiation), Chapter 45 (lasers), Chapter 46 (temperature dependence of resistivity), Chapter 48 (thermonuclear fusion), and Chapter 49 (cosmic background radiation). With a little supplementary material, these sections can be covered even if Chapter 19 is not.

Chapter 48 (Energy from the Nucleus) requires Chapter 47 (Nuclear Physics) for background material, but Chapter 47 need not be followed by Chapter 48. $E = mc^2$ and $E^2 = (pc)^2 + (mc^2)^2$ from relativity theory are also used. The discussion of thermonuclear fusion uses some of the ideas of kinetic theory, chiefly the distribution of molecular speeds. Either Chapter 21 (particularly Section 21–7) should be covered first or you should be prepared to supply a little supplementary material here.

Chapter 49 includes an introduction to high energy particle physics and tells how the ideas of physics are applied to cosmology. Both these topics fascinate many students. In addition, the chapter provides a nice overview of physics. Some knowledge of the Pauli exclusion principle and spin angular momentum (both discussed in Chapter 45) is required. Knowledge of the strong and weak nuclear forces (discussed in Chapter 47) is also required. In addition, beta decay (discussed in Chapter 47) is used several times as an illustrative example. Nevertheless, the chapter can be made to stand alone with the addition of only a small amount of supplementary material.

SUGGESTED COURSES

A bare bones two semester course (about 90 meetings) can be constructed around Chapters 1 through 12, 14, 17, 23 through 33, the first four sections of Chapter 34, and Chapters 37 and 38. The course can be adjusted to the proper length by the inclusion or omission of supplementary material and optional topics. Some time might be available for a few of the modern physics topics. If 4 to 8 additional meetings are available each term, Chapter 15 or 16 (or perhaps both) can be inserted after Chapter 12 and one or more of the optics chapters can be inserted after Chapter 38. As an alternative, you might consider including sections on dielectrics, magnetic properties, semiconductors, and superconductors to emphasize properties of materials. The inclusion of this material must be balanced with the desire to include more of the supplementary modern topics.

A three term course (about 135 meetings) can be constructed by adding the thermodynamics chapters (19 through 22) and some or all of the modern physics chapters (42 through 49) to those mentioned above. If the needs of the class dictate a section on alternating current, two modern physics chapters can be replaced by Chapters 35 and 36.

ESTIMATES OF TIME

The following chart gives estimates of the time required to cover all of each chapter, in units of 50 minute periods. The second and fifth columns of the chart contain estimates of the number of lecture periods needed and includes the time needed to perform demonstrations and discuss the main points of the chapter. The third and sixth columns contain estimates of the number of recitation periods required and includes the time needed to go over problem solutions, answers to end-of-chapter questions, and points raised by students. If your course is organized differently you may wish to add the two numbers to obtain the total estimated time for each chapter.

Use the chart as a rough guide when planning the syllabus for a semester, quarter, or year course. If you omit parts of chapters, reduce the estimated time.

Text Chapter	Number of Lectures	Number of Recitations	Text Chapter	Number of Lectures	Number of Recitations
1	0.3	0.2	26	1.8	1.8
2	2.0	2.0	27	1.5	2.0
3	1.0	1.0	28	1.0	1.0
4	2.0	2.5	29	2.0	2.2
5	2.0	2.0	30	2.0	1.8
6	2.0	2.0	31	2.0	1.2
7	1.8	1.5	32	1.8	1.0
8	2.0	2.0	33	2.0	2.0
9	2.0	1.5	34	0.8	0.3
10	2.0	1.6	35	1.0	1.0
11	2.0	1.5	36	2.0	2.0
12	2.0	2.0	37	1.0	0.5
13	1.0	2.0	38	2.5	2.0
14	2.5	1.8	39	3.0	3.0
15	2.3	2.0	40	2.0	2.0
16	2.0	2.0	41	2.0	2.0
17	2.5	2.0	42	2.5	2.0
18	2.5	2.0	43	2.0	2.0
19	0.7	1.0	44	1.5	1.5
20	2.0	2.0	45	2.2	2.0
21	2.0	1.4	46	2.0	2.0
22	1.5	1.6	47	1.8	2.0
23	1.0	1.0	48	2.0	2.0
24	1.6	1.3	49	2.0	2.0
25	1.8	1.8			

SECTION TWO
LECTURE NOTES

Lecture notes for each chapter of the text are grouped under the headings BASIC TOPICS, SUPPLEMENTARY TOPICS, and SUGGESTIONS.

BASIC TOPICS contains the main points of the chapter in outline form. In addition, one or two demonstrations are recommended to show the main theme of the chapter. You may wish to pattern your lectures after the notes, suitably modified, or simply use them as a check on the completeness of your own notes.

SUPPLEMENTARY TOPICS consists of a list of topics which, although covered in the text, may be safely omitted in order to save time. Many of these are pertinent to laboratory experiments and might be assigned in conjunction with the laboratory or as outside reading. Some deal with modern topics and you may wish to include them to broaden the course.

The SUGGESTIONS sections recommend end-of-chapter questions and problems, films, film loops, video cassettes, computer projects, alternate demonstrations, and other material that might be useful for the course. Section Three of this manual contains a bibliography of references to additional pertinent material. Many of the questions concentrate on points that seem to give students trouble and it is worthwhile dealing with some of them before students tackle a problem assignment. Some questions and problems might be incorporated into the lectures while some might be assigned and used to generate discussion by students in small recitation sections. Answers to some questions appear in Section Four of this manual.

Films and Video cassettes. All of the films, film loops, and video cassettes are short, well done, and highly pertinent to the chapter. It is not possible to review all available material and there are undoubtedly many other fine films and video cassettes that are not listed. Films, film loops, and video cassettes might be incorporated into the lectures, shown during laboratory periods, or set up in a special room for more informal viewing.

An excellent set of video cassettes and disks, *THE MECHANICAL UNIVERSE*, can be obtained from The Mechanical Universe, Caltech, MS 1–70, Pasadena, CA 91125. The set consists of 52 half-hour segments dealing with nearly all important concepts of introductory physics. Historical information and animated graphics are used to present the concepts in an imaginative and engaging fashion. Some physics departments run appropriate segments throughout the course in special viewing rooms. Accompanying textbooks, teacher manuals and study guides are also available.

Many time-tested films originally from Encyclopaedia Britannica, PSSC, Project Physics, and elsewhere have been transferred to video disk and are available under the title *Physics: Cinema Classics* from the American Association of Physics Teachers Instructional Materials Center, Department of Physics and Astronomy, University of Nebraska-Lincoln, Lincoln, NE 68588–0128. The films cover a host of topics in mechanics, thermodynamics, electricity and magnetism, optics, and modern physics. The Center can provide you with tables of contents for the disks.

Computers. Computers have made significant contributions to the teaching of physics. They are widely used in lectures to provide animated illustrations, with parameters under the control of the lecturer; they also provide tutorials and drills that students can work through on their own. The Physics Courseware Laboratory at the University of North Carolina (Raleigh NC 27695–8202) maintains an up-to-date catalog of both commercial and public domain software. It was last published in Computers in Physics **5**, 71 (January/February 1991). Personnel of the laboratory write reviews, which appear in The Physics Teacher and are referenced in the SUGGESTIONS sections of the lecture notes.

Specialized programs are listed in appropriate SUGGESTION sections of the Lecture Notes in this manual. In addition, several available software packages cover large portions of an introductory course. Five of them are:

The *Physics I* and *Physics II* Series (Control Data Company, 3601 West 77th Street, Bloomington, MN 55435). Excellent problem solving tutorials for the IBM PC. Sixteen modules cover important topics in mechanics and twelve cover important topics in electrostatics, magnetostatics, and Faraday's law.

Physics I (Microphysics Programs, 1737 West 2nd Street, Brooklyn, NY 11223). A set of programs that generate problems covering particle dynamics and some aspects of thermodynamics. Different versions are available for IBM PC, Apple II, TRS–80 (models III and IV), Commodore Pet and 64.

Physics Simulations (Kinko's Service Corporation, 4141 State Street, Santa Barbara, CA 93110). A great many demonstrations of important topics in introductory physics. Individual programs are listed under appropriate chapters in the Lecture Notes. These programs are for the Apple Macintosh.

Sensei Physics (Broderbund Software, P.O. Box 12947, San Rafael, CA 94913–2947). Tutorial reviews with animated graphics of most of the major topics of introductory physics. Over 300 problems for student practice. These programs are for Macintosh computers.

Interactive Physics (Knowledge Revolution, 497 Vermont Street, San Francisco, CA 94107). Animations and graphs for a wide variety of mechanical phenomena. User can set up "experiments" with massive objects, strings, springs, dampers, and constant forces. Parameters can easily be changed. For Macintosh computers. Reviewed in The Physics Teacher, September 1991.

You might consider setting aside a room or portion of a lab, equip it with several computers, and make tutorial, drill, and simulation programs available to students. If you have sufficient hardware (and software) you might base some assignments on computer materials.

Computers might also be used by students to perform calculations. Properly selected problems can add greatly to the students' understanding of physics. Problems involving the investigation of some physical system of interest might be assigned as individual projects or might be carried out by a laboratory class.

Commercial spreadsheet programs, of the type used by business, can facilitate problem solving. For a detailed account of how they are used and a collection of informative problems, see *Wondering About Physics ... Using Spreadsheets to Find Out* by D.I. Dykstra and R.G. Fuller (John Wiley & Sons, 1988) and *Dynamic Models in Physics* by F. Potter and C.W. Peck (N. Simpson & Company, 1989). Commercial problem-solving programs such as *Eureka: The Solver* (Borland International, 4585 Scotts Valley Drive, Scotts Valley, CA 95066; IBM PC), *TK Solver!* (Universal Technical Systems Inc., 1220 Rock Street, Rockford, IL 61101; IBM PC, Macintosh), *MathCAD* (MathSoft, Inc., One Kendall Square, Cambridge, MA 02139; IBM PC), and *DERIVE* (Soft Warehouse; 3615 Harding Avenue, Suite 505, Honolulu, HI 96816; IBM PC) can easily be used by students to solve problems and graph results. All these programs allow students to set up a problem generically, then view solutions for various values of input parameters. For example, the range or maximum height of a projectile can be found as a function of initial speed or firing angle, even if air resistance is taken into account. The study guide, *A Student's Companion to Fundamentals of Physics*, that accompanies the text, contains some computer projects. Others are suggested in the Lecture Notes of this manual. A large number of suitable problems and projects can also be found in the book *Introduction to Computational Physics* by Marvin L. De Jong (Addison-Wesley, 1991) and in the two volume calculator supplement *Physics Problems for Programmable Calculators* by J. Richard Christman (John Wiley & Sons, 1981 and 1982, now out of print).

Demonstrations. Notes for most of the chapters are developed around demonstration experiments. Generally speaking, these use relatively inexpensive, readily available equipment, yet clearly demonstrate the main ideas of the chapter. The choice of demonstrations, however, is highly personal and you may wish to substitute others for those suggested here or you may wish to present the same ideas using chalkboard diagrams. Several excellent books give many other examples of demonstration experiments. The following are available from the American Association of Physics Teachers, 5110 Roanoke Place, College Park, MD 20740–4100:

> *Resource Letter PhD-1: Physics Demonstrations*, J.A. Davis and B.G. Eaton, 6 pages (1979). Contains 103 references to books, monographs, indexes, and conference proceedings dealing with physics demonstrations.
>
> *A Demonstration Handbook for Physics*, G.D. Freier and F.J. Anderson, 320 pages (1981). Contains 807 demonstrations, including many which use everyday materials and which can be constructed with minimal expense. Line drawings illustrate every demonstration.
>
> *Physics Demonstration Experiments at William Jewell College*, Wallace A, Hilton, 112 pages (1982). Contains descriptions and photographs of over 300 demonstrations.

The following is available from Robert E. Krieger Publishing Company, Malabar, FL 32950:

> *Physics Demonstration Experiments*, H.F. Meiners, ed. An excellent source of ideas, information, and construction details on a large number of experiments, with over 2000 line drawings and photographs. It also contains some excellent articles on the philosophical aspects of lecture demonstrations, the use of shadow projectors, TV, films, overhead projectors, and stroboscopes.

Appropriate demonstrations described in Freier and Anderson and in Hilton are listed in the SUGGESTIONS sections of the notes. Neither of these books give any construction details, but more information about most demonstrations can be obtained from the book edited by Meiners.

Consider using a computer for data acquisition during demonstrations. Photogate timers, temperature probes, strain gauges, voltage probes, and other devices can be input directly into the computer and results can be displayed as tables or graphs. The screen can be shown to a large class by using a large monitor, a TV projection system, or an overhead projector adapter. Inexpensive software and hardware for the Apple II, Macintosh, and IBM PC can be purchased from Vernier Software, 2920 S.W. 89th Street, Portland, OR 97225. IBM sells software and hardware for its machines. For more sophisticated software consider the commercial packages Assist and Labview.

Laboratories. Hands-on experience with actual equipment is an extremely important element of an introductory physics course. There are many different views as to the objectives of the physics laboratory and the final decision on the types of experiments to be used has to be made by the individual instructor or department. This decision is usually based on financial and personnel considerations as well as on the pedagogical objectives of the laboratory.

Existing laboratories vary widely. Some use strictly cookbook type experiments while others allow the students to experiment freely, with practically no instructions. The equipment ranges from very simple apparatus to rather complex and sophisticated equipment. Physical phenomena may be observed directly or simulated on a computer. Data may be taken by the students or fed into a microprocessor.

Many physics departments have written their own notes or laboratory manuals and relatively few physics laboratory texts are on the market. Two such books, both available from John Wiley & Sons, are

> *Laboratory Physics*, H.F. Meiners, W. Eppenstein, R.A. Oliva, and T. Shannon, 2nd ed. (1987).

Laboratory Experiments in College Physics, C.H. Bernard and C.D. Epp, 6th ed. (1980).

Experiments from these books are listed in the SUGGESTIONS section of the Lecture Notes. MEOS is used to designate the Meiners, Eppenstein, Oliva, Shannon book while BE is used to designate the Bernard and Epp book. Both books contain excellent introductory sections explaining laboratory procedures to students. MEOS also contains a large amount of material on the use of microprocessors in the lab.

Consider having the students use a computer to graph data. Inexpensive software for the Apple II, Macintosh, and IBM PC, called *Graphical Analysis*, can be obtained from Vernier Software (address given above). Data is input from the keyboard and displayed on the screen or printed. Axes can be scaled. Powers or logarithms of values also be plotted automatically. Most spreadsheets can also be used for plotting and scaling.

Student supplements. Several supplements might be recommended to the students. A study guide entitled *A Student's Companion to Fundamentals of Physics*, written to accompany the text, is published by Wiley. The basic concepts of each chapter are reviewed in a workbook format that helps students focus their attention on the important ideas and their relationships to each other. About a dozen problems are solved or partially solved for each chapter.

To those students who might need help with fundamentals such as unit conversion, interpretation of graphs, algebra, trigonometry, geometry, and vector manipulation, you might suggest they purchase a copy of *Prelude to Physics* by C. Swartz (John Wiley & Sons). This book does a nice job of teaching some important techniques and showing their relevancy to the course.

Chapter 1 MEASUREMENT

BASIC TOPICS

I. Base and derived units.

A. Explain that standards are associated with base units and that measurement of a physical quantity takes place by means of comparison with a standard. Discuss qualitatively the SI standards for time, length, and mass. Show a 1 kg mass and a meter stick. Show the simple well-known procedure for measuring length with a meter stick.

B. Explain that derived units are combinations of base units. Emphasize that the speed of light is now a defined unit and the meter is a derived unit. Discuss an experiment in which the time taken for light to travel a certain distance is measured. Example: the reflection of a light signal from the moon. Use a clock and a meter stick to find your walking speed in m/s.

C. This is a good place to review area, volume, and mass density. Use simple geometric figures (circle, rectangle, triangle, cube, sphere, cylinder, etc.) as examples.

II. Systems of units.

A. Explain what a system of units is. Give the 1971 SI base units (Table 1–1). Stress that they will be used extensively.

B. Point out the SI prefixes (Table 1–2). The important ones for this course are mega, kilo, centi, milli, micro, nano, and pico. Stress the simplicity of the powers of ten notation.

C. Most of the students' experience is with the British system. Relate the inch to the centimeter and the slug to the kilogram. Discuss unit conversion. Use speed as an example: convert 50 mph and 3 mph to km/h and m/s. Point out the conversion tables in Appendix F.

III. Properties of standards.

A. Use questions 1, 2, 6, 7, and 18 to discuss choices of standards.

B. Discuss secondary standards such as the meter stick used earlier.

IV. Measurements.

A. Stress the wide range of magnitudes measured. See Tables 1–3, 1–4, and 1–5. Explain the unified mass unit. One atom of ^{12}C has a mass of exactly 12 u. 1 u is approximately 1.661×10^{-27} kg. Assign problem 34 or 35.

B. Discuss indirect measurements. See questions 10 and 11.

V. Skills.

A. Discuss unit arithmetic, unit conversion, powers of ten arithmetic, and significant figures. Assign some of problems 5, 13, 16, 21, and 33.

B. Discuss area and volume calculations. According to the needs of the class assign some of the following problems: 6, 11, 12, 15, 18, and 39.

SUGGESTIONS

1. Demonstrations
Examples of "standards" and measuring instruments: Freier and Anderson Ma1 — 3; Hilton M1.

2. Audio/Visual
Powers of Ten, 16 mm or 3/4" video cassette, b/w, 25 min. Pyramid Films, P.O. Box 1048, Santa Monica, CA 90406. This very popular film is an excellent way to introduce the students to the relative sizes of things in the universe.

3. Laboratory
   a. MEOS Experiment 7-1: *Measurement of Length, Area, and Volume.* Gives students experience using the vernier caliper, micrometer, and polar planimeter. Good introduction to the determination of error limits (random and least count) and calculation of errors in derived quantities (volume and area).
   b. BE Experiments 1 and 2: *Determination of Length, Mass, and Density* and *Determination of π and Density by Measurements and Graphical Analysis.* Roughly the same as the MEOS experiment, but a laboratory balance is added to the group of instruments and the polar planimeter is not included. Graphs of mass vs. radius and radius squared for a collection of disks made of the same material, with the same thickness, are used to establish the quadratic dependence of mass on radius.
   c. MEOS Experiment 7-3: *The Simple Pendulum* and BE Experiment 3: *The Period of a Pendulum — An Application of the Experimental Method.* Students time simple pendula of different lengths, then use the data and graphs (including a logarithmic plot) to determine the relationship between length and period. They calculate the acceleration due to gravity. This is an exercise in finding functional relationships and does not require knowledge of dynamics.

Chapter 2 MOTION ALONG A STRAIGHT LINE

BASIC TOPICS

I. Position and displacement.
   A. Move a toy cart with constant velocity along a table top. Select an origin, place a meter stick and clock on the table, and demonstrate how x(t) is measured in principle. Emphasize that x is always measured from the origin; it is not the cart's displacement during any time interval.
   B. Draw a graph of $x(t)$ and point out it is a straight line. Show what the graph looks like if the cart is not moving. Point out that the line has a greater slope if the cart is going faster. Move the cart so its speed increases with time and show what the curve $x(t)$ looks like. Do with same for a cart that is slowing down.
   C. Some students think of a coordinate as distance. Distinguish between these concepts. Point out that a coordinate defines a position on an axis and can be positive or negative. Demonstrate a negative velocity, both with the cart and on a graph. As another example, throw a ball into the air, pick a coordinate axis (positive in the upward direction, say), and point out when the velocity is positive and when it is negative. Draw the graph of the coordinate as a function of time.
   D. Define the displacement of an object during a time interval. Emphasize that only the initial and final coordinates enter and that an object may have many different positions between these while still having the same displacement. Point out that the displacement is 0 if the initial and final coordinates are the same.

II. Velocity.
   A. Define average velocity over an interval. Stress the meaning of the sign. Go over Sample Problems 2–1 and 2–2. Draw a graph of x vs. t for an object that is accelerating. Pick an interval and draw the line between the end points on the graph. Observe that the average velocity in the interval is the slope of the line. Fig. 2–3 or 2–4 may also be used. Show how to calculate average velocity if the function $x(t)$ is given in algebraic form.
   B. Define instantaneous velocity. To demonstrate the limiting process, go over Table 2–1. Use a graph of x vs. t for an accelerating cart (or Fig. 2–7) to demonstrate that the line

used to find the average velocity becomes tangent to the curve in the limit as Δt vanishes. Remark that the slope of the tangent line gives the instantaneous velocity. Show a plot of v vs. t that corresponds to the x vs. t graph used previously. Show how to calculate the instantaneous velocity if the function $x(t)$ is given in algebraic form. Stress that a value of the instantaneous velocity is associated with each instant of time. Some students think of velocity as being associated with a time interval rather than an instant of time.

C. Define instantaneous speed as the magnitude of the velocity. Compare to the average speed in an interval, which is the total path length divided by the time. Remark that the average speed is not necessarily the same as the magnitude of the average velocity.

D. Note that many calculus texts use a prime to denote a derivative. They also define the derivative of x with respect to time by the limit of $[x(t+\Delta t) - x(t)]/\Delta t$ rather than by the limit of $\Delta x/\Delta t$. Mention the different notations in class so students can relate their physics and calculus texts.

III. Acceleration.

A. Define average and instantaneous acceleration. Show the previous v vs. t graph and point out the line used to find the average acceleration in an interval and the instantaneous acceleration at a given time. Show how to calculate the average and instantaneous acceleration if $x(t)$ or $v(t)$ are given in algebraic form.

B. Interpret the sign of the acceleration. Give examples of objects with acceleration in the same direction as the velocity (speeding up) and in the opposite direction (slowing down). Be sure to include both directions of velocity. Emphasize that a positive acceleration does not necessarily mean speeding up and a negative acceleration does not necessarily imply slowing down.

C. Use graphs of $x(t)$ and $v(t)$ to point out that an object may simultaneously have zero velocity and non-zero acceleration. Explain that if the direction of motion reverses the object must have zero velocity at some instant. Give the position as a function of time as $x(t) = At^2$, for example, and show that the velocity is 0 at $t = 0$ but the acceleration is not 0. Illustrate the function with a graph.

IV. Motion in one dimension with constant acceleration.

A. Derive the kinematic equations for $x(t)$ and $v(t)$. If students know about integration, use methods of the integral calculus. In any event, show that $v(t)$ is the derivative of $x(t)$ and that a is the derivative of $v(t)$.

B. Discuss kinematics problems in terms of a set of simultaneous equations to be solved. Examples: use equations for $x(t)$ and $v(t)$ to algebraically eliminate the time and to algebraically eliminate the acceleration. The equations of constant acceleration motion are listed in Table 2–2. Some instructors teach students to use the table, as is done in Sample Problem 2–8. Others ask students to always start with Eqs. 2–9 and 2–13, then use algebra to obtain the equations needed for a particular problem.

C. To help students see the influence of the initial conditions, sketch graphs of $v(t)$ and $x(t)$ for various initial conditions but the same acceleration. Include both positive and negative initial velocities. Draw a different set of graphs for positive and negative acceleration. Point out where the particle has zero velocity and when it returns to its initial position.

V. Free fall.

A. Give the values for g in SI and British units. Point out that the acceleration due to gravity is directed toward the center of the Earth but that locally the Earth is essentially flat and the acceleration may be taken to be in the same direction at slightly different points. Explain that $a = +g$ if down is taken to be the positive direction and $a = -g$ if up is the positive direction. Do examples using both choices. Throw a ball into the air and emphasize that its acceleration is g throughout its motion, even at the top of its trajectory.

B. Drop a small ball through two photogates, one at the top to turn on a timer and one further down to turn it off. Repeat for various distances and plot the position of the ball as a function of time. Explain that the curve is parabolic and indicates a constant acceleration.

C. Explain that all objects have the same acceleration due to gravity. In reality, different objects may have different accelerations because air influences their motions differently. This can be demonstrated by placing a coin and a wad of cotton in a glass cylinder about 1 m long. Turn the cylinder over and note that the coin reaches the bottom first. Now use a vacuum pump to partially evacuate the cylinder and repeat the experiment. Repeat again with as much air as possible pumped out.

D. Point out that free fall problems are special cases of constant acceleration kinematics and the methods described earlier can be used. Work a few examples. For an object thrown into the air calculate the time to reach the highest point, the height of the highest point, the time to return to the initial height, and its velocity when it returns, all in terms of the initial velocity.

SUGGESTIONS

1. To help students obtain some qualitative understanding of velocity and acceleration, ask them to discuss questions 7, 8, 9, 10, and 11.
2. To make more use of the calculus assign problems 13, 27, 28, 31, 32, and 33. Problems 13, 28, and 31 can also be used to discuss differences between average and instantaneous velocity and between average and instantaneous acceleration.
3. To emphasize the interpretation of graphs assign problems 17, 18, 19, 23, 24, 25, and 26. Many other problems require students to draw graphs after performing a calculation. Assign some.
4. Computer projects
   a. Use a spreadsheet or your own computer program to demonstrate the limiting processes used to define velocity and acceleration. Given the functional form of $x(t)$, have the computer calculate and display the coordinate for some time t_0 and a succession of later times, closer and closer to t_0. For each interval have it calculate and display the average velocity. Be careful to refrain from displaying non-significant figures and be sure to stop the process before all significance is lost.
   b. Have students use the root finding capability of a commercial math program or their own computer programs to solve kinematic problems for which $x(t)$ and $v(t)$ are given functions. Nearly all of them can be set up as problems that involve finding the root of either the coordinate or velocity as a function of time, followed perhaps by substitution of the root into another kinematic equation. Problems need not be limited to those involving constant acceleration. Air resistance, for example, can be taken into account. The same program can be used to solve rotational kinematic problems in Chapter 11.
5. *The Microcomputer Based Lab Project: Motion.* HRM Software, 175 Tompkins Avenue, Pleasantville, NY 10570–9973. Useful for helping students interpret kinematic graphs. As a student moves toward and away from a sonar ranging device, his position, velocity, or acceleration is plotted on a computer monitor. A graph can be designed by the instructor and the student asked to duplicate it by moving. Several sonic rangers are reviewed in TPT January 1988.
6. Demonstrations
   Uniform velocity and acceleration, velocity as a limiting process: Freier and Anderson Mb10 — 13, 15, 18, 21, 22; Hilton M2, M3, M4, M5.
7. Audio/Visual
   a. *Straight-Line Kinematics*, 16 mm, b/w, 34 min. Modern Learning Aids, Division of Ward's Natural Science Establishment, P.O. Box 312, Rochester, NY 14601. Although this film is

over 20 years old, the generation of distance, speed, and acceleration graphs using special test car equipment is an excellent teaching device.

b. *Velocity from Position*; *Position from Velocity I*; *Position from Velocity II*, S8, b/w, 3 min each. American Association of Physics Teachers, 5110 Roanoke Place, College Park, MD 20740–4100. Average and instantaneous velocities are developed for the motion of an automobile. The latter two films discuss the often neglected concept of area under the curve.

c. *Acceleration Due to Lunar Gravity*, S8, color, 4 min. American Association of Physics Teachers (address above). The "Guinea and Feather" experiment is demonstrated on the moon by astronauts, who drop a hammer and a feather in the vacuum of the moon.

d. *Kinematics Graphs*; Bruce Marsh; slide set; American Association of Physics Teachers (address above). Coordinate vs. time and velocity vs. time graphs for one dimensional motion. Use to illustrate lectures.

8. Computer programs

   a. *Motion*, Cross Educational Software, P.O. Box 1536, Ruston, LA 71270. Apple II+. Generates graphs of coordinate, velocity, and acceleration for one dimensional motion. Presents problems in one dimensional dynamics (translational and rotational). Students can be assigned tutorial sections. Some sections can be used to illustrate lectures. Reviewed TPT September 1983.

   b. *Kinematics*, Vernier Software, 2920 S.W. 89th Street, Portland, OR 97225. Apple II. The student is given the parameters of a one-dimensional motion (initial velocity, acceleration, final coordinate, final velocity, time, etc.). The student tries to control the motion of a truck across the screen so its motion corresponds to the given parameters. The computer analyzes the results and provides feedback.

   c. *A Tutorial on Motion*, COMPress, P.O. Box 102, Wentworth, NH 03282. Apple II. Helps with understanding position vs. time and velocity vs. time graphs. Highly interactive; tests students understanding. Reviewed TPT October 1987.

9. Laboratory

   a. MEOS Experiment 7–5: *Analysis of Rectilinear Motion.* Students measure the position as a function of time for various objects rolling down an incline, then use the data to plot speeds and accelerations as functions of time. No knowledge of rotational motion is required. This experiment emphasizes the definitions of velocity and acceleration as differences over a time interval.

   b. MEOS Experiment 8–1: *Motion in One Dimension* (omit the part dealing with conservation of energy). Essentially the same experiment except pucks sliding on a nearly frictionless surface are used. This experiment may be done with dry ice pucks or on an air table or air track.

   c. BE Experiment 7: *Uniformly Accelerated Motion.* The same technique as the MEOS experiments but a variety of setups are described: the standard free fall apparatus, the free fall apparatus with an Atwood attachment, an inclined plane, an inclined air track, and a horizontal air track with a pulley attachment.

Chapter 3 VECTORS

BASIC TOPICS

I. Definition.

   A. Explain that vectors have magnitude and direction, and that they obey certain rules of addition.

B. Example of a vector: displacement. Give the definition of displacement and point out that a displacement does not describe the path of the object. Give the definition and physical interpretation of the sum of two displacements. Demonstrate vector addition by walking along two sides of the room. Point out the two displacements and their sum. Note that the distance traveled is not the magnitude of the displacement. Go back to your original position and point out that the displacement is now zero. Discuss question 1.

C. Compare vectors with scalars and present a list of each.

D. Go over vector notation and insist that students use it to identify vectors clearly.

II. Vector addition and subtraction by the graphical method.

A. Draw two vectors tail to head, draw the resultant, and point out its direction. Explain how the magnitude of the resultant can be measured with a ruler and the orientation can be measured with a protractor. Explain how a scale is used to draw the original vectors and find the magnitude of the resultant.

B. Define the negative of a vector and define vector subtraction as $\mathbf{a} - \mathbf{b} = \mathbf{a} + (-\mathbf{b})$. Graphically show that if $\mathbf{a} + \mathbf{b} = \mathbf{c}$ then $\mathbf{a} = \mathbf{c} - \mathbf{b}$.

C. Show that vector addition is both commutative and associative.

III. Vector addition and subtraction by the analytic method.

A. Derive expressions for the components of a vector, given its magnitude and the angles it makes with the coordinate axes. In preparation for the analysis of forces, find the x component of a vector in the xy plane in terms of the angles it makes with the positive and negative x axis and also in terms of the angles it makes with the positive and negative y axis. Overlays are useful to show vector components.

B. Point out that the components depend on the choice of coordinate system and compare the behavior of vector components with the behavior of a scalar when the orientation of the coordinate system is changed. Find the components of a vector using two differently oriented coordinate systems. Point out that it is possible to orient the coordinate system so that only one component of a given vector is not zero. Remark that a pure translation of a vector (or coordinate system) does not change the components.

C. Define the unit vectors along the coordinate axes. Give the form used to write a vector in terms of its components and the unit vectors. Explain that unit vectors are unitless so they can be used to write any vector quantity. See question 8.

D. Vector addition. Give the expressions for the components of the resultant in terms of the components of the addends. Demonstrate the equivalence of the graphical and analytic methods of finding a vector sum. See the diagram to the right.

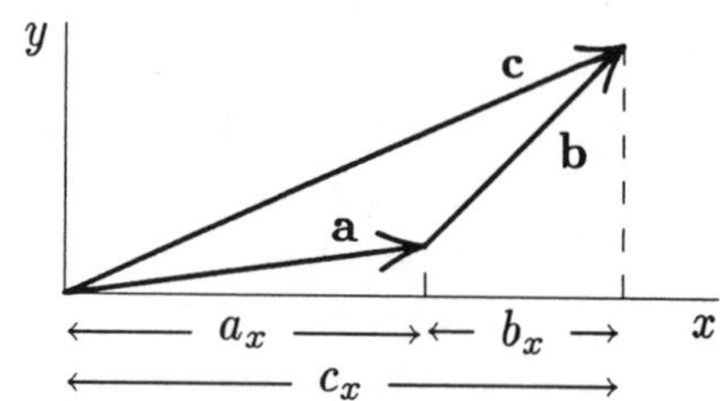

E. Show how to find the magnitude and angles with the coordinate axes, given the components. Explain that calculators give only one of the two possible values for the inverse tangent and show how to determine the correct angle for a given situation.

F. State that two vectors are equal only if their corresponding components are equal. State that many physical laws are written in terms of vectors and that many take the form of an equality between tn two vectors.

IV. Multiplication involving vectors.

A. Multiplication by a scalar. Give examples of both positive and negative scalars multiplying a vector. Give the components of the resulting vector as well as its magnitude and direction. Remark that division of a vector by a scalar is equivalent to multiplication by the reciprocal of the scalar.

B. (May be postponed until Chapter 7.) Scalar product of two vectors. Emphasize that the product is a scalar. Give the expression for the product in terms of the magnitudes of the vectors and the angle between them. To determine the angle, the vectors must be drawn with their tails at the same point. Point out that $\mathbf{a} \cdot \mathbf{b}$ is the magnitude of **a** multiplied by the component of **b** along an axis in the direction of **a**, and vice versa. Explain that $\mathbf{a} \cdot \mathbf{b} = 0$ if **a** is perpendicular to **b**.

C. Either derive the expression for a scalar product in terms of cartesian components or else assign problem 47. Specialize the expression to show that $\mathbf{a} \cdot \mathbf{a} = a^2$ or else assign problem 41. Show how to use the scalar product to calculate the angle between two vectors if their components are known. Consider problem 52.

D. Vector product of two vectors (may be postponed until Chapter 12). Emphasize that the product is a vector. Give the expression for the magnitude of the product and the right hand rule for determining the direction. Explain that $\mathbf{a} \times \mathbf{b} = 0$ if **a** and **b** are parallel. Point out that $|\mathbf{a} \times \mathbf{b}|$ is the magnitude of **a** multiplied by the component of **b** along an axis perpendicular to **a** and in the plane of **a** and **b**, and vice versa. Show that $\mathbf{b} \times \mathbf{a} = -\mathbf{a} \times \mathbf{b}$.

E. Either derive the expression for a vector product in terms of cartesian components or else assign problem 50. First discuss or assign problem 40 and give students the useful mnemonic for the vector products of the unit vectors **i**, **j**, and **k**, written in that order clockwise around a circle. One starts with the first named vector in the vector product and goes around the circle toward the second named vector. If the direction of travel is clockwise the result is the third vector. If it is counterclockwise the result is the negative of the third vector.

SUGGESTIONS

1. Ask students to use graphical representations of vectors to think about questions 2 through 6. Assign problems such as 2, 5, and 7.

2. Problems 10 and 11 cover the fundamentals of vector components. Problems 12, 16, 17, and 19 stress the physical meaning of vector components. Some good problems to test understanding of analytic vector addition are 21, 22, 24, and 25.

3. Multiplication of a vector by a scalar is covered in problem 37. If you cover scalar and vector products assign one or two application problems, such as 48, 52, 56, 58, and 59.

4. Computer project
   Have students use a commercial math program or write their own computer programs to carry out conversions between polar and cartesian forms of vectors, vector addition, scalar and vector products.

5. Demonstrations
   Vector addition: Freier and Anderson Mb2, 3; Hilton M10a, b, c.

6. Audio/visual
   Vector Addition; Bruce Marsh; American Association of Physics Teachers slide set; Publications Sales, AAPT Executive Office, 5112 Berwyn Road, College Park, MD 20740–4100. Graphical and component methods are illustrated.

7. Computer programs
   a. *Vector Addition II*, Vernier Software, 2920 89th Street, Portland, OR 97225. Apple II. User supplies magnitude and direction of up to 19 vectors. Individual vectors and their resultant are drawn on the monitor screen and the magnitude and direction of the resultant are given numerically. Handy for lectures and can also be used to drill students. Reviewed TPT February 1986.

b. *College Physics Series, Vol. I: Vectors and Graphics*, Cross Educational Software, P.O. Box 1536, Ruston, LA 71270. Apple II+. Tutorials on the resolution of vectors into components, vector addition, scalar product, vector product. A quiz follows each tutorial. Reviewed TPT January 1983.

8. Laboratory

BE Experiment 4: *Composition and Resolution of Forces — Force-Table Method.* Students mathematically determine a force that balances 2 or 3 given forces, then check the calculation using a commercial force table. They need not know the definition of a force, only that the forces in the experiment are vectors along the strings used, with magnitudes proportional to the weights hung on the strings. The focus is on resolving vectors into components and finding the magnitude and direction of a vector, given its components.

Chapter 4 MOTION IN TWO AND THREE DIMENSIONS

BASIC TOPICS

I. Definitions.

A. Draw a curved particle path. Show the position vector for several times and the displacement vector for several intervals. Define average velocity over an interval.

B. Define velocity as $d\mathbf{r}/dt$. Write in both vector and component form. Point out that the velocity vector is tangent to the path. Define speed of the magnitude of the velocity.

C. Define acceleration as $d\mathbf{v}/dt$. Write in both vector and component form. Point out that $\mathbf{a}$ is not zero if either the magnitude or direction of $\mathbf{v}$ changes with time.

D. Show that the particle is speeding up only if $\mathbf{a}\cdot\mathbf{v}$ is positive. If $\mathbf{a}\cdot\mathbf{v}$ is negative the particle is slowing down and if $\mathbf{a}\cdot\mathbf{v} = 0$ its speed is not changing.

E. Remark that sometimes the magnitude and direction of the acceleration are given, rather than its components. Remind students how to find the components if such is the case.

F. Go over Sample Problem 5 or a similar problem of your own devising. It shows how to find and use the components of the acceleration.

II. Projectile motion.

A. Demonstrate projectile motion by using a spring gun to fire a ball onto a surface at the firing height. Use various firing angles, including 45°, and point out that the maximum range occurs for a firing angle of 45°. Remark on the symmetry of the range as a function of firing angle. Mention that the maximum range occurs for a different angle when the ball is fired onto a surface at a different height.

B. Draw the trajectory of a projectile, show the direction of the initial velocity, and derive its components in terms of the initial speed and firing angle.

C. Write down the kinematic equations for $x(t)$, $y(t)$, $v_x(t)$, and $v_y(t)$. At first, include both a_x and a_y but then specialize to $a_x = 0$ and $a_y = -g$ for positive y up. Stress that these form two sets of one dimensional equations, linked by the common variable t and are to be solved simultaneously. Note that a_x affects only v_x, not v_y or v_z. Make similar statements about the other components. Throw a ball vertically, then catch it. Repeat while walking with constant velocity across the room. Ask students to observe the motion of the ball relative to the chalkboard and to describe its motion relative to your hand.

D. Point out that the acceleration is the same at all points of the trajectory, even the highest point. Also point out that the horizontal component of the velocity is constant. Remind students how to find the components of the initial velocity, given the initial speed and firing angle.

E. Work examples. Use punted footballs, hit baseballs, or thrown basketballs according to season.
   1. Find the time for the projectile to reach its highest point, then find the coordinates of the highest point.
   2. Find the time for the projectile to hit the ground, at the same level as the firing point. Then find the horizontal range and the velocity components just before landing.
   3. Show that maximum range over level ground is achieved when the firing angle is 45°.
   4. Show how to work problems for which the landing point is not at the same level as the firing point.

F. Point out that all projectiles follow some piece of the full parabolic trajectory. For example, A to D could be the trajectory of a ball thrown at an upward angle from a roof to the street; B to D could be the trajectory of a ball thrown horizontally; C to D could be the trajectory of a ball thrown downward.

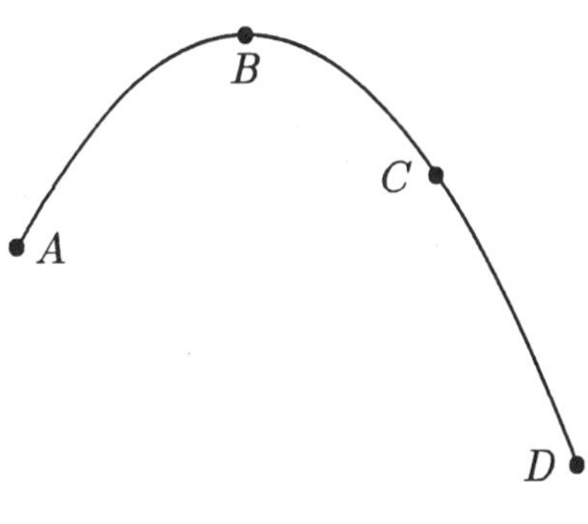

G. Explain how to find the speed and direction of travel for any time. Specialize to the time of impact on level ground and show that the speed is the same as the firing speed but that vertical component of the velocity has changed sign. Remark that this result is true only because air resistance has been neglected.

III. Circular motion.

A. Draw the path and describe uniform circular motion, emphasizing that the speed remains constant. Remind students that the acceleration must be perpendicular to the velocity. By drawing the velocity vector at two times, argue that it must be directed inward. On the diagram show the velocity and acceleration vectors for several positions of the particle.

B. Derive $a = v^2/r$. As an alternative to the derivation given in the text, write the equations for the particle coordinates as functions of time, then differentiate twice.

C. Example: calculate the speed of an Earth satellite, given the orbit radius and the acceleration to due to gravity at the orbit. See Sample Problem 10. Emphasize that the acceleration is toward the Earth.

IV. Relative motion.

A. Material in this section is used in Chapter 5 to discuss inertial frames and in Chapter 12 to discuss rolling without slipping. It is also useful as a prelude to relativity.

B. Relate the position of a particle as given in coordinate system A to the position as given in coordinate system B by $\mathbf{r}_{PA} = \mathbf{r}_{PB} + \mathbf{r}_{BA}$, where $\mathbf{r}_{BA}$ is the position of the origin of B relative to the origin of A. Differentiate to show that $\mathbf{v}_{PA} = \mathbf{v}_{PB} + \mathbf{v}_{BA}$ and $\mathbf{a}_{PA} = \mathbf{a}_{PB} + \mathbf{a}_{BA}$, where $\mathbf{v}_{BA}$ and $\mathbf{a}_{BA}$ are the velocity and acceleration, respectively, of B relative to A.

C. Discuss examples of a ball thrown or rolled in accelerating and non-accelerating trains. The discussion may be carried out for motion in a plane or for one-dimensional motion only.

D. Remark that $\mathbf{a}_{PB} = \mathbf{a}_{PA}$ if the two systems are not accelerating with respect to each other. This is an important point for the discussion of inertial reference frames in Chapter 5.

E. Work several problems dealing with airplanes flying in the wind and boats sailing in moving water. See problems 73, 86, and 88. Emphasize that relative motion problems are chiefly exercises in vector addition. To help students read some of the problems explain that an airplane's "heading" is its direction of motion in a frame attached to the air, while its direction of travel is its direction of motion in a frame attached to the ground.

SUPPLEMENTARY TOPIC

Relative motion at high speeds. Include this topic if you intend to put some modern physics into the course. Write down the relativistic equation for addition of velocities. Show that it reduces to the non-relativistic equation if both the particle and frame B are moving much slower than the speed of light relative to A. Show that if the particle moves with the speed of light in one frame it moves with the speed of light in all frames.

SUGGESTIONS

1. Ask question 1 as part of a general discussion about kinematics.
2. Use questions 2, 3, 7, and 10 to generate discussions of ideal projectile motion. To include air resistance in the discussions, ask questions 5 and 9.
3. Ask question 11 in connection with centripetal acceleration.
4. Ask questions 14, 15, and 18 in connection with relative motion.
5. Assign problems 13, 14, and 15 to have students think about the analysis of motion in two dimensions.
6. Assign problem 22 as the first problem in the study of projectile motion, then have the students go on to several of problems 23 through 27 and 37 through 57.
7. Assign problems 62 and 69 in connection with uniform circular motion.
8. Computer projects
   a. Have students use a commercial math program or their own root finding programs to solve projectile motion problems.
   b. Have students use a spreadsheet or write a computer program to tabulate the coordinates and velocity components of a projectile as functions of time. Have them change the initial velocity and observe changes in the coordinates of the highest point and in the range. Ask them to find the firing angle for the greatest horizontal coordinate when the landing point is above or below the firing point. Some projects can be found in *A Student's Companion to Physics.*
9. Demonstrations
   Projectile motion: Freier and Anderson Mb14, 16, 17, 19, 20, 23, 24, 28; Hilton M13.
10. Audio/Visual
    a. *Accelerated Motion and Angle of Lean*, 3/4" video cassette, color, 9 min. North Texas State University, Department of Physics, Denton, TX 76203. Reviewed AJP 49:383 (1981) and TPT 19:341 (1981). Runners on tracks, accelerating vehicles, and a merry-go-round illustrate the connection between acceleration and the lean of various bodies. Winner of the 1980 AAPT competition.
    b. *Galilean Relativity: Ball Dropped from the Mast of a Ship*, S8, color, approx. 3 min. Kalmia Company, Dept. P1, Concord, MA 01742. One of three titles from the Project Physics Galilean Relativity Series, this loop is best of the three and shows the motion of a falling ball released from the mast of sailboat. Two different reference frames are used to view the motion.
    c. *Velocity in Circular and Simple Harmonic Motion*; *Velocity and Acceleration in Circular Motion*; *The Velocity Vector*; *The Acceleration Vector*, S8, b/w, approx. 4 min each. Kalmia Co. (address given above). Four loops from the PSSC Vector Kinematics series.
    d. *Projectile Motion*; Films for the Humanities and Sciences, P.O. Box 2053, Princeton, NJ 08543. 15 min., color. VHS, Beta, or U-Matic video cassette. Slow motion and stop action pictures of projectiles against scales for distance measurements and with clocks for time measurements. Reviewed TPT November 1989.

e. *Demonstrations of Physics, Volume I, Motion*; Vikas Productions, Inc., P.O. Box 6088, Bozeman, MT 59771. Video cassette; 39 min, color. 6 standard demonstrations. Reviewed TPT 29, 190 (March 1990)

11. Computer programs
    a. *Motion.* See Chapter 2 notes.
    b. *Mechanics*; EduTech, 634 Commonwealth Avenue, Newton Center, MA 02159. Apple IIe, II+. Demonstrates vertical fall with or without air resistance, hunter and monkey experiment, planetary motion.
12. Laboratory
    a. MEOS Experiment 7–9: *Ballistic Pendulum — Projectile Motion* (use only the first method in connection with this chapter). Students find the initial velocity of a ball shot from a spring gun by measuring its range. Emphasizes the use of kinematic equations.
    b. Also see Procedures A and B of BE Experiment 11: *Inelastic Impact and the Velocity of a Projectile.* In addition to using range data to find the initial velocity, students plot range as a function of firing angle.

Chapter 5 FORCE AND MOTION — I

BASIC TOPICS

I. Overview
   A. Explain that objects may interact with each other and as a result their velocities change. State that the strength of an interaction depends on properties of the objects and their relative positions. Gravitational mass is responsible for gravitational interactions, electric charge is responsible for electric and magnetic interactions.
   B. Explain that we split the problem into two parts and say that each body exerts a force on the other and that the net force on a body changes its velocity. Remark that an equation that gives the force in terms of the properties of the objects and their positions is called a force law. Force laws are discussed throughout the course. The dominant theme of this chapter, however, is the relationship between any force and the acceleration it produces.

II. Newton's first law.
   A. State the law: if an object does not interact with any other objects, its acceleration is zero.
   B. Point out that the acceleration depends on the reference frame used to measure it and that the first law can be true for only a select set of frames. Cover the essential parts of the relative motion section of Chapter 4, if they were not covered earlier. Define an inertial frame. Tell students that an inertial frame can be constructed by finding an object that is not interacting with other objects, then attaching a reference frame to it. Any frame that moves with constant velocity relative to an inertial frame is also an inertial frame, but one that is accelerating relative to an inertial frame is not.
   C. Explain that we may take a reference frame attached to the Earth as an inertial frame for the description of most laboratory phenomena but we cannot for the description of ocean and wind currents, space probes, and astronomical phenomena.

III. Newton's second law.
   A. Explain that the environment influences the motion of an object and that force measures the extent of the interaction. The result of the interaction is an *acceleration.* Place a cart at rest on the air track. Push it to start it moving and note that it continues at constant velocity. After it is moving, push it to increase its speed, then push it to decrease its speed. In each case note the direction of the force and the direction of the acceleration. Also give an eraser a shove across a table and note that it stops. Point out that the table top exerts

a force of friction while the eraser is moving. Push the eraser at constant velocity and explain that the force of your hand and the force of friction sum to zero.

B. Define force in terms of the acceleration imparted to the standard 1-kg mass. Explain how this definition can be used to calibrate a spring, for example. Point out that force is a vector, in the same direction as the acceleration. If two or more forces act on the standard mass, its acceleration is the same as when a force equal to the resultant acts. Units: newton, pound. Explain that 1 N is $1\ \text{kg}\cdot\text{m/s}^2$.

C. Have three students pull on a rope, knotted together as shown. Ask one to increase his or her pull and ask the others to report what they had to do to remain stationary.

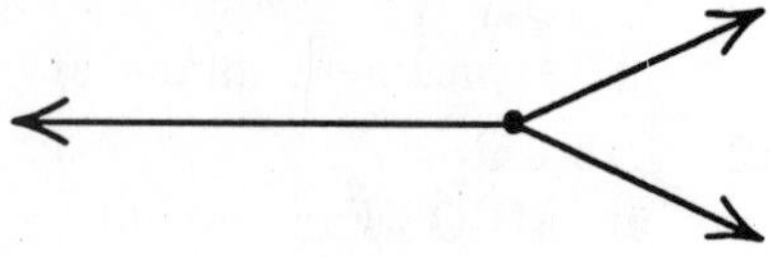

D. Define mass in terms of the ratio of the acceleration imparted to the standard mass and to the unknown mass, with the same force acting. Attach identical springs to two identical carts, one empty and the other containing a lead brick. Pull with the same force (same elongation of the springs) and observe the difference in acceleration. Units: kg, slug.

E. State the second law. Stress that the force that appears is the net or resultant force. Explain that the law holds in inertial frames. Point out that this is an experimentally established law and does not follow as an identity from the definitions of force and mass. Emphasize that $m\mathbf{a}$ is not a force.

F. Discuss examples: calculate the force required to stop an object in a given time, given the mass and initial velocity; calculate the force required to keep an object in uniform circular motion, given its speed and the radius of its orbit. Calculate the acceleration of an object being pushed by two forces in opposite directions and note that the acceleration vanishes if the forces have equal magnitudes. Emphasize that the forces continue to act but their sum vanishes. Some students believe that the forces literally cancel each other and no longer act.

IV. Newton's third law.

A. State the law. Stress that the two forces in question act on different bodies and each helps to determine the acceleration of the body on which it acts. Explain that the third law describes a characteristic of force laws. State that the two forces in an action-reaction pair are of the same type: gravitational, for example.

B. Discuss examples. Hold a book stationary in your hand, identify action-reaction pairs (hand-book, book-Earth). Now allow your hand and the book to accelerate downward with an acceleration less than g and again identify action-reaction pairs. Note that you can control the acceleration of the book by means of the force you exert but once you exert a given force you cannot control the force that the book exerts on you.

V. Applications of Newton's laws involving a single object.

A. Go over the steps used to solve a one-body problem: identify the body and all forces acting on it, draw a free body diagram, choose a coordinate system, write the second law in component form, and finally solve for the unknown.

B. Some special forces should be explained. They are important for many of the problems but are rarely mentioned explicitly. Warn students they must take these forces into account if they act.

1. Explain that weight is the gravitational force on an object and point out that its magnitude is mg, where g is the local acceleration due to gravity and m is the mass of the object. It is directed toward the center of the Earth. Explain that weight varies with altitude and slightly from place to place on the surface of the Earth, but mass does not vary. Emphasize that the appearance of g in the formula for weight does not imply the acceleration of the body is g.

2. Point out that a massless rope transmits force unaltered in magnitude and that the magnitude of the force it exerts on objects at each end is called the tension. If a person pulls an object by exerting a force on a string attached to the object, the motion is as if the person pulled directly on the object. The string serves to define the direction of the force. A frictionless, massless pulley serves to change the direction of a force but does not change the tension in the rope passing over it.
3. Explain that the normal force of a surface on an object originates in elastic and ultimately electric forces. It prevents the object from moving through the surface. State that it is perpendicular to the surface. If the surface is at rest the normal force adjusts so the acceleration component perpendicular to the surface vanishes. More generally, the object and the surface have the same perpendicular acceleration component. Place a book on the table and press on it. State that the normal force is greater than when you were not pressing. Hold the book against the wall by pressing on it and mention that the normal force is horizontal.

C. Set up the situation described in Sample Problem 5–9 using an inclined air track but attach a calibrated spring scale to the support at the top of the incline and tie the other end of the scale to the block. Calculate the tension in the string and compare the result to the reading on the scale. Cut the string, then calculate the acceleration.

D. Consider a person standing on a scale in an elevator. State that the scale measures the normal force and calculate its value for an elevator at rest, one accelerating upward, one accelerating downward with $a < g$, and one in free fall. See Sample Problem 5–12.

VI. Applications of Newton's laws involving more than one object.

A. Explain that when two or more objects are involved, a free body diagram must be drawn for each. A Newton's second law equation, in component form, is also written for each object. Point out that differently oriented coordinate systems may be used for different bodies. Show how to invoke the third law when necessary. Point out that the same symbol can be used for the magnitude of the two forces of an action-reaction pair and that their opposing directions must be taken into account when drawing the free body diagram and in writing the second law equations.

B. Use examples to show how rods, strings, and pulleys relate the motions of bodies in various cases. Explain that, in addition to the second law equations, there will be equations relating the accelerations of the objects. Show that these equations depend on the choice of coordinate systems.

C. Do several examples, carefully explaining each step. If you have not developed an application of your own work one or both of Sample Problems 5–11 in the text. If possible, give a demonstration.

SUGGESTIONS

1. Discuss questions 1 and 2 in connection with Newton's first law.
2. Use questions 4 and 20 to help students think about the influence of forces on bodies. Assign one or two of problems 1, 2, and 3.
3. Discuss questions 15 and 28 in connection with problems 44 and 45.
4. Use questions 11, 13, and 22 and problems 11, 13, and 14 to discuss mass and weight.
5. Use questions 10 and 24 and problems 37, 40, and 73 to discuss Newton's third law.
6. Assign a few applications problems, some from the group 11 through 18 and some from the group 19 through 73.
7. As a prelude to Chapter 9 (the center of mass and conservation of momentum) assign problem 39.

8. Demonstrations
   a. Inertia: Freier and Anderson Mc1 — 5, Me1; Hilton M6.
   b. **F** = *m***a**: Freier and Anderson Md2, Ml1; Hilton M7.
   c. Action-reaction pairs: Freier and Anderson Md1, 3, 4; Hilton M8.
   d. Mass and weight: Freier and Anderson Mf1, 2.
   e. Tension in a string: Freier and Anderson Ml1.
9. Audio/Visual
   a. *Frames of reference*, 16 mm, b/w, 28 min. Modern Learning Aids, Division of Ward's Natural Science Establishment, P.O. Box 312, Rochester, NY 14601. This classic film will never become dated. Frames of reference are illustrated in clever and imaginative ways by Hume and Ivey. Probably the best of the PSSC film series.
   b. *Zero-G*, 16 mm, color, 14 min. National Audiovisual Center, General Services Administration, Washington, DC 20409. (Free loan from NASA regional film library.) The skylab is a "natural" laboratory for "zero g" experiments.
   c. *Inertial Forces — Translational Acceleration*, S8, color, 8 min. Kalmia Company, Dept. P1, Concord, MA 01742. An excellent film showing how apparent weight changes as an elevator moves up and down.
10. Computer programs
   a. *Physics: Elementary Mechanics*, Control Data Company, 3610 West 77th Street, Bloomington, MN 55435. Apple II, IBM PC. Drill problems in collisions, gravitation, satellite motion, rotational dynamics, harmonic motion. Statements of problems are not complete and students must ask the computer for additional data. Helps students think about what information is required to solve mechanics problems. Reviewed TPT May 1986.
   b. *Newton's Laws*, J&S Software, 140 Reed Avenue, Port Washington, NY 11050. Apple II. **F** = *m***a** drill problems. Reviewed TPT February 1984.
   c. *Sir Isaac Newton's Games*, Sunburst Communications, 39 Washington Avenue, Pleasantville, NY 10570. Apple II, IBM PC, Tandy 1000. Games involving the motion of a puck on various surfaces. Gives students an intuitive feel for centripetal force, momentum, impulse, frictional forces, and gravity. Reviewed TPT May 1987.
11. Laboratory
   MEOS Experiment 8–2: *Concept of Mass: Newton's Second Law of Motion.* Students measure the accelerations of two pucks that interact via a spring on a nearly frictionless surface and compare the ratio to the ratio of their masses. This experiment may be done with dry ice pucks or on an air table or air track.

Chapter 6 FORCE AND MOTION — II

BASIC TOPICS

I. Frictional forces.
   A. Place a large massive wooden block on the lecture table. Attach a spring scale, large enough to be read easily. If necessary, tape sandpaper to the table under the block. Pull weakly on the scale and note that the reading is not zero although the block does not move. Pull slightly harder and note that the reading increases but the block still does not move. Remark that there must be a force of friction opposing the pull and that the force of friction increases as the pull increases. Now increase your pull until the block moves and note the reading just before it starts to move. Pull the block at constant speed and note the reading. Have the students repeat the experiment in a qualitative manner, using books resting on their chair arms. To show that the phenomenon depends on the nature of

the surface, the demonstration can be repeated after waxing the wooden block and table top.

B. Give a brief qualitative discussion about the source of frictional forces. Stress that the force of static friction has whatever magnitude and direction are required to hold the two bodies in contact at rest relative to each other, up to a certain limit in magnitude. Define the coefficient of static friction and explain the use of $f_s < \mu_s N$. In particular, explain that if the surface is stationary the force of static friction is determined by the condition that the object on it have zero acceleration. To test if an object remains at rest, the frictional force required to produce zero relative acceleration is calculated and compared with $\mu_s N$.

C. Define the coefficient of kinetic friction and explain that $f_k = \mu_k N$ gives the frictional force as long as the object is sliding on the surface. Also explain that if the surface is stationary the force of kinetic friction is directed opposite to the velocity of the object sliding on it. Remark that this is an example of a force law.

D. Work some examples:

1. Find the angle of an inclined plane for which sliding starts; find the angle at which the body slides at constant speed. These examples can be analyzed in association with a demonstration and the students can use the data to find the coefficients of friction.
2. Analyze an object resting on the floor, with a person applying a force that is directed above the horizontal. Find the minimum applied horizontal force that will start the object moving and point out that it is a function of the angle between the applied force and the horizontal.
3. Consider the same situation but with the object moving. Find its acceleration. This and the previous example demonstrate the dependence of the normal force and the force of friction on the externally applied force.
4. To give an illuminating variant, consider a book being held against the wall by a horizontal force. Calculate the minimum applied force that will keep the book from falling.

II. Drag forces and terminal speed.

A. Make or buy a small toy parachute. Drop two weights side by side and note they reach the floor at the same time. Attach the parachute to one and repeat. Explain that the force of the air reduces the acceleration.

B. State that for turbulent flow of air around an object the magnitude of the drag force is given by $D = \frac{1}{2}C\rho Av^2$, where A is the effective cross-sectional area, ρ is the density of air, and v is the speed of the object relative to the air. C is a drag coefficient, usually determined by experiment. Remark that a parachute increases the cross-sectional area. State that the drag force is directed opposite to the velocity.

C. Explain that as an object falls its speed approaches terminal speed as a limit. Write down Newton's second law for a falling object and point out that the drag and gravitational forces are in opposite directions. Suppose the object is dropped from rest and point out that the acceleration is g at first but as the object gains speed its acceleration decreases in magnitude. At terminal speed the acceleration is zero and remains zero, so the velocity no longer changes. Show that zero acceleration leads to $v_t = \sqrt{2mg/C\rho A}$. Point out Table 6–1, which gives some terminal speeds.

D. Remark that if an object is thrown downward with a speed that is greater than terminal speed it slows down until terminal speed is reached.

E. Qualitatively discuss projectile motion with drag. The horizontal component of the velocity tends to zero while the vertical component tends to the terminal speed. Contrast the trajectory with one in the absence of air resistance.

III. Uniform circular motion.

A. Point out that for uniform circular motion to occur there must be a radially inward force of constant magnitude and that something in the environment of the body supplies the force. Whirl a mass tied to a string around your head and explain that the string supplies the force. Set up a loop-the-loop with a ball or toy cart on a track and explain that the combination of the normal force of the track and the force of gravity supplies the centripetal force. Have students identify the source of the force in examples and problems as they are discussed.

B. Point out that $F = mv^2/r$ is just $F = ma$ with the expression for centripetal acceleration substituted for a.

C. Discuss problem solving strategy. After identifying the forces, find the radial component of the resultant and equate it to mv^2/r.

D. Examples:

1. Find the speed and period of a conical pendulum.
2. Find the speed with which a car can round an unbanked curve, given the coefficient of static friction.
3. Find the angle of banking required to hold a car on a curve without aid of friction.
4. Analyze the loop-the-loop and point out that the ball leaves the track when the normal force vanishes. Show that the critical speed at the top is given by $v^2/r = g$.

SUGGESTIONS

1. To start students thinking about the physical basis of frictional forces, have them discuss question 1.
2. Ways in which the coefficients of friction are used are emphasized in problems 9, and 11. To help students understand the role played by the normal force, assign question 3 in connection with problems 3 and 5. Assign problem 16 in connection with the measurement of coefficients of friction using an inclined plane. Problems 26, 27, 29, 36, and 38 provide interesting applications. Assign a few.
3. Emphasize that terminal speed is a limit that can be reached from either above or below and that the direction of the terminal velocity is opposite that of the force of gravity. Discuss questions 8 and 10. Also consider discussing questions 7 and 11.
4. In the centripetal force section discuss question 12 in connection with problem 51, question 15 in connection with problem 58 or 59, and question 16 in connection with problem 57. Problem 52 is more challenging. Also consider problems 63 and 64. You might also discuss an airplane making a circular turn, then ask question 13 and assign problem 65.
5. Computer projects
   a. Have students use a computer program to investigate objects that are subjected to time dependent forces. To check the program first have them consider a constant force and compare machine generated functions with the known kinematic equations.
   b. Have students modify the program to integrate Newton's second law for velocity dependent forces, then have them investigate the motion of an object subjected to a force that is proportional to v or v^2. It is instructive to have them plot the velocity components as functions of time for a projectile fired straight up or down, subject to air resistance. Consider initial velocities that are both greater and less than the terminal velocity. Also have them study the maximum height and range of projectiles with various coefficients of air resistance. Some projects are given in the Computer Projects section of this manual.
6. Demonstrations
   a. Friction: Freier and Anderson Mk; Hilton M11.

   b. Inclined plane: Freier and Anderson Mj2.
   c. Centripetal acceleration: Freier and Anderson Mb29, 31, Mm1, 2, 4 — 8, Ms5; Hilton M16.
7. Audio/Visual
   Principles of Lubrication, 16 mm, color, 23 min. International Film Bureau, 332 S. Michigan Ave., Chicago, IL 60604. Discusses the laws of friction and presents a variety of examples and models. Reviewed TPT 20:325 (1982).
8. Computer programs
   a. *Mechanics.* See notes for Chapter 4.
   b. *Personal Problems*, Addison-Wesley Publishing Company, Reading, MA 01867. Apple II. The program plots the trajectories of projectiles subjected to air resistance. It also plots the coordinates, velocity components, and acceleration components as functions of time and lists values for points along the trajectory. The user can specify the initial conditions and the coefficient of air resistance. Use this as an alternative to student programming. Reviewed TPT December 1985.
   c. *Physics Simulations I: Ballistic*, Kinko's Service Corporation, 4141 State Street, Santa Barbara, CA 93110. Macintosh. Plots trajectories of projectiles, with or without a drag force proportional to velocity. The drag coefficient can be constant or depend exponentially on altitude. Excellent for illustrating lectures.
9. Laboratory
   a. MEOS Experiment 7–6: *Coefficient of Friction — The Inclined Plane.* Students determine the coefficients of static and sliding friction for 3 blocks on an inclined plane. They devise their own experimental procedures.
   b. MEOS Experiment 7–7: *Radial Acceleration* (Problem I only). The centripetal force and the speed of a ball on a string, executing uniform circular motion, is measured for various orbit radii. Essentially a verification of $F = mv^2/r$.
   c. MEOS Experiment 7–8: *Investigation of Uniform Circular Motion*, or BE Experiment 13: *Centripetal Force.* Students measure the force acting on a body undergoing uniform circular motion, with the centripetal force provided by a spring.
   d. MEOS Experiment 8–3: *Centripetal Force.* Students measure the speed of a puck undergoing uniform circular motion on a nearly frictionless surface. The data is used to calculate the centripetal force.

Chapter 7 WORK AND KINETIC ENERGY

BASIC TOPICS

I. Work done by a constant force.
   A. Write down $W = \mathbf{F} \cdot \mathbf{d} = Fd \cos\phi$ and point out ϕ on a diagram. Explain that this is the work done *on* a particle *by* the constant force $\mathbf{F}$ as the particle undergoes a displacement $\mathbf{d}$. Explain that work can be calculated for each individual force and that the total work done on the particle is the work done by the resultant force. Point out that work is a scalar quantity. Also point out that work is zero for a force that is perpendicular to the displacement and that, in general, only the component of $\mathbf{F}$ tangent to the path contributes to the work. The force does no work if the displacement is zero. Emphasize that work can be positive or negative, depending on the relative orientation of $\mathbf{F}$ and $\mathbf{s}$. For a constant force the work depends only on the displacement, not on details of the path. Units: joule, erg, ft·lb, eV. Give the conversion from electron volts to joules.
   B. Calculate the work done by the force of gravity as a mass falls a distance h and as it rises a distance h. Emphasize the sign. Calculate the work done by a non-horizontal force used

to pull a box across a horizontal floor. Point out that the work done by the normal force and the work done by the force of gravity are zero. Consider both an accelerating box and one moving with constant velocity. Repeat the calculation for a crate being pulled up an incline by a force applied parallel to the incline. Show the work done by gravity is $-mgh$, where h is the change in the height of the crate.

II. Work done by a variable force.

A. For motion in one dimension, discuss the integral form for work as the limit of a sum over infinitesimal path segments. Explain that the sum can be carried out by a computer even if the integral cannot be evaluated analytically.

B. Examples: derive expressions for the work done by an ideal spring and a force of the form k/x^2. If you have not yet discussed the force of an ideal spring, do so now as a preface to the calculation of work. Explain how the spring constant can be found by hanging a mass from the spring and measuring the extension. Demonstrate changes in the spring length during which the spring does positive work and during which the spring does negative work.

C. For motion in more than one dimension, write down the expression for the work in the form $\int_i^f \mathbf{F} \cdot d\mathbf{r}$ and explain its interpretation as the limit of a sum over infinitesimal path segments. Explain that this is the general definition of work. Calculate the work done by the applied force, the force of gravity, and the tension in the string as a simple pendulum is pulled along its arc until it is displaced vertically through a height h by a horizontal applied force $\mathbf{F}$.

III. The work-energy theorem.

A. Define kinetic energy for a particle. Point out that kinetic energy is a scalar and depends on the speed but not on the direction of the velocity. Point out that $v^2 = v_x^2 + v_y^2$ for two-dimensional motion and remark that the appearance of velocity components in the expression does *not* mean K has components.

B. Use Newton's second law to prove the theorem for motion in one dimension. If the students are mathematically sophisticated, extend the theorem to the general case. Stress that it is the total work (done by the resultant force) that enters the theorem.

C. Point out that only the component of a force parallel or antiparallel to the velocity changes the speed. Other components change the direction of motion. Positive total work results in an increase in kinetic energy and speed, negative total work results in a decrease. Remind students of previous examples in which the object moved with constant speed (including uniform circular motion). The total work was zero and the kinetic energy did not change. Avoid quantitative calculations involving frictional forces.

D. Demonstrate the theorem by considering a ball thrown into the air. At first neglect air resistance and point out that during the upward part of the motion the force of gravity does negative work and the kinetic energy decreases. Use energy considerations to compute the maximum height of the ball. Note that $v^2 = v_0^2 + 2as$ (which was derived in the study of kinematics) follows from the work-energy theorem if the force is constant. As the ball falls, the force of gravity does positive work and the kinetic energy increases. Show that the ball returns with its initial speed. Then include air resistance and argue that the ball returns with less than its original speed.

E. As an example consider a stone dropped onto a vertical spring and calculate the maximum compression of the spring, given the mass of the stone, the height from which it is dropped, and the spring constant of the spring.

F. Explain that the work-energy theorem can be applied only to particles and objects that can be treated as particles. To give an example in which it cannot be applied directly, consider a car crashing into a rigid barrier: the barrier does no work but the kinetic energy of the car decreases.

G. Explain that observers in different inertial frames will measure different values of the net work done and for the change in the kinetic energy but both will find $W_{\text{net}} = \Delta K$.

IV. Power.

A. Define power as $P = dW/dt$. Units: watt, horsepower.

B. Show that $P = \mathbf{F} \cdot \mathbf{v}$. Explain that the work done over a time interval is given by $\int P\,dt$.

SUPPLEMENTARY TOPIC

Kinetic energy at high speed. If the course includes modern physics topics, discuss the relativistic expression for kinetic energy. Remark that the work-energy theorem, with the non-relativistic definition of kinetic energy, is not valid if the object has a speed that is an appreciable fraction of the speed of light. Kinetic energy must be redefined. Write down the relativistic expression. Remind students of the value for the speed of light. Use the binomial expansion to show that the relativistic expression reduces to $\frac{1}{2}mv^2$ if $v \ll c$. Calculate the error in the non-relativistic expression for $v/c = 0.99$, 0.1, and 0.01.

SUGGESTIONS

1. Discuss simple machines. Go over Sample Problem 7–4 and problem 3, then ask the students to discuss question 3.
2. For practice in distinguishing between work done by one of several forces acting and the total work, ask students to discuss questions 4, 7, and 14, then have them work problem 2, 3, or 5.
3. Ask question 13 and assign problem 18 on springs. Problem 20 is a good illustration of Hooke's law.
4. Assign problem 23 in connection with the concept of kinetic energy.
5. Assign problems on the work-energy theorem, such as 27, 34, 37, and 38. Also consider question 21.
6. To have students use the calculus, assign problem 33.
7. Assign problems 41, 42, and 45 in connection with the concept of power.
8. Computer project
   Have students use a commercial math program or write a program to numerically evaluate the integral for work, then use the program to calculate the work done by various forces, given as functions of position. Include a non-conservative force and use the program to show the work done on a round trip does not vanish. Some projects can be found in the Computer Projects section of this manual.
9. Demonstrations
   Work: Freier and Anderson Mv1.
10. Computer program
    Work and Energy, J&S Software, 140 Reed Avenue, Port Washington, NY 11050. Apple II. Drill problems on computation of work, work-energy theorem, and conservation of energy.
11. Laboratory
    a. MEOS Experiment 7–16: *Elongation of an Elastomer*. Students measure the elongation of an elastomer for a succession of applied forces and use a polar planimeter to calculate the work done by the force. The experiment may also be done in connection with Chapter 13.
    b. BE Experiment 10: *Mechanical Advantage and Efficiency of Simple Machines*. This experiment can be used to broaden the course to include these topics. A lever, an inclined

plane, a pulley system, and a wheel and axle are studied. In each case the force output is measured for a given force input and the work input is compared to the work output.

Chapter 8 CONSERVATION OF ENERGY

BASIC TOPICS

I. Conservative and non-conservative forces
   A. Definitions. Explain that a force is conservative if either of the following hold:
      1. the work done by the force on every round trip is zero.
      2. the work done by the force is independent of the path.

      Show that the two definitions are equivalent.
   B. Discuss the force of gravity and the force of an ideal spring as examples. For either or both of these show that the work done depends only on the end points and not on the path between, then argue that the work vanishes for a round trip. Point out that on some parts of the path the force does positive work while on other parts it does negative work. Demonstrate that the work done by a spring is independent of path by considering two different motions with the same end points. For the first motion have the mass go directly from the initial point to the final point; for the second have it first go away from the final point before returning to the final point.
   C. Use a force of friction with constant magnitude as an example of a non-conservative force. Consider a block on a horizontal table top and argue that the work done by the force cannot vanish over a round trip since it is negative for each segment. Suppose the block moves around a circular path and friction is the only force that does work. Argue that the object returns to its initial position with less kinetic energy than it had when it started.
   D. Use a cart on a linear air track to demonstrate these ideas. Couple each end of the cart, via a spring, to a support at the corresponding end of the air track. Give the cart an initial velocity and tell students to observe its speed each time it returns to its initial position. Point out that the kinetic energy returns to nearly the same value and that the springs do zero work during a round trip. Reduce or eliminate air flow to show the influence of a non-conservative force. If this is done rapidly and skillfully, you can cause the cart to stop at the opposite end from which it started.

II. Potential energy.
   A. Give the definition of potential energy in terms of work for motion in one dimension. See Eq. 8–5 and emphasize that W is the work done *by* the force responsible for the potential energy.
   B. Discuss the following properties:
      1. The zero is arbitrary. Only potential energy differences have physical meaning.
      2. The potential energy is a *scalar* function of position.
      3. The force is given by $F = -\mathrm{d}U/\mathrm{d}x$ in one dimension.
      4. Units: joule, erg, ft·lb, eV.
   C. Derive expressions for the potential energy functions associated with the force of gravity (uniform gravitational field) and the force of an ideal spring. Stress that the potential energy is a property of the object-Earth or spring-mass system and depends on the *configuration* of the system.
   D. Use the force of friction as an example to explain why a potential energy function cannot be defined for non-conservative forces.
   E. Use the work-energy theorem to show that $W = \Delta U$ is the work that must be applied by an external agent to increase the potential energy by ΔU if the kinetic energy does not

change. Show that ΔU is recovered as kinetic energy when the external agent is removed. Example: raising an object in a gravitational field.

III. Conservation of energy.

A. Explain that if all the forces acting between the objects of a system are conservative then $K + U =$ constant. This follows from the work-energy theorem with the work of the conservative forces represented by the change in potential energy. The negative sign in Eq. 8–5 is essential to obtain this result. Emphasize that U is the sum of the individual potential energies if more than one conservative force acts. Define the total mechanical energy as $E = K + U$.

B. Discuss the conversion of kinetic to potential energy and vice versa. Drop a superball on a rigid table top and point out when the potential and kinetic energies are maximum and when they are minimum. The question of elasticity can be glossed over by saying that to a good approximation the ball rebounds with unchanged speed. Also discuss the energy in a spring-mass system. Return to the cart on the air track and discuss its motion in terms of $K + U =$ constant. To avoid later confusion in the students' minds, start the motion with neither K nor U equal to zero. Emphasize that the energy remains in the system but changes its form during the motion. The agent of the change is the work done by the forces of the springs.

C. Show how to calculate the total energy for a spring-mass system from the initial conditions. Write the conservation principle in the form $\frac{1}{2}mv^2 + U(x) = \frac{1}{2}mv_0^2 + U(x_0)$. Use conservation of energy to find expressions for the maximum speed, maximum extension, and maximum compression, given the total energy.

D. Use the example of a ball thrown upward to demonstrate that conservation of energy must be applied to a system rather than to a single particle. Remark that the Earth does work on the ball, the ball does work on the Earth, and the change in potential energy is the negative of the sum. Show it is mgh, where h is the change in their separation. Remark that both kinetic energies change and the total change is the negative of the change in the potential energy. Explain that because the Earth is so massive the change in its kinetic energy is small and may be neglected.

E. Discuss potential energy curves. Use the curve for a spring-mass system, then a more general one, and show how to calculate the kinetic energy and speed from the coordinate and total energy. Point out the turning points on the curves and discuss their physical significance. Use $F = -\mathrm{d}U/\mathrm{d}x$ to argue that the particle turns around at a turning point. For an object on a frictionless roller coaster track find the speed at various points and identify the turning points.

F. Define stable, unstable, and neutral equilibrium. Use a potential energy curve (a frictionless roller coaster, say) to illustrate. Emphasize that $\mathrm{d}U/\mathrm{d}x = 0$ at an equilibrium point.

IV. Potential energy in 2 and 3 dimensions.

A. Define potential energy as a line integral and explain that it is the limit of a sum over infinitesimal path segments. Remark that conservation of energy leads to $\frac{1}{2}mv^2 + U(x, y, z) =$ constant. Explain that $v^2 = v_x^2 + v_y^2 + v_z^2$ and that v^2 is a scalar.

B. Example: simple pendulum. Since the gravitational potential energy depends on height, in the absence of non-conservative forces the pendulum has the same swing on either side of the equilibrium point and always returns to the same turning points. Demonstrate with a pendulum hung near a blackboard and mark the end points of the swing on the board. For a more adventurous demonstration, suspend a bowling ball pendulum from the ceiling and release the ball from rest in contact with your nose. Stand very still while it completes its swing and returns to your nose.

V. External work and internal energy.
   A. Explain that when forces due to objects external to the system do work W on the system the energy equation becomes $\Delta K + \Delta U = W$ if the internal forces are conservative. K is the total kinetic energy of all objects in the system and U is the total potential energy of their interactions with each other.
   B. Show that if the external force is conservative the system can be enlarged to include the (previously) external agent. Then $W = 0$ and U must be augmented to include the new interactions. Give the example of a ball thrown upward in the Earth's gravitational field.
   C. Explain that if some or all of the internal forces are non-conservative then the total energy must include an internal energy term to take account of energy that enters or leaves some or all of the objects and contributes to the energy (kinetic and potential) of the particles that make up the objects. Distinguish between internal energy and the energy associated with the motion and interactions of the object as a whole. Write $\Delta K + \Delta U + \Delta E_{\text{int}} = W$.
   D. Refer back to the block sliding on the horizontal table top, discussed earlier. Explain that when the block stops all the original kinetic energy has been converted to internal energy.
   E. Explain that the quantity $\int \mathbf{f} \cdot d\mathbf{r}$ along the path of an object does NOT give the work done by friction but it does contribute to the change in kinetic energy of the object, along with similar contributions from other forces, if any. For a block sliding on a table top the work done by friction is algebraically greater than the value of the integral and the difference is the increase in the internal energies of the block and table top.
   F. Explain that a change in the internal energy may be reflected in a change in temperature and that if the temperature of an object is different from the temperature of its environment heat is exchanged between the two. In this case another term is added to the energy equation so it becomes $\Delta K + \Delta U + \Delta E_{\text{int}} = W + Q$, where Q is the heat entering the system. Heat will be discussed more fully in Chapter 20.

SUPPLEMENTARY TOPICS

1. To introduce a topic from modern physics, discuss the equivalence of mass and energy. Say that relativity theory associates an energy with (rest) mass and that the energy is given by $E = mc^2$. State that other forms of energy may be converted to mass energy and vice versa. Thus the mass of an nucleus may decrease, with a corresponding increase in kinetic, potential, or radiation energy.
2. Quantization of energy is another topic from modern physics that can be introduced at this time. Use a potential energy curve to illustrate what is meant by a bound particle. You might use the potential energy of an electron in a hydrogen atom or a harmonic potential energy. At first avoid a situation in which tunneling might occur. Later you might draw a curve with a barrier between the well and the outside and point out that the particle is not bound. Remark that the energy for bound particles is quantized: it can have any one of a set of discrete values, but not a value between. Remark that when the mass is on the order of kilograms the quantum of energy is much too small to measure but when the mass is on the order of an atomic mass the quantum of energy is measurable. Give the expression for a quantum of vibrational energy: hf, where f is the frequency. Give the value for the Planck constant. State that light is a vibrational phenomenon and the energy of light of frequency f is quantized in the same way. A quantum of light energy is called a photon. Emphasize that energy can be emitted or absorbed only in units of the energy quantum. Write the equation for conservation of energy when an atom changes energy from E_i to E_f with the emission of a photon: $E_i - E_f = hf$.

SUGGESTIONS

1. Have the class account for the energy in various situations. See questions 1, 2, 4, 6, 8, 11, 12, and 17.

2. Discuss applications of the conservation of energy principle. See questions 3, 5, 9, 13, and 14 as well as problems 3, 5, 11, 12, 43, and 45.
3. To have students use the calculus to calculate the potential energy assign problems 28, 29, and 39.
4. Draw several potential energy curves and have the class analyze the particle motion for various values of the total energy. This can provide particularly useful feedback as to how well the students have mastered the idea of energy conservation. Also assign problems 7, 38, and 40.
5. Assign some problems dealing with dissipative forces. Consider 47, 48, 57, and 59.
6. Demonstrations
   a. Conservation of energy: Freier and Anderson Mn1 — 3, 6; Hilton M14a, b, e.
   b. Non-conservative forces: Freier and Anderson Mw1.
7. Laboratory
   BE Experiment 9: *Work, Energy, and Friction.* A string is attached to a car on an incline and passes over a pulley at the top of the incline. Weights on the free end of the string are adjusted so the car rolls down the incline at constant speed. The work done by gravity on the weights and on the car is calculated and used to find the change in mechanical energy due to friction. The coefficient of friction is computed. The experiment is repeated for the car rolling up the incline and for various angles of incline. It is also repeated with the car sliding on its top and the coefficients of static and kinetic friction are found.

Chapter 9 SYSTEMS OF PARTICLES

BASIC TOPICS

I. Center of mass.
   A. Spin a chalkboard eraser as you toss it. Point out that, if the influence of air can be neglected, one point (the center of mass) follows the parabolic trajectory of a projectile although the motions of other points are more complicated.
   B. Define the center of mass by giving its coordinates in terms of the coordinates of the individual particles in the system. As an example, consider a system consisting of three discrete particles and calculate the coordinates of the center of mass, given the masses and coordinates of the particles. Point out that no particle need be at the center of mass.
   C. Extend the definition to include a continuous mass distribution. Note that if the object has a point, line, or plane of symmetry, the center of mass must be at that point, on that line, or in that plane. Examples: a uniform sphere or spherical shell, a uniform cylinder, a uniform square, a rectangular plate, or a triangular plate. Show how to compute the coordinates of the center of mass of a complex object comprised of a several simple parts, a table for example. Each part is replaced by a particle with mass equal to the mass of the part, positioned at the center of mass of the part. The center of mass of the particles is then found.
   D. Explain that the general motion of a rigid body may be described by giving the motion of the center of mass and the motion of the object around the center of mass.
   E. Derive expressions for the velocity and acceleration of the center of mass in terms of the velocities and accelerations of the particles in the system.
   F. Derive $\sum \mathbf{F}_{\text{ext}} = M\mathbf{a}_{\text{cm}}$. As an example, consider a two-particle system with external forces acting on both particles and each particle interacting with the other. Invoke Newton's third law to show that the internal forces cancel when all forces are summed.

II. Momentum.

A. Define momentum for a single particle and show that the kinetic energy and momentum of a particle are related by $K = p^2/2m$. If you are including modern physics topics in the course also give the relativistic definitions of kinetic energy and momentum and give the relativistic relation between them.

B. Show that Newton's second law can be written $\sum \mathbf{F} = d\mathbf{p}/dt$ for a particle. Emphasize that the mass of the particle is constant and that this form of the law does not imply that a new term $\mathbf{v}dm/dt$ has been added to $\sum \mathbf{F} = m\mathbf{a}$.

C. State that the total momentum of a system of particles is the *vector* sum of the individual momenta and show that $\mathbf{P} = M\mathbf{v}_{cm}$.

D. Show that Newton's second law for the center of mass can be written $\sum \mathbf{F}_{ext} = d\mathbf{P}/dt$, where $\mathbf{P}$ is the total momentum of the system. Stress that this equation is valid only if the mass of the system is constant.

III. Conservation of linear momentum.

A. Point out that $\mathbf{P}$ = constant if $\sum \mathbf{F}_{ext} = 0$. Stress that one examines the *external* forces to see if momentum is conserved in any particular situation. Point out that one component of $\mathbf{P}$ may be conserved when others are not.

B. Put two carts, connected by a spring, on a horizontal air track and set them in oscillation by pulling them apart and releasing them from rest. Explain that the center of mass does not accelerate and the total momentum of the system is constant. Use the conservation of momentum principle to derive an expression for the velocity of one cart in terms of the velocity of the other. Push one cart and explain that the center of mass is now accelerating and the total momentum is changing.

C. Consider a projectile that splits in two and find the velocity of one part, given the velocity of the other. Point out that mechanical energy is conserved for the cart-spring system but is not for the fragmenting projectile. The exploding projectile idea can be demonstrated with an air track and two carts, one more massive than the other. Attach a brass tube to one cart and a tapered rubber stopper to the other. Arrange so that the tube is horizontal and the stopper fits in its end. The tube has a small hole in its side, through which a firecracker fuse fits. Start the carts at rest and light a firecracker in the tube. The carts rapidly separate, strike the ends of the track, come back together again, and stop. Arrange the initial placement so the carts strike the ends of the track simultaneously. Explain that $\mathbf{P} = 0$ throughout the motion. For a less dramatic demonstration, tie two carts together with a compressed spring between them, then cut the string.

D. State that if $\sum \mathbf{F}_{ext} = 0$ and the center of mass is initially at rest, then it remains at the same point no matter how individual parts of the system move. Refer to the two carts of C above. Consider two skaters with different mass on frictionless ice, separated by a given distance. When they pull on opposite ends of a rope they meet at their center of mass.

E. Explain that observers in two different inertial frames will measure different values of the momentum for a system but they will agree on the conservation of momentum. That is, if the net external force is found to vanish in one inertial frame, it vanishes in all inertial frames.

F. Illustrate the use of conservation of momentum to solve problems by considering the firing of a cannon initially resting on a frictionless surface. Assume the barrel is horizontal and calculate the recoil velocity of the cannon. Explain that muzzle velocity is measured relative to the cannon and that we must use the velocity of the cannonball relative to the Earth.

IV. Variable mass systems.

A. Derive the rocket equations, Eqs. 9–46 and 9–47. Emphasize that we must consider the rocket and fuel to be a single constant mass system. Calculate the momentum before and after a small amount of fuel is expelled and equate the time rate of change of momentum to the net external force.

B. To demonstrate, screw several hook eyes into a toy CO_2 propelled rocket, run a line through the eyes and string the line across the lecture hall. Start the rocket from rest and have the students observe its acceleration as it crosses the hall.

C. As a second example, consider the loading of sand on a conveyor belt and calculate the force required to keep the belt moving at constant velocity.

V. Internal work.

A. Show that if the system is replaced by a single particle with mass equal to the total mass of the system, located at the system center of mass, and acted on by a force equal to the total external force acting on the system, then the work done by the force equals the change in the translational kinetic energy (defined by $\frac{1}{2}mv_{\text{cm}}^2$).

B. Remark that the integral $\int \mathbf{F}_{\text{ext}} \cdot d\mathbf{s}$ over the path of the center of mass is NOT necessarily the work done by the net external force. In particular, it is not when the motions of the center of mass and the point of application of the force are different. In an extreme case the point of application of the external force does not move so the force does zero work, but the center of mass accelerates and the kinetic energy changes. Remind students of the energy equation $\Delta K + \Delta U + \Delta E_{\text{int}} = W$ and stress that W is the true work done by the external force. State that work done by internal forces may contribute to changes in the kinetic energy.

C. Discuss the example of an accelerating car. Point out that the force of the road on the tires accelerates the car but does no work if the tires do not skid. The work that changes the kinetic energy of the car is internal work, done by the engine.

SUGGESTIONS

1. Use questions 1, 3, and 4, along with problems 3, 4, and 7 to generate discussion about the position of the center of mass. To present a challenge, assign problem 12.
2. Question 9 is a good test of understanding of the motion of the center of mass. Discuss it as an introduction to the problems. Assign problem 20.
3. Assign some problems in which the center of mass does not move: 13, 15, 17, and 19, for example.
4. To emphasize the vector nature of momentum, assign problem 29.
5. Assign conservation of momentum problems, such as 33, 36, 39, and 43.
6. To generate a discussion of the role played by internal forces assign problems 58 and 59.
7. Assign problems such as 47, 53, and 55, which are concerned with variable mass systems.
8. Computer project
   Have students use a spreadsheet or write a computer program to follow individual particles in a two or three particle system, given the force law for the forces they exert on each other. The program should integrate Newton's second law for each particle. Have the students use their data to verify the conservation of momentum. See the Computer Projects section of this manual.
9. Demonstrations
   a. Center of mass, center of gravity: Freier and Anderson Mp7, 12, 13; Hilton M18a, b.
   b. Motion of center of mass: Freier and Anderson Mp1, 2, 16 — 19.
   c. Conservation of momentum: Freier and Anderson Mg4, 5, Mi2; Hilton M15.
   d. Rockets: Freier and Anderson Mh.

10. Audio/Visual
    a. *Center of Mass*, S8, color, 3 min. Kalmia Company, Dept. P1, Concord, MA 01742. This film is part of the Mechanics on an Air Track series. Reviewed TPT 16:334 (1976).
    b. *Human Moments*, S8, color, 4 min. American Association of Physics Teachers, 5110 Roanoke Place, College Park, MD 20740–4100. This film is part of the NASA-Skylab film series. Reviewed AJP 44:1021 (1976).
    c. *Bursting Projectile*; Bruce Marsh; American Association of Physics Teachers slide set (address above). Plot shows motions of the center of mass and fragments.
11. Laboratory
    a. MEOS Experiment 7–9: *Ballistic Pendulum — Projectile Motion.* A ball is shot into a trapping mechanism at the end of a pendulum. The initial speed of the ball is found by applying conservation of momentum to the collision and conservation of energy to the subsequent swing of the pendulum. Also see BE Experiment 11: *Inelastic Impact and the Velocity of a Projectile.*
    b. MEOS Experiment 8–6: *Center of Mass Motion.* Two pucks are connected by a rubber band or spring and move toward each other on a nearly frictionless surface. A spark timer is used to record their positions as functions of time. Students calculate and study the position of the center of mass as a function of time. They also find the center of mass velocity. Can be performed with dry ice pucks or on an air table or air track.
    c. MEOS Experiment 8–7: *Linear Momentum.* Essentially the same as 8–6 but data is analyzed to give the individual momenta and total momentum as functions of time. Kinetic energy is also analyzed.

Chapter 10 COLLISIONS

BASIC TOPICS

I. Properties of Collisions
   A. Set up a collision between two carts on an air track. Point out the interaction interval and the intervals before and after the interaction.
   B. Explain that for the collisions considered two bodies interact with each other over a short period of time and that the times before and after the collision are well defined. The force of interaction is great enough that external forces can be ignored during the interaction time. Explain that the identities of the bodies that exit the interaction may be different from those that enter: decays of fundamental particles and nuclei can be included in discussions of collisions.

II. Impulse.
   A. Define the impulse of a force as the integral over time of the force. Note that it is a vector. Clearly distinguish between impulse (integral over time) and work (integral over path). Draw a force vs. time graph for the force of one body on the other during a one-dimensional collision and point out the impulse is the area under the curve.
   B. Define the time averaged force and show that the impulse is the product $\mathbf{F}_{\text{ave}}\Delta t$, where Δt is the time of interaction. Remark that we can use the average force to estimate the strength of the interaction during the collision.
   C. Use Newton's second law to show that the impulse on a body equals the change in its momentum. Refer to the air track collision and point out that it is the impulse of one body on the other that changes the momentum of the second body. Repeat the air track collision. Measure the velocity of one cart before and after the collision and calculate the

change in its momentum. Equate this to the impulse the other cart exerts. Estimate the collision time and calculate the average force exerted on the cart.

D. Use the third law to show that two bodies in a collision exert equal and opposite impulses on each other and show that, if external impulses can be ignored, then total momentum is conserved. Refer to the air track collision. Again stress that external forces are neglected during the collisions considered here.

III. Two-body collisions in one dimension.

A. Define the terms "elastic", "inelastic", and "completely inelastic". Distinguish between the transfer of kinetic energy from one colliding object to the another and the loss of kinetic energy to internal energy.

B. Two-body elastic collisions.

1. Derive expressions for the final velocities in terms of the masses and initial velocities.
2. Specialize the general result to the case of equal masses and one body initially at rest. Demonstrate this collision on the air track using carts with spring bumpers. Point out that the carts exchange velocities.
3. Specialize the general result to the case of a light body, initially at rest, struck by a heavy body. Demonstrate this collision on the air track. Point out that the velocity of the heavy body is reduced only slightly and that the light body shoots off at high speed. Relate to a bowling ball hitting a pin.
4. Specialize the general result to the case of a heavy body, initially at rest, struck by a light body. Demonstrate this collision on the air track. Point out the low speed acquired by the heavy body and the rebound of the light body. Relate to a ball rebounding from a wall. A nearly elastic collision can be obtained with a superball.
5. Point out that, although the total kinetic energy does not change, kinetic energy is usually transferred from one body to the other. Consider a collision in which one body is initially at rest and calculate the fraction that is transferred. Show that the fraction is small if either mass is much greater than the other and that the greatest fraction is transferred if the two masses are the same. This is important, for example, in deciding what moderator to use to thermalize neutrons from a fission reactor.

C. Two-body completely inelastic collisions.

1. Derive an expression for the velocity of the bodies after the collision in terms of their masses and initial velocities.
2. Demonstrate the collision on an air track, using carts with velcro bumpers. Point out that the kinetic energy of the bodies is not conserved and calculate the energy loss. Remark that the energy is dissipated by the mechanism that binds the objects to each other. Some goes to internal energy, some to deformation energy. Note that $\frac{1}{2}(m_1 + m_2)v_{\text{cm}}^2$ is retained. If we use a reference frame attached to the center of mass to describe the collision, we would find the combined bodies at rest after the collision and all kinetic energy lost.

D. Point out that while the greatest energy loss occurs when the interaction is completely inelastic, there are many other inelastic collisions in which less than the maximum energy loss occurs. Note that it is possible to have a collision in which kinetic energy increases (an explosive impact, for example).

IV. Two-body collisions in two dimensions.

A. Write down the equations for the conservation of momentum, in component form, for a collision with one body initially at rest. Mention that these can be solved for two unknowns.

B. Consider an elastic collision for which one body is at rest initially and the initial velocity of the second is given. Write the conservation of kinetic energy and conservation of momentum equations. Point out that the outcome is not determined by the initial velocities but that the impulse of one body on the other must be known to determine the velocities of the two bodies after the collision. State that the impulse is usually not known and in practice physicists observe one of the outgoing particles to determine its direction of motion. Carry out the calculation: assume the final direction of motion of one body is known and calculate the final direction of motion of the other and both final speeds.

C. State, or perhaps prove, that if the particles have the same mass then their directions of motion after the collision are perpendicular to each other.

D. Consider a completely inelastic collision for which the two bodies do not move along the same line initially. Mention that the outcome of this type collision *is* determined by the initial velocities. Calculate the final velocity. Calculate the fraction of energy dissipated.

SUPPLEMENTARY TOPIC

Decay and reaction processes. This topic provides background for Chapters 47 and 49, on nuclear and particle physics and should be covered, now or later, if those chapters are included in the course. Remind students of the concept of mass energy and define the Q value of a decay or reaction. Work several examples. Sample Problem 10–11 which deals with a reaction, illustrates mass to kinetic energy conversion.

SUGGESTIONS

1. To help students with the concept of impulse ask questions 2, 4, and 5.
2. Assign and discuss some problems dealing with inelastic collisions for which the mechanism of kinetic energy loss is given explicitly. See problems 54 and 55. Also discuss questions 14 and 15.
3. Ask questions 9 and 16 in support of the discussion of elastic collisions. For some fun carry out the demonstration described in question 13 and assign problem 38.
4. Assign some problems, such as 29, that ask students to determine if a collision is elastic and some others, such as 44 or 73, that ask them to calculate the change in kinetic energy.
5. Demonstrate the ballistic pendulum and show how it can be used to measure the speed of a bullet. Assign problems 45 and 47. Problem 43 can also be assigned with this group.
6. If you intend to cover the thermodynamics chapters assign and discuss problem 14. Also consider problems 16, 17, and 18. They will prove helpful for the discussion of the microscopic basis of pressure.
7. Assign some problems from the group 60 through 73, which deal with two-dimensional collisions. Include both elastic and inelastic collisions.
8. Computer project
   Have students use a commercial math program or write a program to graph the total final kinetic energy as a function of the final velocity of one object in a two body, one-dimensional collision, given the initial velocities and masses of the two objects. Ask them to run the program for specific initial conditions and identify elastic, inelastic, completely inelastic, and explosive collisions on their graphs. Details of this and other projects are given in the Computer Projects section of this manual.
9. Demonstrations
   Freier and Anderson Mg1 — 3, Mi1, 3, 4, Mw3, 4; Hilton M15d, e, f, M19j.
10. Audio/Visual
    a. *Two Dimensional Collisions, I and II*, S8, color, 3 min. each. Kalmia Company, Dept. P1, Concord, MA 01742. Part of the Project Physics film series. Reviewed TPT 14:56 (1976).

b. *Collisions*, S8, color, 4 min. American Association of Physics Teachers, 5110 Roanoke Place, College Park, MD 20740–4100. Another film from the NASA–Skylab Film Series. Reviewed TPT 14:56 (1976).

11. Computer programs
   a. *Collisions on an Air Track*, Cambridge Development Laboratory, 1696 Massachusetts Avenue, Cambridge, MA 02138. Apple II. Chiefly tutorial but includes segments suitable for lecture illustrations. User specified collisions are simulated and in each case a numerical analysis of dynamic quantities (velocity, momentum, kinetic energy) is given. Reviewed TPT September 1985.
   b. *Collisions and Simple Harmonic Motion*, Educational Materials and Equipment Company, Old Mill Plain Road, Danbury, CT 06811. Apple II (+, e, c, gs). Animated problem-solving tutorials for inelastic collisions (both one and two dimensional) and for harmonic motion. Reviewed TPT May 1988.
12. Interactive videodisk
   Physics and Automobile Collisions by Dean Zollman (John Wiley, 1984). The disk shows collisions of cars with fixed barriers and two car collisions (head-on, at 90°, and at 60°). One sequence shows the influence of bumper design, others show the influence of air bags and shoulder straps on manikins. All are slow motion movies of manufacturers' tests and many show grids and clocks. Students can stop the action to take measurements, then make calculations of momentum and energy transfers. For most exercises a standard player is satisfactory; for a few a computer-controlled player is required.
13. Laboratory
   a. MEOS Experiment 7–10: *Impulse and Momentum.* Students use a microprocessor to measure the force as a function of time as a toy truck hits a force transducer. They numerically integrate the force to find the impulse, then compare the result with the change in momentum, found by measuring the velocity before and after the collision.
   b. BE Experiment 8: *Impulse and Momentum.* In part A a mass is hung on a string that passes over a pulley and is attached to an air track glider. The glider accelerates from rest for a known time and a spark timer is used to find its velocity at the end of the time. The impulse is calculated and compared with the momentum. In part B a glider is launched by a stretched rubber band and a spark record of its position as a function of time is made while it is in contact with the rubber band. A static technique is used to measure the force of the rubber band for each of the recorded glider positions and the impulse is approximated. The result is again compared with the final momentum of the glider.
   c. MEOS Experiment 7–11: *Scattering* (for advanced groups). The deflection of pellets from a stationary disk is used to investigate the scattering angle as a function of impact parameter and to find the radius of the disk.
   d. MEOS Experiment 8–5: *One-Dimensional Collisions.* A puck moving on a nearly frictionless surface collides with a stationary puck. A spark timer is used to record the positions of the pucks as functions of time. Students calculate the velocities, momenta, and energies before and after the collision. May be performed with dry ice pucks or on an air table or track.
   e. MEOS Experiment 8–8: *Two-Dimensional Collisions.* Same as MEOS 8–5 but the pucks are allowed to scatter out of the original line of motion. Students must measure angles and calculate components of the momenta. The experiment may be performed with dry ice pucks or on an air table.
   f. BE Experiment 12: *Elastic Collisions — Momentum and Energy Relations in Two Dimensions.* A ball rolls down an incline on a table top and strikes a target ball initially at rest at the edge of the table. The landing points of the balls on the floor are used to find their

velocities just after the collision. The experiment is run without a target ball to find the velocity of the incident ball just before the collision. Data is used to check for conservation of momentum and energy. Both head-on and grazing collisions are investigated. A second experiment, similar to MEOS 8–8, is also described.

Chapter 11 ROTATION

BASIC TOPICS

I. Rotation about a fixed axis.

A. Spin an irregular object on a fixed axis. A bicycle wheel or spinning platform with the object attached can be used. Draw a rough diagram, looking along the axis. Explain that each point in the body has a circular orbit and that, for any selected point, the radius of the orbit is the perpendicular distance from the point to the axis. Contrast to a body that is simultaneously rotating and translating. See Figs. 11–2 and 3.

B. Define angular position θ (in radians and revolutions), angular displacement $\Delta\theta$, angular velocity ω (in rad/s, deg/s, and rev/s) and angular acceleration α (in rad/s^2, deg/s^2, and rev/s^2). Treat both average and instantaneous quantities but emphasize that the instantaneous quantities are most important for us. Remind students of radian measure.

C. Use Fig. 11–4 to show how an angular displacement is measured. Note that θ is positive for an angle in one direction and negative for an angle in the other. By convention in this text position angles are positive in the counterclockwise direction. Remark that as the body rotates θ continues to increase, even after a revolution.

D. Interpret the signs of ω and α. Give examples of spinning objects for which ω and α have the same sign and for which they have opposite signs.

E. Point out the analogy to one-dimensional linear motion. θ corresponds to x, ω to v, and α to a.

F. Point out that ω and α can be thought of as the components of vectors $\boldsymbol{\omega}$ and $\boldsymbol{\alpha}$, respectively. For fixed axis rotation, the vectors lie along the axis, with the direction of $\boldsymbol{\omega}$ determined by a right hand rule: if the fingers curl in the direction of rotation, then the thumb points in the direction of $\boldsymbol{\omega}$. If $d\omega/dt > 0$ then $\boldsymbol{\alpha}$ is in the same direction; if $d\omega/dt < 0$ then it is in the opposite direction. Use Fig. 11–6 to explain that a vector cannot be associated with a finite angular displacement because two displacements to not add as vectors.

II. Rotation with constant angular acceleration.

A. Emphasize that the discussion here is restricted to rotation about a fixed axis but that the same equations can be used when the axis is in linear translation. This type motion will be discussed in the next chapter.

B. Write down the kinematic equations for $\theta(t)$ and $\omega(t)$. Make a comparison with the analogous equations for linear motion (see Table 11–1).

C. Point out that the problems of rotational kinematics are similar to those for one-dimensional linear kinematics and that the same strategies are used for their solution.

D. Go over examples. Calculate the time and number of revolutions for an object to go from some initial angular velocity to some final angular velocity, given the angular acceleration. If time permits, consider both a body that is speeding up and one that is slowing down. For the latter calculate the time to stop and the number of revolutions made while stopping. Calculate the time to rotate a given number of revolutions and the final angular velocity, again given the angular acceleration.

III. Linear speed and acceleration of a point rotating about a fixed axis.

A. Write down $s = r\theta$ for the arc length. Explain that it is a rearrangement of the defining equation for the radian and that θ must be in radians for it to be valid.

B. Wrap a string on a large spool that is free to rotate about a fixed axis. Mark the spool so the angle of rotation can be measured. Slowly pull out the string and explain that the length of string pulled out is equal to the arc length through which a point on the rim moves. Compare the string length to $r\theta$ for $\theta = \pi/2$, π, $3\pi/2$, and 2π rad. Show that $s = r\theta$ reduces to the familiar result for $\theta = 2\pi$ rad.

C. Differentiate $s = r\theta$ to obtain $v = r\omega$ and $a_t = r\alpha$. Emphasize that radian measure *must* be used. Point out that v gives the speed and a_t gives the acceleration of the string as it is pulled provided it does not slip on the spool. Point out that all points in a rotating rigid body have the same value of ω and the same value of α but points that are different distances from the axis have different values of v and different values of a_t.

D. Point out that the velocity is tangent to the circular orbit but that the total acceleration is not. a_t gives the tangential component while $a_r = v^2/r = r\omega^2$ gives the radial component. The tangential component is not zero only when the point on the rim speeds up or slows down in its rotational motion while the radial component is not zero as long as the object is turning. For students who have forgotten, reference the derivation of $a_r = v^2/r$, given in Chapter 4.

E. Explain how to find the magnitude and direction of the total acceleration in terms of ω, α, and r.

IV. Kinetic energy of rotation and rotational inertia.

A. Show that $K = \frac{1}{2}mr^2\omega^2$ for a particle moving around a circle with angular velocity ω and $K = \frac{1}{2}I\omega^2$, where $I = \sum m_i r_i^2$, for a body rotating about a fixed axis. Explain that I is called the rotational inertia of the body. Mention that many texts call it the moment of inertia.

B. Point out that rotational inertia depends on the distribution of mass and on the position and orientation of the rotation axis. Explain that two bodies may have the same mass but quite different rotational inertias. State that Table 11–2 gives the rotational inertia for various objects and axes. Particularly point out the rotational inertia of a hoop rotating about the axis through its center and perpendicular to its plane. Note that all its mass is the same distance from the axis. Also point out the rotational inertias of a cylinder rotating about its axis and a sphere rotating about a diameter. Note that the mass is now distributed through a range of distances from the axis and the rotational inertia is less than that for a hoop with the same mass and radius.

C. Optional: show how to convert the sum for I to an integral. Use the integral to find the rotational inertia for a thin rod rotating about an axis through its center and perpendicular to its length. If the students have experience with volume integrals using spherical coordinates, derive the expression for the rotational inertia of a sphere.

D. Prove the parallel axis theorem. The proof can be carried out using a sum for I rather than an integral. Explain its usefulness for finding the rotational inertia when the axis is not through the center of mass. Emphasize that the actual axis and the axis through the center of mass must be parallel for the theorem to be valid. Point out the two entries in Table 11–2 for a thin rod and use the parallel axis theorem to obtain the rotational inertia for rotation about one end from the rotational inertia for rotation about the center.

V. Torque.

A. Define torque for a force acting on a single particle. Consider forces that lie in planes perpendicular to the axis of rotation and take $\tau = rF\sin\phi$, where $\mathbf{r}$ is a vector that is perpendicular to the axis and points from the axis to the point of application of the force.

ϕ is the angle between **r** and **F** when they are drawn with their tails at the same point. The definition will be generalized in the next chapter. Explain that $\tau = rF_t = r_\perp F$, where F_t is the tangential component of **F** and $r_\perp$ is the moment arm.

B. Explain that the torque vanishes if **F** is along the same line as **r** and that only the component of **F** that is perpendicular to **r** produces a torque. This is a mechanism for picking out the part of the force that produces angular acceleration, as opposed to the part that produces centripetal acceleration. Also explain that the same force can produce a larger torque if it is applied at a point farther from the axis.

C. Use a wrench tightening a bolt as an example. The force is applied perpendicular to the wrench arm and long moment arms are used to obtain large torques.

D. Explain the sign convention for torques applied to a body rotating about a fixed axis. For example, torques tending to give the body a counterclockwise (positive) angular acceleration are positive while those tending to give the body a clockwise angular momentum are negative. Remark that the convention is arbitrary and the opposite convention may be convenient for some problems.

VI. Newton's second law for rotation.

A. Use a single particle on a circular orbit to introduce the topic. Start with $F_t = ma_t$ and show that $\tau = I\alpha$, where $I = mr^2$. Explain that this equation also holds for extended bodies, although I is then the sum given above.

B. Remark that problems are solved similarly to linear second law problems. Tell students to identify torques, draw a force diagram, choose the direction of positive rotation, and substitute the total torque into $\sum \tau = I\alpha$. Remark that the point of application of a force is important for rotation, so the object cannot be represented by a dot on a force diagram. Tell students to sketch the object and place the tails of force vectors at the application points.

C. Wrap a string around a cylinder, free to rotate on a fixed horizontal axis. Attach the free end of the string to a mass and allow the mass to fall from rest. Note that its acceleration is less than g, perhaps by dropping a free mass beside it. See Sample Problem 11–11.

VII. Work-energy theorem for rotation.

A. Use $dW = \mathbf{F}\cdot d\mathbf{r}$ to show that the work done by a torque is given by $W = \int \tau\, d\theta$ and that the power delivered is given by $P = \tau\omega$.

B. Use $\tau\, d\theta = I\alpha\, d\theta = \frac{1}{2}I\, d(\omega^2)$ to show that $W = \frac{1}{2}I(\omega_f^2 - \omega_i^2)$.

C. For the situation of Sample Problem 11–11 use conservation of energy to find the angular velocity of the cylinder after the mass has fallen a distance h. Use rotational kinematics and the value for the angular acceleration found in the text to check the answer.

SUGGESTIONS

1. Use questions 1, 2, and 6 and problem 1 to discuss radian measure.
2. Use techniques of the calculus to derive the kinematic equations for constant angular acceleration. That is, integrate α = constant twice with respect to time. Assign some problems from the group 2, 4, 5, and 6.
3. To discuss the relationship between angular and linear variables, use questions 13, 14 and 15. Assign problems 32, 33, and 40.
4. Use questions 16, 17, 18, and 20 to guide students through a qualitative discussion of rotational inertia. Also ask questions 19 and 21 to see if they understand the physical significance. Assign problems 49 and 51. Use problem 58 to discuss the radius of gyration, if desired.
5. Use problem 62 or 63 to discuss the calculation of torque.

6. To help students think about $\sum\tau = I\alpha$, discuss problems 67, 68, and 70. To deal with a situation in which the dynamics of more than one object is important, demonstrate the Atwood machine and discuss problem 73. Also assign problem 75.
7. Discuss energy conservation and assign problems 83 and 86.
8. Computer program
   Motion. See notes for Chapter 2.
9. Computer project
   Ask students to use a commercial math program or their own root finding programs to solve rotational kinematic problems.
10. Demonstrations
    a. Rotational variables: Hilton M16a.
    b. Rotational dynamics: Freier and Anderson Ms7, Mt 5, 6, Mo5.
    c. Rotational work and energy: Freier and Anderson Mv2, Mr5, Ms2.
11. Audio/visual
    Non-Commutivity of Angular Displacements; Bruce Marsh; American Association of Physics Teachers, 5112 Berwyn Road, College Park, MD 20740-4100. Slides showing the results of several rotations performed in different order.
12. Laboratory
    MEOS Experiment 7–14: *Rotational Inertia.* The rotational inertia of a disk is measured dynamically by applying a torque (a falling mass on a string wrapped around a flange on the disk). A microprocessor is used to measure the angular acceleration. Small masses are attached to the disk and their influence on the rotational inertia is studied. The acceleration of the mass can also be found by timing its fall through a measured distance. Then $a_t = r\alpha$ is used to find the angular acceleration of the disk. Also see BE Experiment 14: *Moment of Inertia.*

Chapter 12 ROLLING, TORQUE, AND ANGULAR MOMENTUM

BASIC TOPICS

I. Rolling
   A. Remark that a rolling object can be considered to be rotating about an axis through the center of mass while the center of mass moves. The text considers the special case for which the axis of rotation does not change direction. Point out that the rotational motion obeys $\sum\tau = I\alpha$ and the translational motion of the center of mass obeys $\sum\mathbf{F} = m\mathbf{a}$, where $\sum\tau$ is the sum of external torques and $\sum\mathbf{F}$ is the sum of external forces. Emphasize that one of the forces acting is the force of friction produced by the surface on which the object rolls.
   B. Explain that the speed of a point at the top of a rolling object is $v_{\text{cm}} + \omega R$ and the speed of a point at the bottom is $v_{\text{cm}} - \omega R$. Specialize to the case of rolling without slipping. Point out that the point in contact with the ground has zero velocity, so $v_{\text{cm}} = \omega R$. Use Fig. 12–4 as evidence. Also point out that tire tracks in the snow are clean (not smudged) if the tires do not slip.
   C. Use Fig. 12–5 to explain that a wheel rolling without slipping can be viewed as rotating about an axis through the point of contact with the ground. Use this and the parallel axis theorem to show that the kinetic energy is $\frac{1}{2}Mv_{\text{cm}}^2 + \frac{1}{2}I_{\text{cm}}\omega^2$.

D. Consider objects rolling down an inclined plane and show how to calculate the speed at the bottom using energy considerations. If time permits, carry out an analysis using the equations of motion and show how to find the frictional force that prevents slipping.

E. Roll a sphere, a hoop, and a cylinder, all with the same radius and mass, down an incline. Start the objects simultaneously at the same height and ask students to pick the winner. Point out that the speed at the bottom is determined by the dimensionless parameter $\beta = I/MR^2$ and not by I, M, and R alone. All uniform cylinders started from rest reach the bottom in the same time and have the same speed when they get there.

F. Demonstrate the object shown in Fig. 12–26, built from three concentric cylinders with the string wrapped around the smallest, like a yo-yo. Place it on the table and pull the string at various angles. The three different possibilities are shown in the figure. Work out the answers to question 10.

G. Consider a ball striking a bat. Show how to find the point at which the ball should hit so the instantaneous center of rotation is at the place where the bat is held. The striking point is called the center of percussion. When the ball hits there the batter feels no sting.

II. Torque and angular momentum.

A. Define torque as $\boldsymbol{\tau} = \mathbf{r} \times \mathbf{F}$ and explain that this is the general definition. Review the vector product, give the expression for the magnitude ($\tau = rF\sin\phi$), and give the right hand rule for finding the direction. Explain that $\boldsymbol{\tau} = 0$ if $\mathbf{r} = 0$, $\mathbf{F} = 0$, or $\mathbf{r}$ is parallel (or antiparallel) to $\mathbf{F}$.

B. Consider an object going around a circle and suppose a force is applied tangentially. Take the origin to be on the axis but not at the circle center and show the general definition reduces to the expression used in Chapter 11: $\tau = F_t R$, where R is the radius of the circle (not the distance from the origin to the point of application).

C. Define angular momentum for a single particle, using vector notation. Give the expression for the magnitude and the right-hand rule for the direction.

D. Derive the relationship between angular momentum and angular velocity for a particle moving on a circle centered at the origin. Also find the angular momentum if the origin is on a line through the circle center, perpendicular to the circle, but not at the center. Explain that the component along the axis is independent of the position of the origin along the line and that the component perpendicular to the axis rotates with the particle.

E. To show that a particle may have angular momentum even if it is not moving in a circle, calculate the angular momentum of a particle moving with constant velocity along a line not through the origin. Point out that the angular momentum depends on the choice of origin. In preparation for F below you might want to find the time rate of change of $\boldsymbol{\ell}$.

F. Use Newton's second law to derive $\sum \boldsymbol{\tau} = d\boldsymbol{\ell}/dt$ for a particle. Consider a particle moving in a circle, subjected to both centripetal and tangential forces. Take the origin to be at the center of the circle and show that $\sum \boldsymbol{\tau} = d\boldsymbol{\ell}/dt$ reduces to $\sum F_t = ma_t$, as expected. Take the origin be on the line through the center, perpendicular to the circle, but not at the center. Show that the torque associated with the centripetal force produces the change in $\boldsymbol{\ell}$ expected from the discussion of D.

G. Show that the magnitude of the torque about the origin exerted by gravity on a falling mass is mgd, where d is the perpendicular distance from the line of fall to the origin. Write down the velocity as a function of time and show that the angular momentum is $mgtd$. Remark that $\sum \boldsymbol{\tau} = d\boldsymbol{\ell}/dt$ by inspection. See Sample Problem 7.

III. Systems of particles.

A. Explain that the total angular momentum for a system of particle is the vector sum of the individual momenta.

B. Show that $\sum \tau_{\text{ext}} = \mathrm{d}\mathbf{L}/\mathrm{d}t$ for a system of particles for which internal torques cancel. Emphasize that $\sum \tau_{\text{ext}}$ is the result of summing all torques on all particles in the system and that **L** is the sum of all individual angular momenta. Demonstrate in detail the cancellation of internal torques for two particles which interact via central forces. Point out that this equation is the starting point for investigations of the rotational motion of bodies.

C. Show that the component along the axis of the total angular momentum of a rigid body rotating about a fixed axis is $I\omega$. Use the example of a single particle to point out that the angular momentum vector is along the axis if the body is symmetric about the axis but that otherwise it is not. Emphasize that for fixed axis rotation we are chiefly interested in the components of angular momentum and torque along the rotation axis.

D. Make a connection to material of the last chapter by showing that $L = I\omega$ and $\sum \tau = \mathrm{d}L/\mathrm{d}t$ lead to $\sum \tau_{\text{ext}} = I\alpha$ for a rigid body rotating about a fixed axis. Here $\sum \tau_{\text{ext}}$ is the component of the total external torque along the axis.

IV. Conservation of angular momentum.

A. Point out that **L** =constant if $\sum \tau_{\text{ext}} = 0$. State that different objects in a system may change each other's angular momentum but the changes sum vectorially to 0. Also explain that the rotational inertia of an object may change while it is spinning. Then $I_i\omega_i = I_f\omega_f$ if the net external torque vanishes.

B. As examples consider a mass dropped on the rim of a freely spinning platform, a person running tangent to the rim of a merry-go-round and jumping on, and a spinning skater whose rotational inertia is changed by dropping her arms.

C. The third example can be demonstrated easily if you have a rotating platform that can hold a person. Have a student hold weights in each hand to increase the rotational inertia. Start him spinning with arms extended, then have him bring his arms in toward the sides of his body. See Fig. 16. Also carry out the spinning bicycle wheel demonstration described in the text. See Fig. 20.

V. Precession. (optional)

A. Point out that if a torque is always perpendicular to the angular momentum, then the angular momentum vector rotates in the plane of $\boldsymbol{\tau}$ and **L** and maintains a constant magnitude. If you have a bicycle wheel with a leaded rim, set up the demonstration shown in Fig. 23. Before letting go, point out the directions of the torque due to gravity and the angular momentum and ask students to predict what the wheel will do.

B. Use a diagram to point out the direction of the torque. Draw an overhead view and show the angular momentum vector at two times a short interval apart. Note that the change in angular momentum is in the direction of the torque. Remark that the torque remains perpendicular to the angular momentum as the axle turns.

C. Measure the time for one precession and calculate the precessional angular velocity. Add weights to the axle near the wheel and repeat.

D. Geometrically show that if the angular momentum vector turns through the angle $\Delta\phi$ then the magnitude of the change in the angular momentum is $\Delta L = L\Delta\phi$. Equate this to $\tau\Delta t$ and show that the precessional angular velocity is given by $\Omega = \tau/L$. Finally show that the torque is mgr, where m is the mass of the wheel and any attached weights and r is the distance from the support to the center of the wheel.

SUGGESTIONS

1. In connection with rolling on an incline ask questions 1 through 4 and assign problems 1, 3, 6, and 7. To emphasize the condition for no slipping ask questions 5 and 7 and assign problem 9.

2. Assign problems 22 and 34 to stress the importance of the origin in calculations of torque and angular momentum. Problem 29 shows students how to calculate the angular momentum if the cartesian components of the position and momentum vectors are given. Be sure students can calculate the angular momentum of an object moving along a straight line and the angular momentum of an object moving in a circle. See problems 26 and 32. Also consider discussing the angular momentum of a projectile. See problem 41. Include problem 45 in your discussion of the relationship between angular momentum and angular velocity.

3. Use problem 50 to start the discussion of conservation of angular momentum. Problems 56, 58, and 62 include motion along a straight line. Questions 18, 19, and 20 and problems 51, 54, and 55 deal with changes in rotational inertia. Problems 52 and 53 deal with inelastic rotational collisions. Assign one or more from each of these groups.

4. Assign problems 66 and 67 in connection with precession.

5. Computer project
Given the law for the torque between two rotating rigid bodies a computer program or spreadsheet can be used to integrate Newton's second law for rotation and tabulate the angular positions and angular velocities as functions of time. The data can be used to verify the conservation of angular momentum. See the Computer Projects section of this manual.

6. Demonstrations
   a. Rolling: Freier and Anderson Mb4, 7, 30, Mo3, Mp3, Mr1, 4, Ms1, 3, 4, 6; Hilton M10d, M19c.
   b. Conservation of angular momentum: Freier and Anderson Mt1 — 4, 7, 8, Mu1; Hilton M8b, M19i.
   c. Gyroscopes: Freier and Anderson Mu2 — 18; Hilton M19a, b, f, g, h.

7. Audio/Visual
   a. *Angular Momentum: A Vector Quantity*, 16 mm, b/w, 27 min. Modern Learning Aids, Division of Ward's Natural Science Establishment, P.O. Box 312, Rochester, NY 14601. Angular momenta are added vectorially using three spinning wheels.
   b. *Conservation Laws in Zero-G*, 16 mm, color, 18 min. National Audiovisual Center, General Services Administration, Washington, DC 20409. This NASA film contains scenes both from earth-bound and the "zero g" skylab environment in which astronauts are shown spinning in space.
   c. *Human Momenta*; *Moving Astronauts*; *Acrobatic Astronauts*; *Games Astronauts Play*; *Gyroscopes*, S8, color, 4 min. each. American Association of Physics Teachers 5112 Berwyn Road, College Park, MD 20740-4100. These loops are from the skylab series and take advantage of "weightlessness" conditions to demonstrate various rotational motions. Series reviewed AJP 44:1021 (1976).

8. Laboratory
   a. MEOS Experiment 7–12: *Rotational and Translational Motion.* Students measure the center of mass acceleration of various bodies rolling down an incline and calculate the center of mass velocities at the bottom. Results are compared to measured velocities. It is also instructive to use energy methods to find the final speeds.
   b. MEOS Experiment 7–13: *Rotational Kinematics and Dynamics.* Students find the velocity and acceleration of a ball rolling around a loop-the-loop and analyze the forces acting on it.
   c. MEOS Experiment 8–9: *Conservation of Angular Momentum.* Uses the Pasco rotational dynamics apparatus. A ball rolls down a ramp and becomes coupled to the rim of a disk that is free to rotate on a vertical axis. Students measure the velocity of the ball

before impact and the angular velocity of the disk-ball system after impact, then check for conservation of angular momentum.

Chapter 13 EQUILIBRIUM AND ELASTICITY

BASIC TOPICS

I. Conditions for equilibrium.

A. Write down the equilibrium conditions for a rigid body: $\sum \mathbf{F}_{\text{ext}} = 0$, $\sum \tau_{\text{ext}} = 0$ (about any point). Remind students that only external forces and torques enter. Explain that these conditions mean that the acceleration of the center of mass and the angular acceleration about the center of mass both vanish. The body may be at rest or its center of mass may be moving with constant velocity or the body may be rotating with constant angular momentum. Point out that the equilibrium conditions form 6 equations that are to be solved for unknowns, usually the magnitudes of some of the forces or the angles made by some of the forces with fixed lines. Explain that we will be concerned chiefly with static equilibrium for which $\mathbf{P} = 0$ and $\mathbf{L} = 0$. Remark that the subscript "ext" is usually omitted.

B. Show that, for a body in equilibrium, $\sum \tau_{\text{ext}} = 0$ about *every* point.

C. Explain that the gravitational forces and torques, acting on individual particles of the body, can be replaced by a single force acting at a point called the center of gravity. If the gravitational field is uniform over the body the center of gravity coincides with the center of mass and the magnitude of the replacement force is Mg, where M is the total mass.

II. Solution of problems.

A. Give the problem solving steps: isolate the body, identify the forces acting on it, draw a force diagram, choose a reference frame for the resolution of the forces, choose a reference frame for the resolution of the torques, write down the equilibrium conditions in component form, and solve these simultaneously for the unknowns. Point out that the two reference frames may be different and that the reference frame for the resolution of torques can often be chosen so that one or more of the torques vanish.

B. Work examples. Consider an arm and hand holding a mass (Sample Problem 13–2) or a ladder leaning against a wall (Sample Problems 13–3 and 4). In each case show how the situation can be analyzed qualitatively to find the directions of the forces, then solve quantitatively.

C. Use 4 or more bricks to set up the situation described in Fig. 13-37. Point out how large the overhang can be, then assign problem 30 or discuss the solution in the lecture.

III. Elasticity.

A. Point out that you have been considering mainly rigid bodies until now. Real objects deform when external forces are applied. Explain that deformations are often important for determining the equilibrium configuration of a system.

B. Consider a rod of unstrained length L subjected to equal and opposite forces F applied uniformly at each end, perpendicular to the end. Define the stress as F/A, where A is the area of an end. Define strain as the fractional change in length $\Delta L/L$ caused by the stress. Explain that stress and strain are proportional if the stress is sufficiently small. Define Young's modulus E as the ratio of stress to strain and show that $\Delta L = FL/EA$. Explain that Young's modulus is a property of the object and point out Table 13–1.

C. Explain that if the stress is small, the object returns to its original shape when the stress is removed and it is said to be *elastic*. Explain what happens if the stress is large and define *yield strength* and *ultimate strength*.

D. Calculate the fractional change in length for compressional forces acting on rods made of various materials. Use data from Table 13–1.

E. Explain that when the forces are parallel to the ends shearing occurs. Define the stress as F/A and the strain as $\Delta x/L$ where Δx is the displacement of one end relative to the other. Define the shear modulus G as the ratio of stress to strain and show that $\Delta x = FL/GA$.

F. Explain hydraulic compression. Define pressure as the force per unit area exerted by the fluid on the object. Explain that the pressure is now the stress and the fractional volume change $\Delta V/V$ is the strain. Define the bulk modulus B by $p = B\Delta V/V$.

G. To show how elastic properties are instrumental in determining equilibrium go over Sample Problem 9 or a similar problem.

SUGGESTIONS

1. Use questions 6, 7, 9, 10, and 12 to help students gain understanding of the equilibrium conditions in specific situations. Use problem 2 to test for understanding of the equilibrium conditions. Assign a few problems, such as 7, 8, and 10, for which only the total force is important. Assign others, such as 13, 15, 16, and 18, for which torque is also important. To provide a greater challenge assign problems 35 and 44.
2. The fundamentals of elasticity are covered in problems 46 (Young's modulus), 47 (elongation), 50 (shear), and 51 (hydraulic compression). Also assign one or more of problems 54, 55, and 56, in which the laws of elasticity are used in conjunction with the equilibrium conditions to solve for forces and their points of application.
3. Demonstrations
   Freier and Anderson Mo1, 2, 4, 6 — 9, Mp4 — 6, 9, 11, 14, 15, Mq1, 2.
4. Audio/visual
   Equilibrium Examples; Bruce Marsh; American Association of Physics Teachers, 5112 Berwyn Road, College Park, MD 20740-4100. Examples of mechanical equilibrium are shown on slides along with vector diagrams for analysis.
5. Computer program
   Statics; Cross Educational Software, P.O. Box 1536, Ruston, LA 71270. Apple II+. Tutorial on solving equilibrium problems, examples, problems for students to solve. Reviewed TPT February 1983.
6. Laboratory
   a. BE Experiment 5: *Balanced Torques and Center of Gravity.* A non-uniform rod is pivoted on a fulcrum. A single weight is hung from one end and the pivot point moved until equilibrium is obtained. The data is used to find the center of gravity and mass of the rod. Additional weights are hung and equilibrium is again attained. The data is used to check that the net force and net torque vanish.
   b. BE Experiment 6: *Equilibrium of a Crane.* Students study a model crane: a rod attached to a wall pivot at one end and held in place by a string from the other end to the wall. Weights are attached to the crane and the equilibrium conditions are used to calculate the tension in the rod and in the string. The latter is measured with a spring balance.
   c. MEOS Experiment 7–16: *Elongation of an Elastomer* (see Chapter 7 notes).
   d. MEOS Experiment 7–17: *Investigation of the Elongation of an Elastomer with a Microcomputer.* Same as MEOS 7–16 but a microprocessor is used to plot the elongation as a function of applied force. A polar planimeter is used to calculate the work done.

Chapter 14 OSCILLATIONS

BASIC TOPICS

I. Oscillatory motion.

A. Set up an air track and a cart with two springs, one attached to each end. Mark the equilibrium point, then pull the cart aside and release it. Point out the regularity of the motion and show where the speed is the greatest and where it is the least. By reference to the cart define the terms periodic motion, equilibrium point, period, frequency, cycle, and amplitude.

B. Explain that $x(t) = x_m \cos(\omega t + \phi)$ describes the coordinate of the cart as a function of time if $x = 0$ is taken to be the equilibrium point, where the force of the springs on the cart vanishes. State that this type motion is called simple harmonic. Show where $x = 0$ is on the air track, then show what is meant by positive and negative x. Sketch a mass on the end of a single spring and explain that the mass also moves in simple harmonic motion if dissipative forces are negligible.

C. Discuss the equation for $x(t)$.

1. Explain that x_m is the maximum excursion of the mass from the equilibrium point and that the spring is compressed by x_m at one point in a cycle. x_m is called the *amplitude* of the oscillation. Explain that the amplitude depends on initial conditions. Draw several $x(t)$ curves, identical except for amplitude. Illustrate with the air track apparatus.
2. Note that ω is called the angular frequency of the oscillation and is given in radians/s. Define the *frequency* by $f = \omega/2\pi$ and the *period* by $T = 1/f$. Show that $T = 2\pi/\omega$ is in fact the period by direct substitution into $x(t)$; that is, show $x(t) = x(t + T)$. Explain that the angular frequency does not depend on the initial conditions. For the cart on the track, use a timer to show that the period, and hence ω, is independent of initial conditions. Draw several $x(t)$ curves, for oscillations with different periods. Replace the original springs with stiffer springs and note the change in period. Also replace the cart with a more massive cart and note the change in period.
3. Define the phase of the motion and explain that the phase constant ϕ is determined by initial conditions. Draw several $x(t)$ curves, identical except for ϕ, and point out the different conditions at $t = 0$. Remark that the curves are shifted copies of each other. Illustrate various initial conditions with the air track apparatus.

D. Derive expressions for the velocity and acceleration as functions of time for simple harmonic motion. Show that the speed is a maximum at the equilibrium point and is zero when $x = \pm x_m$. Also show that the magnitude of the acceleration is a maximum when $x = \pm x_m$ and is zero at the equilibrium point.

E. Show that the initial conditions are given by $x_0 = x_m \cos\phi$ and $v_0 = -x_m\omega \sin\phi$. Solve for x_m and ϕ: $x_m^2 = x_0^2 + v_0^2/\omega^2$ and $\tan\phi = -v_0/\omega x_0$. Calculate x_m and ϕ for a few special cases: $x_0 = 0$ and v_0 positive, $x_0 = 0$ and v_0 negative, x_0 positive and $v_0 = 0$, x_0 negative and $v_0 = 0$. Tell students how to test the result given by a calculator for ϕ to see if π must be added to it.

II. The force law.

A. State the force law for an ideal spring: $F = -kx$. Point out that the minus sign is necessary for the force to be a restoring force. Hang identical masses on springs with different spring constants, measure the elongations and calculate the spring constants. Remark that stiff springs have larger spring constants than weak springs. Remark that the expression for the force is an idealization. It is somewhat different for real springs.

B. Start with Newton's second law and derive the differential equation for $x(t)$. Show that $x = x_m \cos(\omega t + \phi)$ satisfies the equation if $\omega = \sqrt{k/m}$ and explain that this is the most general solution for a given spring constant and mass.

C. Show a vertical spring-mass system. Point out that the equilibrium point is determined by the mass, force of gravity, and the spring constant. Show, both analytically and with the apparatus, that the force of gravity does not influence the period, phase, or amplitude of the oscillation.

III. Energy considerations.

A. Derive expressions for the kinetic and potential energies as functions of time. Show that the total mechanical energy is constant by adding the two expressions. Remark that the energy is wholly kinetic at the equilibrium point and wholly potential at a turning point. It changes from one form to the other as the mass moves between these points.

B. Show how to use the conservation of energy to find the amplitude, given the initial position and velocity, to find the maximum speed, and to find the speed as a function of position.

IV. Applications.

A. Demonstrate a torsional pendulum and discuss it analytically. Derive the differential equation for the angle as a function of time and compare with the differential equation for a spring to obtain the angular frequency and period in terms of the spring constant and the rotational inertia.

B. Demonstrate a simple pendulum and discuss it analytically in the small amplitude approximation. Derive the differential equation for the angle as function of time and obtain expressions for the angular frequency and period from the equation. Emphasize that the angular displacement must be measured in radians for the small angle approximation to be valid. Have students use their calculators to find the sines of some angles, in radians, starting with large angles and progressing to small angles.

C. Demonstrate a physical pendulum. Use Newton's second law for rotation to obtain the differential equation for the angular displacement. Obtain expressions for its angular frequency and period in the small amplitude approximation. Remind students that the rotational inertia depends on the position of the pivot and show them how to use the parallel axis theorem to find its value.

V. Simple harmonic motion and uniform circular motion.

A. Mount a bicycle wheel vertically and arrange for it to be driven slowly with uniform angular speed. Attach a tennis ball to the rim and project the shadow of the ball on the wall. Note that the shadow moves up and down in simple harmonic motion. Point out that the period of the wheel and the period of the shadow are the same. It is possible to suspend a mass on a spring near the wall and adjust the angular speed and initial conditions so the mass and shadow move together for several cycles. A period of about 1 s works well.

B. Analytically show that the projection of the position vector of a particle in uniform circular motion undergoes simple harmonic motion. Mention the converse: if an object simultaneously undergoes simple harmonic motion in two orthogonal directions, with the same amplitude and frequency, but a $\pi/2$ phase difference, the result is a circular orbit.

VI. Damped and forced harmonic motion.

A. Write the differential equation for a spring-mass oscillator with a damping term proportional to the velocity. Treat the case $(b/2m)^2 < k/m$ and write the solution, including the expression for the angular frequency in terms of k, m, and b. If there is time, prove it is the solution by direct substitution into the differential equation or leave the proof as an exercise for the students. Remark that the natural angular frequency is nearly $\sqrt{k/m}$ if damping is small.

B. Show a graph of the displacement as a function of time. See Fig. 14–19. Point out the exponential decay of the amplitude. Mention that the oscillator loses mechanical energy to dissipative forces.

C. Explain that if $(b/2m)^2 > k/m$ then the mass does not oscillate but rather moves directly back to the equilibrium point. The displacement is a decreasing exponential function of time. To demonstrate under and over damping, attach a vane to a pendulum. Experiment with the size so the pendulum oscillates in air but does not when the vane is in water.

D. Write the differential equation for a forced spring-mass system, including a damping term. Assume an applied force of the form $F_m \cos(\omega_f t)$ and point out that ω_f is not necessarily the same as the natural angular frequency of the oscillator.

E. Mention that when the system is first started transients are present and the motion is somewhat complicated. However, it settles down to a sinusoidal motion with an angular frequency that is the same as that of the impressed force.

F. Also point out that in steady state the amplitude is constant in time but that it depends on the frequency of the applied force. Illustrate with Fig. 14–21, which shows the amplitude as a function of the forcing frequency for various values of the damping coefficient. Mention that the amplitude is the greatest when the forcing frequency nearly matches the natural frequency and say this is the resonance condition. Also mention that at resonance the amplitude is greater for smaller damping and that small damping produces a sharper resonance than large damping.

G. Resonance can be demonstrated with three identical springs and two equal masses, as shown. Fasten the bottom spring to a heavy weight on the floor and drive the upper spring by hand (perhaps standing on a table). Obtain resonance at each of the normal modes (masses moving in the same and opposite directions). After showing the two resonances, drive the system at a low frequency to show a small response, then drive it at a high frequency to again show a small response. Repeat at a resonance frequency to show the larger response. To show pronounced damping effects, attach a large stiff piece of aluminum plate to each mass.

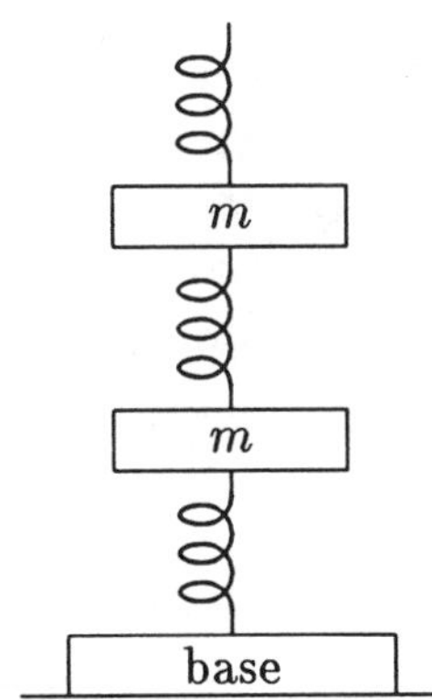

SUGGESTIONS

1. Assign questions 1, 4, and 6 and problems 4 and 13 in support of the spring-mass demonstration and discussion. The maximum force is covered in the first problem and the maximum acceleration is covered in the second. Problems 3, 8, and 10 deal with vertical oscillators. Assign one of these for variety. To include a little more physics consider question 10.

2. Springs in parallel and series test understanding of the spring force law. Consider question 3 and problems 33, 34, and 35.

3. Assign problems 16, 17, 23, and 32 in support of the discussion of the mathematical form of $x(t)$ for simple harmonic motion.

4. Assign problems 42 and 44 in support of the discussion of energy. When assigning problem 42 also ask for the maximum speed of the mass.

5. If oscillators other than a spring-mass system are considered, assign problems 52 and 54 (angular simple harmonic motion), 57 and 61 (simple pendulum), and 63 and 65 (physical pendulum).

6. Computer project
Have students use the numerical integration program developed in connection with Chapter 6 to investigate forced and damped harmonic motion. Have them plot the coordinate as a function

of time to see transients. To study resonance, have them plot the amplitude as a function of forcing frequency. They can also investigate the influence of damping on the amplitude and resonance width. Details are given in the Computer Projects section of this manual.

7. Demonstrations
   a. Simple harmonic motion: Freier and Anderson Mx1, 2, 3, 4, 7.
   b. Pendula: Freier and Anderson Mx6, 9, 10, 11, 12, My1, 2, 3, 8, Mz1, 2, 3, 6, 7, 9; Hilton M14d, f.
8. Audio/Visual
   a. *Simple Harmonic Motion*, 16 mm, color, 17 min. Pennsylvania State University, Applied Research Laboratory, P.O. Box 30, State College, PA 16801. SHM of various phases, frequencies, and amplitudes is shown through computer animation. The relationship of SHM to circular motion is illustrated. Reviewed AJP 47:754 (1979)
   b. *Coupled Oscillators — Equal Masses*; *Coupled Oscillators — Unequal Masses*, S8, color, 3 min. each. Kalmia Company, Dept. P1, Concord, MA 01742. These loops are from the popular Miller series and illustrate the phenomenon of the combination of simple harmonic motion.
9. Computer programs
   a. *Harmonic Motion Workshop*, High Technology Software Products, P.O. Box 60406, 1611 NW 23rd Street, Oklahoma City, OK 73146. Apple II, II+, IIe. Simulation of simple harmonic motion. Displays velocity vector, acceleration vector, kinetic energy, potential energy. Damped and undamped. Useful for lectures. Reviewed TPT October 1983.
   b. *Physics Simulations I: Oscillator*; Kinko's Service Corporation, 4141 State Street, Santa Barbara, CA 93110. Macintosh. Displays a mass in simple harmonic motion, damped or undamped. Plots position, potential energy, and kinetic energy as functions of time.
10. Laboratory
   a. MEOS Experiment 7–2: *The Vibrating Spring.* Students time a vertical vibrating spring with various masses attached, then use the data and a logarithmic plot to determine the relationship between the period and mass.
   b. BE Experiment 15: *Elasticity and Vibratory Motion.* The experiment is much the same as MEOS 7–2 in that a graph is used to determine the relationship between the mass on a spring and the period of oscillation. This measurement is preceded by a static determination of the spring constant.
   c. MEOS Experiment 7–4: *The Vibrating Ring.* Students time the oscillations of various diameter rings, hung on a knife edge, then use the data and a logarithmic plot to determine the relationship between the period and ring diameter. A good example of a physical pendulum.
   d. MEOS Experiment 7–15: *Investigation of Variable Acceleration.* A pendulum swings above a track and a spark timer is used to record its position as a function of time. Its velocity and acceleration are investigated.
   e. MEOS Experiment 7–19: *Harmonic Motion Analyzer.* This apparatus allows students to vary the spring constant, mass, driving frequency, driving amplitude, and damping coefficient of a spring-mass system. They can measure the amplitude, period, and relative phase of the oscillating mass. A variety of experiments can be performed.
   f. MEOS Experiment 8–4: *Linear Oscillator.* A spark timer is used to record the position of an oscillating mass on a spring, moving horizontally on a nearly frictionless surface. The period as a function of mass can be investigated and the conservation of energy can be checked.

g. MEOS Experiment 7–18: *Damped Driven Linear Oscillator.* The amplitude and relative phase of a driven damped spring-mass system are measured as functions of the driving frequency and used to plot a resonance curve.

h. MEOS Experiment 7–20: *Analysis of Resonance with a Driven Torsional Pendulum.* The driving frequency and driving amplitude of a driven damped torsional pendulum is varied and the frequency, amplitude, and relative phase are measured. Damping is electromagnetic and can be varied or turned off. A variety of experiments can be performed.

Chapter 15 GRAVITATION

BASIC TOPICS

I. Newton's law of gravity.

A. This is an important chapter. It is the first chapter devoted to a force law and its ramifications. Students get a glimpse of how a force law and the laws of motion are used together. It reviews the concepts of potential energy, angular momentum, and centripetal acceleration in the context of some important applications. In addition, the discussion of the gravitational fields of continuous mass distributions is a precursor to Gauss' law.

B. Write down the equation for the magnitude of the force of one point mass on another. Explain that the force is one of mutual attraction and is along the line joining the masses. Give the value of G ($6.67 \times 10^{-11}\,\mathrm{N{\cdot}m^2/kg^2}$) and explain that it is a universal constant determined by experiment. If you have a Cavendish balance, show it but do not take the time to demonstrate it. As a thought experiment dealing with the magnitude of G, consider a pair of 100-kg spheres falling from a height of 100 m, initially separated by a bit more then their radii. As they fall, their mutual attraction pulls them only slightly closer together. Air resistance has more influence.

C. Explain that the same mathematical form holds for bodies with spherically symmetric mass distributions (this was tacitly assumed in B) if r is now the separation of their centers. Explain that the force on a point mass anywhere inside a uniform spherical shell is zero. (Optional: use integration to prove that this follows from Newton's law for point masses.) Use this to derive an expression for the force on a point mass inside a spherically symmetric mass distribution. See Sample Problem 15–6.

D. Point out the assumed equivalence of gravitational and inertial mass.

E. Use Newton's law of gravity to calculate g for objects near the surface of the earth and justify the use of a constant acceleration due to gravity in previous chapters. Remark that the acceleration due to gravity is independent of the mass of the body.

F. Optional: Discuss factors that influence g and apparent weight.

II. Gravitational potential energy.

A. Use integration to show that the gravitational potential energy of two point masses is given by $U = -GMm/r$ if the zero is chosen at $r \to \infty$. Demonstrate that this result obeys $F = -\mathrm{d}U/\mathrm{d}r$.

B. Argue that the work needed to bring two masses to positions r apart is independent of the path. Divide an arbitrary path into segments, some along lines of gravitational force and others perpendicular to the gravitational force.

C. Consider a body initially at rest far from the earth and calculate its speed when it gets to the earth's surface. Calculate the escape velocity for the earth and for the moon.

D. Show how to calculate the gravitational potential energy of a collection of discrete masses. Warn the students about double counting the interactions — a term of the sum is associated with each *pair* of masses. Relate this energy to the binding energy of the system.

III. Planetary motion and Kepler's laws.

A. Consider a single planet in orbit about a massive sun. The center of mass for the system is essentially at the sun and it remains stationary.

B. Explain that the orbit is elliptical with the sun at one focus. This is so because the force is proportional to $1/r^2$ and the planet is bound. Draw a planetary orbit and point out the semimajor axis, the perihelion point, and the aphelion point. Define eccentricity. Show that $R_p = a(1-e)$ and $R_a = a(1+e)$, where a is the semimajor axis, R_p is the perihelion distance, and R_a is the aphelion distance.

C. Explain that the displacement vector from the sun to the planet sweeps out equal areas in equal time intervals. Sketch an orbit to illustrate. Show that the torque acting on the planet is zero because the force is along the displacement vector, then show that conservation of momentum leads to the equal area law. Note that the result is true for any central force.

D. For circular orbits show that the square of the period is proportional to the cube of the orbit radius and that the constant of proportionality is independent of the planet's mass. State that the result is also true for elliptical orbits if the radius is replaced by the semimajor axis. Verify the result for planets in nearly circular orbits. The data can be found in Table 15–3.

E. For a body held by gravitational force in circular orbit about another, much more massive body, show that the kinetic energy is proportional to $1/r$ and that the total mechanical energy is $-GMm/2r$. Explain that the energy is zero for infinite separation with the bodies at rest, that a negative energy indicates a bound system, and that a positive energy indicates an unbound system. Describe the orbits of recurring and non-recurring comets. Explain that the expression for the energy is valid for elliptical orbits if r is replace by the semimajor axis. Remark that the energy of a satellite cannot be altered without changing the semimajor axis of its orbit.

F. Remark that the laws of planetary motion hold for moons (including artificial satellites) traveling around planets, binary star systems, stars traveling around the center of a galaxy, and for galaxies in clusters. Explain that when the masses of the two objects are comparable, both objects travel around the center of mass and it is the relative displacement that obeys Kepler's laws. When discussing stars in galaxies you might show how the law of periods has been used to argue for the existence of dark matter.

SUPPLEMENTARY TOPICS

1. Detailed calculations of the gravitational force of a spherical distribution of mass on a point mass.
2. The general theory of relativity. Include a discussion of the distinction between gravitational and inertial mass.

SUGGESTIONS

1. Review the distinction between weight and mass, then discuss questions 2, 3, and 5.
2. Assign problem 1 to stress Newton's force law. Also assign problem 17 to test if students know where the value for g comes from.
3. Discuss problems 2, 10, and 33 in connection with calculations of the gravitational force of a spherically symmetric mass distribution on a point mass. Better students might find problem 30 to be an interesting challenge. Problem 31 is fundamental to the shell theorem.
4. The essentials of gravitational potential energy are covered in problems 37, 46, and 49. Some of these can be used later as models for electrostatic potential energy. Escape velocity and energy are covered in many problems. Consider problems 41 and 42. Problem 43 deals with both the escape speed and the orbital speed.

5. Use questions 11, 14, 16, 21, and 24 in the discussion of planetary orbits. Assign some of problems 56, 62, 71, 72, 78, and 85.
6. Computer project
   Have students use a spreadsheet or write a computer program to integrate Newton's second law for a $1/r^2$ central force and use it to investigate satellite motion. Some projects are described in the Computer Projects section of this manual.
6. Audio/Visual
   a. *Measurement of G, The Cavendish Experiment*, S8, color, 3 min. Kalmia Company, Dept. P1, Concord, MA 01742. The oscillatory motion of a Cavendish balance is shown via time-lapse photography. Another loop from the Miller series.
   b. *Forces (excerpt)*, 16 mm or 3/4" video cassette, b/w, 8 min. Kalmia Company (address above). In this excerpt from the longer PSSC film, a Cavendish balance is constructed of recording tape, medicine bottles, and boxes of sand. Reviewed AJP 31:400 (1963).
   c. *The Determination of the Newtonian Constant of Gravitation*; Films for the Humanities and Sciences, Inc., Box 2053, Princeton, NJ 08543. Experimental setup is explained and data taken. The analysis is left for the student.
7. Computer programs
   a. *Personal Problems.* Addison-Wesley Publishing Company, Reading, MA 01867. Apple II. Program on inverse square orbits plots open and closed orbits, given initial conditions. Numerical values of energy and angular momentum can be obtained for any point on orbit. An impulse can be applied to the object at any point in its motion. Excellent for lecture demonstrations.
   b. *Physics Simulations I: Kepler.* Kinko's Service Corporation, 4141 State Street, Santa Barbara, CA 93110. Macintosh. Plots orbits of one or two planets, with parameters set by user. Use to illustrate lectures or ask students to look at some interesting orbits.
   c. *Intermediate Physics Simulations: Three Bodies in 3-D.* R.H. Good, Physics Department, California State University at Hayward, CA 94542. Apple II. Plots projections of orbits of either 2 or 3 interacting bodies on a user selected plane. Use for lectures.
8. Laboratory
   MEOS Experiment 7–21: *Analysis of Gravitation.* Students use the Leybold-Heraeus Cavendish torsional balance to determine G. Requires extremely careful work and a solid vibration free wall to mount the apparatus.

Chapter 16 FLUIDS

BASIC TOPICS

I. Pressure and density.
   A. Introduce the subject by giving a few examples of fluids, including both liquids and gases. Remark that fluids cannot support shear.
   B. Define density as the mass per unit volume in a region of the fluid. Point out that the limit is a macroscopic limit: the limiting volume still contains many atoms. The density is a scalar and is a function of position in the fluid.
   C. Explain that fluid in any selected volume exerts a force on the material across the boundary of the volume. The boundary may be a mathematical construct and the material on the other side may be more of the same fluid. The boundary may also be a container wall or an interface with another fluid. Explain that, for a small segment of surface area, the force exerted by the fluid is normal to the surface and is proportional to the area. The pressure is the force per unit area and $\mathbf{F} = p\mathbf{A}$, where the magnitude of $\mathbf{A}$ is the area and the

direction of **A** is outward, normal to the surface. Units: Pa (= N/m^2), atmosphere, bar, torr, mm of Hg. Give the conversion or point out the appropriate appendix in the text.

D. Show that in equilibrium with y measured positive above some reference height $dp/dy = -\rho g$, where ρ is the fluid density. Then note that $p_2 - p_1 = -\int \rho g\, dy$, where the integral limits are y_1 and y_2. Point out that the difference in pressure arises because a fluid surface is supporting the fluid above it. Finally, point out that if the fluid is incompressible and homogeneous, then ρ is a constant. If $y_2 - y_1$ is sufficiently small that g is also constant, $p_2 - p_1 = -\rho g(y_2 - y_1)$. Point out that if p_0 is the surface pressure, then the pressure a distance h below the surface is $p = p_0 + \rho g h$. Note that the pressure is the same at all points at the same depth in the fluid. Explain that p_0 is atmospheric pressure if the surface is open to the air and is zero if the fluid is in a tube with the region above the surface evacuated.

E. Connect a length of rubber tubing to one arm of a U-tube partially filled with colored water. Blow into the tube, then suck on it. In each case note the change in water level. Insert the U-tube into a deep beaker of water, with the free end of the tubing out of the water. As the open end is lowered, the change in the level of the colored water will indicate the increase in pressure. Go over Sample Problem 3 to show the equilibrium positions of two immiscible liquids of different densities. Show how to obtain the pressure at the top of one arm in terms of the pressure at the top of the other arm, the densities and quantities of fluids. Point out that the pressures are the same and are the atmospheric pressure if the U-tube is open. Explain that the pressure is always the same at two points which are at the same height *and* can be joined by a line along which neither ρ nor g vary. Use the diagram associated with the problem to point out two places at the same height where the pressure is the same and two places at the same height where the pressures are different.

II. Measurement of pressure.

A. This section not only describes some pressure measuring instruments but also provides some applications of previous material, especially the variation of pressure with depth in a fluid.

B. Show a mercury barometer. A lens system or an overhead projector suitably propped on its side can be used to project an image of the mercury column on a screen for viewing by a large class. Use $p = p_0 + \rho g h$ to show why the height of the column is proportional to the pressure at the mercury pool. Emphasize that the pressure at the top of the column is nearly zero and that this is important for the operation of the barometer.

C. Show a commercial open-tube manometer or explain that such an instrument is similar to the U-tube demonstration done earlier. Explain gauge pressure and emphasize that the instrument measures gauge pressure.

III. Pascal's and Archimedes' principles.

A. State Pascal's principle. Start with $p = p_0 + \rho g h$, consider a change in p_0, and show $\Delta p = \Delta p_0$ if the fluid is incompressible. You can demonstrate the transmission of pressure with a soda bottle full of water, fitted with a tight rubber stopper. Wrap a towel around the neck of the bottle and hit the stopper sharply. With some practice you can blow the bottom out of the bottle cleanly.

B. Apply the principle to a hydraulic jack. Show that $F_1/A_1 = F_2/A_2$. Also explain that if the fluid is incompressible, F_1 and F_2 do the same work. The point of application of the smaller force moves the greater distance. A hydraulic jack can be made from a hot water bottle, fitted with a narrow rubber tube. Put the bottle on the floor and fasten the tube to a tall ringstand so it is vertical. Place a thin wooden board on the bottle to distribute the weight and have a student stand on it. To change the pressure, use a plunger or rubber squeeze ball from an atomizer, or blow into the tube.

C. State Archimedes' principle. Stress that the force is due to the surrounding fluid. Contrast the case of an immersed body surrounded by fluid with one placed on the bottom of the container. Consider a flat board floating on the surface of a liquid, compute the net upward force in terms of the difference in pressure and use $p = p_0 + \rho gh$ to show that this is the weight of the displaced liquid.

D. Explain why some objects sink while others float.

E. Fill a large mouthed plastic vessel with water precisely up to an overflow pipe. Immerse a dense object tied by string to a spring balance. Weigh the object while it is immersed and weigh the displaced water. Observe that the buoyant force is the same as the weight of the displaced water.

F. Explain that for purposes of calculating torque the buoyant force can be taken to act through the center of gravity of the fluid that will be displaced when the object is in place. The force of gravity acts through the center of gravity of the object. Show that these points may not be the same and that the two forces may produce a net torque. Show the relative positions of the center of buoyancy and the center of gravity for stable and unstable equilibrium.

IV. Fluids in motion.

A. Describe:

1. Steady and non-steady flow. Emphasize that the velocity and density fields are independent of time if the flow is steady. They may depend on position, however.
2. Compressible and incompressible flow. Emphasize that the density is independent of both position and time if the flow is incompressible.
3. Rotational and irrotational flow.
4. Viscous and nonviscous flow.

B. Describe streamlines for steady flow and point out that streamlines are tangent to the fluid velocity and that no two streamlines cross. Remark that the velocity is not necessarily constant along a streamline. Describe a tube of flow as a bundle of streamlines. Sketch a tube of flow with streamlines far apart at one end and close together at the other. Explain that since streamlines do not cross the boundaries of a tube of flow they are close together where the tube is narrow and far apart where the tube is wide. Remark that particles do not cross the boundaries of a tube of flow.

V. Equation of continuity

A. Define volume flow rate (volume flux) and mass flow rate (mass flux). Consider a tube of flow with cross-sectional area A at one point and give the physical significance of $A\rho v$ and Av. Remark that the first can be measured in kg/s and the second in m^3/s. Show how to convert m^3/s to gal/s and li/s.

B. State the equation of continuity: $A\rho v =$ constant along a streamline if there are no sources or sinks of fluid and if the flow is steady. Argue that if the equation were not true there would be a build up or depletion of fluid in some regions and the flow would not be steady.

C. Discuss the special case of an incompressible fluid and explain that the fluid speed is great where the tube of flow is narrow and vice versa. Point out that the fluid velocity is great where the streamlines are close together and small where they are far apart. Use the diagram of section IVB above as an example.

VI. Bernoulli's equation.

A. Apply the work-energy theorem to a tube of flow to show that for steady, nonviscous, incompressible flow $p + \frac{1}{2}\rho v^2 + \rho gy =$ constant along a streamline. Point out that this equation also gives the pressure variation in a static fluid ($v = 0$ everywhere).

B. Remark that a typical fluid dynamics problem gives the conditions v, p, y at one point on a streamline and asks for conditions at another. The equation of continuity and Bernoulli's equation can be solved simultaneously for 2 quantities.

C. Work a sample problem. Consider horizontal flow (y = const) through a pipe than narrows. Give the fluid velocity where the pipe is wide and use the equation of continuity to calculate the velocity where it is narrow. Then use Bernoulli's equation to calculate the pressure difference. Emphasize that the pressure must decrease to provide the force that accelerates the fluid as it passes into the narrow region.

D. Now work the same problem but suppose the height of the pipe increases along the direction of flow. Point out the difference in the answers for the pressure.

VII. Applications

A. Section 16–11 of the text serves two purposes. It shows students some practical applications of the continuity and Bernoulli equations and it also provides examples of problem solving techniques.

B. Show a Venturi meter if you have one available. Mention that it can be used to measure fluid velocity and volume flow rate. Draw a diagram of a meter and point out the narrow part of the tube. Explain that the velocity is higher there and the pressure is lower than in the main part of the tube. The manometer measures the pressure difference between the narrow and main parts of the tube. Review the discussion of pressure in U-tubes containing two immiscible fluids and relate the pressure difference to the difference in height of the manometer liquid. Write down Bernoulli's equation for horizontal flow and point out that the pressure difference is directly related to the difference in the squares of the velocities at the two places in the tube. Explain that the velocities are related by the continuity equation and use that equation to eliminate the velocity in the narrow part of the tube. Finally solve for the velocity in the main part of the tube and obtain Eq. 16–24 of the text.

C. Show a pitot tube if one is available and carry out the same type analysis. See problem 69.

D. Consider a water tank with a hole in the bottom and use the Bernoulli and continuity equations to develop an expression for the speed of the fluid as it leaves the hole. Specialize to the case for which the fluid speed is negligible at the top of the tank.

E. Hang two pith balls by strings of equal length, slightly separated. Blow gently between them. Discuss the reduction in pressure in the region between the balls. If you have a high pressure line available, shoot a high velocity air stream vertically upward and hang a smooth handled screw driver in the jet. About 80 psi is needed to lift it. Discuss applications to flying shapes: air foils, baseballs, and golf balls.

SUGGESTIONS

1. Use questions 3 and 7 to discuss pressure. Problems 2, 5, and 7 cover the definition of pressure. Problems 11 and 14 deal with the variation of pressure with depth. Problem 20 deals with the variation of pressure with position in an accelerating fluid.

2. Use problems 29 and 30 in connection with Pascal's principle.

3. Questions 5, 6, and 8 through 23 all provide good examples of Archimedes' principle. Pick several to illustrate applications of the principle. Buoyant stability is covered in questions 18 and 22. Also assign problems 32, 34, and 35 and some of problems 41 through 51.

4. Use problems 53 and 57 as part of the discussion of the equation of continuity.

5. The fundamentals of Bernoulli's equation are covered in question 27 and in problems 58, 59, and 60. Problem 61 requires students to combine the equation of continuity and Bernoulli's equation. Use some of these in the lecture and assign some.

6. Questions 25, 26, 28, 29, 30, and 31 all deal with interesting applications of the continuity and Bernoulli equations. Flow from holes is considered in problems 64, 78, 79, and 84. Lift on an

airplane wing is covered in problems 66, 67, and 68; popping windows are covered in problem 73; and siphons are covered in problem 80. Assign some of these.

7. Demonstrations
   a. Force and pressure: Freier and Anderson Fa, Fb, Fc, Fd, Fe, Ff, Fh; Hilton M20b, M20e, M22b, M22d, M22e, f.
   b. Archimedes' principle: Freier and Anderson Fg; Hilton M20c, M22c.
   c. Bernoulli's principle: Freier and Anderson Fj, Fl1.
8. Audio/visual
   a. *Archimedes' Principle*; Bruce Marsh; American Association of Physics Teachers, 5112 Berwyn Road, College Park, MD 20740-4100. Slide set.
   b. *Demonstrations of Physics: Volume 3: Liquids and Gases*; video cassette, about 30 min. Vikas Productions, Box 6088, Bozeman, MT 59771. Demonstrations of buoyancy and Bernoulli's equation. Reviewed TPT **29**, 403 (September 1991).
9. Laboratory
   a. MEOS Experiment 7–7: *Radial Acceleration* (Problem II only). Students measure the orbit radius of various samples floating on the surface of water in a spinning globe, analyze the forces on the samples. An application of the buoyancy force to rotational motion.
   b. BE Experiment 16: *Buoyancy of Liquids and Specific Gravity.* Archimedes' principle is checked by weighing the water displaced by various cylinders. Buoyant forces are measured by weighing the cylinders in and out of water. The same cylinder is immersed in various liquids and the results used to find the specific gravities of the liquids.

Chapter 17 WAVES — I

BASIC TOPICS

I. General properties of waves.
   A. Explain that wave motion is the mechanism by which a disturbance created at one place travels to another. Use the example of a pulse on a taut string and point out that the displaced string causes neighboring portions of the string to be displaced. Stress that the individual particles have limited motion (perhaps perpendicular to the direction of wave travel), whereas the pulse travels the length of the string. Demonstrate by striking a taut string stretched across the room. Point out that energy is transported by the wave from one place to another. Ask the students to read the introductory section of the chapter for other examples of waves.
   B. Point out that a wave on a string travels in one dimension, water waves produced by dropping a pebble travel in two, and sound waves emitted by a point source travel in three.
   C. Explain the terms longitudinal and transverse. Demonstrate longitudinal waves with a slinky.
   D. State that waves on a taut string of uniform density travel with constant speed and that this course deals chiefly with idealized waves that do not change shape. Take the string to lie along the x axis and draw a distortion in the shape of a pulse, perhaps a sketch of $\exp[-\alpha(x-x_0)^2]$. Remark that the initial displacement of the string can be described by giving a function $f(x)$. Now suppose the pulse moves in the positive x direction and draw the string at a later time. Point out that the maximum has moved from x_0 to x_0+vt, where v is the wave speed. Remark that the displacement can be calculated by substituting $x-vt$ for x in the function $f(x)$. Substantiate the remark by showing that $x-vt=x_0$ if x is the coordinate of the pulse maximum at time t. Explain that $x+vt$ is substituted if the pulse travels in the negative x direction.

II. Sinusoidal traveling waves.

A. Write $f(x) = y_m \sin(kx)$ for the initial displacement of the string and sketch the function. Identify the amplitude as giving the limits of the displacement and point it out on the sketch. Also point out the periodicity of the function and identify the wavelength on the sketch. Show that k must be $2\pi/\lambda$ for $f(x)$ to equal $f(x + n\lambda)$ for all integers n. Remark that k is called the angular wave number of the wave.

B. Substitute $x - vt$ for x in $f(x)$ and explain you will assume the wave travels in the positive x direction. Show that the result is $y(x,t) = y_m \sin(kx - \omega t)$, where $\omega = kv$.

C. State that the motion of the string at any point is simple harmonic and that ω is the angular frequency. Show that at a given place on the string the motion repeats in a time equal to $2\pi/\omega$. This is the period T. Remind students that the frequency is $f = 1/T = \omega/2\pi$.

D. Remark that any given point on the string reaches its maximum displacement whenever a maximum on the wave passes that point. Since the time interval is one period a sinusoidal wave travels one wavelength in one period and $v = \lambda/T = \lambda f = \omega/k$, in agreement with the derivation of $y(x,t)$.

E. Explain that $y(x,t) = y_m \sin(kx + \omega t)$ represents a sinusoidal wave traveling in the negative x direction.

F. Show that the string velocity is $u(x,t) = \partial y/\partial t = -\omega y_m \cos(kx - \omega t)$. Point out that x is held constant in taking the derivative since the string velocity is proportional to the difference in the displacement of the *same* piece of string at two slightly different times. Remark that different points on the string may have different velocities at the same time and the same point may have a different velocity at different times. Contrast this behavior with that of the wave velocity. Point out that for a transverse wave u is transverse.

G. Explain that the wave speed for an elastic medium depends on the inertia and elasticity of the medium. State that, for a taut string, $v = \sqrt{\tau/\mu}$, where τ is the tension in the string and μ is the linear density of the string. Show how to measure μ for a homogeneous, constant radius string. The expression for v may be derived as in Section 17–6 of the text.

H. Point out that the frequency is usually determined by the source and that doubling the frequency for the same string with the same tension halves the wavelength. The product λf remains the same. Remark that if a wave goes from one medium to another the speed and wavelength change but the frequency remains the same. Work an example: given the two densities and the frequency, calculate the wave speed and wavelength in each segment. Draw a diagram of the wave.

III. Energy considerations.

A. Point out that the energy in the wave is the sum of the kinetic energy of the moving string and the potential energy the string has because it is stretched in the region of the disturbance. Energy moves with the disturbance.

B. Show that the kinetic energy of an infinitesimal segment of string is given by $\mathrm{d}K = \frac{1}{2}\,\mathrm{d}m\,v^2 = \frac{1}{2}(\mu\,\mathrm{d}x)(\omega^2 y_m^2)\cos^2(kx - \omega t)$. State that this energy is transported to a neighboring portion in time $\mathrm{d}t = v\,\mathrm{d}x$, so $\mathrm{d}K/\mathrm{t} = \frac{1}{2}\mu v \omega^2 y_m^2 \cos^2(kx - \omega t)$ gives the rate at which kinetic energy is transported past the point with coordinate x, at time t. Explain that when this averaged over a cycle the result is $\overline{\mathrm{d}K/\mathrm{d}t} = \frac{1}{4}\mu v \omega^2 y_m^2$. Remark that this is not zero. Although kinetic energy moves back and forth as the string oscillates, there is a net flow.

C. State without proof that the average rate at which potential energy is transported is exactly the same as the rate for kinetic energy, so the average rate of energy flow is $\overline{P} = \frac{1}{2}\mu v \omega^2 y_m^2$. Note that this depends on the square of the amplitude and on the square of the frequency.

IV. Superposition and interference.

A. Stress that displacements, not intensities, add. State that if y_1 and y_2 are waves that are simultaneously present, then $y = y_1 + y_2$ is the resultant wave. Using diagrams of two similar sinusoidal waves, show that the resultant amplitude can be twice the amplitude of one of them, can vanish, or can have any value between. Mention that the medium must be linear.

B. Start with the waves $y_1 = y_m \sin(kx - \omega t + \phi)$ and $y_2 = y_m \sin(kx - \omega t)$ and show that $y = 2y_m \cos(\phi/2)\sin(kx - \omega t + \phi/2)$. Show that maximum constructive interference occurs if $\phi = 2n\pi$, where n is an integer and maximum destructive interference occurs if $\phi = (2n+1)\pi$, where n is again an integer. Remark that the maximum amplitude is $2y_m$ and the minimum is 0. The derivation depends heavily on the trigonometric identity given as Eq. 17–43. You may wish to verify this identity for the class. Use the expressions for the sine and cosine of the sum of two angles to expand the right side of Eq. 43.

C. Explain that a phase difference can arise if waves start in phase but travel different distances to get to the same point. Show that $\phi = k\Delta x$ and find expressions for the path differences that result in maximum constructive and maximum destructive interference. In the first case Δx is a multiple of λ, while in the second it is an odd multiple of $\lambda/2$.

D. Interference can easily be demonstrated with a monaural amplifier, a signal generator, a microphone, an oscilloscope, and a pair of speakers. Fix the position of speaker S_1 and, with S_2 disconnected, show the wave form on the oscilloscope. Then connect S_2 and show the wave form as S_2 is moved. Because both speakers are driven by the same amplifier, the only phase difference is due to the path difference.

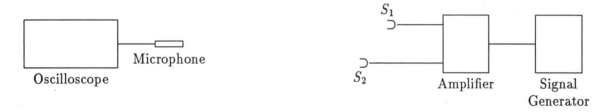

V. Standing waves.

A. Use a vibrating tuning fork (driven, if possible) to set up a standing wave pattern on a string. Otherwise, draw the pattern. Point out nodes and antinodes. Explain that all parts of the string vibrate either in phase or 180° out of phase and that the amplitude depends on position along the string. The disturbance does not travel. If possible, use a stroboscope to show the standing wave pattern. CAUTION: students with epilepsy should not watch this demonstration.

B. Explain that a standing wave can be constructed from two sinusoidal traveling waves of the same frequency and amplitude, traveling with the same speed in opposite directions. Use a trigonometric identity to show that $y_1 + y_2 = 2y_m \sin(kx)\cos(\omega t)$ if the phase constant for each wave is 0. Find the coordinates of the nodes and show they are half a wavelength apart. Also find the coordinates of the antinodes and show they lie halfway between nodes.

C. Point out that standing waves can be created by a wave and its reflection from a boundary. By means of a diagram show how the incident and reflected waves cancel at the fixed end of a string.

D. Remark that for a string fixed at both ends, each end must be a node. Derive the expression for the standing wave frequencies of such a string. Draw diagrams showing the string at maximum displacement for the lowest 3 or 4 frequencies. Remark that for a string with one end fixed and the other free, the fixed end is a node and the free end is an antinode. Derive the expression for the standing wave frequencies. Draw diagrams for the lowest 3 or 4 frequencies.

E. Place two speakers, driven by the same signal generator and amplifier, well apart on the lecture table, facing the class. Standing waves are created throughout the room. Have each student place a finger in one ear and move his head slowly from side to side in an attempt to find the nodes and antinodes. Use a frequency of about 1 kHz.

F. Consider a driven string and describe resonance. Explain that the amplitude becomes large when the driving frequency matches a standing wave frequency. Explain that at resonance the energy supplied by the driving force is dissipated and that off resonance the string does work on the driving mechanism.

G. You may wish to explain that when the string is driven at a non-resonant frequency, each traveling wave and its reflection from an end produce a standing wave, just as at resonance. The standing waves produced by successive reflections, however, do not coincide and a jumble results.

SUPPLEMENTARY TOPIC

The speed of light. This is another interesting topic that is important for modern physics. Cover it if you have time.

SUGGESTIONS

1. Include discussions of questions 4, 5, and 6 when covering the mathematical description of traveling waves. Also assign some of problems 6 through 8 and 11 through 13. Wave speed is covered in problems 27 and 28.
2. Include questions 1 and 2 in a discussion of energy. Also assign problems 33 and 35.
3. The fundamentals of interference are covered in problems 36 and 37. You might also want to include problem 38 to show that path differences might give rise to phase differences. Include question 9 in the discussion of interference.
4. Assign problems 51 and 52 in connection with standing waves and problem 53 in connection with resonant frequencies. Include question 17 in the discussion.
5. Computer project
   Have students use a commercial math program, a spreadsheet, or their own programs to investigate energy in a string carrying a wave. The program should calculate the kinetic, potential, and total energies at a given point and time, given the string displacement as a function of position and time. Use the program to plot the energies as functions of time for a given position. Consider a pulse, a sinusoidal wave, and a standing wave. Demonstrate that energy passes the point in the first two cases but not in the third. For sinusoidal and standing waves, the program should also calculate averages over a cycle. Other projects are described in the Computer Projects section of this manual.
6. Demonstrations
   a. Traveling waves: Freier and Anderson Sa3, 4, 5, 6, 12, 13; Hilton S2a, c, d.
   b. Reflection: Freier and Anderson Sa7, 12, 14.
   c. Standing waves: Freier and Anderson Sa8, 9; Hilton S2b, g.
7. Audio/Visual
   a. *Standing Waves and the Principle of Superposition*, 16 mm, color, 11 min. Encyclopaedia Britannica Educational Corporation, 425 N. Michigan Ave., Chicago, IL 60611. Standing waves by superposition, nodes in standing linear waves, Chladni plates, and soap films are the essential features of this film. Reviewed AJP 41:153 (1973).
   b. *Standing Waves on a String*; *Standing Waves in a Gas*; *Vibrations of a Metal Plate*; *Vibrations of a Drum*, S8, color, 3 – 4 min. each. Kalmia Company, Dept. P1, Concord,

MA 01742. These loops from the Project Physics Series illustrate standing waves in a variety of media and in both one and two dimensions.

c. *Propagation of Waves, III: Interference*, S8, color, 7 min. Walter de Gruyter, Inc., 200 Sawmill River Road, Hawthorne, NY 10532. Interference is illustrated by the ripple tank. Reviewed AJP 45:596 (1977).
d. *Tacoma Narrows Bridge Collapse*, S8, color, 4 min. Kalmia Company, Dept. P1, Concord, MA 01742. This is a classic. It shows the incredible amplitudes that can be built up in a macroscopic structure through resonance effects.
e. *Dynamic Response of a Suspension Bridge*, S8, color, 3.5 min. American Association of Physics Teachers, 5110 Roanoke Place, College Park, MD 20740–4100. To simulate the Tacoma Narrows Bridge collapse, a small suspension bridge is excited in a torsional mode.
f. *Demonstrations of Physics: Volume 5: Waves*; video cassette, about 30 min. Vikas Productions, Box 6088, Bozeman, MT 59771. Demonstrations of waves on a torsion-rod wave machine and Chladni plates. Demonstration of resonance tube apparatus. Reviewed TPT **29**, 403 (September 1991).

8. Interactive videodisk

The Puzzle of the Tacoma Narrows Bridge Collapse by R.G. Fuller, D.A. Zollman, and T.C. Campbell (Wiley, 1982). This videodisk shows the film of the collapse, referenced above. It allows the students to select various demonstration experiments, which are then shown and used to investigate standing waves, resonance phenomena, and the effect of wind on the bridge.

9. Computer programs

a. *Physics Disk 2: Waves*, The 6502 Program Exchange, 2920 Moana, Reno, NV 89509. Apple II. Simulations useful for lectures include the reflection of a pulse at a fixed and at a free end of a string, superposition of two sine waves, standing waves, and beats. Reviewed TPT September 1986
b. *Wave Addition II*, Vernier Software, 2920 S.W. 89 th Street, Portland, OR 97225. Apple II. Simulation of the addition of two waves. In some segments the user chooses the parameters of the second wave. Useful as a lecture demonstration of beats and interference effects. Can also be used to demonstrate Fourier synthesis of sawtooth, square, and triangular waves. Reviewed TPT February 1986.
c. *Animation Demonstration: Animated Waves*, Conduit, The University of Iowa, Oakdale Campus, Iowa City, IA 52242. Apple II. Simulations which can be used to illustrate lectures on standing waves, traveling pulses, Doppler effect for sound, group velocity, and relativistic e-m waves. Reviewed TPT November 1986.
d. *Intermediate Physics Simulations: Sum of Two Waves*, R.H. Good, Physics Department, California State University at Hayward, CA 94542. Apple II. Shows two waves, one with user selected wavelength, and their sum. Use to demonstrate the addition of waves to create beats, standing waves. Can also be used to show group velocity. Illustrations for lectures.
e. *Animation Demonstration*, CONDUIT, The University of Iowa, Oakdale Campus, Iowa City, IA 52242. Apple II. Animated demonstrations of wave motion, particle motion, and quantum mechanics. Use to illustrate lectures. Reviewed TPT November 1986.

10. Laboratory

a. MEOS Experiment 12–1: *Transverse Standing Waves* (Part A). Several harmonics are generated in a string by varying the driving frequency. Frequency ratios are computed and compared with theoretical values. Values of the wave speed found using λf and using $\sqrt{\tau/\mu}$ are compared. The experiment can be repeated for various tensions and various linear densities.

b. BE Experiment 22: *A Study of Vibrating Strings.* A horizontal string is attached to a driven tuning fork vibrator. It passes over a pulley and weights are hung on the end. The weights are adjusted so standing wave patterns are obtained and the wavelength of each is found from the measured distance between nodes. Graphical analysis is used to find the relationship between the wave velocity and the tension in the string and to find the frequency. Several strings are used to show the relationship between the wave velocity and the linear density.

Chapter 18 WAVES — II

BASIC TOPICS

I. Qualitative description of sound waves.

A. Explain that the disturbance that is propagated is a deviation from the ambient density and pressure of the material in which the wave exists. This comes about through the motion of particles. If Chapter 16 was not covered, you should digress to discuss density and pressure briefly. Point out that sound waves in solids can be longitudinal or transverse but sound waves in fluids are longitudinal: the particles move along the line of wave propagation. Waves in crystalline solids moving in low symmetry directions are examples that are neither transverse nor longitudinal. Use a slinky to show a longitudinal wave and point out the direction of motion of the particles. State that sound can be propagated in all materials.

B. Draw a diagram, similar to Fig. 3, to show a compressional pulse. Point out regions of high, low, and ambient density. Also show the pulse at a later time.

C. Similarly, diagram a sinusoidal sound wave in one dimension and draw a rough graph of the pressure as a function of position for a given time. Give the rough frequency limits of audible sound and mention ultrasonic and infrasonic waves.

D. Discuss the idea that the wave velocity depends on an elastic property of the medium (bulk modulus) and on an inertia property (ambient density). Recall the definition of bulk modulus (or introduce it) and show by dimensional analysis that v is proportional to $\sqrt{B/\rho}$. Assert that the constant of proportionality is 1. Point out the wide range of speeds reported in Table 18–1.

II. Interference.

A. Remind students of the conditions for interference. Consider two sinusoidal sound waves with the same amplitude and frequency, traveling in the same direction. Explain that constructive interference occurs if they are in phase and complete destructive interference occurs if they are π rad out of phase.

B. Explain that a phase difference can occur at a detector if two waves from the same source travel different distances. Show that the phase difference is given by $k\Delta x$ $(= 2\pi\Delta x/\lambda)$.

C. Interference of sound waves can be demonstrated by wiring two speakers to an audio oscillator and putting the apparatus on a slowly rotating platform. Students will hear the changes in intensity.

III. Mathematical description of one-dimensional sound waves.

A. If desired, derive $v = \sqrt{B/\rho}$ as it is done in the text.

B. Write $s = s_m \cos(kx - \omega t)$ for the displacement of the material at x. Show how to calculate the pressure as a function of position and time. Relate the pressure amplitude to the displacement amplitude. Explain that a sinusoidal pressure wave traveling in the positive x direction is written $\Delta p(x,t) = \Delta p_m \sin(kx - \omega t)$. State that Δp is the deviation of the pressure from its ambient value. Remind students that $k = 2\pi/\lambda$, $f = \omega/2\pi$, and $\lambda f = v$.

C. Remark that power is transmitted by a sound wave because each element of fluid does work on neighboring elements. Show that the kinetic energy in an infinitesimal length dx of a sinusoidal sound wave traveling along the x axis is $dK = \frac{1}{2}A\rho\omega^2 s_m^2 \sin^2(kx - \omega t)\,dx$, where A is the cross-sectional area. Show that its average over a cycle is $\overline{dK} = \frac{1}{4}A\rho\omega^2 s_m^2$. Argue that this energy moves to a neighboring segment in time $dt = dx/v$ and show that the rate of kinetic energy flow is, on average, $\overline{dK/dt} = \frac{1}{4}A\rho v\omega^2 s_m^2$, where v is the speed of sound. Tell students that the rate of flow of potential energy is exactly the same, so the rate of energy flow is $\overline{P} = \frac{1}{2}A\rho v\omega^2 s_m^2$.

D. Define intensity as the average rate of energy flow per unit area and show that it is given by $I = \frac{1}{2}\rho v\omega^2 s_m^2$. Show that conservation of energy implies that the intensity decreases as the reciprocal of the square of the distance as a spherical wave moves outward from an isotropic point source.

E. Show a scale of the range of human hearing in terms of intensity. Introduce the idea of sound level and define the bel and decibel. Discuss both absolute (relative to $10^{-12}\,\mathrm{W/m^2}$) and relative intensities. Remark that an increase in intensity by a factor of 10 means an increase in sound level by 10 db. If you have a sound level meter, use an oscillator, amplifier, and speaker to demonstrate the change of a few db in sound level.

IV. Standing longitudinal waves and sources of sound.

A. Use a stringed instrument or a simple taut string to demonstrate a source of sound. Point out that the wave pattern on the string is very nearly a standing wave, produced by a combination of waves reflected from the ends. If the string is vibrating in a single standing wave pattern then sound waves of the same frequency are produced in the surrounding medium. Demonstrate the same idea by striking a partially filled bottle, then blowing across its mouth. Also blow across the open end of a ball point pen case. If you have them, demonstrate Chladni plates.

B. Derive expressions for the natural frequencies and wavelengths of air pipes open at both ends and closed at one end. Stress that pressure nodes occur near open ends and that pressure antinodes occur at closed ends. Define the terms fundamental, overtone, and harmonic.

C. Optional: Discuss the quality of sound for various instruments in terms of harmonic content. If possible, demonstrate the instruments.

D. Demonstrate voice patterns by connecting a microphone to an oscilloscope and keeping the setup running through part or all of the lecture. This is particularly instructive in connection with part C.

V. Beats.

A. Demonstrate beats using two separate oscillators, amplifiers, and speakers, operating at nearly, but not exactly, the same frequency. If possible, show the time dependence of the wave on an oscilloscope. Remark that the sound is like that of a pure note but the intensity varies periodically. Explain that this technique is used to tune instruments in an orchestra.

B. Remark that you will consider displacement oscillations at a point in space when two sound waves of the same amplitude and nearly the same frequency are present. Write the expression for the sum of $s_1 = s_m \cos(\omega_1 t)$ and $s_2 = s_m \cos(\omega_2 t)$, where $\omega_1 \approx \omega_2$, but the two frequencies are not exactly equal. Show that $s_1 + s_2 = 2s_m \cos(\omega' t)\cos(\omega t)$, where $\omega' t = (\omega_1 - \omega_2)/2$ and $\omega = (\omega_1 + \omega_2)/2$. Remark that because the difference in frequencies is much smaller than either constituent frequency we can think of the oscillation as having an angular frequency of $\omega = (\omega_1 + \omega_2)/2$ and a time dependent amplitude. Note that the angular frequency of the amplitude is $\omega' = |\omega_1 - \omega_2|/2$ but the angular frequency of the intensity is $\omega_{\text{beat}} = |\omega_1 - \omega_2|$. The latter is the beat angular frequency.

VI. Doppler effect.

A. Explain that the frequency increases when the source is moving toward the listener, decreases when the source is moving away, and that similar effects occur when the listener is moving toward or away from the source. Use Figs. 18–17 and 20 to illustrate the physical basis of the phenomenon.

B. Derive expressions for the frequency when the source is moving and for the frequency when the listener is moving. Point out that the velocities are measured relative to the medium carrying the sound.

C. The effect can be demonstrated by placing an auto speaker and small audio oscillator (or sonalert type oscillator) on a rotating table. The sonalert can also be secured to a cable and swung in a circle. Show the effect of a passive reflector by moving a hand-held sonalert rapidly toward and away from the blackboard.

SUGGESTIONS

1. In connection with the discussion of sound sources, consider questions 6, 7, 8, and 10.
2. The speed of sound is emphasized in problems 1, 3, and 4 while its dependence on the bulk modulus and density is covered in problem 5.
3. Use problems 29 and 33 to discuss sound intensity and problems 30 and 31 to discuss sound level. They will help students with the concepts of bel and decibel. Al so consider problem 32.
4. Assign problems 21 and 25 in connection with interference.
5. Assign problems 48 and 61 when discussing standing waves. If you use a Knudt's tube in the lab, assign problem 49.
6. Tuning stringed instruments is covered in problems 53 and 56.
7. Assign problems 65 and 66 in connection with beats.
8. Use problem 69 in a discussion of the Doppler effect. Assign problems 75 and 85. Also assign problem 80 in connection with sonic booms.
9. Computer projects
   A spreadsheet or computer program can be used to add waves. Have students use it to investigate beats. See the Computer Projects section of this manual for details.
10. Demonstrations
    a. Wavelength and speed of sound in air: Freier and Anderson Sa16, 17, 18, Sh1; Hilton S3e, f.
    b. Sound not transmitted in a vacuum: Freier and Anderson Sh2; Hilton S3a.
    c. Sources of sound, acoustical resonators: Freier and Anderson Sd3, Se, Sf, Sj6; Hilton S4, S7.
    d. Harmonics: Freier and Anderson Sj2 — 5
    e. Beats: Freier and Anderson Si4 — 6; Hilton S5.
    f. Doppler shift: Freier and Anderson Si1 — 3; Hilton S6.
11. Audio/Visual
    a. *Propagation of Waves II: Standing Waves and the Doppler Effect*, S8, color, 4 min. Walter de Gruyter, Inc., 200 Sawmill River Road, Hawthorne, NY 10532. A toy train moves the source in this ripple tank demonstration of the Doppler effect.
    b. *Demonstrations in Acoustics*, 3/4" video cassette, color, (various lengths). University of Maryland, Department of Physics and Astronomy, College Park, MD 20742. 29 different demonstrations in elementary acoustics are presented. A good source of demonstrations. Reviewed AJP 49:608 (1981). See the review for a complete list of the experiments.

   c. *Plates, Bars, Membranes, and Beats*; Thomas D. Rossing; American Association of Physics Teachers, 5112 Berwyn Road, College Park, MD 20740–4100. Slides showing standing wave patterns.
   d. *Experiments on the Doppler Effect*; Films for the Humanities and Sciences, Inc., Box 2053, Princeton, NJ 08543. The effect is discussed, experimental setup is explained, and data taken. The analysis is left for the student.

12. Computer programs
   a. *Animation Demonstration: Animated Waves.* See Chapter 19 notes.
   b. The microcomputer Based Lab Project *Sound*, HRM Software, 175 Tompkins Avenue, Pleasantville, NY 10570-9973. Sound is picked up by a microphone and intensity is plotted as a function of time on the monitor screen. Use this as an alternative to an oscilloscope. It has the advantages that sound patterns can be stored on disk and recalled for later use and two patterns can be displayed simultaneously for comparison. Any portion of a pattern can be magnified for closer study.

13. Laboratory
   a. MEOS Experiment 12–2: *Velocity of Sound in Air* and BE Experiment 23: *Velocity of Sound in Air — Resonance-Tube Method.* Resonance of an air column is obtained by holding a tuning fork of known frequency at the open end of a tube with one closed end. The length of the column is changed by adjusting the amount of water in the tube. The wavelength and speed of sound are found.
   b. MEOS Experiment 12–3: *Velocity of Sound in Metals* and BE Experiment 24: *Velocity of Sound in a Metal — Kundt's-Tube Method.* A Kundt's tube is used to find the frequency of sound excited in a rod with its midpoint clamped and its ends free. The wavelength is known to be twice the rod length and λf is used to find the speed of sound. In another experiment, a transducer and oscilloscope are used to time a sound pulse as it travels the length of a rod and returns.
   c. MEOS Experiment 12–4: *Investigation of Longitudinal Waves.* The amplitude and phase of a sound wave are investigated as functions of distance from a speaker source. To do this, Lissajous figures are generated on an oscilloscope screen by the source signal and the signal picked up by a microphone. To eliminate noise, the speaker and microphone should be in a large sound-proof enclosure with absorbing walls. Use MEOS Experiment 10–10 to familiarize students with the oscilloscope and Lissajous figures.

Chapter 19 TEMPERATURE

BASIC TOPICS

I. The zeroth law of thermodynamics.
   A. Explain that if two bodies, not in thermal equilibrium, are allowed to exchange energy then they will do so and one or more of their macroscopic properties will change. When no further changes take place the bodies are in thermal equilibrium. Explain that two bodies in thermal equilibrium are said to have the same temperature.
   B. For gases the properties of interest include pressure, volume, internal energy, and the quantity of matter. Other properties may be included for other materials. The quantity of matter may be given as the number of particles or as the number of moles.
   C. Explain what is meant by diathermal and adiabatic walls and remark that diathermal walls are used to obtain thermal contact without an exchange of particles. Adiabatic walls are used to thermally isolate a system.

D. State the zeroth law: if body A and body B are each in thermal equilibrium with body C, then A is in thermal equilibrium with B. Discuss the significance of the zeroth law. State that it is the basis for considering the temperature to be a property of an object. If it were not true then, at best, an object might have a large number of temperatures, depending on what other objects were in thermal equilibrium with it.

E. Explain that the temperature of a body is measured by measuring some property of a thermometer in thermal equilibrium with it. Illustrate by reminding students that the length of the mercury column in an ordinary household thermometer is a measure of the temperature. Explain that the zeroth law guarantees that the same temperature, as measured by the same thermometer, will be obtained for two substances in thermal equilibrium with each other.

II. Temperature measurements.

A. Mention that the value of the temperature obtained depends on the substance used for the thermometer and on the property measured but that several techniques exist that allow us to define temperature independently of the thermometric substance and property.

B. Describe a constant-volume gas thermometer. If one is available, demonstrate its use. If not, show Fig. 19–5. The gas is placed in thermal contact with the substance whose temperature is to be measured and the pressure is adjusted so that the volume has some standard value (for that thermometer). After corrections are made, the temperature is taken to be proportional to the pressure: $T = ap$, where a is the constant of proportionality.

C. Describe the triple point of water and explain that water at the triple point is assigned the temperature $T = 273.16\,\mathrm{K}$. Solve for a and show that $T = 273.16(p/p_3)$.

D. Point out that thermometers using different gases give different values for the temperature when used as described. Explain the limit used to obtain the Kelvin temperature. See Fig. 19–6.

E. Define the Celsius and Fahrenheit scales. Give the relationships between the degree sizes and the zero points. Give equations for conversion from one scale to another and give the temperature value for the ice and steam points in each system. Use Fig. 19–7 and Table 19–2.

F. Define the Kelvin scale and explain the kelvin as a unit of temperature. Give the relationship between the Celsius and Kelvin scales. Give the ice and steam points on the Kelvin scale.

III. Thermal expansion.

A. Describe linear expansion and define the coefficient of linear expansion: $\alpha = \Delta L/L\Delta T$. Point out Table 19–3. Obtain a bimetallic strip and use both a bunsen burner and liquid nitrogen (or dry ice) to show bending. After the students see the strip bend ask which of the metals has the greater coefficient of linear expansion. Explain that these devices are often used in thermostats.

B. Discuss area and volume expansion. Consider a plate and show that the coefficient of area expansion is 2α. Consider a rectangular solid and show that the coefficient of volume expansion is 3α. In each case apply the equation for linear expansion to each dimension of the object and find ΔA or ΔV to first order in ΔT.

C. Explain that the length of a scratch on the flat face of an object increases as the temperature increases. The area of a hole also increases. Carefully drill a 1/2 inch hole in a piece of aluminum, roughly $1\frac{1}{4}$ inch thick. Obtain a 13 mm diameter steel ball bearing and place it in the hole. It will not pass through. Heat the plate on a bunsen burner and the ball passes through easily.

D. Demonstrate volume expansion of a gas using a flat bottomed flask, a bulbed tube, a two hole stopper, and some colored water. Partially evacuate the bulb so the colored water stands in the tube somewhat above the stopper. Place your hand on the bulb to warm the air inside and the water in the tube drops in response.

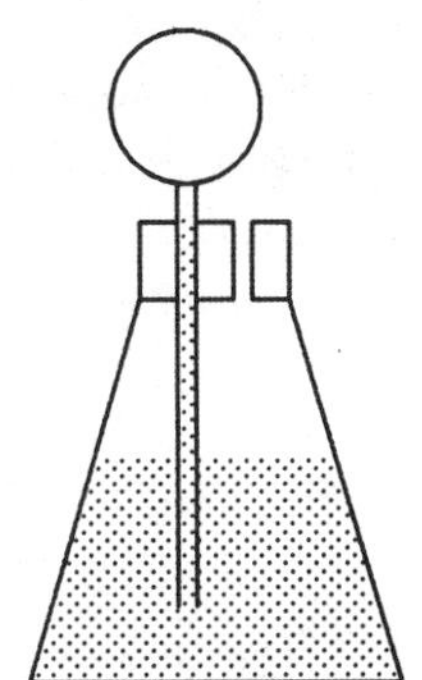

SUPPLEMENTARY TOPIC

Use Fig. 19–13 to discuss thermal expansion from a microscopic viewpoint. Point out the importance of the asymmetry in the potential energy function for the interaction between two molecules.

SUGGESTIONS

1. After discussing thermal equilibrium, ask question 2.
2. After discussing gas thermometers, ask questions 3, 4, and 12, assign problems 6 and 7.
3. Use questions 11, 13, and 14 to discuss the general problem of defining and measuring temperature. Note that different properties are used to measure temperatures in different situations. To emphasize the great variety of temperature measuring techniques, assign problems 1, 2, 5, and 8. Problem 13 asks students to compare temperature readings on various scales.
4. Use question 17 and problem 36 in connection with the ball and hole demonstration, question 18 and problem 46 in connection with the bimetallic strip demonstration. Assign problem 53 or 54 to good students.
5. Introduce Newton's law of cooling and assign problems 17 and 18.
6. Demonstrations
   a. Thermometers: Freier and Anderson Ha1 — 4; Hilton H1.
   b. Thermal expansion: Freier and Anderson Ha5 — 12; Hilton H2.
7. Audio/visual
   Demonstrations of Physics: Volume 4: Thermal effects; video cassette, about 30 min. Vikas Productions, Box 6088, Bozeman, MT 59771. Demonstrations of thermal expansion, heat capacity, heat of fusion, simultaneous boiling and freezing of water. Reviewed TPT **29**, 403 (September 1991).
8. Laboratory
   MEOS Experiment 9–3: *Linear Expansion* and BE Experiment 18: *Linear Coefficient of Expansion of Metals.* The length of a metal rod is measured at room temperature and at 100°C (in a steam jacket), then the data is used to compute the coefficient of thermal expansion. The experiment can be repeated for several different metals and the results compared.

Chapter 20 HEAT AND THE FIRST LAW OF THERMODYNAMICS

BASIC TOPICS

I. Heat.

A. Explain that when thermal contact is made between two bodies at different temperatures, a net flow of energy takes place from the higher temperature body to the lower temperature body. The temperature of the hotter body decreases, the temperature of the cooler body increases, and the net flow continues until the temperatures are the same. Energy also flows from warmer to cooler regions of the same body. State that heat is energy that is transferred because of a temperature difference. Distinguish between heat and internal energy. Emphasize that the idea of a body having heat content is not meaningful. Also emphasize that heat is not a new form of energy. The energy transferred may be the kinetic energy of molecules or the energy in an electromagnetic wave. Examples: a bunsen burner flame, radiation across a vacuum. State that heat is usually measured in Joules but calories and British thermal units are also used. 1 kcal = 3.969 Btu = 4187 J. Remark that the unit used in nutrition, a Calorie (capitalized) is 1 kcal.

B. Remind students of the energy equation studied in Chapter 8. Tell them that for the systems considered here the center of mass remains at rest (or has a constant velocity) and changes in potential energy are ignored. Processes considered change only the internal energy. A new term, however, must be added since the environment can exchange heat with the system, as well as do macroscopic work on the system. Write $\Delta E_{\text{int}} = Q - W$, where Q is the heat absorbed *by* the system and W is the work done *by* the system.

C. Stress the sign convention for heat and work: Q is positive if the system absorbs heat, W is positive if the system does positive work.

D. Stress that heat and work are alternate means of transferring energy and explain that, for example, temperature changes can be brought about by both heat and mechanical work. To demonstrate this, connect a brass tube, fitted with a rubber stopper, to a motor as shown. Make a wooden brake or clamp which fits tightly around the tube. Put a few drops of water into the tube, start the motor, and exert pressure on the tube with the clamp. Soon the stopper will fly off. Note that mechanical work was done and steam was produced.

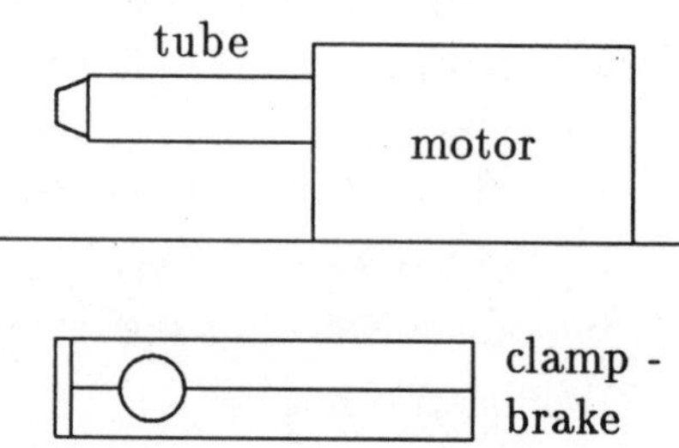

II. Heat capacity.

A. Define the heat capacity of a body as the amount of heat absorbed per degree of temperature change: for a small temperature change $C = Q/\Delta T$. Point out that it depends on the temperature and on the constraints imposed during the transfer. The heat capacity at constant volume is different from the heat capacity at constant pressure because positive work is done by the system when the temperature is increased at constant pressure. More heat is therefore required to obtain the same increase in internal energy and temperature.

B. Point out that the heat capacity depends on the amount of material. Define the specific heat c and the molar specific heat. Explain they are independent of the amount of material. Point out Table 20–1. You might use C' to denote a molar specific heat.

C. Do a simple calorimetric calculation (see Sample Problem 20–3). Stress the fundamental idea: the heat that leaves one body enters another, so the sum of the heats absorbed by all objects in a closed system vanishes.

D. Explain that energy must be transferred to or from a body when it changes phase (liquid to gas, etc.). The energy per unit mass is called the heat of transformation or latent heat. Point out Table 20–2. If time permits work a calorimetric problem that involves a change in phase (see Sample Problem 20–2).

E. (Optional) Discuss the heat capacities of a solid and explain qualitatively why they eventually decrease as the temperature is lowered. Show Fig. 20–2.

III. Heat, work, and the first law of thermodynamics.

A. Describe a gas in a cylinder fitted with a piston. Remind students that as the piston moves the gas volume changes and the gas does work $W = \int p\,dV$ on the piston. Explain that when heat enters the cylinder the gas exchanges energy with its environment through both work and heat.

B. Draw a p-V diagram (such as Fig. 20–4) and mark initial and final states, with $V_f > V_i$. Explain that p and V are thermodynamic state variables and have definite, well defined values for a given thermodynamic state. They can be used to specify the state. Point out there are many paths from the initial to the final state. Define the term "quasi-static process" and explain that the various paths on the diagram represent quasi-static processes, for which the system is infinitesimally close to equilibrium states. Point out that for different paths p is a different function of V and different amounts of work are done by the gas. Also explain that different amounts of heat are transferred for different paths. Work and heat are not thermodynamic state variables.

C. Explain that $Q - W$ is independent of the process. Define the internal energy by $\Delta E_{\text{int}} = Q - W$ and point out that ΔE_{int} is the same for any two selected states regardless of the path used to get from one to the other. State that ΔE_{int} is the change in mechanical energy (kinetic and potential energy) of all the particles that make up the system. Stress that the first law $\Delta E_{\text{int}} = Q - W$ is an expression of the conservation of energy.

D. Show that $\Delta E_{\text{int}} = mc_V\,\Delta T$ for any process and explain that $Q = mc_V\,\Delta T$ only if the volume remains constant.

III. Applications of the first law.

A. Adiabatic process. Explain that $Q = 0$ and $\Delta E_{\text{int}} = -W$. As an example, consider a gas in a thermally insulated cylinder and allow the volume to change by moving the piston. Explain that when the internal energy increases the temperature goes up for most materials. This can be achieved by compressing the gas. The opposite occurs when the piston is pulled out. Stress that no heat has been exchanged. Illustrate an adiabatic process on a p-V diagram.

B. Constant volume process. Explain that $\Delta E_{\text{int}} = Q$ and $\Delta T = Q/mc_V$ since $W = 0$. Illustrate on a p-V diagram.

C. Isobaric process. Explain that $W = p(V_f - V_i)$ and $Q = mc_p\Delta T$ for a quasi-static isobaric process. If the changes in temperature and volume are both known $\Delta E_{\text{int}} = mc_p\Delta T - p\Delta V$ can be used to find the change in internal energy, provided no phase change takes place. For a change in phase, show that $\Delta E_{\text{int}} = mL - p\Delta V$. Illustrate the two processes on a p-V diagram.

D. Describe adiabatic free expansion and note that $\Delta E_{\text{int}} = 0$. Explain that this process is not quasi-static and cannot be shown on a p-V diagram. The end points, however, are well defined thermodynamic states and are points on a p-V diagram.

E. Cyclical process. Explain that all state variables return to their original values at the end of each cycle and, in particular, $\Delta E_{\text{int}} = 0$. Thus $Q = W$. Illustrate on a p-V diagram. For later reference stress that heat may be absorbed (or rejected) and work done during a cyclic process.

IV. Transfer of heat.

A. Explain that steady state heat flow can be obtained if both ends of a slab are held at different temperatures. Define the thermal conductivity k of the material using $H = -kA\,dT/dx$ for a slab of uniform cross section A. Here H is the rate of heat flow. Emphasize that the negative sign appears because heat flows from hot to cold. Stress that H and T are constant in time in the steady state. Explain that $H = kA(T_H - T_C)/L$ for a uniform bar of length L, with the cold end held at temperature T_C and the hot end held at temperature

T_H.

B. A demonstration that shows both thermal conductivity and heat capacity can be constructed from three rods of the same size, one made of aluminum, one made of iron, and one made of glass. Use red wax to attach small ball bearings at regular intervals along each rod. Clamp the rods so that each has one end just over a bunsen burner. The rate at which the wax melts and the ball bearings drop off is mostly dictated by the thermal conductivity of the rods, but it is influenced a bit by the specific heats.

C. For a practical discussion, introduce the idea of R value and discuss home insulation.

D. Qualitatively discuss radiation as a means of energy transfer. Place a heating element at the focal point of one spherical reflector and some matches, stuck in a cork, at the focal point of the another. Place the reflectors several meters apart and adjust the positions so that the heater is imaged at the matches. Use a 1 kW or so heater. The matches will ignite in about a minute.

SUGGESTIONS

1. Emphasize the definitions of heat, work, and internal energy. Distinguish between heat and internal energy, a source of confusion to many students. Ask questions 2, 4, 24, and 25.
2. Following the discussion of heat exchange and heat capacity, ask question 9. To test the fundamental concepts of heat capacity and heat of transformation assign some of problems 2 through 7. Problems 18 and 19 are good examples of calorimetry problems. Problems 27 and 29 involve changes of phase.
3. Problem 37 is a good test of understanding of the first law. Also assign problems such as 36, 38, 39, and 42, which involve the interpretation of p-V diagrams. Tell students to pay attention to signs. Also consider question 28.
4. Use question 32 as a review of energy transfer mechanisms. Use question 11 to stimulate discussion of convection and cooling.
5. Following the discussion of thermal conductivity, ask questions 10 and 19. Assign problems 47, 48, and 56 in connection with heat conduction.
6. Demonstrations.
   a. Heat capacity and calorimetry: Freier and Anderson Hb1, 2.
   b. Work and heat: Freier and Anderson He1 — 6.
   c. Heat transfer: Freier and Anderson Hc, Hd1 — 7, Hf; Hilton H3.
   d. p-V relations: Freier and Anderson Hg1 — 3; Hilton H5f.
7. Computer program
   Physics Vol. 6: Thermodynamics, Cross Educational Software, P.O. Box 1536, Ruston, LA 71270. Apple II. Tutorial programs on calorimetry, p-V, p-T, and V-T diagrams, thermodynamic cycles, heat engines, and molecular motion. Reviewed TPT April 1985.
8. Laboratory
   a. MEOS Experiment 9–1: *Calorimetry — Specific Heat and Latent Heat of Fusion.* Students use a calorimeter to find the specific heat of water and a metal sample. They also measure the latent heat of fusion of ice. Since the specific heat of the stirring rod and the calorimeter must be taken into account, this is a good exercise in experimental design.
   b. MEOS Experiment 9–2: *Calorimetry — Mechanical Equivalent of Heat* and BE Experiment 30: *The Heating Effect of an Electric Current.* A calorimeter is used to find the relationship between the energy dissipated by a resistive heating element and the temperature rise of the water in which it is immersed. Students must accept $P = i^2 R$ for the power output of the heating element. With slight revision these experiments can also be used in conjunction with Chapter 28.

c. BE Experiment 19: *Specific Heat and Temperature of a Hot Body.* A calorimeter is used to obtain the specific heat of metal pellets. In a second part, a calorimeter and a metal sample with a known specific heat are used to find the temperature of a Bunsen burner flame.

d. BE Experiment 20: *Change of Phase — Heat of Fusion and Heat of Vaporization.* A calorimeter is used to measure the heat of fusion and heat of vaporization of water. If the lab period is long or writeups are done outside of lab, experiments 19 and 20 may be combined nicely.

e. MEOS Experiment 9–6: *Calorimetry Experiments* (with a microprocessor).

f. MEOS Experiment 9–4: *Thermal Conductivity.* The sample is sandwiched between a thermal reservoir and a copper block. The rate at which energy passes through the sample is found by measuring the rate at which the temperature of the copper increases. Temperature is monitored by means of a thermocouple.

g. MEOS Experiment 9–5: *Thermal Conductivity with Microprocessor.*

Chapter 21 THE KINETIC THEORY OF GASES

BASIC TOPICS

I. Macroscopic description of an ideal gas.

A. Explain that kinetic theory treats the same type problems as thermodynamics but from a microscopic viewpoint. It uses averages over the motions of individual particles to find macroscopic properties. Here it is used to clarify the microscopic basis of pressure and temperature.

B. Define the mole. Define Avogadro's number N_A and give its value, $6.02 \times 10^{23}\,\text{mol}^{-1}$. Explain the relationships between the mass of a molecule, the mass of the sample, the molar mass, the number of moles, the number of molecules, and Avogadro's number. These often confuse students.

C. Write down the ideal gas equation of state in the form $pV = nRT$ and in the form $pV = NkT$. Here N is the number of molecules and n is the number of moles. Give the values of R and k and state that $k = R/N_A$. Explain that for real gases at low density pV/T is nearly constant. Point out that the equation of state connects the thermodynamic variables n (or N), p, V, and T. Draw some ideal gas isotherms on a p-V diagram.

D. To show how the equation of state is used in thermodynamic calculations, go over Sample Problems 21–1 and 2.

E. Derive expressions for the work done by an ideal gas during an isothermal process and during an isobaric process.

II. Kinetic theory calculations of pressure and temperature.

A. Go over the assumptions of kinetic theory for an ideal gas. Consider a gas of molecules with only translational degrees of freedom. Assume the molecules are small and are free except for collisions of negligible duration. Also assume collisions with other molecules and with walls of the container are elastic. At the walls the molecules are specularly reflected.

B. Discuss a gas in a cubic container and explain that the pressure at the walls is due to the force of molecules as they bounce off. By considering the change in momentum at the wall per unit time, show that the pressure is given by $p = nMv_{\text{rms}}^2/3V$, where M is the molar mass. Define the rms speed. Use Table 21–1 to give some numerical examples of v_{rms}^2 and calculate the corresponding pressure. For many students the rms value of a quantity needs clarification. Consider a system of 5 or so molecules and select numerical values for their speeds, then calculate v_{rms}^2 numerically.

C. Substitute $p = nMv_{\rm rms}^2/3V$ into the ideal gas equation of state and show that $v_{\rm rms} = \sqrt{3RT/M}$. Remark that this equation can be used to calculate the rms speed for a particular (ideal) gas at a given temperature.

D. Rearrange the equation for the rms speed to obtain $\frac{1}{2}Mv_{\rm rms}^2 = \frac{3}{2}RT$ and use $M/m = N_A$ to show this can be written $\frac{1}{2}mv_{\rm rms}^2 = \frac{3}{2}kT$, where m is the mass of a molecule. Remark that the left side is the mean kinetic energy of the molecules and point out that the temperature is proportional to the mean kinetic energy.

III. Internal energy and equipartition of energy.

A. Explain that the internal energy of a monatomic ideal gas is the sum of the kinetic energies of the molecules and write $E_{\rm int} = \frac{1}{2}Nmv_{\rm rms}^2 = \frac{3}{2}NkT = \frac{3}{2}nRT$, where N is the number of molecules and n is the number of moles. Stress that for an ideal gas the internal energy is a function of temperature alone, not the pressure and volume individually. This is an approximation for a real gas. Emphasize that the velocities used in computing the internal kinetic energy are measured relative to the center of mass and that the internal energy does not include the kinetic energy associated with motion of the system as a whole.

B. Point out that if adiabatic work W is done on the gas the internal energy increases by W and the temperature increases by $\Delta T = 2W/3nR$.

C. Point out that the expression obtained above for ΔT agrees closely with experimental values for monatomic gases but gives values that are too high for gases of diatomic and polyatomic molecules. Draw diagrams of these types of molecules and explain that they have two and three degrees of rotational freedom, respectively. Some of the energy goes into motions other than the translational motion of the molecules. Define the term degree of freedom and show how to count the number for monatomic, diatomic and polyatomic molecules.

D. State the equipartition theorem: in thermal equilibrium the energy is distributed equally among all degrees of freedom, with each receiving $\frac{1}{2}kT$ for each molecule. Point out that this agrees with the previous result for monatomic gases: there are three degrees of freedom per molecule and an energy of $\frac{1}{2}kT$ is associated with each.

E. Discuss diatomic molecules and explain there are 2 new degrees of freedom, rotational in nature. Show that $E_{\rm int} = \frac{5}{2}nRT = \frac{5}{2}NkT$. Explain that $\frac{3}{2}nRT$ is in the form of translational kinetic energy and nRT is in the form of rotational kinetic energy.

F. Discuss polyatomic molecules. State that there are now 3 rotational degrees of freedom and show that $E_{\rm int} = 3nRT = 3NkT$. Explain that $\frac{3}{2}nRT$ is in the form of translational kinetic energy and $\frac{3}{2}nRT$ is in the form of rotational kinetic energy.

G. Explain that vibrational motions may also contribute to the internal energy and that, since a vibration has both kinetic and potential energy, there are two degrees of freedom and energy kT associated with each vibrational mode. Explain that, in fact, for most materials vibrational modes generally do not contribute to the internal energy except at extremely high temperatures. Quantum mechanics is required to explain why vibrational modes are frozen out.

IV. Heat capacities of ideal gases.

A. Use equations previously derived for $\Delta E_{\rm int}$ to obtain expressions for the molar specific heat at constant volume C_V. Point out the different results for monatomic, diatomic, and polyatomic molecules. Remark that C_V is used to denote *molar* specific heats in this and the next chapter. The symbol is not used for heat capacity as it was in the last chapter.

B. Show that the molar specific heat at constant pressure is related to the molar specific heat at constant volume by $C_p = C_v + R$ and derive the formulas for C_p for monatomic, diatomic, and polyatomic ideal gases.

C. For each type ideal gas obtain the value for the ratio of molar specific heats $\gamma = C_p/C_v$. Point out these values are independent of T.

D. Derive pV^γ = constant for an ideal gas undergoing an adiabatic quasi-static process. Also derive the expression $W = -(p_fV_f - p_iV_i)/(\gamma - 1)$ for the work done by the gas during an adiabatic change of state. Draw an ideal gas adiabat on a p-V diagram. Suppose the initial pressure and volume and the final volume are given. Show how to calculate the final pressure and temperature.

SUPPLEMENTARY TOPICS

1. Mean free path. This topic emphasizes the collisions of molecules and adds depth to the kinetic theory discussion but it is not crucial to subsequent chapters. Discuss as much as time allows.
2. Distribution of molecular speeds. This section deals with the Maxwell distribution and provides a deeper understanding of average speed and root-mean-square speed. Include it if you intend to cover thermonuclear fusion later in the course.

SUGGESTIONS

1. To start students thinking about the postulates of kinetic theory, ask questions 1, 2, 8, and 11.
2. Assign a problem, such as 7, that is a straightforward application of the ideal gas law. Then assign problems that show how the law is used to compute changes in various quantities when the gas changes state: 10 and 11, for example. Problems 18 through 23 deal with real-life applications. If possible, assign a few.
3. Use problem 31 in a discussion of the kinetic basis of pressure. Also assign problem 28 or 29. You may wish to discuss mixtures of gases and partial pressures; if so, consider problem 17.
4. Assign problem 32 when you deal with the kinetic basis of temperature. Question 7 and problem 35 deal with the neglect of gravity.
5. After discussing the various specific heats, ask question 36 and assign problem 61. Assign problem 62 to emphasize the dependence of the heat capacity on the process. Problem 59 illustrates an isothermal process.
6. Problems 12 and 15 provide illustrations of the work done by an ideal gas and problem 58 illustrates a change in internal energy.
7. Demonstrations
   a. Avogadro's number: Hilton H4a.
   b. Kinetic theory models: Freier and Anderson Hh1, 2, 4, 5.

9. Audio/Visual
   a. *Pressure, Volume, and Boyle's Law*, S8, color, 4 min. Kalmia Company, Dept. P1, Concord, MA 01742. Computer-animated particles in a box are used to illustrate the pressure-volume relationship. Reviewed Science Books & Films 121:169 (1976).
   b. *The Determination of Boltzmann's Constant*; Films for the Humanities and Sciences, Inc., Box 2053, Princeton, NJ 08543. Experimental setup is explained and data taken. The analysis is left for the student.
   c. *pV Isotherms of Carbon Dioxide*; Films for the Humanities and Sciences, Inc., Box 2053, Princeton, NJ 08543.
   d. *Ideal Gas Law*; *Gravitational Distribution*; *The Maxwell-Boltzmann Distribution*; *Brownian Motion and Random Walk, Deviations from an Ideal Gas*, S8 or 16 mm reels, color, 3 — 4 min. each. Kalmia Company, Dept. P1, Concord, MA 01742. Selected loops from the FITCH-MIT series dealing with various topics related to the kinetic theory of gases. Reviewed AJP 44:810 (1976).

10. Computer programs
    a. *Physics Vol. 6: Thermodynamics.* See Chapter 20 notes.
    b. *Animation Demonstration: Animated Particles*; CONDUIT, The University of Iowa, Oakdale Campus, Iowa City, IA 52242. The influence of gravitational and magnetic fields are simulated. Reviewed TPT November 1986.
    c. *Intermediate Physics Simulations: Gas — with Collisions and Speed Distributions*, R.H. Good, Physics Department, California State University at Hayward, CA 94542. Apple II. Large and small mass particles are represented on the screen. They are segregated to start and all particles of the same mass have the same velocity. Thermal equilibrium is achieved through collisions with the walls and with each other. Graphs of the velocity distributions are shown and students see them approach Maxwellian distributions. Excellent demonstration for showing the approach to thermal equilibrium.
    d. *Physics Simulations III: Gas*, Kinko's Service Corporation, 4141 State Street, Santa Barbara, CA 93110. An excellent simulation of gas molecules in a box. Use for illustration of lectures.
11. Laboratory
    a. BE Experiment 17: *Pressure and Volume Relations for a Gas.* The volume of gas in a tube is adjusted by changing the amount of mercury in the tube and a U-tube manometer is used to measure pressure. A logarithmic plot is used to determine the relationship between pressure and volume.
    b. MEOS Experiment 9–8: *Kinetic Theory Model.* The Fisher kinetic theory apparatus, consisting of a large piston-fitted tube of small plastic balls, is used to investigate relationships between pressure, temperature, and volume for a gas. A variable-speed impeller at the base allows changes in the average kinetic energy of the balls; the piston can be loaded to change the pressure. A variety of experiments can be performed.

Chapter 22 ENTROPY AND THE SECOND LAW OF THERMODYNAMICS

BASIC TOPICS

I. Engines and refrigerators.
   A. Distinguish between reversible and irreversible processes. Remark that reversible processes are quasi-static but not all quasi-static processes are reversible (*i.e.* quasi-static processes involving friction). Also mention that for a gas the path of a reversible process can be plotted on a p-V diagram. As examples consider reversible and irreversible compressions of an ideal gas.
   B. Discuss heat engines and refrigerators in general, from the point of view of the first law only. Explain that they run in cycles and that an engine absorbs heat at a high temperature, rejects heat at a low temperature, and does work. Describe a refrigerator in similar terms. Define the efficiency of an engine and the coefficient of performance of a refrigerator. Remark that heat engines and refrigerators may be reversible or irreversible.
   C. Remind students that a cycle is a process for which the system starts and ends in the same equilibrium state and that $\Delta E_{\text{int}} = 0$, $\Delta p = 0$, $\Delta T = 0$, and $\Delta V = 0$ for a cycle.
   D. Consider a gas undergoing a reversible cycle consisting of two constant volume processes and two constant pressure processes. Illustrate with a p-V diagram. Mention that, when run as a heat engine, heat enters the gas during one of the constant volume processes and one of the constant pressure processes; heat leaves the gas during the other two processes. The gas does positive work during the isobaric expansion and negative work during the

isobaric compression. Refer to the p-V diagram and show that the net work done by the gas is positive. Show that the efficiency is given by $e = (|Q_{\text{in}}| - |Q_{\text{out}}|)/|Q_{\text{in}}|$. Remark that the cycle operating in the reverse direction is a refrigerator.

II. The second law of thermodynamics.

A. Give both forms: in terms of heat engines and in terms of refrigerators. Remark that both statements imply the use of cycles, for otherwise the internal energy would be different for the initial and final states. The first form rules out the perfect heat engine, for which $Q_{\text{out}} = 0$; the second form rules out the perfect refrigerator, for which no work need be done. State that the two statements are equivalent since if either is violated an engine or refrigerator that violates the other can be constructed. Optional: prove the equivalence.

B. Define a Carnot cycle. Assume the working substance is an ideal gas and draw the cycle on a p-V diagram. Stress that the working substance passes through a succession of equilibrium states. Show that the work done by the system during a cycle is $W = |Q_H| - |Q_C|$, where H denotes the higher temperature reservoir and C denotes the lower temperature reservoir. Stress that $|Q_H|$ and $|Q_C|$ are magnitudes: for an engine Q_H is positive and Q_C is negative. Remark that the cycle can be run in reverse as a refrigerator.

C. Show that for a Carnot cycle $|Q_H|/T_H = |Q_C|/T_C$. Also show that the efficiency of an ideal gas Carnot heat engine is given by $e = 1 - |Q_C|/|Q_H| = 1 - T_C/T_H$ and that the coefficient of performance of an ideal gas Carnot refrigerator is $K = T_C/(T_H - T_C)$. Remark that $e < 1$ unless $T_C = 0$. Explain that an engine with $e = 1$ has never been built.

D. Use the second law to show that the efficiencies of all reversible heat engines operating between the same two temperatures are the same and that the efficiency of an irreversible engine is less than that of a reversible engine. In particular, show the converse allows the construction of an engine or refrigerator that violates the second law. Remark that $e = 1 - T_C/T_H$ for *every* Carnot engine, regardless of the working substance.

III. Entropy.

A. Remind students that Q_H and Q_C have opposite signs for a Carnot cycle so $Q_H/T_H + Q_C/T_C = 0$. Argue that any reversible cycle can be constructed from a series of alternating isotherms and adiabatic lines and, in the limit of infinitesimal differences between isotherms, $\oint dQ/T = 0$. See Figs. 22–11 and 12.

B. Define the entropy difference between two infinitesimally close states as $dS = dQ/T$ and between any two states as $\Delta S = \int dQ/T$. Explain that the integral is independent of path and that S is therefore a thermodynamic state variable. Stress that a reversible path must be used to evaluate the integral but that entropy differences are defined regardless of whether the actual process is reversible or irreversible. The end points must be equilibrium states, however.

C. Consider the adiabatic free expansion of an ideal gas. Point out that the process is irreversible, $Q = 0$, and $\Delta E_{\text{int}} = 0$. Since the gas is ideal, $T_f = T_i$. Find the change in entropy by evaluating $\int dQ/T$ over a reversible isotherm through the initial and final states. Point out that the isothermal path does not represent the actual process. Show that $\Delta S = nR\ln(V_f/V_i)$ and state this is positive.

D. Derive expressions for the change in entropy for an ideal gas undergoing processes at constant volume, constant pressure, and constant temperature.

E. Consider two identical rigid containers of ideal gas, at different temperatures, T_H and T_C. Place them in contact in an adiabatic enclosure. Show they reach equilibrium at temperature $T_m = (T_H + T_C)/2$. Then consider a reversible, constant volume process which connects the initial and final states and show that $\Delta S = C_V \ln(T_m^2/T_H T_C)$. Remark that this is positive.

F. Remark that for reversible processes the total entropy of the system and its environment does not change. This is because, for the combination of system and environment, the process is adiabatic and $dQ = 0$ for each segment of the reversible path. On the other hand, entropy increases for an adiabatic *irreversible* process.

G. State that the second law is equivalent to the statement that for processes which proceed from an initial equilibrium state to a final equilibrium state the total entropy of a closed system (system + environment) does not decrease. State that if the process is reversible the total entropy does not change and if the process is irreversible it increases. Point out that the previous two examples are consistent with this statement. Optional: show this statement is equivalent to the first and second forms of the second law.

SUGGESTIONS

1. In order to make the ideas of thermodynamic equilibrium, reversible process, and irreversible process more concrete, ask a few of the questions in the group 6 through 11.
2. Consider practical engines and their efficiencies by approximating their operation by Carnot cycles. For a gasoline engine $T_H \approx 1000°F$ and $T_L \approx 400°F$. Compare actual efficiencies with the Carnot efficiency. Actual efficiencies can be obtained by considering the fuel energy available and the work actually obtained.
3. Consider practical refrigerators. Look in a catalog for typical values of the coefficient of performance and compare with the Carnot coefficient of performance.
4. Use questions 12 through 20 to discuss real and Carnot engines. Problems 4, 5, and 7 cover the fundamentals of cycles. Problem 10 covers work and heat transfer in a Carnot cycle.
5. To start students thinking about entropy changes as they occur in common processes, ask a few of the questions in the group 21 through 31. Assign problems 34, 37, 46, and 50. To include entropy changes in calorimetry experiments, ask problems 43 and 44.
6. Ask bright students to draw a Carnot cycle on a T-S diagram. See question 24.
7. Demonstrations
   Engines: Freier and Anderson Hm5, Hn; Hilton H5a, b.
8. Audio/Visual
   a. *The Reversibility of Time*, S8, color, 4 min. Kalmia Company, Dept. P1, Concord, MA 01742. To illustrate that reverse motion is not always obvious, various scenes are run forward and backward. Part of the Project Physics Series.
   b. *Entropy of Mixing*; Philip R. Pennington; American Association of Physics Teachers, 5112 Berwyn Road, College Park, MD 20740–4100. Slides use dot patterns to show mixing.

Chapter 23 ELECTRIC CHARGE

BASIC TOPICS

I. Charge.

A. Explain that there are two kinds of charge, called positive and negative, and that like charges repel each other, unlike charges attract each other. Give the SI unit (coulomb) and explain that it is defined in terms of current, to be discussed later. Optional: explain that current is the flow of charge and is measured in amperes. One coulomb of charge passes a cross section each second in a wire carrying a steady current of 1 A.

B. Carry out the following sequence of demonstrations. They work best in dry weather.

1. Suspend a pith ball by a string. Charge a rubber rod by rubbing it with fur, then hold the rod near the pith ball. The ball is attracted, touches the rod, then flies away after a short time. Use the rod to push the ball around without touching it. Explain

that the rod and ball carry the same type charge. Hold the fur near the pith ball and explain that they are oppositely charged.

2. Repeat using a second pith ball and a wooden rod charged by rubbing it on a plastic sheet (this replaces the traditional glass rod – silk combination and works much better). Place the two pith balls near each other and explain they are oppositely charged.
3. Suspend a charged rubber rod by a string. Use another charged rubber rod to push it around without touching it. Similarly, pull it with the charged wooden rod. Also show that only the rubbed end of the rubber rod is charged.

II. Conductors and insulators.

A. Explain the difference between a conductor and an insulator as far as the conduction of charge is concerned. Explain that excess charge on a conductor is free to move and generally does so when influenced by the electric force of other charges. Excess charge on a conductor is distributed so the net force on any of it is zero. Any excess charge on an insulator does not move far from the place where it is deposited. Remind students of the demonstration which showed that only the rubbed end of the rubber rod remains charged. Metals are conductors. The rubber rod is an insulator. Mention semiconductors and superconductors.

B. Use an electroscope to demonstrate the conducting properties of conductors. Charge the electroscope by contact with a charged rubber rod and explain why the leaves diverge. Discharge it by touching the top with your hand. Explain why the leaves converge. Recharge the electroscope with a charged wooden rod, then bring the charged rubber rod near the electroscope, but do not let it touch. Note the decrease in deflection and explain this by pointing out the attraction of the charge on the rod for the charge on the leaves. Throughout, emphasize the motion of the charge through the metal leaves and stem of the electroscope.

C. Demonstrate charging by induction. Bring a charged rubber rod near to but not touching an uncharged electroscope. Touch your finger to the electroscope, then remove it. Remove the rubber rod and note the deflection of the leaves. Bring the rubber rod near again and note the decrease in deflection. Observe that the electroscope and rod are oppositely charged. Confirm this with the wooden rod. Explain the process.

III. Coulomb's law.

A. Assert that experimental evidence convinces us that there are only two kinds of charge and that the force between a pair of charges is along the line joining them, has magnitude proportional to the product of the magnitudes of the charges and is inversely proportional to the square of the distance between them. Further, the force is attractive for unlike charges and repulsive for like charges.

B. Write down Coulomb's law for the magnitude of the electric force exerted by one point charge on another. Give the SI value for ϵ_0 and for $1/4\pi\epsilon_0$. Stress that the law holds for point charges. Note in detail that the mathematical form of the law contains all the qualitative features discussed previously. If Chapter 15 was covered, point out the similarity with Newton's law of gravity and mention that, unlike charge, there is no negative mass.

C. Explain that a superposition law holds for electric forces and illustrate by finding the resultant force on a charge due to two other charges. Use the analogy with the law of gravity to show that the force of one spherical distribution of charge on another obeys the same law as two point charges and that the force on a charge inside a uniformly charged spherical shell is zero.

IV. Quantization and conservation of charge.

A. State that all measured charge is an integer multiple of the charge on a proton: $q = ne$. Give the value of e: 1.60×10^{-19} C. State that the charge on the proton is $+e$, the charge on the electron is $-e$, and the neutron is neutral.

B. Remark that macroscopic objects are normally neutral; they have the same number of protons as electrons. Stress that the word "neutral" describes the algebraic sum of the charges and does not indicate the absence of charged particles. Remark that when an object is charged the charge imbalance is usually slight but significant. Go over Sample Problem 23–3.

C. State that charge is conserved in the sense that for a closed system the sum of all charge before an event or process is the same as the sum after the event or process. Stress that the charges in the sum must have appropriate signs. Example: rubbing a rubber rod with fur. The rod and fur are oppositely charged afterwards and the magnitude of the charge is the same on both. Also discuss the conservation of charge in the annihilation and creation of fundamental particles and note that the identity of the particles may change in an event but charge is still conserved. Examples: beta decay, electron-positron annihilation.

SUPPLEMENTARY TOPIC

The constants of physics. Include this section to add some breadth to the course.

SUGGESTIONS

1. Discuss questions 1, 2, 3, and 6, perhaps in connection with demonstrations or lab experiments. Also see problems 7 and 12. Discuss charging by induction and go over question 5.
2. Use questions 7 and 8 to see if students can distinguish conductors from insulators.
3. Use questions 12 and 14 to test for qualitative understanding of polarization and the decrease in electric force with increasing distance.
4. Question 13 is important for understanding the vanishing of the electrostatic field in a conductor, discussed later.
5. Problems 8, 9, and 13 deal with the addition of electric forces in one dimension, problems 10 and 14 deal with the addition of electric forces in two dimensions. Also see problem 17. All cause students to think about the direction of the force.
6. Demonstrations
   a. Charging, electroscopes: Freier and Anderson Ea1, 2, 11; Hilton E1a — f.
   b. Electric force: Freier and Anderson Ea5, 6, 8, 12, 15, 17, Eb3, 4, 9, 10, 12, Ec4 — 6.
   c. Induction: Freier and Anderson Ea12, 13, 14; Hilton E1g.
   d. Touch a grounded wire to several places within a small area of a wall. Rub a balloon with fur and place it in contact with that area. Ask students to explain why the balloon sticks.
7. Audio/Visual
   The following film and loop series have material which is pertinent for Chapters 23 through 27. The films are listed in groups rather than divided among the various chapters. Appropriate films for a particular chapter can be selected by title.
   a. *Electric Fields and Moving Media*, 16 mm or 3/4" video cassette, color, 32 min. Educational Development Corporation, Distribution Center, 39 Chapel Street, Newton, MA 02160. Both static and time-varying electric fields and their effects on various media are illustrated. Natural phenomena and technological applications are shown. Reviewed SBF 11:104 (1975).
   b. *Introduction to Electrostatics*; *Insulators and Conductors*; *Electrostatic Induction*; *The Electroscope*; *Charge Distribution: The Faraday Ice-Pail Experiment*; *Charge Distribution: Concentration and Point Discharge*; *The Van de Graaff Generator*; *The Photoelectric Effect*; *Capacitors and Dielectrics*; *Problems in Electrostatics*, S8, color, 4 min. each. Kalmia Company, Dept. P1, Concord, MA 01742. These loops make up the Electrostatic Series developed by A.E. Walters of Rutgers University. Large lecture hall apparatus is utilized

to illustrate the various phenomena. Student notes are included with each film. Parts of this series are reviewed in TPT 12:507, 13:254, 13:371, and 14:58.

c. *Capacitance of Capacitor Combinations — Parallel*; *Capacitance of Capacitor Combinations — Series*; *Charge on the Outside of a Conductor*; *Conductors, Insulators, and Capacitors*; *Coulomb's Law*; *Discharging the Electroscope — Conduction and Ionization*; *Discharging the Electroscope — the Photoelectric Effect*; *Electric Field and Induced Charges*; *Electrostatic Attraction*; *Electrostatic Repulsion*; *Increasing the Potential of a Capacitor*; *Polarity*; *Variation of Charge with Curvature*; *A Working Model of a Van de Graaff Generator*, S8, color, 4 min. each. Encyclopaedia Britannica Educational Corporation, 425 N. Michigan Ave., Chicago, IL 60611. This popular Electrostatic Series produced by Albert Baez illustrates a variety of electrostatic phenomena and is a good source of demonstrations.

d. *Millikan's Oil-Drop Experiment*; Films for the Humanities and Sciences, Inc., Box 2053, Princeton, NJ 08543. Experimental setup is explained and data taken. The analysis is left for the student.

e. *Demonstrations of Physics: Volume 6: Electricity and Magnetism*; video cassette, about 30 min. Vikas Productions, Box 6088, Bozeman, MT 59771. Demonstrations of electrostatics, series and parallel DC circuits, magnetic fields, electromagnetic induction. Reviewed TPT **29**, 403 (September 1991).

8. Computer program
Basic Concepts of Electricity, Series I: Basic Concepts, Merlan Scientific Ltd., 247 Armstrong Avenue, Georgetown, Ontario L7G 4X6, Canada. Apple II+, IIe. Introduction and drill on charging by rubbing, current in simple circuits, electric potential difference. Reviewed TPT November 1983.

9. Laboratory
MEOS Experiment 10–2: *The Electrostatic Balance.* A coulomb torsional balance is used to find the functional relationship between the electrostatic force of one small charged ball on another and the separation of balls. An electrostatic generator is used to charge the balls.

Chapter 24 THE ELECTRIC FIELD

BASIC TOPICS

I. The electric field.

A. Use a fluid to introduce the idea of a field. The temperature of the fluid $T(x, y, z, t)$ is an example of a scalar field and the velocity $\mathbf{v}(x, y, z, t)$ is an example of a vector field. Point out that these functions give the temperature and velocity at the place and time specified by the dependent variables.

B. Explain that charges may be thought to create an electric field at all points in space and that the field exerts a force on another charge, if present. The important questions to be answered are: given the charge distribution what is the field? given the field what is the force on a charge?

C. Consider two point charges and remark that each creates a field and that the field of either one exerts a force on the other. Explain that the two together produce a field that is the superposition of the individual fields and that this field exerts a force on a third charge, if present.

D. Define the field at any point as the force per unit charge on a positive test charge at the point, in the limit of a vanishingly small test charge. Mention that the limiting process eliminates the influence of the test charge on the charge creating the field. SI units: N/C.

E. Use Coulomb's law to obtain the expression for the field of a point charge. Explain that the field of a collection of charges is the vector sum of the individual fields.

II. Electric field lines.

A. Explain that field lines are useful for visualizing the field. Draw field lines for a point charge and explain that, in general, the field at any point is tangent to the line through that point and that the magnitude of the field is proportional to the number of lines per unit area that pass through a surface perpendicular to the lines.

B. By considering a sphere around a point charge and calculating the number of lines per unit area through the sphere, show that the $1/r^2$ law allows us to associate lines with a charge and to take the number of lines to be proportional to the charge. Explain that lines can be thought of as directed, that they originate at positive charge and terminate at negative charge.

C. Show Figs 24–2, 3, 4, and 5 or similar diagrams that illustrate the field lines of some charge distributions.

D. Field lines can be illustrated by floating some long seeds in transformer oil in a shallow, flat bottomed dish. Place two metal plates in the dish and connect them to an electrostatic generator. The seeds line up along the field lines. You can place the apparatus on an overhead projector and shadow project the seeds.

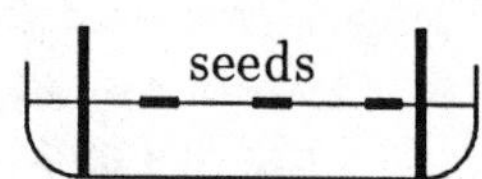

III. Calculation of the electric field.

A. Remind the students of the expression for the field of a point charge. State that the field is radially outward for a positive charge and radially inward for a negative charge. Also remind them that the total field is the vector sum of the individual fields of the charge being considered.

B. Derive an expression for the field of an electric dipole by considering the field of two charges with equal magnitudes and opposite signs. Consider a field point on a line perpendicular to the dipole moment, on a line along the dipole moment, or a general point. Evaluate the expression in the limit of vanishingly small separation and finite dipole moment. Define the dipole moment and stress that it points from the negative toward the positive charge. Point out that the field is proportional to $1/r^3$ for points far from the dipole.

C. Consider a small set of discrete charges and calculate the electric field by evaluating the vector sum of the individual fields. Example: the field at the center of a square with various charges on its corners.

D. As an introduction to the fields of continuous charge distributions go over the ideas of linear, area, and volume charge densities. Graphically show how a line of charge is divided into infinitesimal segments and point out that a segment of length ds contains charge $dq = \lambda\, ds$. Explain that for purposes of calculating the field each segment can be treated as a point charge and that the fields of all segments are summed vectorially to find the total field.

E. Derive an expression for the field on the axis of a continuous ring of charge. Carefully explain how the integral is set up and how the vector nature of the field is taken into account by dealing with components. Explain in detail the symmetry argument used to show the field is along the axis.

F. Extend the calculation to find an expression for the field on the axis of a charged disk and for an infinite sheet of charge. Remark that the field of a sheet is perpendicular to the sheet and is independent of distance from the sheet. This will be useful later when parallel plate capacitors are studied.

IV. Motion of a charge in an electric field.

A. Point out that the electric force on a charge is $q\mathbf{E}$ and explain that the electric field used is that due to all *other* charges (except q). Substitute the force into Newton's second law and remind the students that once the acceleration and initial conditions are known, kinematics can be used to find the subsequent motion of the charge.

B. Find the trajectory of a charge moving into a region of uniform field, perpendicular to its initial velocity. Compare to projectile motion problems studied in Chapter 4. See Sample Problem 24–8.

C. Show that the force on a dipole in a uniform field is zero and that the torque is $\mathbf{p} \times \mathbf{E}$. Also show that the potential energy of a dipole is $-\mathbf{p} \cdot \mathbf{E}$. Emphasize that the potential energy minimum occurs when the dipole moment is aligned with the field. To review oscillatory rotational motion calculate the angular frequency of small angle oscillations for a dipole with rotational inertia I in an electric field. Assume no other forces act.

SUGGESTIONS

1. Assign problem 40 to illustrate the measurement of an electric field.
2. Center a qualitative discussion of electric field lines on questions 3 through 6. Have students sketch field lines for various charge distributions. See problems 2, 3, and 4.
3. Have students work some of problems 14, 15, 16, and 17. These deal with the superposition of fields.
4. Use question 10 to test for understanding of the idea that a charge does not exert a force on itself.
5. Problems 25 and 26 are a good tests of understanding of the derivation of the dipole field. Ask question 12.

7. Assign problems 37 and 38 to support the discussion of the field of a uniformly charged disk. Assign problems 33 and 34 to give students practice in deriving expressions for the field of a continuous charge distribution.

7. Assign problems 47, 52, and 56 to help students with the motion of point charges in fields. Assign questions 17 and 18 and problems 58 (torque) and 59 (energy) in connection with the discussion of a dipole in a field. Also consider question 19.

8. To include the Milliken oil drop experiment, assign problems 49 and 50.

9. Computer projects

   a. Have students use a commercial math program or write their own programs to calculate the electric fields of discrete charge distributions. Have them use the programs to plot the magnitude of the field at various distances from a dipole, along lines that are perpendicular and parallel to the dipole moment.

   b. Have students write programs to trace field lines for discrete charge distributions.

   These and other projects are described in the Computer Projects Section of this manual.

10. Demonstrations

    Electric field lines: Freier and Anderson Eb1, Ec2 — 4.

11. Computer programs

    a. *Physics Disk 3: Electric Fields and Potentials*, The 6502 Program Exchange, 2920 Moana, Reno, NV 89509. Apple II. Generates field lines and equipotential surfaces for user supplied distribution of discrete charges. Diagrams can be stored for later display. Chiefly for lecture illustrations. Reviewed TPT September 1986.

    b. *Physics Simulations II: Coulomb*, Kinko's Service Corporation, 4141 State Street, Santa Barbara, CA 93110. Macintosh. User gives up to 15 charges and their positions, then the program displays electric field lines.

c. *Laboratory Simulations in Atomic Physics*, Norwalk High School Science Department, County Street, Norwalk, CT 06851. Apple II. Simulations of the deflection of an electron by an electric field, the Thompson e/m experiment, the Millikan oil drop experiment, and a mass spectrometer. Parameters are selected by the user. Excellent for illustrating lectures. Some parts can be used in connection with this chapter, some in connection with Chapter 30. Reviewed TPT March 1984.

Chapter 25 GAUSS' LAW

BASIC TOPICS

I. Electric flux.

A. Start by discussing some of the important concepts in a general way. Define a vector surface element. Define the flux of a vector field through a surface. Distinguish between open and closed surfaces and explain that for the latter the surface normal is taken to be *outward.* Interpret the surface integral for the flux as a sum over surface elements. If you covered Chapter 16 use the velocity field of a fluid as an example.

B. Define electric flux. Point out that it is the normal component of the field which enters. Also point out that the sign of the contribution of any surface element depends on the choice for the direction of $d\mathbf{A}$.

C. Interpret electric flux as a quantity that is proportional to the net number of field lines penetrating the surface. Remind students that the number of lines through a small area perpendicular to the field is taken to be proportional to the magnitude of the field. By considering surfaces with the same area but different orientations, show that the net number of penetrating lines is proportional to the cosine of the angle between the field and the normal to the surface. Conclude that $\mathbf{E}\cdot d\mathbf{A}$ is proportional to the number of lines through $d\mathbf{A}$.

D. Stress that lines roughly in the same direction as the normal contribute positively to the flux, lines roughly in the opposite direction contribute negatively, and lines that pass completely through a volume do not contribute to the flux through its boundary. Point out that zero flux through a surface does not imply zero field at points on the surface.

E. As an example, calculate the flux through each side of a cube in a uniform electric field.

II. Gauss' law.

A. Write down the law. Stress that the surface is closed and that the charge appearing in the law is the net charge enclosed. Interpret the law as a statement that the number of (signed) lines crossing the surface is proportional to the net charge inside and make the statement plausible by reminding students that the field of each charge is proportional to the charge and its direction depends on the sign of the charge.

B. Illustrate by considering the surface of a sphere with positive charge inside, with negative charge inside, with both positive and negative charge inside, and with charge outside. In each case draw representative field lines with the number of lines proportional to the net charge. Stress that the position of the charge inside is irrelevant for the flux through the surface. Also use Gauss' law to calculate the flux.

C. Use Gauss' law and symmetry arguments to obtain an expression for the electric field of a point charge.

III. Gauss' law and conductors.

A. Argue that the electrostatic field vanishes inside a conductor and use Gauss' law to show that there can be no net charge at interior points under static conditions. Point out that exterior charge and charges on the surface separately produce fields in the interior but that the resultant field vanishes. For contrast, point out that an insulator may have charge distributed throughout.

B. Demonstrate that any excess charge on a conductor resides on the exterior surface. Use a hollow metal sphere with a small hole cut in it. As an alternative, solder shut the top on an empty metal can and drill a small hole in it. This will not work as well because of the sharp edges. Charge a rubber rod by rubbing it with fur and touch it to the inside of the sphere, being careful not to touch the edge of the hole. Repeat several times to build up charge. Now scrape at the interior with a metal transfer rod, again being careful not to touch the edge of the hole. Touch the transfer rod to an uncharged electroscope and note the lack of deflection. Scrape the exterior of the sphere with the transfer rod and touch the electroscope. Note the deflection.

C. Show how to calculate the charge on the inner and outer surfaces of neutral and charged conducting spherical shells when charge is placed in the cavities. See problem 19.

D. Use Gauss' law to show that the magnitude of the field just outside a charged conductor is given by $E = \sigma/\epsilon_0$, where σ is the surface charge density.

IV. Applications of Gauss' law.

A. Derive expressions for the electric field at various points for a uniformly charged sphere and for a uniformly charged thick spherical shell. Remark that such distributions are possible if the sphere or shell is not conducting. Carefully give the symmetry argument to show the field is radial and has the same magnitude at all points on a concentric sphere.

B. Derive an expression for the electric field at a point outside an infinite sheet with a uniform charge distribution. Contrast with the field outside an infinite conducting sheet with the same area charge density on one surface. Point out that for the conductor the field is not due only to the charge on the surface being considered. Another field must be present to produce a net field of zero in the interior and this doubles the field in the exterior.

C. Consider a point charge at the center of a neutral spherical conducting shell and derive expressions for the electric field in the various regions. Repeat for a charged shell.

D. Work one problem with cylindrical symmetry. For example, consider charge distributed uniformly throughout a cylinder and find the field in all regions.

E. Note that Gauss' law can be used to find **E** only if there is adequate symmetry.

SUPPLEMENTARY TOPIC

Experimental verification of Gauss' law. Verification of the $1/r^2$ law.

SUGGESTIONS

1. Use questions 4 through 9 to help students understand the flux integral and charge that appear in Gauss' law. Assign it if you covered Chapter 16. Problem 1 is a good example of a fluid flux calculation. Use problems 2 and 3 to introduce electric flux. The latter problem also demonstrates the vanishing of the total flux for a closed surface in a uniform field.
2. Problems 5 and 6 illustrate the fundamental idea of Gauss' law. Problems 10 and 12 are also instructive.
3. A more detailed discussion of the symmetry arguments used in connection with Gauss' law can be given with the aid of questions 10, 11, 18, 20, and 22. Choose one or more according to the applications you stressed in lecture.
4. Use questions 13, 14, 15, and 17 and problems 18, 19, and 20 to discuss the electrostatic properties of conductors.
5. Assign a variety of problems dealing with applications: 23, 24, 27, and 30 (cylinders of charge); 31, 37, and 39 (planes of charge); 42, 44, and 47 (spheres of charge). Assign problem 54 to challenge good students.

6. Computer project
Have students use a commercial math program or their own programs to evaluate the flux integral in Gauss' law. Have them separately calculate the flux through each face of a cube containing a point charge. Consider various positions of the charge within the cube to show that the flux through individual faces may change as the charge changes position but the total flux remains the same and obeys Gauss' law. Repeat for a point charge outside the cube. Details are given in the Computer Projects Section of this manual.
7. Demonstrations
Charges on conductors: Freier and Anderson Ea7, 18, 23, Eb7; Hilton E1h.

Chapter 26 ELECTRIC POTENTIAL

BASIC TOPICS

I. Electric potential
   A. Define the potential difference of two points as the negative of the work per unit charge done by the electric field when a positive test charge moves from one point to the other. Stress the sign of the potential: the potential of the end point is higher than that of the initial point if the work is negative. The electric field points from regions of high potential toward regions of low potential and positive charge tends to be repelled from regions of high potential. The region near an isolated positive charge has a higher potential than regions far away. The opposite is true for a negative charge. Unit: volt. Define electron volt as a unit of energy.
   B. If you covered Chapter 15, use the similarity of Coulomb's law and Newton's law of gravity to argue that the electrostatic force is conservative and that the work is independent of path. If you did not cover Chapter 15, either derive or state these results.
   C. Show that the definition is equivalent to $V_b - V_a = -\int \mathbf{E} \cdot d\mathbf{s}$, where the integral is along a path from a to b. Point out that the potential is constant in regions of zero field. Note that the unit N/C is the same as V/m and the latter is a more common unit for **E**.
   D. Point out that the potential is a scalar and that only potential differences are physically meaningful. One point can be chosen arbitrarily to have zero potential and the potential at other points is measured relative to the potential there. Often the potential is chosen to be zero where the field (or force) is zero. For a finite distribution of charge the potential is usually chosen to be zero at a point far away (infinity). Show a voltmeter and remark that the meter reads the potential difference between the leads.
   E. Show that the potential a distance r from an isolated point charge is given by $V = q/4\pi\epsilon_0 r$. Remark that this is the potential energy per unit test charge of a system consisting of the point charge q and the test charge. Explain that the equation is valid for both positive and negative charge. Show how to calculate the potential due to a collection of point charges. Derive the expression for the potential of an electric dipole.
   F. Give some examples of calculations of the potential from the electric field. Start with a uniform electric field, like that outside a uniform plane distribution of charge, and show that potential is given by $-Ex+C$, where C is a constant. Since the distribution is infinite the point at infinity cannot be picked as the zero of potential.
   F. As a more complicated example, consider one of the configurations discussed in the last chapter, a point charge at the center of a spherical conducting shell, say. Take the potential to be zero at infinity and compute its value at points outside the outer surface, within the shell, and inside the inner surface. As an alternative you might find expressions for the potential in various regions around and inside a sphere with a uniform charge distribution.

G. Write down the integral expressions for the potential due to a line of charge and for a surface of charge, in terms of the linear and area charge densities. Work an example, such as the potential of a uniform finite line of charge or a uniform disk of charge.

II. Equipotential surfaces.

A. Define the term equipotential surface. Show diagrams of equipotential surfaces for an isolated point charge and for the region between two uniformly charged plates. Equipotential surfaces of a dipole are shown in Fig. 26–3.

B. Point out that the field does zero work if a test charge is carried between two points on the same surface and note that this means that the force, and hence **E**, is perpendicular to the equipotential surfaces. Note further that the work done by the field when a charge is carried from any point on one surface to any point on another is the product of the charge and the negative of the potential difference.

III. Calculation of **E** from V.

A. Remind students that $\Delta V = -E\Delta x$ for a uniform field in the positive x direction. Note that E has the form $-\Delta V/\Delta x$ and **E** is directed from high to low potential. Use this result to reenforce the idea of an equipotential surface and the fact that **E** is perpendicular to equipotential surfaces.

B. Generalize the result to $E = -\mathrm{d}V/\mathrm{d}s$, where s is the distance along a normal to an equipotential surface. Then broaden this further to $E_x = -\partial V/\partial x$, $E_y = -\partial V/\partial y$, and $E_z = -\partial V/\partial z$. Verify that the prescription works for a point charge and for a dipole.

IV. Electrostatic potential energy.

A. Remark that when a charge Q moves from point a to point b the potential energy of the system changes by $Q(V_b - V_a)$, where V is the potential due to the other charges. When charge Q is brought into position from infinity (where the potential is zero), the potential energy changes by QV, where V is the potential at the final position of Q due to charge already in place.

B. Show that the potential energy of two point charges is given by $q_1q_2/4\pi\epsilon_0 r$, where r is their separation and the zero of potential energy is taken to be infinite separation. Point out that the potential energy is positive if the charges have like signs and negative if they have opposite signs. Explain that the potential energy decreases if two charges of the same sign move apart or two charges of opposite sign move closer together.

C. Remind students that potential energy can be converted to kinetic energy. Explain what happens if the charges used in the last example are released from their positions. Consider a proton fired directly at a heavy nucleus with charge Ze and find the distance of closest approach in terms of the initial speed.

D. Calculate the potential energy of a simple system: charges at the corners of a triangle or square. Assume the charges are brought in from infinity one at a time and sum the potential energies. Explain that the total is the sum over charge pairs. Show how to calculate the potential energy of any collection of point charges.

E. Explain that the potential energy of a system of charges is the work an agent must do to assemble the system from rest at infinite separation. This is the negative of the work done by the field.

V. An isolated conductor.

A. Recall that the electric field vanishes at points in the interior of a conductor. Argue that the surface must be an equipotential surface and that V at all points inside must have the same value as on the surface. State this is true whether or not the conductor is charged and whether or not an external field exists.

B. Consider two spherical shells of different radii, far apart and connected by a very fine wire. Explain that $V_1 = V_2$ and show that $q_1/R_1 = q_2/R_2$, then show that the surface charge density varies inversely with the radius: $\sigma_1/\sigma_2 = R_2/R_1$. Recall that E is proportional to σ just outside a conductor and argue that σ and E are large near places of small radius of curvature and small near places of large radius of curvature. Use an electrostatic generator to show discharge from a sharp point and from a rounded (larger radius) ball. Discuss the function of lightning rods and explain their shape.

SUPPLEMENTARY TOPICS

The electrostatic generator. Explain how it works. This might be done in lab if they are used there. Spend some time explaining safety precautions.

SUGGESTIONS

1. Questions 1, 2, 7, 8, and 12 can be used to help students think about the arbitrariness of the zero of potential.
2. The relationship between the electric field and potential is explored in questions 13, 14, 21, and 22. Use them to test students' understanding.
3. Use questions 9 and 15 and problems 4, 10, 16, and 20 to test for understanding of equipotential surfaces.
4. Ask students to calculate potential differences for various situations: see problems 8, 9, 13, 14, and 32.
5. Use question 23 and problems 58, 59, and 61 in connection with the discussion of electrostatic potential energy. Also assign some conservation of energy problems, such as 64 and 67.
6. Use questions 16 and 18 to aid in a discussion of the field and potential of a conductor. Assign problems 74, 76, and 81.
7. Computer project
   Have students use a commercial math program or their own root finding programs to plot equipotential surfaces for a discrete charge distribution. It is instructive to consider two unequal charges (any combination of signs). Details are given in the Computer Projects portion of this manual.
8. Demonstrations
   Electrostatic generators: Freier and Anderson Ea22, Ec1; Hilton E1i, j.
9. Computer program
   Physics Disk 3: Electric Fields and Potentials. See Chapter 24 notes.
10. Laboratory
    MEOS Experiment 10–1: *Electric Fields* and BE Experiment 25: *Mapping of Electric Fields.* Students map equipotential lines on sheets of high resistance paper with metallic electrodes at two sides. In the MEOS experiment an audio oscillator generates the field and an oscilloscope or null detecting probe is used to find points of equal potential. If students are not familiar with oscilloscopes you might want to preface this experiment with Part A of MEOS Experiment 10–10. In the BE experiment the field is generated by a battery and a galvanometer is used as a probe.

Chapter 27 CAPACITANCE

BASIC TOPICS

I. Capacitance.

A. Describe a generalized capacitor. Draw a diagram showing two separated, isolated conductors. Assume they carry charge q and $-q$ respectively, draw representative field lines, and point out that all field lines start on one conductor and terminate on the other. Explain that there is a potential difference V between the conductors and that the positively charged conductor is at the higher potential. Define capacitance as $C = q/V$. Explain that V is proportional to q and that C is independent of q and V. C does depend on the shapes, relative positions, and orientations of the conductors and on the medium surrounding them. Unit: 1 farad = 1 C/V.

B. Show a radio tuning capacitor and some commercial fixed capacitors. Mention that one usually encounters μF and pF capacitors. Capacitors on the order of 1 F have been developed for the electronics industry.

C. Remark that in circuit drawings a capacitor is denoted by ⊣⊢.

D. Remark that a battery can be used to charge a capacitor. The battery transfers charge from one plate to the other until the potential difference of the plates is the same as the terminal potential difference of the battery. Calculate the charge given the battery potential difference and the capacitance.

E. Show how to calculate capacitance in principle. Put charge q on one conductor, $-q$ on the other, and calculate the electric field due to the charge, then the potential difference between the conductors. Except for highly symmetric situations, the charge is not uniformly distributed over the surfaces of the conductors and fairly sophisticated means must be used to calculate V. The text deals with symmetric situations for which Gauss' law can be used to calculate the electric field.

F. Examples: derive expressions for the capacitance of two parallel plates (neglect fringing) and two coaxial cylinders or two concentric spherical shells. Use Gauss' law to find the electric field, then evaluate the integral for the potential difference. Emphasize that the field is due to the charge on the plates.

G. Large demonstration parallel plate capacitors with variable plate separation are available commercially. You can also make one using two $\approx$ 1-ft diameter circular plates of 1/8 inch aluminum sheet. Attach an aluminum disk to the center of each with a hole drilled for a support rod. Use an insulating rod on one and a metal rod on the other. By sliding the two conductors closer together, you can show the effect of changing d while holding q constant. An electroscope serves as a voltmeter.

H. Explain how the equivalent capacitance of a device can be measured. Consider a black box with two terminals. State that a potential difference V is applied and the total charge q deposited is measured. The capacitance is q/V.

I. Derive $1/C_{eq} = 1/C_1 + 1/C_2$ for the equivalent capacitance of two capacitors in series and $C_{eq} = C_1 + C_2$ for the equivalent capacitance of two capacitors in parallel. Emphasize that two capacitors in parallel have the same potential difference, two in series have the same charge. Explain the usefulness of these equations for circuit analysis.

II. Energy storage.

A. Derive the expression $W = \frac{1}{2}q^2/C$ for the work required to charge a capacitor. Explain that, as an increment of charge is transferred, work is done by an external agent (a battery, for example) against the electric field of the charge already on the plates. Show that this expression is equivalent to $W = \frac{1}{2}CV^2$. Interpret the result as the potential energy stored in the charge system and explain that it can be recovered when the capacitor is discharged.

B. Remark that if two capacitors are in parallel the larger stores the greater energy. If two capacitors are in series the smaller stores the greater energy.

C. Show that the energy density in a parallel plate capacitor is $\frac{1}{2}\epsilon_0 E^2$. State that this result is quite general and that its volume integral gives the work required to assemble charge to create the electric field E. Explain that the energy may be thought to reside in the field or it may be considered to be the potential energy of the charges.

D. Integrate the energy density to find an expression for the energy stored in the electric field of a charged spherical capacitor or a charged cylindrical capacitor. Compare the result with $\frac{1}{2}q^2/C$.

III. Dielectrics.

A. Explain that when the region between the conductors of a capacitor is occupied by insulating material the capacitance increases by a factor $\kappa > 1$, called the dielectric constant of the material. Remark that $\kappa = 1$ for a vacuum.

B. Use a large commercial or home-made capacitor to show the effect of a dielectric. Charge the capacitor, then isolate it and insert a glass plate between the plates. The electroscope shows that V decreases and, since q is fixed, the capacitance increases.

C. Calculate the change in stored energy that occurs when a dielectric slab is inserted between the plates of an isolated parallel plate capacitor (see Sample Problem 27–9). Also calculate the change in stored energy when the slab is inserted while the potential difference is maintained by a battery. Explain that the battery now does work in moving charge from one plate to the other.

D. Explain that dielectric material between the plates becomes polarized, with the positively charged ends of the dipoles attracted toward the negative conductor. The field of the dipoles opposes the external field, so the electric field is weaker between the plates than it would be if the material were not there. This reduces the potential difference between the conductors for a given charge on them. Since the potential difference is less for the same charge on the plates, the capacitance is greater.

E. Explain that if the polarization is uniform, the material behaves like neutral material with charge on its surfaces.

F. Optional: Show how Gauss' law can be written in terms of $\kappa\mathbf{E}$ and the free charge. Show how to compute the polarization charge for a parallel plate capacitor with dielectric material between its plates.

SUGGESTIONS

1. Use questions 1, 2, and 4 to discuss charging a capacitor.
2. Use questions 3, 6, and 10 to emphasize the dependence of capacitance on geometry.
3. The fundamental idea of capacitance is illustrated by problem 2. Assign problem 8 to have students compare spherical and plane capacitors. Problem 10 covers the dependence of the capacitance of a parallel plane capacitor on area and separation.
4. Problem 22 is a good test of understanding of the derivations of the expressions for the equivalent capacitance of parallel and series combinations. Problems 19 and 20 cover equivalent capacitance, charge, and potential difference for series and parallel combinations. Also consider assigning some problems in which students must find the equivalent capacitance of various combinations. See problems 15, 16, and 17, for example.
5. Problem 36 covers most of the important points discussed in connection with energy storage. Also assign problem 40, which deals with the energy density around a charged metal sphere and problem 46, which deals with the energy needed to separate the plates of a parallel plate capacitor.

6. Include questions 17 and 18 in the discussion of the influence of a dielectric on capacitance.
7. To test understanding of induced polarization charge, ask question 12. Be sure to consider charge, potential difference, and electric field.
8. Audio/Visual
Capacitor I: Voltage and Force, *Capacitor II: Dipoles and Dielectrics*, S8, color, 4 min. each. Walter de Gruyter, Inc., 200 Sawmill River Road, Hawthorne, NY 10532. Shows the relationship of the voltage on a capacitor to the plate separation and uses a beam balance to show the force-voltage relationship. Reviewed AJP 45:1014 (1977).
9. Demonstrations
   a. Charge storage: Freier and Anderson Eb8, Ed3, 7; Hilton E4b.
   b. Capacitance and voltage: Freier and Anderson Ed1; Hilton E4c, d.
   c. Energy storage: Freier and Anderson Ed8
   d. Dielectrics: Freier and Anderson Ed2, 4.
10. Laboratory
   a. MEOS Experiment 10–7 (Part B): *Measuring Capacitance with a Ballistic Galvanometer.* A ballistic galvanometer is used to measure the capacitance of individual capacitors and capacitors in series and parallel. Students must temporarily accept on faith that the deflection of the galvanometer is proportional to the total charge that passes through it.
   b. MEOS Experiment 11–2 (Part C): *Coulomb Balance Attachment (to the current balance).* Students use gravitational force to balance the force of one capacitor plate on the other. The voltage and plate separation are used to find the charge on the plates, then ϵ_0 is calculated.

Chapter 28 CURRENT AND RESISTANCE

BASIC TOPICS

I. Current and current density.
   A. Explain that electric current is moving charge. Draw a diagram of a long straight wire with positive charge moving in it. Consider a cross section and state that the current is dq/dt if charge dq passes the cross section in time dt. Give the sign convention: both positive charge moving to the right and negative charge moving to the left constitute currents to the right. Early on it is good to use the words "conventional current" quite often. Later "conventional" can be dropped. Many high school courses now take the current to be in the direction of electron flow and it is worthwhile making the effort to reduce confusion in students' minds. Unit: 1 ampere = 1 C/s.
   B. Explain that under steady state conditions, in which no charge is building up or being depleted anywhere in the wire, the current is the same for every cross section. Remark that current is a scalar, but arrows are used to show the direction of positive charge flow.
   C. Explain that current is produced when charge is free to move in an electric field. For most materials it is the negative electrons that move and their motion is opposite to the direction of the electric field. Current is taken to be in the direction opposite to that of electron drift, in the direction of the field.
   D. Distinguish between the drift velocity and the velocities of individual charges. Note that the drift velocity of electrons in an ordinary wire is zero unless an electric field is turned on. Also note that the drift speed is many orders of magnitude smaller than the average electron speed.
   E. Explain that current density is a microscopic quantity used to describe current flow at a point. Use the same diagram but now consider a small part of the cross section and state

that $J = i/A$ in the limit as the area diminishes to a point. State that current density is a vector in the direction of the drift velocity for positive charge and opposite the drift velocity for negative charge. Explain that $i = \int \mathbf{J} \cdot d\mathbf{A}$ is the current through a finite surface, where $d\mathbf{A}$ is normal to the surface. This reduces to $J = i/A$ for uniform current density and for an area that is perpendicular to the current. Unit: A/m^2.

F. Derive $\mathbf{J} = en\mathbf{v}_d$ and show how to calculate the drift speed from the free-electron concentration and current in the wire, assuming uniform current density. You may want to go over the calculation of the free-electron concentration from the mass density of the sample and the atomic or molecular mass numbers of its constituents.

II. Resistance and resistivity.

A. Define resistance by $R = V/i$ and point out that R may depend on V. Unit: $1\,\text{ohm} = 1\,\text{V/A}$; abbreviation: Ω. Also define resistivity ρ and conductivity σ. Point out Table 28–1. Explain that the latter quantities are characteristic of the material while resistance also depends on the sample shape and the positions of the current leads.

B. Make a sketch similar to the one shown here. Indicate that $V_a - V_b = iR$ is algebraically correct, even if i is negative, and effectively defines the resistance of the sample with the leads connected at a and b. Emphasize that the point at which the current enters is iR higher in potential than the point at which it leaves.

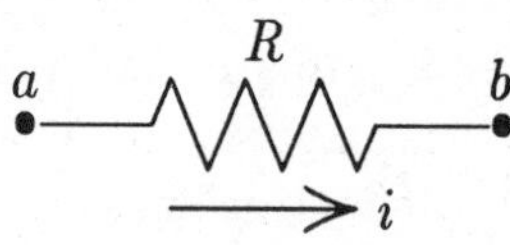

C. Show that $R = \rho L/A$ for a conductor with uniform cross section A and length L, carrying a current that is uniformly distributed over the cross section.

D. Point out that for many samples the current is proportional to the potential difference and the resistance is independent of the voltage applied. These materials are said to obey Ohm's law. Also point out that many important materials do not obey Ohm's law. Show Fig. 28–11.

E. Use a variable-voltage power supply and connect, in turn, samples of ohmic (carbon resistor) and non-ohmic (solid state diode) material across the terminals. Use analog meters to display the current and potential difference and vary the supply smoothly and fairly rapidly. For the ohmic material it will be apparent that i is proportional to V, while for the non-ohmic material it will be apparent that i is not proportional to V.

F. Give a qualitative description of the mechanism that leads to Ohm's law behavior. Explain that collisions with atoms cause the drift velocity to be proportional to the applied field. Assume the electrons have zero velocity after each collision and that they accelerate for a time τ between collisions. Show that an electron goes the same distance on the average during the first five collisions as it does during the second five so the drift velocity is proportional to the field even though the electron accelerates between collisions. Now consider the quantitative aspects: derive the expression for the drift velocity in terms of $\mathbf{E}$ and the mean free time τ, then derive $\rho = m/ne^2\tau$. Emphasize that the mean free time is determined by the electron speed and since drift is an extremely small part of the speed, τ is essentially independent of the electric field. Point out that a long mean free time means a small resistivity because the electrons accelerate for a longer time between collisions and thus have a higher drift speed.

G. Remark that the resistivity of a sample depends on the temperature. Define the temperature coefficient of resistivity and point out the values given in Table 28–1.

III. Energy considerations.

A. Point out that when current flows from the high to the low potential side of any device, energy is transferred from the current to the device at the rate $P = iV$. Reproduce Fig. 13 and note that $P = i(V_a - V_b)$ is algebraically correct if P is the power supplied to the device. Note that if P is negative the device is supplying energy at the rate $-P$.

B. Give examples: Energy may be converted to mechanical energy (a motor), to chemical energy (a charging battery), or to internal energy (a resistor). Also note the converse: mechanical energy (a generator), chemical energy (a discharging battery), and internal energy (a thermocouple) may be converted to electrical energy.

C. Explain that in a resistor the electrical potential energy of the free electrons is converted to kinetic energy as the electric field does work on them and that the kinetic energy is lost to atoms in collisions. This increases the thermal motion of the atoms. Show that the rate of energy loss in a resistor is given by $P = i^2 R = V^2/R$.

SUPPLEMENTARY TOPICS

1. Semiconductors
2. Superconductors

Both topics are important for modern physics and technology. Say a few words about them if you have time or encourage students to read about them on their own.

SUGGESTIONS

1. Use question 2 to expand the discussion of the sign convention for currents. Also assign problem 9 or 10.
2. Use questions 6 and 7 in a discussion of the current density and drift velocity. Use questions 1 and 5 in a discussion of the relationship between current and current density. Definitions are covered in problems 2 (current), 5 (current density), and 6 (drift speed).
3. Use questions 9, 11, and 12 in a discussion of Ohm's law.
4. The dependence of resistance on length and cross section is emphasized in problems 27 and 28.
5. As part of the coverage of energy dissipation by a resistor, ask students to think about questions 19, 21, and 22. Assign problems 47 and 48. Problems 61 and 62 illustrate some practical applications.
6. Demonstrations
   a. Model of resistance: Freier and Anderson Eg1.
   b. Thermal dissipation by resistors: Freier and Anderson Eh3.
   c. Fuses: Freier and Anderson Eh5.
   d. Ohm's law: Freier and Anderson Eg2, Eo1; Hilton E2c.
   e. Measurement of resistance, values of resistance: Freier and Anderson Eg3, 6; Hilton E3b.
   f. Temperature dependence of resistance: Freier and Anderson Eg4, 5.
7. Laboratory
   a. MEOS Experiment 10–3: *Electrical Resistance.* An ammeter and voltmeter are used to find the resistance of a light bulb and wires of various dimensions, made of various materials. The dependence of resistance on length and cross section is investigated. Resistivities of the substances are calculated and compared.
   b. BE Experiment 29: *A Study of the Factors Affecting Resistance.* A Wheatstone bridge and a collection of wire resistors is used to investigate the dependence of resistance on length, cross section, temperature, and resistivity.
   c. MEOS Experiment 10–8: *Temperature Coefficient of Resistors and Thermistors.* A Wheatstone bridge is used to measure the resistances of a resistor and thermistor in a water-filled thermal reservoir. The temperature is changed by an immersion heater. Students see two different behaviors. A voltmeter-ammeter technique can replace the bridge if desired.

d. Also see MEOS Experiment 9–2 and BE Experiment 30, described in the Chapter 20 notes. These experiments can be revised to emphasize the power dissipated by a resistor. In several runs the students measure the power dissipated for different applied voltages.

Chapter 29 CIRCUITS

BASIC TOPICS

I. Emf devices.

A. Explain that an emf device moves positive charge from its negative to its positive terminal or negative charge in the opposite direction and maintains the potential difference between its terminals. Emf devices are used to drive currents in circuits. Example: a battery is an emf device with an internal resistance. Note the symbol used in circuit diagrams to represent an ideal emf device (no internal resistance).

B. Explain that a direction is associated with an emf and that it is from the negative to the positive terminal, inside the device. This is the direction current would flow if the device acted alone in a completed circuit. Point out that when current flows in this direction the device does positive work on the charge and define the emf of an ideal device as the work per unit positive charge: $\mathcal{E} = \mathrm{d}W/\mathrm{d}q$. Also point out that the positive terminal of an ideal device is $\mathcal{E}$ higher in potential than the negative terminal, regardless of the direction of the current. Unit: volt.

C. Point out that the rate at which energy is supplied by an ideal device is $i\mathcal{E}$. State that for a battery the energy comes from a store of chemical energy. Mention that a battery is charging if the current and emf are in opposite directions.

II. Single loop circuits.

A. Consider a circuit containing a single emf and a single resistor. Use energy considerations to derive the steady state loop equation (Kirchhoff's loop rule): equate the power supplied by the emf to the power loss in the resistor.

B. Derive the loop equation by picking a point on the circuit, selecting the potential to be zero there, then traversing the circuit and writing down expressions for the potential at points between the elements until the zero potential point is reached again. Tell the students that if the current is not known a direction must be chosen for it and used to determine the sign of the potential difference across the resistor. When the circuit equation is solved for i, a negative result will be obtained if the current is actually opposite in direction to the arrow. As you carry out the derivation remind students that current enters a resistor at the high potential end and that the positive terminal of an emf is at a higher potential than the negative terminal.

C. Consider slightly more complicated single loop circuits. Include the internal resistance of the battery and solve for the current. Place two batteries in the circuit, one charging and the other discharging. Once the current is found, calculate the power gained or lost in each element.

D. For the circuits considered, show how to calculate the potential difference between two points on the circuit and point out that the answer is independent of the path used for the calculation. Explain the difference between the closed and open circuit potential difference across a battery.

III. Multiloop circuits.

A. Explain Kirchhoff's junction rule for steady state current flow. State that it follows from the conservation of charge and the fact that charge does not build up anywhere when the steady state is reached.

B. Using an example of a two-loop circuit, go over the steps used to write down the loop and junction equations and to solve for the currents. Explain that if the current directions are unknown an arbitrary choice must be made in order to write the equations and that if the wrong choice is made, the values obtained for the current will be negative.

C. Warn students not to write duplicate junction equations. Define a branch and state that different symbols must be used for currents in different branches. State out that the total number of equations will be the same as the number of branches, that the number of independent junction equations equals one less than the number of junctions, and that the remaining equations are loop equations. Also state that each current must appear in at least one loop equation.

D. Derive expressions for the equivalent resistance of two resistors in series and in parallel. Contrast with the expressions for the equivalent capacitance of two capacitors in series and in parallel. Show how to calculate potential differences across resistors in series and currents in resistors in parallel. Show how series and parallel combinations can sometimes be used to solve complicated circuits. Mention that not all circuits can be considered combinations of series and parallel connections.

IV. RC circuits.

A. Consider a series circuit consisting of an emf, a resistor, a capacitor, and a switch. Suppose the switch is closed at time $t = 0$ with the capacitor uncharged. Use the loop rule and $i = dq/dt$ to show that $R(dq/dt) + (q/C) = \mathcal{E}$. By direct substitution, show that $q(t) = C\mathcal{E}[1 - e^{-t/RC}]$ satisfies this equation and yields $q = 0$ for $t = 0$. Also find expressions for the potential differences across the capacitor and across the resistor. Show that $q = C\mathcal{E}$ for times long compared to RC.

B. Explain that $\tau = RC$ is called the time constant for the circuit and that it is indicative of the time required to charge the capacitor. If RC is large the capacitor takes a long time to charge. Show that $q/C\mathcal{E} \approx 0.63$ when $t = \tau$.

C. Show that the current is given by $i(t) = (\mathcal{E}/R)e^{-t/RC}$. Point out that $i = \mathcal{E}/R$ for $t = 0$ and that the potential difference across the capacitor is zero at that time because the capacitor is uncharged. Thus the potential difference across the resistor is $\mathcal{E}$. Also point out that the current tends toward zero for times that are long compared to τ. Then the potential difference across the resistor is zero and the potential difference across the capacitor is $\mathcal{E}$.

D. Derive the loop equation for a series circuit consisting of a capacitor and resistor. Suppose the capacitor has charge q_0 at time $t = 0$ and show that $q = q_0 e^{-t/RC}$. Again find expressions for the potential differences across the capacitor and resistor. Point out that RC is indicative of the time for discharge.

E. Write the expression for the energy initially stored in the capacitor: $U = \frac{1}{2}q_0^2/C$. Evaluate $\int_0^\infty i^2 R\,dt$ to find the energy dissipated in the resistor as the capacitor discharges. Show that these energies are the same.

SUPPLEMENTARY TOPIC

Electrical measuring instruments. This material can be covered, as needed, in conjunction with the laboratory.

SUGGESTIONS

1. Use questions 2, 4, and 5 in a discussion of emfs and batteries. Also ask question 8. Problem 1 covers the fundamental idea of emf. Use problems 9 and 21 to discuss the distinction between the emf and terminal potential difference of a battery.
2. Assign some problems dealing with multiloop circuits. Consider problems 29, 30, 34, 41, and 46.

3. Assign problems 58 and 59 if voltmeters and ammeters are discussed in lecture or lab. Also consider problems 60 and 61.
4. A computer can easily be programmed to solve simultaneous linear equations. Have students use such a program to solve multiloop circuit problems. Details of a program and some specific projects are given in the Computer Projects Section of this manual.
5. Demonstrations
   a. Seats of emf: Freier and Anderson Ee2, 3, 4; Hilton E3f.
   b. Measurement of emf: Freier and Anderson Eg7.
   c. Resistive circuits: Freier and Anderson Eh1, 2, 4, Eo2 — 8; Hilton E2b, E3a, c, d, g.
6. Computer programs
   a. *Circuit Lab*, Mark Davids, 21825 O'Conner, St. Clair Shores, MI 48080. Atari 800 and 800XL, Apple II. One of four basic circuits can be selected. Light bulbs, switches, resistors, ammeters, and voltmeters are placed in the circuit by the user, who also selects values for the circuit elements. Ammeters and voltmeters then show correct values. Use as a drill or to illustrate circuits in lectures. Reviewed TPT April 1986.
   b. *Basic Electricity*, Programs for Learning, Inc., P.O. Box 954, New Milford, CT 06776. Apple II. Drill on circuits containing batteries and resistors. Reviewed TPT April 1984.
7. Laboratory
   a. MEOS Experiment 10–7 (Part A): *Measuring Current with a d'Arsonval Galvanometer.* Students determine the characteristics and sensitivity of a galvanometer. To expand this lab, ask the students to design an ammeter and a voltmeter with full scale deflections prescribed by you. Students practice circuit analysis while trying to understand design considerations.
   b. MEOS Experiment 10–9: *The emf of a Solar Cell.* Students study a slide wire potentiometer and use it to measure the emf of a solar cell. Another experiment that gives them practice in circuit analysis.
   c. BE Experiment 28: *Measurements of Potential Difference with a Potentiometer.* Students study a slide wire potentiometer and use it to investigate the emf and terminal voltage of a battery and the workings of a voltage divider.
   d. BE Experiment 26: *A Study of Series and Parallel Electric Circuits.* Students use ammeters and voltmeters to verify Kirchhoff's laws and investigate energy balance for various circuits. They also experimentally determine equivalent resistances of resistors in series and parallel. This experiment can be extended somewhat by having them consider a network of resistors that cannot be reduced by applying the rules for series and parallel resistors. Also see BE Experiment 27: *Methods of Measuring Resistance.* Two voltmeter-ammeter methods and a Wheatstone bridge method are used to measure resistance and to check the equivalent resistance of series and parallel connections.
   e. BE Experiment 31: *Circuits Containing More Than One Potential Source.* Similar to BE Experiment 26 described above but circuits with more than one battery are considered. The two experiments can be done together, if desired.
   f. MEOS Experiment 10–4: *The R-C Circuit.* Students connect an unknown resistor to a known capacitor, charged by a battery. The battery is disconnected and a voltmeter and timer are used to measure the time constant. The value of the resistance is calculated. In a second part an unknown capacitor is charged by means of a square wave generator and the decay is monitored on an oscilloscope. Again the time constant is measured, then it is used to calculate the capacitance. A third part explains how to use a microprocessor to collect data. Also see BE Experiment 32: *A Study of Capacitance and Capacitor Transients.*

Chapter 30 THE MAGNETIC FIELD

BASIC TOPICS

I. Definition of the field and force on a moving charge.

A. Explain that moving charges create magnetic fields and that a magnetic field exerts a force on a moving charge. Both the field of a moving charge and the force exerted by a field depend on the velocity of the charge involved. The latter property distinguishes it from an electric field.

B. Define the magnetic field: the force on a moving test charge is $q_0\mathbf{v} \times \mathbf{B}$ after the electric force is taken into account. Review the rules for finding the magnitude and direction of a vector product. Point out that the force must be measured for at least two directions of $\mathbf{v}$ since the component of $\mathbf{B}$ along $\mathbf{v}$ cannot be found from the force. The direction of $\mathbf{B}$ can be found by trying various directions for $\mathbf{v}$ until one is found for which the force vanishes. The magnitude of $\mathbf{v}$ can be found by orienting $\mathbf{v}$ perpendicular to $\mathbf{B}$. Units: $1\,\text{tesla} = 1\,\text{N/A·m}$, $1\,\text{gauss} = 10^{-4}\,\text{T}$. Point out the magnitudes of the fields given in Table 30–1.

C. Explain that the magnetic force on any moving charge is given by $\mathbf{F}_B = q\mathbf{v} \times \mathbf{B}$. Point out that the force is perpendicular to both $\mathbf{v}$ and $\mathbf{B}$ and is zero for $\mathbf{v}$ parallel or antiparallel to $\mathbf{B}$. Also point out that the direction of the force depends on the sign of q. Remark that the field cannot do work on the charge and so cannot change its speed or kinetic energy. A magnetic field can change the direction of travel of a moving charge. It can, for example, be used to produce a centripetal force and can cause a charge to move in a circular orbit.

D. To show a magnetic force qualitatively, slightly defocus an oscilloscope so the central spot is reasonably large. Move a bar magnet at an angle to the face of the scope and note the movement of the beam.

E. Point out that the total force on a charge is $q(\mathbf{E} + \mathbf{v} \times \mathbf{B})$ when both an electric and a magnetic field are present.

II. Magnetic field lines.

A. Explain that field lines can be associated with a magnetic field. At any point the field is tangent to the line through that point and the number of lines per unit area that pierce a plane perpendicular to the field is proportional to the magnitude of the field.

B. To show field lines project Fig. 30–3 or place a sheet of clear plastic over a bar magnet and place iron filings on the sheet. Place the arrangement on an overhead projector. Explain that the filings line up along field lines. Also show Fig. 30–7

C. Point out that magnetic field lines form closed loops; they continue into the interior of the magnet, for example. Contrast with electric field lines and remark that no magnetic charge has yet been found. Mention that magnetic field lines would start and stop at magnetic monopoles, if they exist. Remark that lines enter at the south pole of a magnet and exit at the north pole.

III. Motion of charges in magnetic fields.

A. Derive $v = E/B$ for the speed of a charge passing through a velocity selector.

B. Outline the Thompson experiment and derive Eq. 30–12 for the charge-to-mass ratio.

C. Show how the Hall effect can be used to determine the sign and concentration of charge carriers in a conductor. Mention that these measurements are important for the semiconductor industry. Also mention that the Hall effect is used to measure magnetic fields. Show a Hall effect teslameter.

D. Consider a charge with velocity perpendicular to a constant magnetic field. Show that the orbit radius is given by $r = mv/qB$ and the period of the motion is given by $T = 2\pi m/qB$ (independently of v) for non-relativistic speeds. If you are covering modern topics state that $r = p/qB$ is relativistically correct but $p = mv/\sqrt{1 - v^2/c^2}$, where c is the speed of light, must be used for the momentum. Remark that the orbit is a helix if the velocity of the charge has a component along the field. Mention that cyclotron motion is used in cyclotrons and synchrotrons. If you have time, explain how a cyclotron works.

IV. Force on a current loop.

A. Run a flexible non-magnetic wire near a strong permanent magnet. Observe that the wire does not move. Turn on a power supply so about 1 A flows in the wire and watch the wire move. Remark that magnetic fields exert forces on currents.

B. Consider a thin wire carrying current, with all charges moving with the drift velocity. Start with the force on a single charge and derive $d\mathbf{F}_B = i\,d\mathbf{L} \times \mathbf{B}$ for an infinitesimal segment and $\mathbf{F}_B = i\mathbf{L} \times \mathbf{B}$ for a finite straight segment in a uniform field. Stress that $d\mathbf{L}$ and $\mathbf{L}$ are in the direction of the current.

C. Consider an arbitrarily shaped segment of wire in a uniform field. Show that the force on the segment between a and b is $\mathbf{F}_B = i\mathbf{L} \times \mathbf{B}$, where $\mathbf{L}$ is the vector joining the ends of the segment. This expression is valid only if the field is uniform.

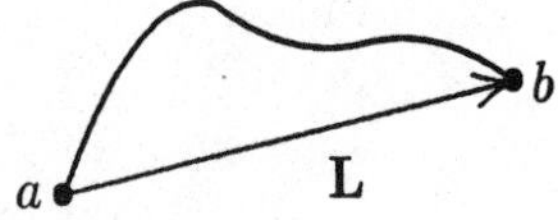

D. Point out that the force on a closed loop in a uniform field is zero since $\mathbf{L} = 0$.

E. Calculate the force of a uniform field on a semicircular loop of wire, in the plane perpendicular to $\mathbf{B}$. See Sample Problem 30–7. Do this as in the text, then repeat using the result given in C above.

V. Torque on a current loop.

A. Calculate the torque exerted by a uniform field on a rectangular loop of wire arbitrarily oriented with two opposite sides perpendicular to $\mathbf{B}$. See Fig. 30–26.

B. Define the magnetic dipole moment of a current loop ($\mu = NiA$) and give the right hand rule for determining its direction. For a rectangular loop in a uniform field, show that $\boldsymbol{\tau} = \boldsymbol{\mu} \times \mathbf{B}$. State that the result is generally valid for any loop in a uniform field. Mention that other sources of magnetic fields, such as bar magnets and the Earth, have dipole moments. Mention that many fundamental particles have intrinsic dipole moments.

C. Note that this is a restoring torque and that if the dipole is free to rotate it will oscillate about the direction of the field. If damping is present it will line up along the field direction. Remark that this is the basis of magnetic compasses.

D. Explain how analog ammeters and voltmeters work. To demonstrate the torque on a current carrying coil, remove the case from a galvanometer and wire it to a battery and resistor so that it fully deflects.

E. Remark that a potential energy cannot be associated with a moving charge in a magnetic field but can be associated with a magnetic dipole in a magnetic field. Show that $U = -\boldsymbol{\mu} \cdot \mathbf{B}$. Find the work required to turn a dipole through 90° and 180°, starting with it aligned along the field. Point out that U is a minimum when $\boldsymbol{\mu}$ and $\mathbf{B}$ are parallel and is a maximum when they are antiparallel.

SUGGESTIONS

1. Use questions 2, 5, and 6 to help in understanding the definition of the magnetic field. Use questions 1, 3, and 16 to discuss the magnetic force on a moving charge. The dependence of magnetic force on velocity and charge is emphasized in problems 2 and 3.

2. Use question 9 to test for understanding of the motion of charges in magnetic fields. Problems 23, 28, and 29 deal with the circular orbit of a charge in a uniform magnetic field. Crossed electric and magnetic fields, used as a velocity filter, are explored in problems 10, 12, and 13. Problems 31, 32, 33, and 34 deal with mass spectrometers. Problems 41 and 42 deal with cyclotrons. Use some of these problems to include practical applications.

3. Use problems 14, 15, and 16 to help students study the Hall effect.
4. Use questions 13, 14, and 17 to include more detail in the discussion of the magnetic force on a current carrying wire. Use problems 45 and 47 to stress the importance of the angle between the magnetic field and the current carrying wire on which it exerts a force. Use problems 53 and 58 to emphasize that the force of a uniform magnetic field on a closed loop is zero. Problems 48 and 61 ask students about the dynamics of current-carrying wires in magnetic fields.
5. Magnetic dipoles and the torques exerted on them by magnetic fields are explored in problems 62 and 64. Problem 67 deals with the energy of a dipole in a field. Also consider problem 68.
6. Computer project
   Have students use numerical integration of Newton's second law to investigate the orbits of charges in magnetic and electric fields.
7. Audio/Visual
   Because Chapters 30 and 31 are highly interrelated, the following films may be used profitably with either chapter.
   a. *The Magnetic Field*; *The Field from a Steady Current*; *Field vs. Current*; *Uniform and Non-Uniform Fields*, S8, color, 3 min. each. Kalmia Company, Dept. P1, Concord, MA 01742. These loops are part of the Adler series produced at MIT.
   b. *Magnetic Fields and Electric Currents, I*, 16 mm, color, 14.5 min. BFA Educational Media, Division of Phoenix Films, 468 Park Avenue, New York, NY 10016. Various phenomena associated with magnetic fields are presented. Includes a wire carrying current and the magnetic properties of iron.
   c. *Magnetic Fields and Electric Currents, II*, 16 mm, color, 12.5 min. BFA Educational Media (see address above). A companion film to the one listed above, it deals with the interaction of magnetic fields and illustrates the simple electric motor.
8. Demonstrations
   a. Magnets and compasses: Hilton E6a, b, c, d.
   b. Force on an electron beam: Freier and Anderson Ei18, Ep8, 11.
   c. Forces and torques on wires: Freier and Anderson Ei7, 12, 13 — 15, 19, 20; Hilton E7a, b(1), c.
   d. Meters: Freier and Anderson Ej1, 2.
   e. Hall effect: Freier and Anderson Ei16.
9. Computer programs
   a. *Charged Particle Workshop*, High Technology Software Products, P.O. Box 60406, 1611 NW 23rd Street, Oklahoma City, OK 73146. Apple II. Shows trajectories of charged particles in a uniform electric field, a uniform magnetic field, and crossed electric and magnetic fields. Velocity components can be displayed. Can be used to illustrate lectures.
   b. *Laboratory Simulations in Atomic Physics.* See Chapter 24 notes.
10. Laboratory
   a. BE Experiment 33: *A Study of Magnetic Fields.* A small magnetic compass is used to map field lines of various permanent magnets, a long straight current carrying wire, a single loop of current carrying wire, a solenoid, and the Earth. Parts of this experiment might be performed profitably in connection with Chapter 31.
   b. MEOS Experiment 11–3: *Determination of e/m.* Students use the accelerating potential and the radius of the orbit in a magnetic field to calculate the charge-to-mass ratio for the electron.
   c. MEOS Experiment 11–5: *The Hall Effect.* Students measure the Hall voltage and use it to calculate the drift speed and carrier concentration for a bismuth sample. The influence

of the magnetic field on the Hall voltage is also investigated. Values of the magnetic field are given to them by the instructor.

Chapter 31 AMPERE'S LAW

BASIC TOPICS

I. Magnetic field of a current.

A. Place a magnetic compass near a wire carrying a DC current of several amperes, if possible. Turn the current on and off, reverse the current. Note the deflection of the compass needle and remark that the current produces a magnetic field and that the field reverses when the current reverses.

B. Write the Biot-Savart law for the field produced by an infinitesimal segment of a current carrying wire. Give the value for μ_0. Draw a diagram to show the direction of the current, the displacement vector from the segment to the field point, and the direction of the field. Explain that dB is in the direction of $d\mathbf{s}\times\mathbf{r}$. Point out the angle between **r** and **ds**. Mention that the integral for the field of a finite segment must be evaluated one component at a time. Point out that the angle between dB and a coordinate axis must be used to find the component of dB.

C. Example: Show how to calculate the magnetic field of a straight finite wire segment. See the text, but use finite limits of integration. State that magnetic fields obey a superposition principle and point out that the result of the calculation can be used to find the field of a circuit composed of straight segments. Specialize the result to an infinite straight wire. Demonstrate the right hand rule for finding the direction of **B** due to a long straight wire.

D. Explain that the field lines around a straight wire are circles in planes perpendicular to the wire, centered on the wire. Draw a diagram to illustrate. Use symmetry to argue that the magnitude of the field is uniform on a field line. Point out that for other current configurations B is not necessarily uniform on a field line.

E. Show how to find the force per unit length of one long straight wire on another. Treat currents in the same and opposite directions. Lay two long automobile starter cables on the table. Connect them in parallel to an auto battery, with a $0.5\,\Omega$, 500 W resistor and an "anti-theft" switch or starter relay in each circuit. Close one switch and note that the wires do not move. Close the other switch and note the motion. Show parallel and antiparallel situations. It is better to reconnect the wires or rearrange them rather than to use a reversing switch.

F. Give the definition of the ampere and remind students of the definition of the coulomb.

II. Ampere's law.

A. Write the law in integral form. Explain that the integral is a line integral around a closed contour and interpret it as a sum over segments. Point out that it is the tangential component of **B** that enters. Explain that the current that enters is the net current through the contour. Two currents in opposite directions tend to cancel, for example. Illustrate by considering a contour with 5 or 6 wires passing through, with some currents in each direction. Also consider a wire passing through the plane of the contour but outside the contour. Mention that this current produces a magnetic field at all points on the contour but the integral of its tangential component is zero.

B. Explain the right hand rule that relates the direction of integration around the contour and the direction of positive current through the contour.

C. Pick a functional form for the magnetic field ($B_x = 2xy$, $B_y = -y^2$, $B_z = 0$, for example). Be sure the divergence is zero and the curl is not. Now consider a simple contour, such as a square in the xy plane. Integrate the tangential component of the field around the contour and calculate the net current through it.

D. Use Ampere's law to calculate the magnetic field *outside* a long straight wire. Either use without proof the circular nature of the field lines or give a symmetry argument to show that **B** at any point is tangent to a circle through the point and has constant magnitude around the circle. Point out that the integration contour is taken tangent to **B** in order to evaluate the integral in terms of the unknown magnitude of **B**.

E. Use Ampere's law to calculate the field *inside* a long straight wire with a uniform current distribution. Note that the use of Ampere's law to find B has the same limitations as Gauss' law when used to find E: there must be sufficient symmetry.

F. Use Ampere's law to calculate the field inside a solenoid. First argue that, for a long tightly wound solenoid, the field at interior points is along the axis and nearly uniform while the field at exterior points is nearly zero.

G. Similarly, use Ampere's law to calculate the field inside a toroid.

III. Magnetic dipole field.

A. Use the Biot-Savart law to derive an expression for the field of a circular current loop at a point on its axis. Stress the resolution of d**B** into components.

B. Take the limit as the radius becomes much smaller than the distance to the field point and write the result in terms of the dipole moment. Explain that the result is generally true for loops of any shape as long as the field point is far from the loop. Remind students that the dipole moment of a loop is determined by its area and the current it carries.

SUGGESTIONS

1. Use question 3 as part of the discussion of magnetic field lines.
2. Use questions 10 and 12 to discuss forces between current carrying wires.
3. The field of a long straight wire is considered in problem 3, while the superposition of fields due to two or more long straight wires is considered in problems 27, 28, 29, and 33. Assign at least one of these. Combinations of long straight segments and circular segments are considered in problems 10 through 13 and finite straight segments are considered in problems 17, 18, 19, 20, 22, 23, and 24. If you have not discussed them in class, have students work problems 17 and 22 before attempting the others.
4. Use questions 13 and 14 to help students think about the interpretation of Ampere's law. Problems 40 and 44 are good tests of fundamental understanding of the law. Also have students work problems 42, 46, or 50, which deal with practical calculations. Have better students attempt problem 48 or 51.
5. Solenoids and toroids are the subjects of problems 53 and 56. These configurations will also be studied in the chapter on inductance. You might also include problem 65, on Helmholtz coils.
6. Fields of dipoles and torques on dipoles are considered in problem 71. Assign it for its own value, but especially if you intend to cover Chapter 34.
7. Computer projects
   a. Have students use the Biot-Savart law and numerical integration to calculate the magnetic field due to a circular current loop at off-axis points. They can use a commercial math program or their own programs.
   b. Use numerical integration to verify Ampere's law for several long straight wires passing through a square contour. Have them show the result of the integration is independent of the positions of the wires, as long as they are inside the square. Also have them consider a wire outside.

8. Demonstrations
   a. Magnetic fields of wires: Freier and Anderson Ei8 — 11; Hilton E7b, d, E9b, c.
   b. Magnetic forces between wires: Freier and Anderson Ei1 — 6; Hilton E7e, f, g, E9a.
9. Computer program
   Physics Simulations II: Ampere, Kinko's Service Corporation, 4141 State Street, Santa Barbara, CA 93110. Macintosh. Positions and currents of up to 9 coaxial loops are specified by the user, then the program displays magnetic field lines. Use to illustrate lectures.
10. Laboratory
   a. MEOS Experiment 11–1: *The Earth's Magnetic Field.* A tangent galvanometer is used to measure the Earth's magnetic field. The dip angle is calculated.
   b. MEOS Experiment 11–2: *The Current Balance.* The gravitational force on a current carrying wire is used to balance the magnetic force due to current in a second wire. The data can be used to find the value of μ_0 or to find the current in the wires. The second version essentially defines the ampere. Part B describes how a microprocessor can be used to collect and analyze the data.
   c. BE Experiment 34: *Measurement of the Earth's Magnetic Field.* The oscillation period of a small permanent magnet suspended inside a solenoid is measured with the solenoid and the Earth's field aligned. The reciprocal of the period squared is plotted as a function of the current in the solenoid and the slope, along with calculated values of the solenoid's field, is used to find the Earth's field.
   d. MEOS Experiment 11–3: *Determination of e/m.* Students find the speed and orbit radius of an electron in the magnetic field of a pair of Helmholtz coils and use the data to calculate e/m. Information from this chapter is used to compute the field, given the coil radius and current. If you are willing to postulate the field for the students, this experiment can be performed in connection with Chapter 30.

Chapter 32 FARADAY'S LAW OF INDUCTION

BASIC TOPICS

I. The law of induction.
   A. Connect a coil (50 to 100 turns) to a sensitive galvanometer and move a bar magnet in and out of the coil. Note that a current is induced only when the magnet is moving. Show all possibilities: the north pole entering and exiting the coil, the south pole entering and exiting the coil. In each case point out the direction of the induced current. With a little practice you might also demonstrate effectively that the deflection of the galvanometer depends on the speed of the magnet.
   B. To show the current produced by changing the orientation of a loop, align the loop axis with the earth's magnetic field and rapidly rotate the loop once through 180°. Note the deflection of a galvanometer in series with the loop. Explain that this forms the basis of electric generators.
   C. Connect a coil to a switchable DC power supply. Connect a voltmeter (digital, if possible) to the supply to show when it is on. Place a second coil, connected to a sensitive galvanometer, near the first. Show that when the switch is opened or closed current is induced in the second coil, but that none is induced when the current in the first coil is steady.
   D. Define the magnetic flux through a surface. Unit: 1 weber = 1 T·m^2. Point out that Φ_B measures the number of magnetic field lines that penetrate the surface. Remark that $\Phi_B = BA\cos\theta$ when **B** is uniform over the surface and makes the angle θ with its normal.

E. Give a qualitative statement of the law: an emf is generated around a closed contour when the magnetic flux through the contour changes. Stress that the law involves the flux through the surface bounded by the contour. Point out the surface and contour for each of the demonstrations done, then remark that the contour may be a conducting wire, a physical boundary of some material, or it may be a purely geometric construction. Remark that if the contour is conducting then current flows.

F. Give the equations for Faraday's law: $\mathcal{E} = -\mathrm{d}\Phi_B/\mathrm{d}t$ for a single loop and $\mathcal{E} = -N\,\mathrm{d}\Phi_B/\mathrm{d}t$ for N tightly packed loops. Note that the emf's add.

II. Lenz' law.

A. Explain Lenz' law in terms of the magnetic field produced by the current induced if the contour is a conducting wire. Stress that the induced field must re-enforce the external field in the interior of the loop if the flux is decreasing and must tend to cancel it if the flux is increasing. This gives the direction of the induced current, which is the same as the direction of the emf. Review the right-hand rule for finding the direction of the field produced by a loop of current-carrying wire. State that Lenz' law can be used even if the contour is not conducting. The current must then be imagined.

B. Optional: Give the right-hand rule for finding the direction of positive emf. When the thumb points in the direction of $\mathrm{d}\mathbf{A}$, then the fingers curl in the direction of positive emf. If Faraday's law gives a negative emf, then it is directed opposite to the fingers. Stress that the negative sign in the law is important if the equation, with the right-hand rule, is to describe nature.

C. Consider a rectangular loop of wire placed perpendicular to a magnetic field. Assume a function $B(t)$ and calculate the emf and current. Show how the directions of the emf and current are found. Point out that an *area* integral is evaluated to find Φ_B and a *time* derivative is evaluated to find the emf. Some students confuse the variables and integrate with respect to time.

III. Motional emf.

A. Consider a rectangular loop being pulled with constant velocity past the boundary of a uniform magnetic field. Calculate the emf and current.

B. Consider a rod moving with a constant velocity that is perpendicular to a uniform magnetic field. Show how to complete the loop and calculate the emf. Mention that the emf exists only in the moving rod, whether the rest of the contour is conducting or not.

C. Consider a rectangular loop of wire rotating with constant angular velocity about an axis that is in the plane of the loop and through its center. Take the magnetic field to be uniform and point out that now the flux is changing because the angle between the field and the normal to the loop is changing. Derive the expression for the emf and point out it is time dependent.

IV. Energy considerations.

A. Point out that an emf does work at the rate $\mathcal{E}i$, where i is the current. Explain that for a current induced by motion the energy comes from the work done by an external agent or from the kinetic energy of the moving portion of the loop.

B. Consider four conducting rails that form a rectangle, three fixed and the fourth riding on two of them. Take the magnetic field to be uniform and normal to the loop. Assume most of the electrical resistance of the loop is associated with the moving rail. First suppose the moving rail has constant velocity and derive expressions for the emf, current, and magnetic force on the rail. Next derive expressions for the rate at which an external agent must do work to keep the velocity constant and for the rate at which energy is dissipated by the resistance of the loop. Point out that all the energy supplied by the agent is dissipated.

C. Now suppose the rail is given an initial velocity and thereafter it is acted on by the magnetic field alone. Use Newton's second law to derive an expression for the velocity as a function of time. Compare the rate at which the kinetic energy is decreasing with the rate of energy dissipation in the resistance. Remark that this phenomenon finds practical application in magnetic braking.

D. Mention that energy is also dissipated when a current is induced by a changing magnetic field and it comes from the agent that is changing the field. Remark that more details will be given in the next chapter.

V. Induced electric fields.

A. Explain that a changing magnetic field produces an electric field, which is responsible for the emf. The emf and electric field are related by $\mathcal{E} = \oint \mathbf{E} \cdot d\mathbf{s}$, where the integral is around the contour. Remind students that this integral is the work per unit charge done by the field as a charge goes around the contour. Write Faraday's law as $\oint \mathbf{E} \cdot d\mathbf{s} = -\frac{d}{dt} \int \mathbf{B} \cdot d\mathbf{A}$. Note that $d\mathbf{s}$ and $d\mathbf{A}$ are related by a right-hand rule: fingers along $d\mathbf{s}$ implies thumb along $d\mathbf{A}$. This is consistent with Lenz' law.

B. State that the induced electric field is like an electrostatic field in that it exerts a force on a charge but that it is unlike an electrostatic field in that it is not conservative. For an electrostatic field the integral defining the emf vanishes.

C. Consider a cylindrical region containing a uniform magnetic field, along the axis. Assume a time dependence for **B** and derive expressions for the electric field inside the region and outside the region. See Sample Problem 4. Point out that the lines of **E** form closed circles concentric with the cylinder and that the magnitude of **E** is uniform around a circle.

SUPPLEMENTARY TOPICS

The betatron. If you have time, cover this topic to add a modern flavor to the course. It is a nice application of Faraday's law.

SUGGESTIONS

1. Answers to questions 10 through 16 and 18 through 21 depend on understanding Lenz' law. Use several as examples and several to test the students.
2. Questions 6 through 8 and 23 through 25 deal with some important applications of Faraday's law. Use them in discussions or assign them for students to think about.
4. Assign problems 1, 12, and 13 to give students some practice in carrying out calculations of magnetic flux.
5. Assign some of problems 2, 4, 5, 6, 9, 11, 14, and 15 to cover the emf's generated by various time dependent magnetic fields. Addition of emf's is covered in problem 19. This is a good problem to test for understanding of the sign of an induced emf.
6. Motional emf is covered in problems 23 through 32. Problem 22 is a particularly good test of understanding of Faraday's law. AC generators are considered in problems 29 through 31. Also consider rods sliding on rails in magnetic fields. See problems 23, 24, 25, and 27. If you use a flip coil in the lab, assign problems 17 and 18.
7. Assign problems 40, 41, and 43 in connection with the discussion of induced electric fields.

8. Demonstrations
   a. As a supplementary demonstration, take a large, long coil, mount it vertically, insert a solid soft iron rod with a foot or so sticking out, and connect the coil via a switch to a large DC power supply. Place a solid aluminum ring around the iron rod. The ring should fit closely but be free to move. Close the switch and the ring will jump up, then settle down. Repeat with a ring that has a gap in it. Finally, use an AC power supply. The effect can be enhanced by cooling the ring with liquid nitrogen.

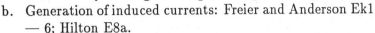

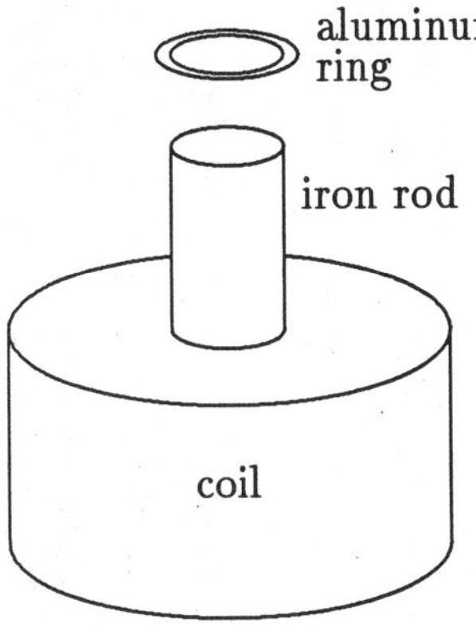

   b. Generation of induced currents: Freier and Anderson Ek1 — 6; Hilton E8a.
   c. Eddy currents: Freier and Anderson Ei1 — 6; Hilton E8d.
   d. Generators: Freier and Anderson: Eq4 — 7, Er1; Hilton E8b, c.
9. Audio/Visual
   a. *The Concept of a Changing Flux*; *Faraday's Law of Induction*, S8, color, 3 min. each. Kalmia Company, Dept. P1, Concord, MA 01742. Two more loops in the Adler-MIT electromagnetism series.
   b. *Lenz' Law*; *Large Inductance: Current Buildup*, S8, color, 3 min. each. American Association of Physics Teachers, 5110 Roanoke Place, College Park, MD 20740. Demonstrations of Lenz' law using different conductors, an aluminum disk, and a hollow aluminum conductor.
10. Laboratory
   a. BE Experiment 35: *Electromagnetic Induction.* Students measure the magnitude and observe the direction of current induced by a changing magnetic flux in a simple galvanometer circuit. Changing flux is produced by moving permanent magnets, by moving current carrying coils, and by changing current in a coil.
   b. MEOS Experiment 11–4: *The Magnetic Field of a Circular Coil.* The emf generated in a small search coil when a low frequency AC current flows in a given circuit (a circular coil in this case) is used to determine the magnetic field produced by the circuit. The field is investigated as a function of position, specified in spherical coordinates.

Chapter 33 INDUCTANCE

BASIC TOPICS

I. Definition of inductance.
   A. Connect a light bulb and choke coil in parallel across a switchable DC supply. Close the switch and note that the lamp is initially brighter than when steady state is reached. Open the switch and note that the light brightens before going off. Remark that this behavior is due to the changing magnetic flux through the coil and that the flux is created by the current in the coil itself.
   B. Point out that when current flows in a loop it generates a magnetic field and the loop contains magnetic flux due to its own current. If the current changes so does the flux and an emf is generated around the loop. The total emf, due to all sources, determines the current. Remark that the self flux is proportional to the current and the induced emf is proportional to the rate of change of the current.
   C. Define the inductance by $L = N\Phi_B/i$, where N is the number of turns, Φ_B is the magnetic flux through each turn, and i is the current in the circuit. Unit: 1 henry = 1 V·s/A.
   D. Remark that Faraday's law yields $\mathcal{E} = -L\, di/dt$ for the induced emf.

E. Inductors are denoted by ꝏꝏ in circuit diagrams. Point out that if the circuit element looks like a ꝏꝏ b (with current i directed from a to b), then $V_b - V_a = -L\,di/dt$ is algebraically correct. As an example use $i(t) = i_m \sin(\omega t)$. Note that i is positive when it is directed from a to b and negative when it is directed from b to a. Compute $V_a - V_b = Li_m\omega \cos(\omega t)$. Graph i and the potential difference as functions of time to show the phase relationship. Remark that a real inductor can be regarded as a pure inductance in series with a pure resistance.

F. Show how to calculate the inductance of an ideal solenoid. Use the current to calculate the field, then the flux, and finally equate $N\Phi_B$ to Li and solve for L. Point out that L is independent of i but depends on geometric factors such as the cross-sectional area, length, and the number of turns per unit length.

G. Optional: Show how to calculate the inductance of a toroid.

II. An LR circuit.

A. Derive the loop equation for a single loop containing a source of emf (an ideal battery), a resistor, and an inductor in series: $\mathcal{E} - iR - L\,di/dt = 0$, where the current is positive if it leaves the positive terminal of the seat of emf. Use the prototypes developed earlier:

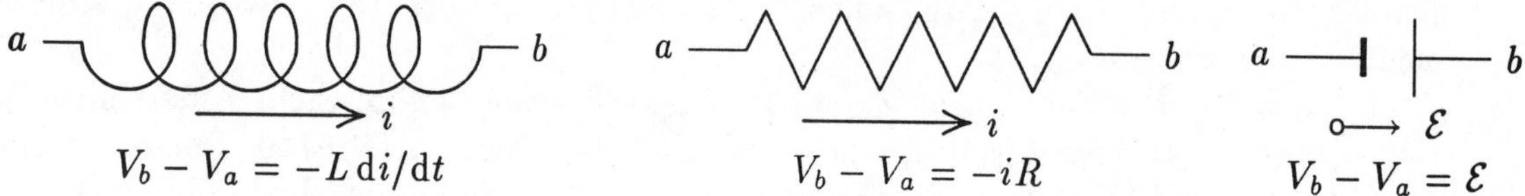

Remark that these are correct no matter whether the current is positive or negative or whether it is increasing or decreasing. Write down the solution for the current as a function of time for the case $i(0) = 0$: $i = (\mathcal{E}/R)[1 - e^{-Rt/L}]$. Show that the expression satisfies the loop equation and meets the initial conditions. Show a graph of $i(t)$; point out the asymptotic limit $i = \mathcal{E}/R$ and the time constant $\tau_L = L/R$. Remark that if L/R is large the current approaches its limit more slowly than if L/R is small.

B. Explain the qualitative physics involved. When the battery is turned on and the current increases, the emf of the coil opposes the increase and the current approaches its steady state value more slowly than if there were no inductance. At long times the current is nearly constant so di/dt and the induced emf are small. The current is nearly the same as it would be in the absence of an inductor. Just after the battery is turned on the potential difference across the resistor is 0 and the potential difference across the inductor is $\mathcal{E}$. After a long time the potential difference across the resistor is $\mathcal{E}$ and the potential difference across the inductor is 0.

C. Repeat the calculation for a circuit with an inductor and resistor, but no battery. Take the initial current to be i_0 and show that $i(t) = i_0 e^{-t/\tau_L}$. Graph the solution and show the position of τ_L on the time axis. Point out that the emf of the coil opposes the decrease in current.

D. Demonstrate the two circuits by connecting a resistor and coil in series to a square wave generator. Observe the current by placing oscilloscope leads across the resistor. Observe the voltage drop across the coil. Vary the time constant by varying the resistance.

III. Energy considerations.

A. Consider a single loop circuit containing an ideal battery, a resistor, and an inductor. Assume the current is increasing. Write down the loop equation, multiply it by i, and identify the power supplied by the battery and the power lost in the resistor. Explain that the remaining term describes the power being stored by the inductor, in its magnetic field. Point out the similarity between $i\mathcal{E}$ and $-iL\,di/dt$ for the rate at which work is being done by an ideal battery and by an inductor (with emf $-L di/dt$).

B. Integrate $P = iL\,di/dt$ to obtain $U_B = \frac{1}{2}Li^2$ for the energy stored in the magnetic field (relative to the energy for $i = 0$).

C. Consider the energy stored in a long current carrying solenoid and show that the energy density is $u_B = B^2/2\mu_0$. Explain that this gives the energy density at a point in any magnetic field and that the energy required to establish a given magnetic field can be calculated by integrating the expression over the volume occupied by the field.

IV. Mutual induction.

A. Repeat the demonstration experiment discussed in note IC for Chapter 32. Explain it in terms of the concept of mutual induction. Point out that the flux through the second coil is proportional to the current in the first. Define the coefficient of the mutual induction of the second coil with respect to the first by $M_{12} = N_2\Phi_{12}/i_1$. Show that $\mathcal{E}_2 = -M_{12}\,di_1/dt$ is the emf induced in the second coil when the current in the first changes. State without proof that $M_{12} = M_{21}$.

B. Example: derive the coefficient of mutual inductance for a small coil placed at the center of a solenoid or for a small tightly wound coil placed at the center of a larger coil.

C. Show that two inductors connected in series and well separated have an equivalent inductance of $L = L_1 + L_2$. Then show that if their fluxes are linked $L = L_1 + L_2 \pm 2M$, where the minus sign is used if the field lines have opposite directions. Also consider inductors in parallel. See problems 5, 6, and 49.

SUGGESTIONS

1. Use questions 2, 5, 6, and 15 when discussing the calculation of inductance. Assign problem 2 (coil), 3 (solenoid), or 8 (two parallel wires) as an example of a typical inductance calculation.
2. Use some of questions 7 through 14 when discussing the *LR* circuit. Assign problem 18. *LR* time constants are considered in problems 14, 15, and 17.
3. After discussing energy flow in a simple *LR* circuit with increasing current assign problems 32 and 35.
4. Problem 38 deals with energy storage and energy density in an inductor.
5. Demonstrations
   a. Self-inductance: Freier and Anderson Eq1 — 3; Hilton E12d.
   b. LR circuit: Freier and Anderson Eo11, En5 — 7; Hilton E12c.

Chapter 34 MAGNETISM AND MATTER

BASIC TOPICS

I. Gauss' law for magnetism.

A. Explain that a magnetic monopole is a particle that produces a magnetic field even while at rest. Remark that no magnetic monopole has been observed yet but it is currently being sought. Write down Gauss' law for the magnetic field and state that magnetic field lines form closed contours so the flux through any closed surface vanishes. If monopoles were found to exist, the law would be modified to include them. Compare with Gauss' law for the electric field.

B. To show that the ends of a magnet are not monopoles, magnetize a piece of hard iron wire. Use a compass to locate and mark the north and south poles. Break the wire into pieces and again use the compass to show that each piece has a north and a south pole. Repeat a few times using smaller pieces each time. Remark that the same results would be obtained if the breaking process were continued to the atomic level. Individual atoms and particles are magnetic dipoles, not monopoles.

II. Magnetic dipoles in matter.

A. Explain that current loops and bar magnets produce magnetic fields which, for points far away, are dipole fields. Review the expressions for the magnetic field of a dipole and for the dipole moment of a loop in terms of the current and area. Place a bar magnet under a piece of plastic sheet on an overhead projector. Sprinkle iron filings on the sheet and show the field pattern of the magnet. Remind students that field lines emerge from the north pole and enter at the south pole.

B. Explain that electrons in atoms create magnetic fields by virtue of their orbital motions. Derive Eq. 34–11, which gives the relationship between orbital angular momentum and dipole moment for a negative particle, such as an electron. State that the Bohr magneton ($\mu_B = eh/4\pi m$) is the natural unit for atomic dipole moments. Give its value (9.27×10^{-24} J/T).

C. Explain that the electron and many other fundamental particles have intrinsic dipole moments, related to their intrinsic spin angular momentum. Give the magnitude of the electron's intrinsic dipole moment: 9.27×10^{-24} J/T.

D. Remark that it is chiefly the orbital and spin dipole moments of electrons that are responsible for the magnetic properties of materials. Explain how to calculate the dipole moment of an atom: $\boldsymbol{\mu} = (-e/2m)\mathbf{L} + (-e/m)\mathbf{S}$, where $\mathbf{L}$ is the total orbital angular momentum and $\mathbf{S}$ is the total spin angular momentum of the electrons of the atom.

E. Explain that protons and neutrons also have intrinsic dipole moments but that these are much smaller than that of an electron because the masses are so much larger. Remark that nuclear magnetism has found medical applications.

III. Magnetization.

A. Define magnetization as the dipole moment per unit volume. Although only uniformly magnetized objects are considered in the text you may wish to state the definition as the limiting value as the volume shrinks to zero.

B. State that a magnetized object produces a magnetic field both in its exterior and interior and write $\mathbf{B} = \mathbf{B}_0 + \mathbf{B}_M$ for the total field. Here $\mathbf{B}_0$ is the applied field and $\mathbf{B}_M$ is the field due to dipoles in the material.

IV. Paramagnetism and diamagnetism.

A. Give a qualitative discussion of paramagnetism. Explain that paramagnetic substances are composed of atoms with net dipole moments and, in the absence of an external field, the moments have random orientations, so that they produce no net magnetic field. An external field tends to align the moments and the material produces its own field. Since the moments, on average, are aligned with the external field, the total field is stronger than the external field alone. Alignment is opposed by thermal agitation and both the net magnetic moment and magnetic field decrease as the temperature increases.

B. Remind students that the potential energy of a dipole $\boldsymbol{\mu}$ in a magnetic field $\mathbf{B}$ is given by $U = -\boldsymbol{\mu} \cdot \mathbf{B}$ and show that the energy required to turn a dipole end for end, starting with it aligned with the field, is $2\mu B$. Calculate U for $\mu = \mu_B$ and $B = 1\,\mathrm{T}$. Calculate the mean translational kinetic energy for an ideal gas at room temperature ($\frac{3}{2}kT$) and remark that there is sufficient energy for collisions to reorient the dipoles. Calculate the temperature for which $2\mu B = \frac{3}{2}kT$.

C. Give the Curie law for small applied fields. Point out that M is proportional to B and inversely proportional to T. Describe saturation and explain that there is an upper limit to the magnetization. The limit occurs when all atomic dipoles are aligned. Use a teslameter or flip coil to measure the magnetic field just outside the end of a large, high current coil. Put a large quantity of manganese in the coil and again measure the field.

D. Give a qualitative discussion of diamagnetism. Explain that an external field changes the electron orbits so there is a net dipole moment and that the induced moment is directed opposite to the field. This tends to make the total field weaker than the external field alone. Bismuth is an example of a diamagnetic substance.

E. Explain that diamagnetic effects are present in all materials but are overshadowed by paramagnetic or ferromagnetic effects if the atoms have dipole moments.

V. Ferromagnetism.

A. Explain that, for iron and other ferromagnetic substances (such as Co, Ni, Gd, and Dy), the atomic dipoles are aligned by an internal mechanism (exchange coupling) so the substance can produce a magnetic field spontaneously, in the absence of an external field. At temperatures above its Curie temperature, a ferromagnetic substance becomes paramagnetic. Gadolinium is ferromagnetic with a Curie temperature of about 20° C. Put a sample in a beaker of cold water ($T < 20°$ C) and use a weak magnet to pick it up from the bottom of the beaker but not out of the water. Add warm water to the beaker and the sample will drop from the magnet.

B. Describe ferromagnetic domains and explain that the dipoles are aligned within any domain but are oriented differently in neighboring domains. The magnetic fields produced by the various domains cancel for an unmagnetized sample. When the sample is placed in a magnetic field domains with dipoles aligned with the field grow in size while others shrink. The dipoles in a domain may also be reoriented somewhat as a unit.

C. Define hysteresis (see Fig. 34–15) and explain that the growth and shrinkage of domains are not reversible processes. Domain size is dependent not only on the external field but also on the magnetic history of the sample. When the external field is turned off, the material remains magnetized. Explain the difference between soft and hard iron in terms of hysteresis. Use a large, high current coil to magnetize a piece of hard iron and show that it remains magnetized when the current is turned off. Also magnetize a piece of soft iron and show it is magnetized only as long as the current remains on. When the current is turned off, very little permanent magnetization remains. Soft iron is used for transformer coils.

SUPPLEMENTARY TOPIC

The earth's magnetic field. Section 34–5 describes the magnetic field of the earth. The shape, cause, and some of the ramifications of the earth's field are important topics and should be covered if you have the time. If not, you might intersperse some of the information in your other lectures. Explain that the field can be approximated by a magnetic dipole field. Draw a sphere, label the north and south geographic poles, draw a dipole moment vector at the center (pointing roughly from north to south, about 10° away from the axis of rotation), draw some magnetic field lines. Remark that the north pole of the dipole is at the south geographic pole. Define declination and inclination.

SUGGESTIONS

1. To test for understanding of Gauss' law for magnetism, assign problem 6 or 8. Problem 9 is more challenging.
2. Ask students to think about a permanent bar magnet that pierces the surface of a sphere and explain why the net magnetic flux through the surface is zero. Also ask them about the flux as a single charge as it crosses the surface and the flux of a single magnetic monopole as it crosses the surface.
3. Ask students to think about questions 1 and 2 in connection with induced magnetism.
4. In order to emphasize the different mechanisms for paramagnetism and diamagnetism, ask question 10.

5. Magnetization in a paramagnetic substance is covered in problems 21, 22, and 23. If Chapter 21 was included in the course also consider assigning problem 25. The attraction of a bar magnet for a paramagnetic substance is covered in problem 20.
6. The repulsion of a diamagnetic substance by a bar magnet is covered in problem 26. This makes a nice companion to problem 20.
7. The Curie temperature of a ferromagnet is covered in problem 28 and the magnetization of a ferromagnet is covered in problem 31. Use problem 29 to show that magnetic interactions are not responsible for ferromagnetism.
8. Demonstrations
   a. Field of a magnet: Freier and Anderson Er4
   b. Gauss' law: Freier and Anderson Er12
   c. Paramagnetism: Freier and Anderson Es3, 4
   d. Diamagnetism: Hilton E10b.
   e. Ferromagnetism: Freier and Anderson Es1, 2, 6 — 10; Hilton E10a, E10c, d.
   f. Levitation: Freier and Anderson Er10, 11
9. Films
   a. *Monopoles and Dipoles*, S8, color, 3 min. The Kalmia Company, Dept. P1, Concord, MA 01742. A film loop in the Adler-MIT electromagnetism series.
   b. *Ferromagnetic Domain Wall Motion*, S8, color, 4 min. The Kalmia Company (address given above). Domain boundary movements are illustrated in this film loop by Franklin Miller.
   c. *Magnetic Domains and Magnetization Processes in a Gadolinium Iron Garnet*, S8, color, 4 min. American Association of Physics Teachers, 5110 Roanoke Place, College Park, MD 20740. A transparent ferromagnetic material is subjected to an alternating external magnetic field.
10. Laboratory
    MEOS Experiment 11-6: *Magnetization and Hysteresis.* Faraday's law is used to measure the magnetic field inside an iron toroid for various applied fields. A plot of the field as a function of the applied field shows hysteresis. A method for obtaining the hysteresis curve as an oscilloscope trace is also given.

Chapter 35 ELECTROMAGNETIC OSCILLATIONS

BASIC TOPICS

I. LC oscillations.
   A. Draw a diagram of an LC series circuit and assume the capacitor is charged. Explain that as charge flows, energy is transferred from the electric field of the capacitor to the magnetic field of the inductor and back again. When the capacitor has maximum charge, the current (dq/dt) vanishes, so no energy is stored in the inductor. When the current is a maximum the charge on the capacitor vanishes and no energy is stored in that element.
   B. Write down the loop equation, then convert it so the charge q on the capacitor is the dependent variable. If the direction of positive current is into the capacitor plate with positive charge q, then $i = dq/dt$. If it is out of that plate, then $i = -dq/dt$.
   C. Write down the solution: $q(t) = Q\cos(\omega t + \phi)$. Show by direct differentiation that this is a solution if $\omega^2 = 1/LC$. Show that ϕ is determined by the initial conditions and treat the special case for which $q = Q$, $i = 0$ at $t = 0$.

D. Once the solution is found, derive expressions for the current, the energy stored in the capacitor, and the energy stored in the inductor, all as functions of time. Sketch graphs of these quantities. Show that the total energy is constant.

E. In preparation for the next chapter it is also worthwhile deriving expressions for the potential differences across the capacitor and the inductor. Draw graphs of them as well. Mention that the charge on the capacitor is proportional to the potential difference across its plates and that the time rate of change of the current is proportional to the potential difference across the terminals of the inductor.

F. Note that the form of the differential equation for q is the same as that for the displacement x of a block oscillating on the end of a spring. Make the analogy concrete by explaining that if q is replaced by x, L is replaced by m, and C is replaced by $1/k$ the equation for q becomes the equation for x. Also point out that the current is analogous to the velocity of the block, that the energy stored in the inductor is analogous to the kinetic energy of the block, and that the energy stored in the capacitor is analogous to the potential energy stored in the spring.

II. Damped and forced oscillations.

A. Write down the loop equation for a single RLC loop, then convert it so q is the dependent variable. State that $q(t) = Q\,e^{-Rt/2L}\cos(\omega' t + \phi)$ satisfies the differential equation. Here ω' is somewhat less than $1/\sqrt{LC}$. If time permits, the expression for ω' can be found by substituting the assumed solution into the differential equation.

B. Draw a graph of $q(t)$ and point out that the envelope decreases exponentially. Each time the capacitor is maximally charged, the charge on the positive plate is less than the previous time. Explain that this does violate the conservation of charge principle since the total of the charge on both plates of the capacitor is always zero. Energy is dissipated in the resistor.

C. To show the oscillations, wire a resistor, inductor, and capacitor in series with a square wave generator and connect an oscilloscope across the capacitor. The scope shows a function proportional to the charge. Also connect the oscilloscope across the resistor to show a function that is proportional to the current. If you have a dual-trace scope show the functions simultaneously. Show the effect of varying C (use a variable capacitor), R (use a decade box), and L (insert an iron rod into the coil). If time permits, show that oscillations occur only if $1/LC > (R/2L)^2$.

D. Consider an RLC circuit with a sinusoidal oscillator and write down the loop equation. You may also want to write down the solution. In any event, show the solution by sketching graphs of the current amplitude as a function of the impressed frequency for several values of the resistance (see Fig. 35–6). Point out that the current amplitude is greatest when the impressed frequency matches the natural frequency of the circuit and that the peak becomes larger as the resistance is reduced.

E. Demonstrate resonance phenomena by wiring an RLC loop in series with a sinusoidal audio oscillator. Look at the current by putting the leads of an oscilloscope across the resistor. Use a decade box for the resistor and measure the current amplitude for various frequencies and for several resistance values. Be sure the amplitude of the oscillator output remains the same. Explain that similar circuits are used to tune radio and TV's.

F. Use a sweep generator to show the current amplitude. Set the oscilloscope sweep rate to accommodate that of the generator and put a small diode in series with the scope leads. Usually this will have enough capacitance that only the envelope will be displayed.

SUPPLEMENTARY TOPIC

Other oscillators. Section 35–7 deals chiefly with feedback oscillators. Include it to add some breadth to the course.

SUGGESTIONS

1. Questions 1, 3, 4, 5, and 6 can be used to help students think about the LC circuit discussions.
2. Discuss the initial conditions and the determination of the phase constant for an LC circuit, then ask questions 2 and 7.
3. If you compare an oscillating LC circuit to an oscillating mass on a spring, ask question 12. To fully answer the question you will need to consider a circuit with resistance and a mechanical oscillator subjected to a drag force. Assign problem 25.
4. Assign problems 1, 2, 4, 5, and 15 to test for understanding of the fundamentals of LC oscillations. The frequency of oscillation is covered in problems 9, 10, 11, and 24 and the relationship of the frequency to the inductive and capacitive time constants is explored in problem 12.
5. Demonstrations

   LCR series circuit: Freier and Anderson En12, Eo13; Hilton E13c, d, e. By making the resistance small you can demonstrate many of the ideas of this chapter.

Chapter 36 ALTERNATING CURRENTS

BASIC TOPICS

I. Elements of circuit analysis.

   A. Consider a sinusoidally varying current in a resistor and state that the potential difference across the resistor is in phase with the current and that the amplitudes are related by $I_R = V_R/R$. Emphasize that $v_R(t)$ gives the potential of one end of the resistor relative to the other. Draw a phasor diagram: 2 arrows along the same line with length proportional to I_R and V_R respectively. Both make the angle ωt with the horizontal axis and rotate in the counterclockwise direction. Point out that the vertical projections represent $i_R(t)$ and $v_R(t)$ and these vary in proportion to $\sin(\omega t)$ as the arrows rotate.

   B. Consider sinusoidally varying current in a circuit branch containing a capacitor. Start with $i_C = dq/dt = C\,dv_C/dt$, then show that v_C lags i_C by 90° and that the amplitudes are related by $I_C = V_C/X_C$, where $X_C = 1/\omega C$ is the capacitive reactance. Draw a phasor diagram to show the relationship. Mention that the unit of reactance is the ohm.

   C. Consider a sinusoidally varying current in a circuit branch containing an inductor. Start with $v_L = L\,di_L/dt$, then show that v_L leads i_L by 90° and that the amplitudes are related by $I_L = V_L/X_L$, where $X_L = \omega L$ is the inductive reactance. Draw a phasor diagram to show the relationship.

   D. Wire a small resistor in series with a capacitor and a signal generator. Use a dual trace oscilloscope with one set of leads across the resistor and the other set across the capacitor. Remind students that the potential difference across the resistor is proportional to the current, so the scope shows i_C and v_C. Point out the difference in phase. Repeat with an inductor in place of the capacitor.

II. An RLC series circuit.

   A. Draw the circuit. Assume the generator emf is given by $\mathcal{E}(t) = \mathcal{E}_m \sin(\omega t)$ and the current is given by $i(t) = I\sin(\omega t - \phi)$. Pick consistent directions for positive emf and positive current. Construct a phasor diagram step by step (see Fig. 36–6). First draw the current and resistor voltage phasors, in phase. Remind students that the current is the same in every element of the circuit so voltage phasors for the other elements can be drawn using the phase relations between voltage and current developed earlier. Draw the capacitor voltage phasor lagging by 90° and the inductor voltage phasor leading by 90°. Make $V_L > V_C$. Their lengths are IX_C and IX_L respectively. Draw the projections of the phasors on the vertical axis and remark that the algebraic sum must be $\mathcal{E}(t)$.

B. Draw the impressed emf phasor. Remark that its projection on the inductance phasor must be $V_L - V_C$ and that its projection on the resistance phasor must be V_R. Make the analogy to a vector sum.

C. Use the phasor diagram to derive the expression for the current amplitude: $I = \mathcal{E}_m/Z$, where $Z = \sqrt{R^2 + (X_L - X_C)^2}$ is the impedance of the circuit. Show that the impedance is frequency dependent by substituting the expressions for the reactances. Also show that I is greatest for $X_C = X_L$ or $\omega = 1/\sqrt{LC}$ and remark that this is the resonance condition discussed in the last chapter.

D. Use the phasor diagram to derive the expression for the phase angle of i relative to $\mathcal{E}$: $\tan\phi = (X_L - X_C)/R$. Point out that the phase angle vanishes at resonance and $\mathcal{E}$ leads i if $X_L > X_C$, but $\mathcal{E}$ lags i if $X_L < X_C$. For later use show that $\cos\phi = R/Z$.

III. Power considerations.

A. Discuss average values over a cycle. Show that the average of $\sin^2(\omega t + \phi)$ is $\frac{1}{2}$ and that the average of $\sin(\omega t)\cos(\omega t)$ is 0. Define the rms value of a sinusoidal quantity. Point out that AC meters are usually calibrated in terms of rms values.

B. Derive the expression for the power input of the AC source: $P = i\mathcal{E} = i_m\mathcal{E}_m \sin(\omega t + \phi)\sin(\omega t)$. Show that the average over a cycle is given by $\overline{P} = \mathcal{E}_{\rm rms} i_{\rm rms}\cos\phi$. Do the same for the power dissipated in the resistor. In particular, show that its average value can be written $i^2_{\rm rms}R$ or $\mathcal{E}_{\rm rms} i_{\rm rms} R/Z$. Recall that $R/Z = \cos\phi$ and then use this relationship to show that the average power input equals the average power dissipated in the resistor.

C. Show that the average rate of energy flow into the inductor and capacitor are each zero.

D. Explain that $\cos\phi$ is called the power factor. If it is 1 the source delivers the greatest possible power for a fixed generator amplitude. Remark that the power factor is 1 at resonance.

SUPPLEMENTARY TOPIC

The transformer. Use Faraday's law to show how the potential difference across the secondary is related to the potential difference across the primary. Explain what step-up and step-down transformers are. A dual trace oscilloscope can be used to demonstrate transformer voltages. Assume a purely resistive load and show how to find the primary and secondary currents. Show that, as far as the primary current is concerned, the transformer and secondary circuit can be replaced by a resistor with $R_{\rm eq} = (N_p/N_s)^2 R$, where N_p is the number of turns in the primary coil, N_s is the number of turns in the secondary coil, and R is the load resistance. Explain impedance matching.

SUGGESTIONS

1. When discussing solutions to the RLC loop equation, include questions 1, 4, 11, and 12.
2. Use questions 2 and 3 in the discussion of phasor diagrams.
3. Use questions 5, 7, 8, and 10 in the discussion of reactance and the relative phases of the potential and current.
4. Include question 13 in the discussion of power. More detail can be added to the discussion of the power factor by including some of questions 14, 15, 17, and 18.
5. Assign problems 8 and 9 in connection with discussions of the phase and amplitude of separate inductive and capacitive circuits. Phase is also covered in problems 10, 11, 22, and 23.

6. Assign problem 20 to have students think about voltages around an LCR circuit.
7. Power in an RLC circuit is covered in problems 37 and 38 and the power factor in problem 40.
8. Demonstrations
   a. See the RLC circuit demonstrations listed in Chapter 35 notes.
   b. Measurements of reactance and impedance: Freier and Anderson Eo9.
   c. Transformers: Freier and Anderson Ek7, Em1, 2, 4, 5, 7, 8, 10; Hilton E11.
9. Films
   Electromagnetic Oscillator I: Free Oscillations, *Electromagnetic Oscillator II: Forced Oscillations*; S8, color, 3.3 min. (I) and 7.4 min. (II). Walter de Gruyter, Inc., 200 Sawmill River Road, Hawthorne, NY 10532. I. Damped, critically damped, and over damped oscillations are illustrated in an RLC circuit. II. Using a double beam oscilloscope, the phase, voltage, and current of weakly damped resonant circuits are shown. Reviewed AJP 45:1014 (1977).
10. Laboratory
   a. MEOS Experiment 10–11: *A.C. Series Circuits.* Students use an oscilloscope and AC meters to investigate voltage amplitudes, phases, and power in RC and RLC circuits. Voltage amplitudes and phases are plotted as functions of the driving frequency to show resonance. Reactances and impedances are calculated from the data.
   b. BE Experiment 37: *A Study of Alternating Current Circuits.* An AC voltmeter is used to investigate the voltages across circuit elements in R, RC, RL, and RLC circuits, all with 60 Hz sources. Reactances and impedances are computed. If possible, oscilloscopes should be used. A section labelled optional describes their use. This experiment is pedagogically similar to the text and can be used profitably to reenforce the ideas of the chapter. Warning: the lab book uses the word vector rather than phasor.

Chapter 37 MAXWELL'S EQUATIONS

BASIC TOPICS

I. The Ampere-Maxwell law.
   A. Use Table 37–I to review the equations of electricity and magnetism discussed so far. Note the absence of any counterpart to Faraday's law, i.e. the creation of magnetic fields by changing electric flux. Tell students it should be there and you will now discuss its form.
   B. Consider the charging of a parallel plate capacitor. Remind students that in Ampere's law $d\mathbf{s}$ and $d\mathbf{A}$ are related by a right-hand rule and the surface integral is over any surface bounded by the closed contour.

      In the diagram, surfaces A, B, and C are all bounded by the contour which forms the left end of the figure. If we choose surface A or C then Ampere's law as we have taken it gives $\oint \mathbf{B}\cdot d\mathbf{s} = \mu_0 i$, but if we choose surface B, it gives $\oint \mathbf{B}\cdot d\mathbf{s} = 0$. Since the integral on the left side is exactly the same in all cases, something is wrong.

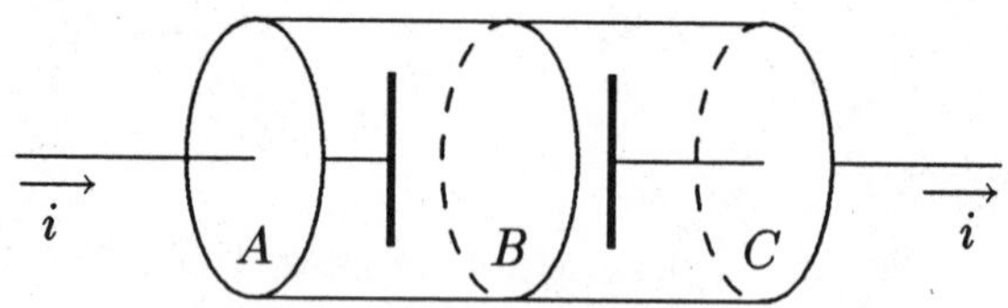

   C. Note that the situation discussed and the lack of symmetry in the electromagnetic equations suggests that Ampere's law as used so far must be changed. Experiment confirms this conjecture.
   D. Explain that if the electric flux through an open surface changes with time then there is a magnetic field and the magnetic field has a tangential component at points on the

boundary. Write down the Ampere-Maxwell law: $\oint \mathbf{B} \cdot d\mathbf{s} = \mu_0 i + \mu_0 \epsilon_0 \, d\Phi_E/dt$, where Φ_E is the electric flux through the surface. Compare to Faraday's law and point out the interchange of **B** and **E**, the change in sign, and the appearance of the factor $\mu_0 \epsilon_0$.

E. Give the right-hand rule that relates the normal to the surface used to calculate Φ_E and the direction of integration around its boundary. State that the surface may be a purely mathematical construction and that the law holds for any surface.

II. Displacement current.

A. Define the displacement current $i_d = \epsilon_0 \, d\Phi_E/dt$. Explain that it does not represent the flow of charge and is not a true current, but that it enters the Ampere-Maxwell law in the same way as a true current. Discuss the direction of i_d. Consider a region in which the electric field is uniform and is changing. Find the direction for both an increasing and a decreasing field.

B. Refer to the Ampere-Maxwell law. Explain that there are no changing electric fields in the examples of Chapter 31 so only true currents were considered. Explain that in the region between the plates of a charging capacitor there is no true current but there is a displacement current.

C. Consider a parallel plate capacitor with circular plates, for which dE/dt is given. Show that the total displacement current in the interior of the capacitor equals the true current into the capacitor. Explain that the sum of the true and displacement currents is continuous. Optional: discuss a leaky capacitor.

D. Derive expressions for **B** at various points along the perpendicular bisector of the line joining the plate centers. Consider points between the plates and outside them. See Sample Problem 37–1.

III. Maxwell's equations.

A. Write down the four equations in integral form and review the physical processes that each describes. See Table 37–2.

B. Carefully distinguish between the line and surface integrals that appear in the equations and give the right-hand rules that relate the direction of integration for the contour integrals and the normal to the surface for the surface integrals.

C. Review typical problems: the electric field of a point charge, the magnetic field of a uniform current in a long straight wire, the magnetic field at points between the plates of a capacitor with circular plates, the electric field accompanying a changing uniform magnetic field with cylindrical symmetry.

D. State that in the absence of dielectric and magnetic materials these equations describe all electromagnetic phenomena to the atomic level and the natural generalizations of them provide valid descriptions of electromagnetic phenomena at the quantum level. They are consistent with modern relativity theory. Optional: for completeness you may want to rewrite the equations and include magnetization and electric polarization terms.

SUGGESTIONS

1. To test for understanding of the direction of the magnetic field induced by a changing electric field, assign questions 2 and 5. The directions of the displacement current, **E**, and **B**, are also covered in question 7. It is worthwhile asking this question before question 5.

2. Problems 13 and 15 help students think carefully about displacement current. The continuity of current and displacement current through a charging capacitor is covered in problem 9.

3. To give students some physical intuition about the magnitudes of displacement currents and induced magnetic fields, ask questions 3 and 10 and assign problems 4 and 11.

4. To give students practice in associating Maxwell's equations with various electromagnetic phenomena, assign problem 16.

Chapter 38 ELECTROMAGNETIC WAVES

BASIC TOPICS

I. Qualitative features of electromagnetic waves.

A. Explain that an electromagnetic wave is composed of electric and magnetic fields. The disturbance, analogous to the string shape that moves on a taut string, is made up of the fields themselves, moving through space or a material medium. Also explain that electromagnetic waves carry energy and momentum.

B. State that the wave speed in a vacuum is given by $c = 1/\sqrt{\mu_0\epsilon_0}$ and is about 3.00×10^8 m/s. The existence of waves and this expression for the wave speed in vacuum are predicted by Maxwell's equations. Since the values of c and μ_0 are fixed, this fixes ϵ_0.

C. Show the electromagnetic spectrum (Fig. 38–1) and point out the visible, ultraviolet, infrared, x-ray, microwave, and radio regions. Remark that all the waves are fundamentally the same, differing only in wavelength and frequency.

D. Restate that the visible spectrum extends from just over 400 nm to just under 700 nm. Remark that while color is largely subjective, violet is at the short wavelength end while red is at the high wavelength end. Use a prism to display the spectrum. Show Fig. 38–2 of the text and remark that human eyes are most sensitive in the green–yellow portion of the spectrum and that sensitivity falls off rather rapidly on either side.

E. State that an accelerating charge creates electromagnetic radiation. Show diagrams of an oscillating electric dipole antenna and its fields (see Figs. 3, 4, 5, and 6). Point out that **E** and **B** are perpendicular to each other and to the direction of propagation and that they oscillate in phase with each other at any point. Explain the term polarization.

II. Traveling sinusoidal waves.

A. Take $E(x,t) = E_m \sin(kx - \omega t)$, along the y axis, and $B(x,t) = B_m \sin(kx - \omega t)$, along the z axis. Remark that both fields travel in the positive x direction and that they are in phase. Remind students that the minus sign in the argument becomes a plus sign for a wave traveling in the negative x direction.

B. Consider a rectangular area in the xy plane, with infinitesimal width Δx and length h (along y). Evaluate $\oint \mathbf{E}\cdot d\mathbf{s}$ and Φ_B, then show that Faraday's law yields $\partial E/\partial x = -\partial B/\partial t$. Substitute the expressions for E and B to show that $E = cB$, where $c = \omega/k$. Stress that the magnitudes of **E** and **B** are related. Remark that **E** is different at different points because **B** changes with time.

C. Consider a rectangular area in the xz plane, with infinitesimal width Δx and length h (along z). Evaluate $\oint \mathbf{B}\cdot d\mathbf{s}$ and Φ_E, then show that the Ampere-Maxwell law yields $-\partial B/\partial x = \mu_0\epsilon_0\, \partial E/\partial t$. Combine this with the result of part B to show that $c = 1/\sqrt{\mu_0\epsilon_0}$. Remark that **B** is different at different points because **E** changes with time. Emphasize the role played by the displacement current.

III. Energy and momentum transport.

A. Define the Poynting vector $\mathbf{S} = (1/\mu_0)\mathbf{E}\times\mathbf{B}$ and explain that it is in the direction of propagation and that its magnitude gives the electromagnetic energy per unit area that crosses an area perpendicular to the direction of propagation per unit time. Remark that for a plane wave $S = EB/\mu_0 = E^2/\mu_0 c = cB^2/\mu_0$.

B. Consider the plane wave of Section II, propagating in the positive x direction. Consider a volume of width Δx and cross section A (in the yz plane) and show that the electric and magnetic energies in it are equal and that the total energy is $\Delta U = (EBA/\mu_0 c)\Delta x$, for small Δx. This energy passes through the area A in time $\Delta t = \Delta x/c$ so the rate of energy flow per unit area is EB/μ_0, as previously postulated.

C. Explain that most electromagnetic waves of interest oscillate rapidly and we are not normally interested in the instantaneous values of the energy or energy density. Explain how to find the average over a period of the square of a sinusoidal function. Define the intensity as the time average of the magnitude of the Poynting vector and write expressions for it in terms of the average energy density and in terms of the field amplitudes.

D. Explain that electromagnetic waves transport momentum and that S/c gives the momentum that crosses a unit area per unit time. The momentum is in the direction of **S**. Also explain that if an object absorbs energy U then it receives momentum U/c. If the object reflects energy U then it receives momentum $2U/c$.

E. As an example of radiation pressure, you may wish to consider solar pressure. S can be determined from the solar constant $1.38\,\text{kW/m}^2$ (valid just above the earth's atmosphere).

IV. Polarization.

A. Remind students that a linearly polarized electromagnetic wave is one for which the electric field is everywhere parallel to the same line. As the wave passes by any point the field oscillates along the line of polarization.

B. Explain that a linearly polarized wave can be resolved into two other linearly polarized waves with mutually orthogonal polarization directions. Take the original polarization direction to be at the angle θ to one of the new directions and show that the amplitudes are given by $E_1 = E_m \cos\theta$ and $E_2 = E_m \sin\theta$, where E_m is the original amplitude.

C. Explain that the electric field associated with unpolarized light does not remain in the same direction for more than about 10^{-8} s and the new direction is unrelated to the old.

D. Shine unpolarized light through crossed Polaroid sheets and note the change in intensity as the second sheet is rotated. Show that the intensity does not change if the first sheet is rotated. Remark that for an ideal polarizing sheet the transmitted intensity is half the incident intensity.

E. Derive the law of Malus. Explain that the light emerging from the first Polaroid sheet is linearly polarized in a direction determined by the orientation of the sheet. Remark that this direction is called the polarizing direction of the sheet. Draw a diagram of the electric field amplitude as the light enters the second sheet, at an angle θ to the polarizing direction of the second sheet. Resolve the amplitude into components along the polarizing direction and perpendicular to it. Explain that the first component is transmitted through the sheet while the second is absorbed. The amplitude of the transmitted wave is proportional to $\cos\theta$ and the intensity is proportional to $\cos^2\theta$.

F. Assign Problem 51 and carry out the demonstration. Shine unpolarized light onto two crossed Polaroid sheets and remark that no light is transmitted. Then slide another sheet between the two and point out the change in transmitted intensity as you rotate the sheet in the middle. The sheets can be taped to ringstands to hold them.

SUGGESTIONS

1. In discussing the speed of light in terms of ϵ_0 and μ_0, remind students that these constants enter electrostatics and magnetostatics respectively and were first encountered in situations that had nothing to do with wave propagation. Have them answer question 20.
2. Remind students that, when the relationship between **E** and **B** for a plane wave was discussed, the Faraday and Ampere-Maxwell laws were used. Then ask them to discuss question 7.
3. The relationship between frequency, wavelength, and speed is explored in problems 3 and 4. These also give some examples of high and low frequency electromagnetic radiation and ask

students to interpret Fig. 38–2, which graphs the sensitivity of the human eye as a function of wavelength.

4. To stress the relationship between **E** and **B** in an electromagnetic wave, assign problem 12 or 18.

5. To stress that electromagnetic waves need not be sinusoidal, assign problems 13 and 14. Remark that the wave associated with any function obeys the wave equation and travels with speed c in empty space.

6. To emphasize the magnitude of the energy and momentum carried by an electromagnetic wave, discuss questions 8 and 9 and assign problems 27 and 39. Problem 22 deals with the average energy in terms of the magnetic field amplitude. Also consider some problems that deal with point sources: 21 and 25, for example.

7. Use problem 48 to test for understanding of the term "polarization". Also consider question 11.

8. The fundamentals of polarizing sheets are covered in problems 49 and 50. In the first the incident light is unpolarized while in the second it is polarized. Discuss question 13 and assign problem 59.

9. Demonstrations

   Radiation: Freier and Anderson Ep4, 5.

10. Computer programs
    a. *Physics Simulations II: Radiation*; Kinko's Service Corporation, 4141 State Street, Santa Barbara, CA 93110; Macintosh. Shows electric field lines of an accelerating charge in linear, circular, or oscillatory motion. User selects the velocity and can view either the near or far field.
    b. *Intermediate Physics Simulations: Moving Charge*; R.H. Good, Physics Department, California State University at Hayward, CA 94542; Apple II. Shows electric field lines of a moving charge with user selected velocity. The user can change the velocity and radiation fields are shown when the charge accelerates.
    c. *Intermediate Physics Simulations: Radiating Dipole*; R.H. Good (address given above); Apple II. Animated diagrams showing the electric field lines of a radiating electric dipole. Excellent illustration for a lecture on sources of radiation.

11. Laboratory
    a. MEOS Experiment 13–7; *Polarization of Light.* Polaroid sheets are first investigated and the law of Malus is verified. Then a polaroid sheet is used to investigate polarization by reflection, by refraction, and by scattering. Brewster's angle is found. Rotation of the direction of polarization by a sugar solution is also studied and crossed polarizers are used to check various objects for stresses. Essentially a series of demonstrations performed by students.
    b. BE Experiment 46; *Polarized Light.* Similar to MEOS Experiment 13–7 except light transmitted by a calcite crystal is also investigated. A photodetector is used to obtain quantitative data.

Chapter 39 GEOMETRICAL OPTICS

BASIC TOPICS

I. Wave and geometrical optics
   A. Explain that optical phenomena outside the quantum realm can be understood in terms of Maxwell's equations and that the wave nature of electromagnetic radiation must be

taken into account to explain many important phenomena. State that some of these will be discussed later.

B. Explain that if the wavelength of the light is much smaller than any obstacles it meets or any slits through which it passes, then the important property is the direction of motion, not details of the wave nature. This is the realm of geometrical optics.

C. Define a ray as a line that gives the direction of travel of a wave. It is perpendicular to the wave fronts (surfaces of constant phase). Explain that geometrical optics deals largely with tracing rays as light is reflected from surfaces or passes through materials.

D. Explain that when light traveling in one medium strikes a boundary with another medium, some is reflected and some is transmitted into the second medium. Draw a plane boundary between two media and show an incident, a reflected, and a refracted ray. Label the angles these rays make with the normal to the surface.

E. Tell students that the speed of light may be different in different materials and state that the speeds of light in the two media are crucial for determination of the amplitudes of the reflected and refracted light and for determination of the angle of refraction. Define the index of refraction of a medium as the ratio of the wave speed in vacuum to the wave speed in the medium and write $v = c/n$. Remark that the index of refraction is a property of the medium and depends on the wavelength. Point out Table 39–1, which gives the indices of refraction of various materials. Note that the index of refraction for a vacuum is 1 and is nearly 1 for air. State it is wavelength dependent and point out Fig. 39–2.

II. Reflection at a plane surface.

A. Consider a plane wave incident on a plane surface. Write down the law of reflection: $\theta_1 = \theta_1'$.

B. Consider a point source in front a plane mirror. Draw both incident and reflected rays and show that the reflected rays appear to come from a point behind the mirror. Show that the object and image lie on the same normal to the mirror and that they are the same distance from the mirror. Remark that no light comes from the image and that the image is said to be virtual.

C. Define the object distance p and image distance i and explain that the latter is taken to be negative for virtual images. The law of equal distance is written $p = -i$.

D. Give the condition for being able to see an image. Draw a mirror, an eye, a source, and its image. Draw the line from the image to the eye and state that the image can be seen if this line intersects the mirror. Show that length of a wall mirror with its top edge at eye level need reach only halfway to the floor for a person to see his feet. Demonstrate with a mirror resting on the floor and half-covered with a cloth. Have a student stand in front of the mirror and start with the cloth about shoulder height and lower it until the student can see his feet.

III. Polarization by reflection.

A. Reflect a well collimated beam of unpolarized light from a plane glass surface. A slide projector beam does nicely. Darken the room and obtain a reflection spot on the ceiling. Place a Polaroid sheet in the reflected beam and note the change in intensity of the spot as you rotate it. Remark that the reflected light is partially polarized.

B. Orient the incident beam so the angle of incidence is Brewster's angle and use the Polaroid sheet to show the reflected light is now entirely polarized.

C. Discuss Brewster's law. Explain that unpolarized light incident on a boundary is partially or completely polarized on reflection. When the angle of incidence and the angle of refraction sum to 90° the reflected light is completely polarized, with **E** perpendicular to the plane of the incident and reflected rays. Show that the angle of incidence θ_B for completely polarized reflected light is given by $\tan\theta_B = n_2/n_1$, where medium 1 is the medium of the reflected ray.

IV. Refraction.

A. Write down the law of the law of refraction: $n_1 \sin\theta_1 = n_2 \sin\theta_2$.

B. Explain that light rays are bent toward the normal when light enters a more optically dense medium (higher index of refraction) and are bent away from the normal when it enters a less optically dense medium.

C. Consider light striking a water surface from air and trace a few rays. Consider light from an underwater source and trace a few rays as they enter the air. Consider a slab of glass with parallel sides and show that the emerging ray has the same direction as the entering ray but is displaced along the slab. Optional: derive the expression for the displacement.

D. Trace a ray through a prism and derive the expression for the angle of deviation: $\psi = \theta_1 + \theta_2 + \phi$, where θ_1 is the angle of incidence, θ_2 is the angle of emergence, and ϕ is the prism angle. Explain that ψ is different for different colors because n depends on wavelength.

E. Shine an intense, monochromatic, well-collimated beam on a prism and point out the reflected and refracted beams. A laser works reasonably well but it is difficult for the class to see the beam. Use smoke or chalk dust to make it visible. To avoid the mess, use an arc beam or the beam from a 35 mm projector, filtered by red glass. Make a $\frac{1}{2}$ in. hole in a 2 in. by 2 in. piece of aluminum and insert it in the film gate. Use white light from the projector and the prism to show that different wavelengths are refracted through different angles.

F. Explain total internal reflection. Show that no wave is transmitted when the angle of incidence is greater than the critical angle and derive the expression for the critical angle in terms of the indices of refraction. Stress that the index for the medium of incidence must be greater than the index for the medium of the refracted light. Total internal reflection can be demonstrated with some pieces of solid plastic tubing having a diameter larger than that used for fiber optics. The beam inside is quite visible. If time permits, discuss fiber optics and some of its applications.

V. Spherical mirrors.

A. Consider a point source in front of a concave spherical mirror. Draw a diagram that shows the optic axis (called the central axis in this text), the center of curvature, and the source on the axis, outside the focal point. Show that small-angle rays form an image and that object and image distances are related by $1/p + 1/i = 2/r$. To emphasize the small-angle approximation, consider the case $p = 2r$ and use a full hemispherical concave surface. The small-angle formula predicts all rays cross the axis at $i = (2/3)r$, but the ray that strikes the edge of the mirror crosses at the vertex.

B. Explain that the mirror equation is also valid for convex mirrors and for any position of the object, even virtual objects for which incoming rays converge toward a point behind the mirror. Give the sign convention: p and i are positive for real objects and images (in front of the mirror) and are negative for virtual objects and images (behind the mirror); r is positive for concave mirrors (center of curvature in front of the mirror) and negative for convex mirrors (center of curvature behind mirror). Remark that a surface is concave or convex according to its shape as seen from a point on the incident ray.

C. Define the focal point as the image point when the incident light is parallel to the axis. By considering a source far away, show that $f = r/2$. Consider a concave mirror and show that for $p > f$ the image is real, for $p < f$ the image is virtual. Also show that for $p = f$ parallel rays emerge after reflection.

D. Describe a geometric construction for finding the image of an extended source. Trace rays from an off-axis point: one through the center of curvature, one through the focal point, and one parallel to the axis. Use both concave and convex mirrors as examples. Explain that the geometric construction gives the same result as the small-angle approximation if reflection is assumed to take place at a plane through the mirror vertex and perpendicular to the optic axis. The law of reflection cannot be applied at this plane, of course.

E. Define lateral magnification and show that $m = -i/p$. Explain the sign: m is positive for erect images and negative for inverted images. Virtual images of real objects are erect and real images of real objects are inverted.

F. Take the limit $r \to \infty$ and show that the mirror equation makes sense for a plane mirror.

VI. Spherical refracting surfaces.

A. Draw a convex spherical boundary between two media, use the law of refraction to trace a small-angle ray from a source on the axis, and show that $n_1/p + n_2/i = (n_2 - n_1)/r$, where n_1 is the index of refraction for the region of incident light and n_2 is the index of refraction for the region of refracted light. You can demonstrate the bending of the light using a laser and a round bottom flask. Use a little smoke or chalk dust to make the beam visible in air and just a pinch of powdered milk in the water to make it visible inside the flask.

B. Explain the sign convention. Point out that real images are on the opposite side of the boundary from the incident light and virtual images are on the same side. Explain that p and i are positive for real objects and images, negative for virtual objects and images. r is positive for convex surfaces, negative for concave. With this sign convention the equation holds for concave or convex surfaces and for $n_2 > n_1$ or $n_1 > n_2$.

C. Consider the limit $r \to \infty$, which yields $i = -pn_2/n_1$. This is the solution to the apparent depth problem. For water 4 inches deep, a ball on the bottom appears to be at a depth of about 3 inches. Use an aquarium filled with water and a golf ball to make a hallway display.

VII. Thin lenses.

A. Explain that a lens consists of two refracting surfaces close together in vacuum. State or derive the thin lens equation: $1/p + 1/i = (n - 1)(1/r_1 - 1/r_2)$, where n is the index of refraction for the lens material. Stress that the equation holds for small-angle rays. State that it also holds to a good approximation for a lens in air. r_1 is the radius of the first surface struck by the light and r_2 is the radius of the second. They are positive or negative according to whether the surfaces are convex or concave when viewed from a point on the incident ray. You may wish to generalize the equation by retaining the indices of refraction. The result is $1/p + 1/i = (n_2/n_1 - 1)(1/r_1 - 1/r_2)$. This allows you to consider a thin glass or air lens in water.

B. By considering $p \to \infty$ show that the focal length is given by $1/f = (n - 1)(1/r_1 - 1/r_2)$ or more generally by $1/f = (n_2/n_1 - 1)(1/r_1 - 1/r_2)$. Show that the same value, including sign, is obtained no matter which surface is struck first by light. Then show that $1/p + 1/i = 1/f$. Point out that there are two focal points, the same distance from the lens but on opposite sides. For a converging lens rays from a point source at f on one side are parallel on the other side; incident parallel rays converge to f on the other side. For a diverging lens rays that converge toward f on the other side emerge parallel; rays that are parallel emerge as diverging from f on the incident side.

C. Show how to locate the image of an extended object by tracing a ray parallel to the axis, a ray through the lens center, and a ray along a line through the first focal point (on the incident side for a converging lens and on the other side for a diverging lens).

D. Define lateral magnification and show that $m = -i/p$. Explain that the sign tells whether the image is erect or inverted.

E. Consider all possible situations: converging lens with $p > f$, $p < f$, and $p = f$; diverging lens with $p > f$, $p < f$, and $p = f$. In each case show whether the image is real or virtual, erect or inverted, and find its position relative to the focal point.

F. Note that most optical instruments are constructed from a combination of two or more lenses. Point out that to analyze them, one considers one lens at a time, with the image of the previous lens as the object of the lens being considered. This sometimes leads to virtual objects. Note that the overall magnification is given by $m = m_1 m_2 m_3 \ldots$ and that the sign of m tells whether the image is erect or inverted. If the image lies on the opposite side of the system from the object and is outside the system then it is real, otherwise it is virtual.

SUPPLEMENTARY TOPIC

Optical instruments. This section may be studied in the laboratory. Ask students to experiment with the image forming properties of positive and negative lenses, then construct one or more optical instruments. Display several instruments in the lab.

SUGGESTIONS

1. Problems 1 and 2 cover the laws of reflection and refraction. Problems 6, 7, 11, and 12 present some situations that are more challenging.
2. Assign problems 21 and 22 in connection with total internal reflection. Problems 26 and 27 deal with optical fibers. Assign either problem 29 or 31 in connection with polarization by reflection.
3. Assign problem 35 in connection with images in plane mirrors. Interesting applications are covered in problem 41 (rotation of mirror), 42 (area of mirror used to form image), and 40 (can observer see an image?). Question 9 deals with the idea of a virtual image and questions 6 and 7 deal with the apparent reversal of right and left. Problems 36, 37, 38, and 39 deal with images in multiple mirrors. Assign at least one of them.
4. Use questions 10 and 11 to discuss images in spherical mirrors. Problem 48 covers nearly all possibilities. Lateral magnification is covered in problem 46.
5. Assign problem 51 in connection with spherical refracting surfaces.
6. Use questions 14 and 16 to discuss images formed by thin lenses. For comprehensive coverage of nearly all relationships, assign problem 65. Also assign problem 70, which deals with a compound system and includes a ray tracing exercise. Problems 57, 59, and 60 test understanding of the lensmaker's equation.
7. Ask questions 15 and 18 in connection with problem 61. These deal with the immersion of a thin lens in a medium.
8. Consider expanding the course a little by including problems 79 and 80, which deal with the human eye.
9. Demonstrations
   a. Plane mirrors: Freier and Anderson Ob1 - 6, Ob8, 9; Hilton O1c, d.
   b. Refraction at a plane surface: Freier and Anderson Od1 - 7
   c. Prisms: Freier and Anderson Of1 - 4; Hilton O2a, b, O3a.
   d. Total internal reflection: Freier and Anderson Oe1 - 7; Hilton O2c, d.
10. Audio/visual

    Demonstrations of Physics: Volume 7: Light; video cassette, about 30 min. Vikas Productions, Box 6088, Bozeman, MT 59771. Demonstrations of reflection and refraction, image formation by mirrors and lenses. Uses laser beams to trace rays. Reviewed TPT **29**, 403 (September 1991).

11. Computer program
    Optics and Light, Focus Media, Inc., 839 Stewart Avenue, Garden City, NY 11530. Apple II. Demonstration and tutorial on Snell's law and thin lenses. User selects the parameters, then the program draws a ray diagram. Reviewed TPT January 1985.
12. Laboratory
    a. MEOS Experiment 13-3: *Prism Spectrometer.* Helium lines are used to determine the index of refraction as a function of wavelength for a glass prism. A good example of dispersion and excellent practice in carrying out a rather complicated derivation involving Snell's law. Also see BE Experiment 43: *Index of Refraction with the Prism Spectrometer* and BE Experiment 44: *The Wavelength of Light.* In the second of these experiments students use a prism spectrometer to determine the wavelength of lines from a sodium source.
    b. MEOS Experiment 13–1: *Laser Ray Tracing.* A laser beam is used to investigate the laws of reflection and refraction and to observe total internal reflection and the formation of images by spherical mirrors. Measurements are used to calculate the index of refraction of several materials, including liquids, and the focal length of mirrors. Tracing is done by arranging the apparatus so the laser beam grazes a piece of white paper on the lab table. Much the same set of activities are described in BE Experiment 38: *Reflection and Refraction of Light*, but pins are used as objects rather than a laser source and rays are traced by positioning other pins along them. The technique can be used if you do not have sufficient lasers for the class.
    c. BE Experiment 39: *The Focal Length of a Concave Mirror.* Several methods are described, including a technique which involves finding the radius of curvature. Others involve finding the image when the object distance is extremely long, when it is somewhat greater than $2f$, and when it is somewhat less than $2f$. Then the mirror equation is used to solve for f.
    d. MEOS Experiment 13–2: *Lenses.* A light source and screen on an optical bench are used to find the focal lengths and magnifications of both convex and concave lenses. Chromatic and spherical aberrations are also studied. Also see BE Experiment 40: *Properties of Converging and Diverging Lenses*, a compendium of techniques for finding focal lengths.
    e. BE Experiment 41: *Optical Instruments Employing Two Lenses.* Students construct simple two lens telescopes and microscopes on optical benches, then investigate their magnifying powers. By trying various lens combinations they learn the purposes of the objective and eyepiece lenses.

Chapter 40 INTERFERENCE

BASIC TOPICS

I. Huygens' principle.
   A. Shine monochromatic light through a double slit and project the pattern on the wall. Either use a laser or place a single slit between the source and the double slit. Use a diagram to explain the setup. Point out the appearance of light in the geometric shadow and the occurrence of dark and bright bands. You can make acceptable double slits by coating a microscope slide with lamp black or even black paint. Tape a pair of razor blades together and draw them across the slide. By inserting various thicknesses of paper or shim stock between the blades you can obtain various slit spacings.
   B. Explain that Huygens' principle will be used to understand the pattern, then state the principle. Describe plane wave propagation in terms of Huygen wavelets: draw a plane wave front, construct spherical wave fronts of the same radius centered at several points along the plane wave front, then draw the plane tangent to these.

C. Use Huygens' principle to derive the law of refraction. Assume different wave speeds in the two media and show that the wavelengths are different. Consider wavefronts one wavelength apart and show that $\sin\theta_1/\sin\theta_2 = v_1/v_2$. Explain that $n = c/v$ and obtain the law of refraction.

D. Go back to the double slit pattern and explain that those parts of an incident wave front that are within the slit produce spherical wavelets that travel to the screen while wavelets from other parts are blocked. Some wavelets reach the geometric shadow. The spreading of the pattern beyond the shadow is called diffraction and will be studied in the next chapter. Wavelets from different slits arrive at the same point on the screen and interfere to produce the bands. This phenomena will be studied in this chapter.

II. Two-slit interference patterns.

A. Draw a diagram of a plane wave incident normally on a two-slit system and draw a ray from each slit to a screen far away. Remark that the waves are in phase at the slits but they travel different distances to get to the same point on the screen and may have different phases there. The electric fields sum to the total electric field. At some points the two fields cancel, at other points they reenforce each other. Remind students that the intensity is proportional to the square of the total field, not to the sum of the squares of the individual fields.

B. Point out that if the screen is far away the two rays are nearly parallel, then show that the difference in distance traveled is $d\sin\theta$, where d is the slit separation and θ is the angle the rays make with the forward direction. Explain the condition $d\sin\theta = m\lambda$ for a maximum of intensity and the condition $d\sin\theta = (m+\frac{1}{2})\lambda$ for a minimum.

C. Show that a lens can be used to obtain the same pattern, even if the screen is not far away.

III. The intensity.

A. Take the two fields to be $E_1 = E_0\sin(\omega t)$ and $E_2 = E_0\sin(\omega t+\phi)$, where $\phi = (2\pi/\lambda)d\sin\theta$. This is easily shown by remarking that $\phi = k\Delta d$, where $k = 2\pi/\lambda$ and $\Delta d = d\sin\theta$ (derived earlier).

B. Explain how the fields can be represented on a phasor diagram. If you did not cover Chapter 36 explain that a phasor has a length proportional to the amplitude and makes the angle ωt or $\omega t+\phi$ with the horizontal axis. Its projection on the vertical axis is proportional to the field. Sum the phasors to obtain the total field. Show that the amplitude E_θ of the total field is $2E_0\cos(\phi/2)$. Plot the intensity $4E_0^2\cos^2(\phi/2)$ as a function of ϕ. Point out that $\phi = 0$ produces a maximum, that maxima occur at regular intervals, and that the minima are halfway between adjacent maxima.

C. Show that the intensity at a maximum is 4 times the intensity due to one source alone. Remark that no energy is gained or lost. All energy through the slits arrives at the screen. The presence of the slitted barrier, however, redistributes the energy.

D. Note the half-width of each maximum, at half the peak, is given by $\sin\theta = \lambda/4d$. The smaller λ/d the sharper the maximum. Near the central maximum, where $\sin\theta \approx \tan\theta \approx \theta$, the linear spread on the screen is $y \approx (\lambda/2d)D$, where D is the distance from the slits to the screen.

E. It is also worth noting that since $\sin\theta = m\lambda/d \le 1$ for a maximum, the smaller λ/d, the more maxima occur.

F. For completeness you might mention the amplitude of the wavelets fall off as $1/r$ and are not quite the same at the screen. Show this is a negligible effect for the patterns considered here.

IV. Coherence

A. Explain that two waves are coherent if their relative phase does not change with time.

B. Explain that the two interfering waves must be coherent to obtain an interference pattern. The phase difference at the observation point must be constant over the observation time. Explain why two incandescent lamps, for example, do not produce a stable interference pattern. The light is from many atoms and the emission time for a single atom is about 10^{-8} s. The phase difference changes in a random way over times that are short compared to the observation time. State that in this case the intensities add.

C. Explain that an extended source can be used to obtain an interference pattern. Light from each atom goes through both slits and forms a pattern, but the patterns of different atoms are displaced from each other, according to the separation of the atoms in the source. No pattern is seen unless the incident light is restricted to that from a small region of the source. If you did not use a laser in the demonstration, explain the role of the single slit in front of the double slit.

D. Explain that a laser produces coherent light even though many atoms are emitting simultaneously. Because emission is stimulated, light from any atom is in phase with light from all other atoms. A laser can be used to form an interference pattern without restricting the incident beam.

V. Thin film interference.

A. Cut a 1 to 2 mm slit in a 2" square piece of aluminum and insert it in the film gate of a 35 mm projector. Let the beam impinge on a soap bubble to show the effect.

B. Consider normal incidence on a thin film of index n_1 in a medium of index n_2 and suppose the medium behind the film has index n_3. Explain that a wave reflected at the interface with a medium of higher index undergoes a phase change of π. If $n_1 < n_2 < n_3$, waves reflected at both surfaces undergo phase changes of π. Consider all other possibilities and then specialize to a thin film of index n in air. Give the conditions for maxima and minima for both the reflected light and the transmitted light, assuming near normal incidence. Note that the wavelength in the medium must be used to calculate the phase change on traveling through the medium. Define optical path length and point out its importance for thin film interference.

C. Broaden the discussion qualitatively by including non-normal incidence. Note that for some angles conditions are right for destructive interference of a particular color while at other angles conditions are right for constructive interference of the same color. Also note that these angles depend on λ. Hence the soap bubble colors.

D. If time permits, discuss Newton's rings. Use a plano-convex lens and a plane sheet of glass together with a laser. Use a diverging lens to spread the beam.

SUPPLEMENTARY TOPIC

The Michelson interferometer. This is an excellent example of an application of interference effects. Set up a hallway demonstration and give a brief explanation.

SUGGESTIONS

1. In the discussion of coherence, give a more detailed explanation of the single slit placed between the source and the double slit. Also use questions 4 and 16.
2. Questions 6 and 7 are good tests of understanding. Use them in the discussion or ask students to answer them for homework.
3. Problems 1 through 14 deal with the basics of interference. Assign one or two. Also consider problem 35. Use problem 28 or 30 to test for understanding of the derivation of the double-slit equation.
4. Assign problems 19, 20, 21, and 22 in connection with the double-slit interference pattern.

5. Use problems 47 and 55 to help with the discussion of thin films.
6. Problems 78, 79, and 80 illustrate some applications of a Michelson interferometer.
7. Film
   The Michelson Interferometer, S8, color, 4 min. The Kalmia Company, Dept. P1, Concord, MA 01742. Part of the Miller Demonstrations in Physics series. Reviewed TPT 14:253 (1976).
8. Demonstrations
   a. Double slit interference: Freier and Anderson O14, 5, 9; Hilton O7a.
   b. Thin film interference: Freier and Anderson O115 — 18; Hilton O7b, c, d.
   c. Michelson interferometer: Freier and Anderson O119.
9. Computer program
   Light Waves, Educational Materials and Equipment Company, P.O. Box 17, Pelham, NY 10803. Apple II, TRS-80 I, III, & W; IBM PC. Simulations of Young's experiment with user selected parameters. Students can view either a graph of the intensity or a simulated intensity pattern. Useful for lecture demonstrations. Reviewed TPT September 1984.
10. Laboratory
    a. MEOS Experiment 13–4: *Interference and Diffraction.* Students observe double-slit patterns of water waves in a ripple tank, sound waves, microwaves, and visible light. In each case except water waves, they measure and plot the intensity as a function of angle, then use the data to calculate the wavelength. A microcomputer can be used to take data and plot the intensity of a visible light pattern.
    b. MEOS Experiment 13–6: *The Michelson Interferometer.* An interferometer is used to measure the wavelengths of light from mercury and a laser and to find the index of refraction of a glass pane and air. Good practical applications.

Chapter 41 DIFFRACTION

BASIC TOPICS

I. Qualitative discussion of single-slit diffraction.
   A. Shine coherent monochromatic light on a single slit and project the pattern on the wall. Point out the broad central bright region and the narrower, less bright regions on either side, with dark regions between. Also point out that light is diffracted into the geometric shadow.
   B. Remark that diffraction can be discussed in terms of Huygens wavelets emanating from points in the slit. Explain that they not only spread into the shadow region but that they arrive at any selected point with a distribution of phases and interfere to produce the pattern. Explain that for quantitative work this chapter deals with Fraunhofer diffraction, with the screen far from the slit.
   C. Draw a single slit with a plane wave incident normal to it. Also draw parallel rays from equally spaced points within the slit, all making the same angle θ with the forward direction. Point out that all wavelets are in phase at the slit. The first minimum can be located by selecting θ so that, at the observation point, the ray from the top of the slit is 180° out of phase with the ray from the middle of the slit. All wavelets then cancel in pairs. Show that this leads to $a \sin\theta = \lambda$, where a is the slit width. Point out that this value of θ determines the width of the central bright region and that this region gets wider as the slit width narrows. Use $\sin\theta \approx \tan\theta \approx \theta$ (in radians) to show that the linear width of the central region on a screen a distance D away is $2D\lambda/a$. Use a variable width slit or a series of slits to demonstrate the effect.

D. By dividing the slit into fourths, eighths, etc. and showing that in each case the wavelets cancel in pairs if θ is properly selected, find the locations of other minima. Show that $a \sin\theta = m\lambda$ for a minimum.

E. Explain that for $a < \lambda$ the central maximum covers the whole forward direction. No point of zero intensity can be observed. Also remark that the intensity becomes more uniform as a decreases from λ. This was the assumption made in the last chapter when the interference of only one wavelet from each slit was considered.

F. Qualitatively discuss the intensity. Draw a phasor diagram showing 10 or so phasors representing wavelets from equally spaced points in the slit. Show that each wavelet at the observation point is out of phase with its neighbor by the same amount. First show the phasors with zero phase difference ($\theta = 0$), then show them for a larger value of θ. Show that they approximate a circle at the first minimum and then, as θ increases, they wrap around to form another maximum, with less intensity than the central maximum. Point out that as θ increases the pattern has successive maxima and minima and that the maxima become successively less intense.

II. The intensity.

A. Draw a diagram showing 10 or so phasors along the arc of a circle and let ϕ be the phase difference between the first and last. See Fig. 41–6. Explain that you will take the limit as the number of wavelets increases without bound and draw the phasor addition diagram as an arc. Use geometry to show that $E_\theta = E_m(\sin\alpha)/\alpha$, where $\alpha = \phi/2$. Point out that the intensity can be written $I_\theta = I_m(\sin^2\alpha)/\alpha^2$. By examining the path difference for the rays from the top and bottom of the slit, show that $\alpha = (\pi a/\lambda)\sin\theta$. Explain that these expressions give the intensity as a function of the angle θ.

B. Sketch the intensity as a function of θ (see Fig. 41–7) and show mathematically that the expression just derived predicts the positions of the minima as found earlier.

C. (Optional) Set the derivative of $(\sin\alpha)/\alpha$ equal to 0 and show that $\tan\alpha = \alpha$ at an intensity maximum. State that the first two solutions are $\alpha = 4.493\,\text{rad}$ and $7.725\,\text{rad}$. Use these results to show that the intensity at the first two secondary maxima are 4.72×10^{-2} and 1.65×10^{-2}, relative to the intensity for $\theta = 0$. You might also want to pick a wavelength and slit width, then find the angular positions of the first two secondary maxima. Remark that they are *not* precisely at midpoints between zeros of intensity.

III. Double-slit diffraction.

A. Consider the double-slit arrangement discussed in the previous chapter. Point out that the electric field for the light from each of the slits obeys the equation developed for single-slit diffraction and these two fields are superposed. They have the same amplitude, $E_m(\sin\alpha)/\alpha$ and differ in phase by $(2\pi d/\lambda)\sin\theta$, where d is the center-to-center slit separation. The result for the intensity is $I_\theta = I_m(\cos^2\beta)(\sin^2\alpha)/\alpha^2$, the product of the single-slit diffraction equation and the double-slit interference equation. Here $\beta = (\pi d/\lambda)\sin\theta$.

B. Sketch I_θ vs. θ for a double slit and point out that the single-slit pattern forms an envelope for the double-slit interference pattern. Remark that this is so because d must be greater than a.

C. Show how to calculate the number of interference fringes within the central diffraction maximum and remark that the result depends on the ratio d/a but not on the wavelength.

D. Discuss missing maxima. Point out that the first minimum on either side of the central single-slit diffraction maximum might coincide with a double-slit interference maximum, in which case the maximum would not be seen. Show that the maximum of order m is missing if $d/a = m$.

IV. Multiple-slit patterns.

A. Make or purchase a set of multiple-slit barriers with 3, 4, and 5 slits, all with the same slit width and spacing. Multiple slits can be made using razor blades and a lamp blackened microscope slide. Use a laser to show the patterns in order of increasing number of slits. Finish with a commercial grating.

B. Qualitatively describe the pattern produced as the number of slits is increased. Point out the principle maxima and, if possible, the secondary maxima. Remark that the principle maxima narrow and that the number of secondary maxima increases as the number of slits increases. Remark that for gratings with a large number of rulings the principal maxima are called lines. For each barrier sketch a graph of the intensity as a function of angle. Explain that the single-slit diffraction pattern forms an envelope for the pattern.

C. Remark that you will assume the slits are so narrow that the patterns you will consider lie well within the central maximum of the single-slit diffraction pattern and you need to consider only one wave from each slit. Explain that lines occur whenever the path difference for rays from two adjacent slits is an integer multiple of the wavelength: $d\sin\theta = m\lambda$. Remark that m is called the order of the line. Also remark that the angular positions of the lines depend only on the ratio d/λ and not on the number of slits or their width.

D. Consider N phasors of equal magnitude that form a regular polygon and remark this is the configuration for an interference minimum adjacent to a principal maximum. Show that for one of these minima the phase difference for waves from adjacent slits is $2\pi(m + 1/N)$ and the path difference is $d\sin\theta = \lambda(m + 1/N)$. Replace θ with $\theta + \delta\theta$, where $d\sin\theta = m\lambda$, to derive the expression $\delta\theta = \lambda/Nd\cos\theta$ for the angular half-width of the principal maximum at angle θ. Explain that this predicts narrowing of the principal maxima as the number of slits is increases. Also explain that principal maxima at large angles are wider than those at small angles.

E. Show a commercial transmission grating and tell students a typical grating consists of tens of thousands of lines ruled over a few centimeters. Explain that light is transmitted through both the rulings and the regions between but since these represent different thickness of material the phase of the waves leaving the rulings is different from that of waves leaving the regions between. As a result the diffraction pattern is the same as that of a multiple-slit barrier. Say that a diffraction pattern is also produced by lines ruled on a reflecting surface.

F. Put a grating in front of a white-light source and point out the spectrum. Put a grating in front of a discharge tube to display the emission spectrum of hydrogen or mercury. Note the separation of the lines corresponding to the same principal maximum produced by different frequency light. Explain that atoms produce light with certain discrete frequencies and that these are separated by the grating. Remark that measurements of the angles can be used to compute the wavelengths present if the ruling separation is known. Point out the colors of a compact disk.

V. X-ray diffraction (optional).

A. Explain that x rays are electromagnetic radiation with wavelength on the order of 10^{-10} m (1 Å). Point out that crystals are regular arrays of atoms with spacings on that order and so can be used to diffract x rays.

B. Consider a set of parallel crystalline planes and explain that reflection of the incident beam occurs at each plane, with the angle of reflection equal to the angle of incidence. Draw a diagram like Fig. 41–22 and state that x-ray diffraction is conventionally described in terms of the angle between the ray and the plane, rather than the normal to the plane. Show that waves reflected from the planes interfere constructively if $2d\sin\theta = m\lambda$.

C. Explain that for a given set of planes intense diffracted waves are produced only if waves are incident at an angle θ that satisfies the Bragg condition, given above. Measurements of these angles can be used to investigate the crystal structure. Show how to calculate the distance between planes, given the wavelength and the scattering angle. Explain that a crystal with a known structure can be used as a filter to obtain x rays of a given wavelength from a source with a broad range of wavelengths.

SUPPLEMENTARY TOPICS

1. Diffraction from a circular aperture. This topic is important for its application to diffraction patterns of lenses and the diffraction limit to the resolution of objects by a lens system. Show a diagram or picture (like Fig. 41–9) and point out the bright central disk and the secondary rings. Tell students that the angular position of the smallest ring of zero intensity occurs for $\theta = 1.22\lambda/d$, where d is the diameter of the aperture. If you intend to discuss the resolving power of a grating, the Rayleigh criterion for a circular aperture should be covered first since it is easier to present and understand. You can demonstrate the Rayleigh criterion by drilling two small holes, closely spaced, in the bottom of a tin can. Place the can over a light bulb and let students view it from various distances. See problem 41–18. Also use red and blue filters to show the dependence on wavelength.
2. Dispersion and resolving power of a grating. Define the dispersion of a grating and show it is $m/d\cos\theta$ for a line of order m, occurring at angle θ. Note that dispersion can be increased by decreasing the ruling separation but dispersion does not depend on the number of rulings. If you have gratings with different ruling separations use them to show the hydrogen spectrum and point out the difference. Define the resolving power of a grating and show it is Nm for the line of order m. Remark that the resolving power does depend on the number of rulings and that the greater this number, the greater the resolving power. Show the sodium spectrum with a grating for which the two D lines cannot be resolved, then show it with one for which they can. Explain that dispersion and resolving power measure different aspects of the pattern produced by a grating. The lines produced by two different wavelengths may be fairly well separated in angle (large dispersion) but cannot be resolved because the principal maxima are so wide (small resolving power).

SUGGESTIONS

1. To test for understanding of the derivation of the single-slit diffraction equation, ask question 13. Basic relationships are explored in problems 1 through 6 and 8. Assign a few of these.
2. Following the discussion of the equation for the double-slit pattern, ask question 14. Characteristics of the pattern are explored in problems 35, 36, 38, and 40.
3. Diffraction from a circular aperture with application to the Rayleigh criterion for resolution is covered in problems 18 through 34. Assign one or two if you cover this topic.
4. After discussing diffraction patterns of multiple slits ask questions 18 and 20. Problems 43, 44, 46, and 47 cover the fundamental equation for an intensity maximum. Problem 51 deals with the single-slit diffraction pattern that forms the envelope of the multiple-slit interference pattern and accounts for missing orders. Problem 50 deals with line width.
5. Ask questions 21, 22, and 25 in connection with resolving power and dispersion. Assign some of problems 55, 60, 63, and 65.
6. After discussing x-ray diffraction by crystals assign problems 71 and 72. Problems 79 and 80 are a little more challenging. Problem 78 deals with the geometry of a square lattice.
7. Computer projects
   Assign some of the computer projects described in the Computer Projects section of this manual. A computer is used to plot the intensity pattern for various situations including the case when the screen is not far from the sources.

8. Films
   a. *Joseph Fraunhofer: Diffraction*, 16 mm, color, 16 min. Office of Instructional Media, Rensselaer Polytechnic Institute, Troy, NY 12181. This historic film by Professor Leitner illustrates a series of diffraction experiments that were performed by Fraunhofer circa 1820. Reviewed AJP 44:116 (1976).
   b. *Interference and Diffraction*, 16 mm, b/w, 19 min. Modern Learning Aids, Division of Ward's Natural Science Establishment, P.O. Box 312, Rochester, NY 14601. This PSSC film, although nearly 30 years old, provides an excellent visual illustration of the phenomena of interference and diffraction through the use of a ripple tank. Reviewed AJP 32:62 (1964).
   c. *Shadow of a Hole*, S8, color, 3.5 min. American Association of Physics Teachers, 5110 Roanoke Place, College Park, MD 20740. A changing diffraction pattern is shown as the observation screen is moved away from the observer. Reviewed TPT 15:564 (1977) and AJP 46:197 (1978).
   d. *Diffraction — Single Slit*; *Diffraction — Double Slit*, S8, color, 4 min. each. The Kalmia Company, Dept. P1, Concord, MA 01742. These excellent loops are part of the Franklin Miller Demonstrations in Physics series. Reviewed TPT 14:253 (1976).
9. Computer program
   Physics Simulations III: Diffraction, Kinko's Service Corporation, 4141 State Street, Santa Barbara, CA 93110. Macintosh. Program shows intensity plots for single slits, double slits, and other apertures.
10. Demonstrations
   a. Single-slit diffraction: Freier and Anderson O12, 3, 6, 7; Hilton O7a.
   b. Multiple-slit diffraction: Freier and Anderson O110, 13; Hilton O7a; g.
   c. Diffraction by circular and other objects: Freier and Anderson O121 — 23; Hilton O7g.
   d. Diffraction by crystals: Freier and Anderson O114.
11. Laboratory
   a. MEOS Experiment 13–4: *Interference and Diffraction*. See Chapter 40 notes.
   b. MEOS Experiment 13–5; *Diffraction Gratings*. Wavelengths of the helium spectrum are found using a grating spectrometer and the influence of the number of grating rulings is investigated.
   c. BE Experiment 44; *The Wavelength of Light*. Wavelengths of the sodium spectrum are found using a grating spectrometer. The wavelength of a laser is also found.
   d. BE Experiment 45; *A Study of Spectra with the Grating Spectrometer*. Sources used are a sodium lamp, an incandescent bulb, a mercury lamp, and lamp containing an unknown element. The limits of the visible spectrum are determined and the unknown element is identified.

Chapter 42 RELATIVITY

BASIC TOPICS

I. Introduction
   A. Consider a wave on a string and remind students that its speed relative to the string is given by $v_w = \sqrt{\tau/\mu}$, where τ is the tension and μ is the linear mass density. Explain that, according to non-relativistic mechanics, an observer running with speed v_0 with the wave measures a wave speed of $v_w - v_o$ and an observer running against the wave measures a wave speed of $v_w + v_o$. Remark that these results are *not* valid for light (or fast moving waves and particles). The speed of light in a vacuum is found to be the same regardless of the speed of the observer (or the speed of the source).

B. Remark that this fact has caused us to revise drastically our idea of time. If, for example, two observers moving at high speed with respect to each other both time the interval between two events they obtain different results.

C. Explain that special relativity is a theory which relates measurements taken by two observers who are moving with respect to each other. Although it sometimes seems to contradict everyday experience, it is extremely well supported by experiment.

D. State the postulates: the laws of physics are the same for observers in all inertial frames; the speed of light in a vacuum is the same for all directions and in all inertial frames. Remind students what an inertial frame is. Explain that the laws of physics are relationships between measured quantities, not the quantities themselves. Newton's laws and Maxwell's equations are examples. State that relativity has forced us to revise Newton's second law but not Maxwell's equations.

II. Time measurements.

A. Explain the term *event* and note that three space coordinates and one time coordinate are associated with each event. Explain that each observer may think of a coordinate system with clocks at all places where events of interest occur and that the clocks are synchronized. Outline the synchronization process involving light. State that the coordinate system and clock used by an observer are at rest with respect to the observer and may be moving from the viewpoint of another observer.

B. State that two observers in relative motion cannot both claim that two events at different places are simultaneous if their motion is not perpendicular to the line joining the coordinates of the events. To illustrate show Fig. 42–4 and explain that the events are simultaneous in Sam's frame but the Red event occurs before the Blue event in Sally's frame. Show that signals from the events met at the midpoint of Sam's spaceship but the signal from the Red event gets to the midpoint of Sally's spaceship before the signal from the Blue event. Stress the importance of the second postulate for reaching these conclusions.

C. Explain the light flasher used to measure time, in principle. Consider a flasher at rest in one frame, take two events to be a flash and the subsequent reception of reflected light back at the instrument, then remark that the time interval is $\Delta t_0 = 2D/c$, where D is the separation of the mirror from the flash bulb. Consider the events as viewed in another frame, moving with speed v perpendicularly to the light ray, and show the interval is $\Delta t = 2D/c\sqrt{1 - v^2/c^2} = \Delta t_0/\sqrt{1 - v^2/c^2}$. This is also written $\Delta t = \gamma\Delta t_0$, where γ $(= 1/\sqrt{1 - v^2/c^2})$ is called the Lorentz factor. State that $v/c < 1$ and $\gamma > 1$.

D. Remark that Δt_0 is the *proper time interval* and that both events occur at the same coordinate in the frame in which it is measured. Point out that Δt is larger than Δt_0. Explain that the same result is obtained no matter what clocks are used for the measurement (as long as they are accurate and each is at rest in the appropriate frame). Ask students to identify a frame to measure the proper time interval for a ball thrown from third to first base. Note that $\Delta t \approx \Delta t_0$ if $v \ll c$.

E. State that time dilation has been observed by comparing clocks carried on airplanes to clocks remaining behind and by comparing the average decay time of fast moving fundamental particles to their decay time when at rest. You might want to discuss the twin paradox here.

III. Length measurements.

A. Point out the problem with measuring the length of an object that is moving relative to the meter stick: the position of both ends must be marked *simultaneously* (in the rest frame of the meter stick) on the meter stick. If the speed v of the object is known another method can be used to measure its length: put a mark on a coordinate axis along the line of motion of the object, then measure the time Δt_0 taken by the object to pass the mark.

The length is given by $L = v\Delta t_0$. Note that Δt_0 is a proper time interval but L is not the proper length.

B. Explain that the length of the object, as measured in its rest frame is $L_0 = v\Delta t$, where Δt is the time interval measured in that frame. Substitution of $\Delta t = \gamma\Delta t_0$ leads to $L = L_0/\gamma$. State that L_0, the length as measured in the rest frame of the object, is called the ***proper length.*** Since $\gamma > 1$, all observers moving with respect to the object measure a length that is less than the rest length. The same result is obtained no matter what method is used to measure length. Note that $L \approx L_0$ if $v \ll c$.

IV. The Lorentz transformation.

A. Consider two reference frames: S' moving with speed v in the positive x direction relative to S. Remark that the coordinates of an event as measured in S are written x, y, z, t while the coordinates as measured in S' are written x', y', z', t'. Write down the Lorentz transformation for the coordinate differences of two events: $\Delta x' = \gamma(\Delta x - v\Delta t)$, $\Delta y' = \Delta y$, $\Delta z' = \Delta z$, $\Delta t' = \gamma(\Delta t - v\Delta x/c^2)$. Remark that these equations reduce to the Galilean transformation if $v \ll c$: $\Delta x' = \Delta x - v\Delta t$, $\Delta y' = \Delta y$, $\Delta z' = \Delta z$, $\Delta t' = \Delta t$.

B. Explain that the transformation equations can be solved for Δx and Δt, with the result $\Delta x = \gamma(\Delta x' + v\Delta t')$, $\Delta t = \gamma(\Delta t' + v\Delta x'/c^2)$. From the viewpoint of an observer in S', S is moving in the negative x' direction, so the two sets of equations are obtained from each other when v is replaced by $-v$ and the primed and unprimed symbols are interchanged.

C. Discuss some consequences of the Lorentz transformation equations:

1. Simultaneity. Take $\Delta t = 0$, $\Delta x \neq 0$ and show that $\Delta t' = -\gamma v\Delta x/c^2$ ($\neq 0$). If two events are simultaneous and occur at different places in S, then they are not simultaneous in S'. Point out that $\Delta t'$ is positive for Δx negative and is negative for Δx positive. Similarly, take $\Delta t' = 0$, $\Delta x' \neq 0$ and show $\Delta t = \gamma v\Delta x'/c^2$ ($\neq 0$).
2. Time dilation. Consider two events that occur at the same place in S and show that $\Delta t' = \gamma\Delta t$. Point out that Δt is the proper time interval. Also show that the events do not occur at the same place in S': $\Delta x' = -\gamma v\Delta t$. Work the same problem for two events that occur at the same place in S'.
3. Length measurement. Suppose the object is at rest in S' and the meter stick is at rest in S. Marks are made simultaneously in S on the meter stick at the ends of the object. Thus $\Delta t = 0$. Show that $\Delta x' = \gamma\Delta x$ and point out that $\Delta x'$ is the rest length. Work the same problem with the object at rest in S and the meter stick at rest in S'.
4. Causality. Consider two events, the first of which influences the second. For example a particle is given an initial velocity along the x axis and collides with another particle. Remark that t_2 (the time of the collision) must be greater than t_1 (the time of firing). Take $\Delta t = t_2 - t_1$ and $\Delta x > 0$, then show that the Lorentz transformation predicts $\Delta t'$ is positive for every frame for which $v < c$. The collision cannot happen before the firing in any frame moving at less than the speed of light.
5. Velocity transformation. Tell students that u represents the velocity of frame S' relative to S, $\mathbf{v}$ and $\mathbf{v}'$ represent the velocity of a particle, as measured in S and S', respectively. Divide the Lorentz equation for Δx by the Lorentz equation for Δt to show that the x component of the particle velocity in S is $v = (v' + u)/(1 + uv'/c^2)$. Show this reduces to the Galilean transformation $v = v' + u$ for $u \ll c$. Take $v' = c$ and show that $v = c$. If $v' < c$ then $v < c$ for all frames moving at less than the speed of light.

V. Relativistic momentum and energy.

A. Explain that the non-relativistic definition of momentum must be generalized if momentum is to be conserved in collisions involving particles moving at high speeds. State that the proper generalization is $\mathbf{p} = m\mathbf{v}/\sqrt{1 - v^2/c^2}$. Remark that $\mathbf{p}$ is unbounded as the particle speed approaches the speed of light.

B. Remark that $m/\sqrt{1 - v^2/c^2}$ is sometimes referred to as the relativistic mass of the particle; m is then called its rest mass and is represented by m_0. In this text m is used for the rest mass and is called simply the mass. The concept of relativistic mass is not used in this text.

C. Remark that the definition of energy must be changed if the work-energy theorem is to hold for particles at high speeds. State that the relativistically correct expression for the energy of a free particle is $E = mc^2/\sqrt{1 - v^2/c^2}$. Take the limit as v/c becomes small and show that E can then be approximated by $mc^2 + \frac{1}{2}mv^2$. Thus the correct relativistic definition of the kinetic energy is $K = E - mc^2$. Point out that the particle has energy mc^2 when it is at rest and remark that mc^2 is called the rest energy.

D. Explain that mass and rest energy are not conserved in many interactions involving fundamental particles but that total energy E is; rest energy can be converted to kinetic energy and vice versa.

E. Derive $E^2 = (pc)^2 + (mc^2)^2$ and explain that this expression replaces $E = p^2/2m$ ($= mv^2/2$). Remark that $E = pc$ for a massless particle, such as a photon.

SUPPLEMENTARY TOPIC

The Doppler effect for light. The expression for the frequency transformation can be derived easily by considering the measurement of the period in two frames. Suppose an observer in S obtains T for the interval between successive maxima at the same place. This is a proper time interval and the interval in another frame S' is γT. If S' is moving parallel to the wave, however, the two events do not occur at the same place in S' and γT is not the period in that frame. An observer in S' must wait for a time $|\Delta x'|/c$ longer before the next maximum is reached at the place of the first. Thus $T' = \gamma T + |\Delta x'|/c$ or since $\Delta x' = -\gamma T v/c$, $T' = \gamma T(1 + v/c) = T\sqrt{(1+\beta)/(1-\beta)}$. Thus $f' = f\sqrt{(1-\beta)/(1+\beta)}$. If S' is moving perpendicularly to the wave the two events occur at the same place in both frames and $T' = \gamma T$, so $f' = f/\gamma$.

SUGGESTIONS

1. Use questions 1 and 2 to broaden the discussion of inertial frames.
2. Simultaneity and time measurements are the issues in questions 7, 8, and 9. Ask them to test understanding. Also assign problems 7 and 20.
3. Length contraction is the issue in question 10. Also consider question 11. Assign problems 11, 13, and 16.
4. Use questions 15 and 16 to broaden the discussion of mass and rest energy.
5. Assign problems 17 and 18 in support of the discussion of the Lorentz transformation. Assign problems 26, 28, and 32 in connection with the relativistic velocity transformation. Assign problems 33 and 37 in connection with the relativistic Doppler effect. Assign problems 43, 46, 47, and 53 in connection with relativistic energy and momentum. If you covered cyclotrons in Chapter 30, assign problem 57.
6. Computer project
   Have students write a computer program or design a spreadsheet to evaluate the Lorentz transformation equations. Then have them use it to investigate simultaneity, length contraction, and time dilation. Specific projects are given in the Computer Projects Section of this manual.
7. Computer programs
   a. *Spacetime Software*, Spacetime Software, 20 Davis Road, Belmont, MA 02178. Allows students to view events, motion of objects, and collisions from various reference frames. Includes interesting exercises and projects.

b. *Physics Simulations I: Einstein*, Kinko's Service Corporation, 4141 State Street, Santa Barbara, CA 93110. Macintosh. The screen is split to shown the views of events as seen in two frames which are moving relative to each other. Clocks time intervals between events. Use to demonstrate time dilation, length contraction, twin paradox.

c. *Intermediate Physics Simulations: Relativistic motion*, R.H. Good, Physics Department, California State University at Hayward, CA 94542. Apple II. User adjusts the velocity (in two dimensions) of a moving clock, which ticks at uniform intervals and lays down a marker at each tick. The screen shows the time in the observer's frame and in the rest frame of the clock. Use to demonstrate time dilation and length contraction.

Chapter 43 QUANTUM PHYSICS — I

BASIC TOPICS

I. Introduction.

A. Explain that this chapter deals with some of the fundamental results of quantum mechanics, as applied to electromagnetic radiation. The first few sections describe experimental results that can be understood only if light is regarded as made up of particles. Remark that interference and diffraction phenomena require waves for their explanation. Reconciliation of these opposing views will be discussed in the next chapter.

B. Explain that the energy of a photon is related to the frequency of the wave through $E = hf$ and the momentum of a photon is related to the wavelength of the wave through $p = h/\lambda$. Show these equations predict $p = E/c$, the classical relationship. Also explain that the energy density is nhf, where n is the photon concentration, and that the intensity is Rhf, where R is the rate per unit area with which photons cross a plane perpendicular to their direction of motion. Recall the discussion of the Poynting vector in Chapter 38. Explain that the Planck constant is a constant of nature and pervades quantum mechanics. Give its value (6.63×10^{-34} J·s) and calculate the photon energy and momentum for visible light, radio waves, and x-rays.

C. Point out that classically monochromatic electromagnetic radiation can have any value of energy. Quantum mechanically this is not true but since h is so small the discreteness of the energy values is important only at the atomic level.

II. The photoelectric effect.

A. Sketch a schematic of the experimental setup. Explain that monochromatic light is incident on a sample. It is absorbed and part of the energy goes to electrons, some of which are emitted. The energy of the most energetic electron is found by measuring the stopping potential V_0.

B. Point out that the stopping potential is independent of the light intensity. As the intensity is increased, more electrons are emitted but they are not more energetic. Show a plot of the stopping potential as a function of frequency and point out that the relationship is linear and that as the frequency is increased the electrons emitted are more energetic. Also state that electrons are emitted promptly when the light is turned on. If the radiation energy were distributed throughout the region of a wave it would take a noticeable amount of time for an electron to accumulate sufficient energy to be emitted, since an electron has a small surface area. This argument can be made quantitative (see Sample Problem 43–3).

C. Give the Einstein theory. Electromagnetic radiation is concentrated in photons, with each photon having energy hf. The most energetic electrons after emission are those with the greatest energy while in the material and, in the interaction with a photon, receive energy hf. If the light intensity is increased without changing the frequency, there are more photons and hence more electrons emitted, but no single electron can receive more energy. Furthermore, the electron receives energy immediately and need not wait to absorb the proper amount.

D. Show that this analysis leads to $hf = \phi + K_m$, where ϕ is the work function, the energy needed to remove the most energetic electron from the material. It is characteristic of the material. Remark that $K_m = eV_0$ and that the Einstein theory predicts a linear relationship between V_0 and f and predicts a minimum frequency for emission: $hf = \phi$. Remark that the emitted electrons have a distribution of speeds if $hf > \phi$ because they come from states with different energies.

III. The Compton effect.

A. Note that in the explanation of the photoelectric effect a photon is assumed to give up all its energy to an individual electron. The photon then ceases to exist. Explain that a photon might transfer only part of its original energy in an interaction with an electron. Since a lower energy means a lower frequency, the scattered light has a longer wavelength than the incident light.

B. Discuss the experiment. Light is scattered from electrons in matter and the intensity of the scattered light is measured as a function of wavelength for various scattering angles. Show Fig. 43–5. Stress that the experimental data can be explained by considering the interaction to be a collision between two particles, with energy and momentum conserved. Relativistic expressions, however, must be used for energy and momentum.

C. Remark that the situation is exactly like a two-dimensional collision between 2 particles. Write down the relativistic expressions for the momentum and energy of a particle with mass (the electron) and remind students of the rest energy. Assume the electron is initially at rest and that the photon is scattered through the angle ϕ. The electron leaves the interaction at an angle θ to the direction of the incident photon. Write down the equations for the conservation of energy and the conservation of momentum in two dimensions. Write down the momentum and energy of the photon in terms of the wavelength and solve for the change on scattering of the wavelength.

D. Note that the change in wavelength is independent of wavelength and that the change is significant only for short wavelength light, in the x-ray and gamma ray regions. Also state that the theoretical results successfully predict experimental data. The widths of the curves are due chiefly to moving electrons, for which $\Delta\lambda$ is slightly different, and the peak near $\Delta\lambda = 0$ is due to scattering from more massive particles (atoms as a whole). Stress that the particle picture of light accounts for experimental data.

IV. Cavity radiation and the quantization of energy.

A. Describe a radiation cavity as a hollow block of material with a small hole to the outside. The material is kept at a uniform constant temperature and the electromagnetic radiation in the cavity is studied by observing some that leaks out of the hole. Explain that the interior walls absorb and emit radiation and that the distribution of energy among the various wavelengths depends on the temperature but is independent of the material, the size of the cavity, and the shape of the cavity. Write down the Stefan-Boltzmann law for the radiant intensity: $I(T) = \sigma T^4$ and give the value of the Stefan-Boltzmann constant ($5.670 \times 10^{-8}\,\mathrm{W/m^2 \cdot K^4}$). Remind students of the absolute temperature scale.

B. The quantity of interest is the spectral radiancy $S(\lambda)$, defined so that $S(\lambda)\,d\lambda$ is the radiation rate per unit area for electromagnetic energy in the wavelength range from λ to $\lambda + d\lambda$. Remark that $I(T) = \int_0^\infty S(\lambda)\,d\lambda$. Explain that the electromagnetic energy in the cavity is in thermal equilibrium with the material and the spectral distribution should be predicted when thermodynamics is applied to the radiation. Show Fig. 43–8, which compares experimental results with classical theory. Point out that the experimental curve reaches a peak in the infrared or red and falls off on either side. As the temperature increases the peak becomes sharper and moves toward the blue end of the spectrum. Classical theory does not predict a peak. Write down the Wien displacement law: $\lambda_{max}T = 2898\,\mu\text{m}\cdot\text{K}$.

C. Give the Planck law and explain that it fits the data at all wavelengths. Describe the assumptions used by Planck to derive the law: radiation is emitted and absorbed by atoms in the walls and the energy absorbed or emitted is quantized. Thus the radiation energy in the cavity is also quantized, in units of hf, where f is the frequency of the radiation. The law is derived by assuming the quanta of radiation are in thermodynamic equilibrium with the cavity walls. Explain that in the current view all electromagnetic energy is quantized.

D. Briefly discuss the correspondence principle: quantum mechanical results reduce to classical results when the situation is such that classical results are valid. Show that the Planck expression for $R(\lambda)$ reduces to the classical expression if $hc/\lambda kT << 1$.

V. The hydrogen atom and line spectra.

A. Use a commercial hydrogen tube to show the visible hydrogen spectrum. Since the intensity is low you will not be able to project this but you can purchase inexpensive $8'' \times 10''$ sheets of plastic grating material, which can be cut into pieces and passed out to the students. Point out Fig. 43–10.

B. Explain that an electron bound in an atom, molecule, or macroscopic matter can have an energy which is one of a discrete set of values, not a continuum. The energy of any bound particle is quantized. It radiates or absorbs energy by changing its state and the frequency of the radiation is given by $hf = |E_f - E_i|$. Show Fig. 43–9.

C. Give the expression $E = -me^4/8\epsilon_0^2h^2n^2$ for the energy levels of a hydrogen atom. Show Fig. 43–13 and point out the Lyman, Balmer, and Paschen series. Develop the formula $1/\lambda = R(1/\ell^2 - 1/u^3)$ for the wavelengths of emitted radiation. Here ℓ is the quantum number of the lower energy level and u is the quantum number of the upper. R is the Rydberg constant, given by $R = me^4/8\epsilon_0^2h^3c = 1.097 \times 10^7\ \text{m}^{-1}$. Remind students of the lines they saw and explain that $\ell = 2$ for the visible lines. For the red line $u = 3$ and for the bluest visible line $u = 6$.

D. Point out that classical physics predicts a continuous spectrum. Since the electron is accelerating as it goes around the nucleus, it should radiate continuously and fall into the nucleus. Stress that it does not, the atom is stable, and the spectrum is discrete. Quantum mechanically the atom does not radiate as long as it remains in a state. Radiation occurs only when the atom makes a transition to a lower energy state.

E. (Optional). Go over the Bohr derivation of the hydrogen atom energies, as in Section 43–10.

SUGGESTIONS

1. Emphasize that the energy in a light beam is the product of the number of photons and the energy of each photon. Assign some of problems 8, 9, 10, and 11. Ask questions 1, 2, 3, and 4 as part of a discussion of photon properties.
2. After discussing the photoelectric effect, ask questions 9, 10, 12, and 14. Assign problems 24 and 26. After covering the Compton effect you may wish to go over problem 28.
3. After discussing the Compton effect, ask questions 15, 16, and 17. Also assign problem 31. Consider problems 37 and 39.
4. To help in understanding the Planck radiation law, assign problem 42 or go over it in class, then ask the students to do problem 47. Also consider problems 49 and 50.
5. After discussing the hydrogen spectrum, ask questions 22, 23, and 27. Also assign problems 58 and 63. If you have emphasized the terms *binding energy* and *excitation energy*, assign problem 68. Problem 76 is a good review of the Bohr theory. Assign it if you covered Section 10.

6. Question 28 is a good review question. Ask it when the chapter has been studied.
7. A commercial math program or a student-generated root-finding program can be used to solve the equations for the photoelectric and Compton effects. Students may be interested, for example, in seeing how the Compton lines broaden when the electrons are not initially at rest. Assign some exercises as homework or set aside some laboratory time for a more detailed investigation.
8. Demonstrations
   Photoelectric effect: Freier and Anderson MPb1; Hilton A4a, b.
9. Computer program
   Atoms and Matter, Focus Media, Inc., 839 Stewart Avenue, Garden City, NY 11530. Apple II. A series of programs that simulate various modern experiments. One plots radiative intensity vs. frequency for a blackbody at a temperature chosen by the user. This display can be used to illustrate the lecture. The tutorial material can be used by the students. Reviewed TPT December 1986.

10 Laboratory
   a. MEOS Experiment 14–2: *The Photoelectric Effect.* Students investigate the characteristics of various photocells, then use a plot of stopping potential vs. frequency to determine the Planck constant. A mercury source and optical filters are used to obtain monochromatic light of various frequencies.
   b. MEOS Experiment 14–3: *Analysis of Spectra.* A spectroscope is used to obtain the wavelengths of hydrogen and helium lines. Hydrogen lines are compared with predictions of the Balmer equation.

Chapter 44 QUANTUM PHYSICS — II

BASIC TOPICS

I. Matter waves.
   A. Explain that electrons and all other particles have waves associated with them, just as photons have electromagnetic waves associated with them. State that the waves exhibit interference and diffraction effects. Draw a diagram of a single-slit barrier with a beam of monoenergetic electrons incident on it and a fluorescent screen or other mechanism for detecting electrons behind it. Explain that an intense central maximum is obtained and that many electrons arrive in this region. Secondary maxima are also obtained.
   B. State that the width of the central maximum depends on the speed of the electrons and narrows if the speed is increased. The maximum also narrows if more massive particles are used, at the same speed. Remind students that when they studied the single-slit diffraction of electromagnetic waves they found the width of the central maximum narrowed as the wavelength decreased. Conclude that the momentum of the particle is related to the wavelength of the wave and that one is proportional to the reciprocal of the other.
   C. State that the particle energy and the wave frequency are related by $E = hf$ and that the particle momentum and the wavelength are related by $p = h/\lambda$. Calculate the wavelengths of a 1-eV electron and a 35-m/s baseball.
   D. By way of example, state that crystals diffract electrons of appropriate wavelength ($\approx 10^{-10}$ m) and the angular positions of the scattering maxima can be found using Bragg's law, suitably modified to account for changes in the propagation direction that occur when matter waves enter the crystal.

II. The wave function.

A. State that a one-dimensional matter wave is denoted by $\psi(x)$ and that $|\psi|^2$ gives the probability density for finding the particle near x. Similarly, E^2 is proportional to the probability density for finding a photon. In the limit of a large number of particles $|\psi|^2$ is proportional to the particle concentration. Explain that, at the atomic and particle level, physics deals with probabilities. What can be analyzed is the probability for finding a particle, not its certain position.

B. Explain that, for a particle confined to the region between 0 and L on the x axis, possible wave functions are given by $\psi_n(x) = A\sin(n\pi x/L)$, where $n = 1, 2, \ldots$. These are standing waves and vanish at $x = 0$ and $x = L$. You might want to include the time dependence by writing $\Psi = A\sin(n\pi x/L)f_n(t)$ and explaining that $f_n(t)$ is a function of time with magnitude 1. The wavelength of one of the traveling waves in the standing wave is $\lambda = 2L/n$ and the particle momentum has magnitude $nh/2L$. Use $E = p^2/2m$ to show that $E_n = n^2h^2/8mL^2$ gives the quantized energy levels. Explain that confinement of the particle leads to energy quantization and that energy is quantized for any bound particle.

C. Explain that the particle is certainly between $x = 0$ and $x = L$, so $\int_0^L |\psi_n|^2\,dx = 1$. The wave function is said to be *normalized* if it obeys this condition. Show that the normalization condition leads to $A = \sqrt{2/L}$ for a particle in a box.

D. Use the particle confined to a one-dimensional box as an example and explain that $\psi_n^2\,dx = (2/L)\sin^2(n\pi x/L)\,dx$ gives the probability that the particle will be found between x and $x+dx$ when it is in the state with the given wave function. Sketch several of the probability density functions and point out that there are several places where the probability density vanishes. See Fig. 44–10.

E. Explain that experimentally the probability can be found, in principle, by performing a large number of position measurements and calculating the fraction for which the particle is found in the designated segment of the x axis. Since a position measurement changes the state of the particle, it must be restarted in the same state each time.

III. The hydrogen atom.

A. Graph the potential energy as a function of distance from the proton for the electron in a hydrogen atom and explain that the electron is bound. Give the expression for the energy levels in terms of the principal quantum number. State that quantum mechanics predicts these allowed values.

B. Give the ground state wave function and obtain the expression for the probability density. Show that the volume of a spherical shell with thickness dr is $4\pi r^2\,dr$ and define the radial probability density as $P(r) = 4\pi r^2|\psi|^2$. Sketch $P(r)$ for the ground state and point out there is a range of radial distances at which the electron might be found. Contrast with the Bohr model. Locate the most probable radius and the average radius.

IV. Barrier tunneling.

A. Show Fig. 44–13 and explain that the wave function penetrates a finite barrier. It is oscillatory (in position) outside the barrier, where $E > V$, and exponential inside, where $E < V$. The figure shows the probability density.

B. Explain that the particle has a probability of being found on either side of the barrier. Contrast to the behavior of a classical particle.

C. Write down Eqs. 44–20 and 21 for the transmission coefficient and explain that this measures the probability of transmission through the barrier. Remark that transmission is small for high, wide barriers and becomes larger as the barrier height decreases and as the barrier width narrows. Also define the reflection coefficient R by $R = 1 - T$.

V. The uncertainty principle.

A. Because a different answer might result each time the position of the electron is measured, there is an uncertainty in the position. It can be defined similarly to the standard deviation of a large collection of experimental results. Similar statements can be made about momentum measurements. Explain that the uncertainties in position and momentum are both determined by the particle wave function. Explain that if the electron is placed in a state for which the uncertainty in position is small then the uncertainty in momentum is large and vice versa.

B. As an example, explain that for a particle in a one-dimensional infinite square well the uncertainty in position is about L and the uncertainty in momentum is about $2p = nh/L$. Remark that the uncertainty in position can be reduced by reducing L and that this increases the uncertainty in momentum. Show that $(\Delta x)(\Delta p) = nh$.

C. State the uncertainty principle: $(\Delta x)(\Delta p_x) \approx h$ and give similar expressions for the other cartesian components. Explain that Δx and Δp_x can be changed by changing the state (wave function) of the particle, but that their product is always greater than h.

D. Draw a barrier with a single slit and ask students to consider a plane electron wave impinging on the barrier from one side. Explain that diffraction takes place and sketch a graph of the probability density on a screen placed on the other side of the barrier. Mention that an electron picks up a component of momentum parallel to the slit and consequently may arrive at a point on the screen that is not directly behind the slit. Remark that the electron has been localized to within the slit at the time of passage but, as a result, its momentum is uncertain and so its position of arrival at the screen is uncertain. Remind students that if the slit is narrowed the diffraction pattern flares. The uncertainty in position has been decreased but the uncertainty in momentum has been increased.

E. Discuss the energy-time uncertainty relationship: $\Delta E \cdot \Delta t \approx h$. Show how it can be applied to the lifetime of an electron in an excited state. Because the lifetime is not infinite the uncertainty in the energy of the state cannot be zero.

V. Waves and particles.

A. Photons, electrons, and other quantum entities have both particle and wave properties and these complement each other. Explain that particle properties are detected in some experiments and wave properties in others.

B. Consider electrons incident on a double-slit barrier and explain that an interference pattern is obtained even if only one electron is in the system at any time. Now suppose a particle detector is placed at each slit and assume each electron is detected as it passes through one slit or the other. Graph the intensity pattern at the screen and remark that interference effects are missing. Explain that the act of detecting a particle collapses the wave function so it is transmitted through one slit only (the slit at which the particle was detected).

SUGGESTIONS

1. In the discussion of wave-particle duality, include questions 1 and 2.
2. Include questions 4, 5, 7, and 9 in a discussion of de Broglie wavelength. Also assign problems 10 and 16.
3. After discussing a particle confined between rigid walls, ask questions 14 and 15. Assign problems 21, 23, and 25. Assign problem 29 for the better students. To emphasize probability densities, ask questions 13 and 23.
4. Discuss problems 33 and 34 in connection with the ground state of a hydrogen atom.
5. Ask questions 17 and 18 in connection with tunneling. Also assign problems 37 and 38.

6. Following the discussion of the uncertainty principles, ask questions 24 and 25. Assign problems 42 and 47.
7. Films
   a. *Matter Waves*, 16 mm, b/w, 28 min. Modern Learning Aids, Division of Ward's Natural Science Establishment, P.O. Box 312, Rochester, NY 14601. A comparison of electron and optical diffraction patterns are shown in a modern version of the original experiment that showed the wave nature of the electron. Reviewed AJP 33:63 (1965).
   b. *Electron Diffraction*; Films for the Humanities and Sciences, Inc., Box 2053, Princeton, NJ 08543. Experimental setup is explained and data taken. The analysis is left for the student. Nice demonstration of how powder patterns are formed by a collection of randomly oriented crystals. Reviewed TPT 27, 541 (November 1989).
8. Demonstrations
   Electron diffraction: Hilton 13b.
9. Laboratory
   MEOS Experiment 14–5: *Electron Diffraction.* The Sargent-Welch electron diffraction apparatus is used to investigate the diffraction of electrons by aluminum and graphite. Since powder patterns (rings) are obtained you will need to explain their origin.

Chapter 45 ALL ABOUT ATOMS

BASIC TOPICS

I. The Schrodinger equation and the structure of hydrogen.
   A. Remind students of the hydrogen spectrum and write the expression for the allowed values of the energy. Tell students that the integer n that appears is called the principal quantum number.
   B. Explain that the Schrodinger equation is a differential equation for the wave function of a particle and that the main ingredient that causes two identical particles to have different wave functions is their potential energy function. For an electron in a hydrogen atom the potential energy function is $U(r) = -e^2/4\pi\epsilon_0 r$, where r is the distance from the proton to the electron. Mention that when this potential energy function is used in the Schrodinger equation and the reasonable condition that the wave functions remain finite everywhere is applied, then the allowed energy values are predicted. Draw a graph of $U(r)$ and draw lines across it to indicate the values of the first few energy levels.
   C. Say that for a given allowed energy value the Schrodinger equation may predict several wave functions and that different wave functions are associated with different angular momentum states.

II. Orbital and spin angular momentum.
   A. Remark that orbital angular momentum is quantized and that the allowed values of its magnitude are given by $L = \sqrt{\ell(\ell+1)}\hbar$, where $\hbar = h/2\pi$. The orbital quantum number ℓ can take on the values 0, 1, 2, ..., $n-1$ for a given value of n. Emphasize that for a hydrogen atom the $n = 2$, $\ell = 0$ and $n = 2$, $\ell = 1$ states, for example, have the same energy but different angular momenta.
   B. State that the z component of the angular momentum is given by $L_z = m_\ell \hbar$, where $m_\ell = 0, \pm 1, \pm 2, \ldots, \pm \ell$. m_ℓ is called the magnetic quantum number. The z axis can be in any direction, perhaps defined by an external magnetic field. Point out that the angle θ between the angular momentum vector and the z axis is given by $\cos\theta = m_\ell/\sqrt{\ell(\ell+1)}$. The smallest value of θ occurs when $m_\ell = \ell$ and it is not zero. Explain that the angles

L makes with the x and y axes cannot be known if the angle between **L** and the z axis is known. Discuss this in terms of the precession of **L** about the z axis.

C. Explain that the electron and some other particles have intrinsic angular momentum, as if they were spinning on axes. The magnitude of the electron spin angular momentum is $\sqrt{3/4}\hbar$ and the z component is either $-\frac{1}{2}\hbar$ or $+\frac{1}{2}\hbar$ (there are two possible states). You might want to remark that spin is not predicted by the Schrodinger equation but that it is predicted by relativistic modifications to quantum mechanics.

III. Magnetic dipole moments.

A. Explain that the electron has a magnetic dipole moment because of its orbital motion and write down $\boldsymbol{\mu}_\ell = -(e/2m)\mathbf{L}$ and $\mu_{\ell z} = -(e/2m)L_z = -(e\hbar/2m)m_\ell$. Give the value of the Bohr magneton ($\mu_B = e\hbar/2m = 9.28 \times 10^{-24}$ J/T). Remind students that because of its motion the electron experiences a torque in an external magnetic field and produces its own magnetic field (provided $m_\ell \neq 0$).

B. State that the spin magnetic moment is $\mu_{sz} = -2m_s\mu_B$. Stress the appearance of the factor 2. The electron produces a magnetic field and experiences a torque in a magnetic field because of this moment.

C. Remark that the energy of an electron is changed by$-\mu_z B$ when an external field **B** is applied in the positive z direction. Thus states with the same n but different m_ℓ have different energies in a magnetic field. State that this is called the Zeeman effect. Photons with an energy equal to the energy difference of the two spin states cause the spin to flip. The phenomenon can be detected by measuring the absorption of the beam.

D. Briefly describe the Stern-Gerlach experiment. Explain that a magnetic dipole in a *non-uniform* magnetic field experiences a force and that $F_z = \mu_z \, dB/dz$ for a field in the z direction that varies along the z axis. Atoms with different values of m_ℓ experience different forces and arrive at different places on a screen. That discrete regions of the screen receive atoms is experimental evidence for the quantization of the z component of angular momentum.

E. To emphasize the practical, qualitatively explain NMR and its use in diagnostic medicine. You might also explain how local magnetic fields in solids, for example, can be measured using magnetic resonance techniques.

IV. Hydrogen wave functions.

A. Explain that states for hydrogen are classified using 4 quantum numbers:

1. The principal quantum number n, which determines the energy.
2. The orbital quantum number ℓ, which determines the magnitude of the orbital angular momentum.
3. The magnetic quantum number m_ℓ, which the determines the z component of the orbital angular momentum.
4. The spin quantum number m_s, which determines the z component of the spin angular momentum..

B. Explain that traditionally each value of n is said to label a shell and the shells are named $K, L, M, N, \ldots$, in order of increasing n. Remark that a shell may consist of many states, but each is associated with the same value of the energy.

C. Remind students that for a given shell ℓ may take on the values $0, 1, 2, \ldots, n-1$. There are n different values in all. Explain that all the states with given values of n and ℓ are said to form an orbital. Remind students that for a given value of ℓ, m_ℓ may take on any integer value from $-\ell$ to $+\ell$, $2\ell + 1$ values in all. Since m_s can have either of 2 values, an orbital consists of $2(2\ell + 1)$ states. Either state or prove that the shell with principal quantum number n has $2n^2$ states.

D. List all the states for $n = 1$, 2, and 3. Group them according to n and remark that all states with the same n have the same energy, all states with the same ℓ have the same magnitude of orbital angular momentum, and all states with the same m_ℓ have the same z component of orbital angular momentum. Remark that states with different values of n, ℓ, and m_ℓ have different wave functions.

E. Give the ground state wave function and obtain the expression for the probability density. Remark that ψ has spherical symmetry and explain that this is true of all $\ell = 0$ wave functions. Show that the volume of a spherical shell with thickness dr is $4\pi r^2\,dr$ and define the radial probability density as $P = 4\pi r^2|\psi(r)|^2$. Sketch P for the ground state (Fig. 45–7*a*) and point out there is a range of radial distances at which the electron might be found. Contrast this with the Bohr model. Locate the most probable radius and the average radius.

F. Show a graph of P for $n = 2$, $\ell = 0$ (Fig. 45–7*b*). Write down the expression for P and point out that it is spherically symmetric. Note that the average radius and most probable radius both increase with n.

V. X rays and the numbering of the elements.

A. Explain that x rays are produced by firing energetic electrons into a solid target. Show Fig. 45–15 and point out the continuous part of the spectrum and the peaks. Also point out that there is a sharply defined minimum wavelength to the x-ray spectrum. Explain that the continuous spectrum results because the electrons lose some or all of their kinetic energy in close (decelerating) encounters with nuclei. This energy appears as photons and $\Delta K = hf$. Explain that a photon of minimum wavelength is produced when an electron loses all its kinetic energy in a single emission. Derive the expression for the minimum wavelength in terms of the original accelerating potential and point out it is independent of the target material.

B. Explain that the line spectrum in Fig. 45–15 appears because incident electrons interact with atomic electrons and knock some of the deep-lying electrons out of the atoms. Electrons in higher levels drop to fill the holes, emitting photons with energy equal to the difference in energy of the initial and final atomic levels. The K_α line is produced when electrons drop from the L ($n = 2$) shell to the K ($n = 1$) shell and the K_β line is produced when electrons drop from the M ($n = 3$) shell to the K shell. Explain Fig. 45–17.

C. Show Fig. 45–18 and state that when the square root of the frequency for any given line is plotted as a function of the atomic number of the target atom, the result is nearly a straight line. Argue that the innermost electrons have an energy level scheme close to that of hydrogen but with an effective nuclear charge of $(Z - 1)e$, where the 1 accounts for screening by electrons close to the nucleus. Z is the number of protons in the nucleus, the atomic number. Use the expression for hydrogen energy levels and for K_α put $n = 2$ for the initial state and $n = 1$ for the final state, then show that $\sqrt{f}$ is proportional to $(Z - 1)$.

D. Remark that this relationship was used to position the chemical elements in the periodic table independently of their chemical properties. This technique was particularly important for elements in the long rows of the periodic table, which contain many elements with similar chemical properties. Today the technique is used to identify trace amounts of impurities in materials.

VI. Atom building and the periodic table.

A. Give the "rules" for atom building.

1. The 4 quantum numbers n, ℓ, m_ℓ, m_s can be used to label states. They have the same restrictions on their values as for hydrogen. Remark that wave functions and energies are different for electrons with the same quantum numbers in different atoms. Also remark that the energy depends on ℓ.

2. No more than one electron can have any given set of quantum numbers. This is a general principle of quantum mechanics, called the Pauli exclusion principle.

B. Explain that as more protons are added to the nucleus the electron wave functions pull in toward regions of low potential energy. This and the dependence of the energy on ℓ means that states associated with one principal quantum number may not be filled before states associated with the next principal quantum number are started. For example, a 5*s* state is lower in energy than a 4*d* state, in different atoms. It also accounts for the fact that all atoms are nearly the same size.

C. Show a periodic table. Point out the inert gas atoms and explain they all have filled shells. Point out the alkali metal and alkaline earth atoms and state they have one and two electrons, respectively, outside closed shells. Remark that electrons in partially filled shells are chiefly responsible for chemical activity. Point out the atoms in which d and f states are being filled and finally those in which p states are being filled.

VII. The laser.

A. List the characteristics of laser light: monochromatic, coherent, directional, can be sharply focused. See the text for quantitative comparisons with light from other sources.

B. Explain the mechanism of light absorption: an incident photon is absorbed if hf corresponds to the energy difference of two electron states of the material and the upper state is initially empty. An electron makes the jump from the lower to the upper state. Explain spontaneous emission: an electron spontaneously (without the aid of external radiation) makes the transition from one state to a lower state (if that state is empty) and a photon with hf equal to the energy difference is emitted. Emphasize that in most cases the electron remains in the upper state for a time on the order of 10^{-9} s but that there are metastable states in which the electron remains for a longer time ($\approx 10^{-3}$ s). Explain stimulated emission: with the electron in an upper state an incident photon with the proper energy can cause it to make the jump to a lower state. The result is two photons of the same energy, moving in the same direction, with waves having the same phase and polarization. Remark that laser light is produced by a large number of such events, each triggered by a photon from a previous event. Hence all laser photons are identical. Explain that metastable states are important since the electron must remain in the upper state until its transition is induced. Compare with light produced by random spontaneous transitions.

C. Explain that, in thermodynamic equilibrium, upper levels are extremely sparsely populated compared to the ground state. To obtain laser light the population of an upper level must be increased; otherwise absorption events would equal or exceed stimulated emission events. A laser must be pumped. Write down the expression for the thermal equilibrium number of atoms in the state with energy E: $n(E) = Ce^{-E/kT}$ Explain that C is independent of energy but depends on the number of atoms present. State that the temperature T is on the Kelvin scale.

D. Use Fig. 45–21 to describe the three level laser. First describe the equilibrium distribution of electrons among the states, then describe pumping from the ground state to the highest level and the fast decay to the metastable state. Describe the distribution when the population has been inverted. Finally describe the stimulation of emission and the build-up of the number of photons with the same energy and phase.

E. Discuss the helium-neon laser, paying particular attention to the roles of the walls and mirror ends. Go over the four characteristics of laser light discussed earlier and tell how each is achieved.

SUGGESTIONS

1. To emphasize the difference between the Bohr and quantum models of the hydrogen atom and

to get students to think about wave functions, ask questions 1, 2, and 13.

2. To test for understanding of the angular momentum quantum numbers, assign some of problems 3, 5, 7, and 8. To stress the connection between angular momentum and magnetic dipole moment, assign problems 10 and 11. To discuss the Stern-Gerlach experiment in more detail, include questions 14, 15, and 17. Also assign problems 31 and 32.
3. When you discuss the enumeration of hydrogen states assign problem 9. Hydrogen wave functions and radial probability densities are covered in problems 20, 22, and 26. Ask one or more of these.
4. Use problem 40 to test for understanding of the Pauli exclusion principle. To emphasize the role played by spin in the building of the periodic table, ask questions 20 and 23. Also include questions 19 and 22 in the discussion of the periodic table and assign problem 39.
5. The existence of a minimum wavelength in the continuous x-ray spectrum provides an argument for the particle nature of light. Either discuss this or see if the students can devise the argument. Assign question 24 and problem 46. After discussing characteristic x-ray lines and Moseley plots, ask questions 26, 28, and 31. Assign problems 50 and 55.
6. Ask questions 33 and 34 to see if students understand the properties of laser light. Ask question 37 to test for understanding of the laser mechanism. Also assign problems 65 and 70.
7. Demonstrations
   a. Thompson and Bohr models of the atom: Hilton A5a, b.
   b. Zeeman effect: Freier and Anderson MPc1; Hilton A20a.
   c. X-ray apparatus: Hilton A7a, b, c.
   d. Lasers: Hilton A12.
8. Audio/Visual
   a. A.F. Burr and Robert Fisher; *Electron Distribution in the Hydrogen Atom*; American Association of Physics Teachers slide set; Publications Sales, AAPT Executive Office, 5112 Berwyn Road, College Park, MD 20740-4100. Probability distributions for n =1 to n = 6.
   b. *Introduction to Lasers*, 16 mm, color film, 17 min., Encyclopaedia Britannica Educational Corporation, 425 N. Michigan Ave., Chicago, IL 60611. This film discusses the development and uses of the laser. Three of the principal developers are featured. The phenomenon of lasing is illustrated through animated sequences and laboratory demonstrations. Reviewed AJP 42:525 (1974).
9. Computer program
   Animation Demonstration: Electron Waves in an Atom, CONDUIT, The University of Iowa, Oakdale Campus, Iowa City, IA 52242. Compares classical electron orbits and quantum wave patterns. Shows quantization of orbits by applying boundary conditions and simulates radiative transitions. Reviewed TPT November 1986

Chapter 46 THE CONDUCTION OF ELECTRICITY IN SOLIDS

BASIC TOPICS

I. Electron energy bands.
   A. Explain that energy levels for electrons in crystalline solids are grouped into bands with the levels in any band being nearly continuous and with gaps of unallowed energies between. Remark that bands are produced when atoms are brought close together. Wave functions then overlap and extend throughout the solid. Show Fig. 46–3 and remark that low energy bands are narrow since the wave functions are highly localized around nuclei and overlap

is small. High energy bands are wide because overlap is large. When the atoms are close together outer shell electrons are influenced by many atoms rather than just one.

B. Remind students that since the Pauli exclusion principle holds the lowest total energy is achieved when electrons fill the lowest states with one electron in each state. Thus at $T = 0\,\text{K}$ all states are filled up to a maximum energy.

C. Remark that for a metal the highest occupied state is near the middle of a band, while for an insulator or semiconductor it is at the top of a band.

D. Write down the Fermi-Dirac probability function $p(E)$, given by Eq. 46–7, and state that it gives the thermodynamic probability that a state with energy E is occupied. Show that for $T = 0\,\text{K}$, $p(E) = 1$ for $E < E_F$ and $p(E) = 0$ for $E > E_F$. To give a numerical example, calculate the probabilities of occupation for states 0.1 and 1 eV above the Fermi energy, then 0.1 and 1 eV below, at room temperature. Graph E vs. $p(E)$ for $T = 0$ and for $T > 0$. See Figs. 46–6*b* and 7*b*. Also show the graph for a still higher temperature and point out that the central region (from $p = 0.9$ to $p = 0.1$, say) widens. This quantitatively describes the thermal excitation of electrons to higher energy states. Remark that the Fermi-Dirac probability function is valid for any large collection of electrons, including the collections in metals, insulators, and semiconductors.

II. Metallic conduction.

A. Write down Eq. 46–1 for the resistivity and remark that n is the concentration of conduction electrons and τ is the mean time between collisions of electrons with atoms. Ask students to review Section 28–6. Remark that a low resistivity results if the electron concentration is large or the mean free time is long. In a rough way, if there are few collisions per unit time then the mean free time is long and the electrons are accelerated by the electric field for a long time before colliding, so the drift velocity is large. Remark that quantum mechanics must be used to determine n and τ.

B. Explain that for metals the energies of conduction electrons (those in partially filled bands) are primarily kinetic and to a first approximation we may take the electrons to be trapped in a box the size of the sample. The so-called free electron model of a metal takes the potential energy to be zero in the box.

C. Define the density of states function $n(E)$ and the density of occupied states function $n_0(E)$. Explain that $n_0(E) = n(E)p(E)$ and that the total electron concentration in a metal is given by $n = \int n(E)p(E)\,\mathrm{d}E$. In principle, this equation can be solved for the Fermi energy as a function of temperature. State that for nearly free electrons in a metal $n(E)$ is given by Eq. 46–3 and that the Fermi level is given by Eq. 46–6. Evaluate the expression for copper and show that E_F is about 7 eV above the lowest free electron energy. Strictly this is the result for $T = 0$ but the variation of E_F and n with temperature is not important in a first approximation for metals.

D. Explain that the electric current is zero when no electric field is present because states for which the velocities are $+\mathbf{v}$ and $-\mathbf{v}$, for example, have the same energy. If one is filled then so is the other. Thus the average velocity of the electrons vanishes. A current arises in an electric field because the electrons accelerate: they tend to make transitions within their band to other states such that the changes in their velocities are opposite to the field.

E. Explain that the acceleration caused by an electric field does not continue indefinitely because the electrons are scattered by atoms of the solid. As a result, the electron distribution distorts only slightly. Some states with energy slightly greater than E_F and velocity opposite the field become occupied while some states with energy slightly less than E_F and velocity in the direction of the field become vacant. Electrons with energy E_F have speeds v_F given by $E_F = \frac{1}{2}mv_F^2$ but the average speed (the drift speed) is considerably less because most electrons can be paired with others moving with the same speed in the opposite direction.

F. Explain that a steady state is reached and that the drift velocity is then proportional to the applied electric field. Only electrons near the Fermi energy suffer collisions and the additional velocities they obtain from the field between collisions are insignificant compared to their velocities in the absence of the field. Thus the mean free time is essentially independent of the field and Ohm's law is valid.

G. State that electrons in a perfectly periodic lattice do not suffer collisions, a result that is predicted by quantum mechanics. Collisions with the atoms occur because they are vibrating. Collisions also occur if the solid contains impurities or other imperfections. As the temperature increases vibrational amplitudes of the atoms increases and so does the number of collisions per unit time. As a result, the mean free time becomes smaller. This explains the increase with temperature in the resistivity of a metal.

III. Insulators and semiconductors.

A. Explain that a filled band cannot contribute to an electric current because the average electron velocity is always zero, even in an electric field. State that insulators and semiconductors have just the right number of electrons to completely fill an integer number of bands and that, in the lowest energy state, all bands are either completely filled or completely empty. For metals, on the other hand, the highest occupied state is near the middle of a band. Metals always have partially filled bands. Show Fig. 46–8 and identify the valence and conduction bands for an insulator.

B. Explain that as the temperature is raised from $T = 0\,\mathrm{K}$ a small fraction of the electrons in the valence band of an insulator or semiconductor are thermally excited across the gap into the conduction band. For a semiconductor the gap is small (about $1\,\mathrm{eV}$) and at room temperature both bands can contribute to the current. The conductivity, however, is still small compared to that of a metal. For an insulator the gap is large (more than $5\,\mathrm{eV}$), so the number of promoted electrons is extremely small and the current is insignificant for laboratory fields. Explain that silicon and germanium are the only elemental semiconductors although there are many semiconducting compounds. Carbon is a prototype insulator, with a gap of $5.5\,\mathrm{eV}$. Compare with silicon, which has a gap of $1.1\,\mathrm{eV}$. Resistivities of metals and semiconductors are compared in Table 46–1.

C. When electrons are promoted across the gap they contribute to the current in an electric field. The valence band becomes partially filled and electrons there also contribute. It is usually convenient to think about the few empty states in this band rather than the large number of electrons there. These are called holes and behave as if they were positive charges. In contrast to electrons, holes drift in the direction of $\mathbf{E}$. Compare the carrier concentrations of metals and semiconductors at room temperature. See Table 46–1.

D. Explain the different signs for the temperature coefficients of resistivity, also given in Table 46–1. Explain that for both metals and semiconductors near room temperature the mean free time decreases with increasing temperature. For metals the electron concentration is essentially constant but for semiconductors n increases dramatically with temperature as electrons are thermally promoted across the gap. This effect dominates and the resistivity of an intrinsic semiconductor decreases with increasing temperature.

E. Explain that the proper kind of replacement atoms (donors) can increase the number of electrons in the conduction band and another kind (acceptors) can increase the number of holes in the valence band. They produce n and p type semiconductors, respectively. By considering the number of electrons in their outer shells explain why phosphorus is a donor and aluminum is an acceptor. Point out that wave functions for impurity states are highly localized around the impurity and so do not contribute to the conductivity. Go over Sample Problem 46–5, which shows that only a relatively small dopant concentration can increase the carrier concentration enormously. Doped semiconductors are used in nearly all semiconducting devices.

IV. Semiconducting devices.

A. Show a commercial junction diode and draw a graph of current vs. potential difference (Fig. 46–13). Include both forward and back bias. Explain that it is a rectifier, with high resistance for current in one direction and low resistance for current in the other direction. Demonstrate the *i-V* characteristics by placing a diode across a variable power supply and measuring the current for various values of the potential. Reverse the potential to show the rectification.

B. Describe a *p-n* junction and remark that the diffusion of carriers leaves a small depletion region, nearly devoid of carriers, straddling the metallurgical junction. Explain the origin of the electric field in the depletion region and the origin of the contact potential. Stress that the field is due to uncovered impurity atoms, positive donors on the *n* side and negative acceptors on the *p* side.

C. Describe a diffusion current as one that arises because particles diffuse from regions of high concentration toward regions of low concentration. Explain that this motion results from the random motion of the particles. More particles leave a high concentration region simply because there are more particles there, not because they are driven by any applied force. State that the diffusion current for both electrons and holes in an unbiased *p-n* junction is from the *p* to the *n* side, against the contact electric field. Point out that the drift current is from the *n* toward the *p* side and that the diffusion and drift currents cancel when no external field is applied. Point out the depletion zone and the currents on Fig. 46–12.

D. Draw a circuit with a battery across a *p-n* junction, the positive terminal attached to the *n* side. Explain that this is a back bias. The internal electric field is now larger, the barrier to diffusion is higher, and the reverse current is extremely small. See Fig. 46–15(*a*). Also explain that the width of the depletion zone is increased by application of a reverse bias.

E. Draw the circuit for forward bias. The internal electric field is now smaller, the barrier to diffusion is lower, and the current increases dramatically. The depletion zone narrows. Explain Fig. 46–15(*b*).

F. Explain how diodes are used for rectification and how light emitting diodes work.

G. Optional. Explain how a junction transistor works. Explain the mechanism by which the gate voltage of a MOSFET controls current through the channel. Remove the covers from a few chips and pass them around with magnifying glasses for student inspection.

SUGGESTIONS

1. To test for understanding of the conduction process, ask questions 5 and 7.
2. Use questions 2, 4, and 6 to help in the discussion of the distribution of electrons among the states. Also assign problems 6 and 9. Consider problems 18 and 19 together. After discussing holes, assign problem 17.
3. After discussing the differences between conductors, semiconductors, and insulators, ask questions 11, 12, 18, and 23. Also ask question 21 in connection with the temperature coefficient of resistivity.
4. Use questions 19 and 20 in a discussion of doping. Also assign problems 33 and 34.
5. After discussing *p-n* junctions, ask questions 30 and 32. Also assign problem 37.
6. Computer project

   Ask students to use a root finding program to carry out calculations of the electron concentration in the conduction band and hole concentration in the valence band of both intrinsic and doped semiconductors. Then ask them to calculate the contact potential for a p-n junction with given dopant concentrations.

Chapter 47 NUCLEAR PHYSICS

BASIC TOPICS

I. Nuclear properties.

A. Explain that the nucleus of an atom consists of a collection of tightly bound neutrons, which are neutral, and protons, which are positively charged. A proton has the same magnitude charge as an electron. Define the term nucleon and state that the number of nucleons is called the mass number and is denoted by A, the number of protons is called the atomic number and is denoted by Z, and the number of neutrons is denoted by N. Point out that $A = Z + N$. Remark that nuclei with the same Z but different N are called *isotopes.* The atoms have the same chemical properties and the same chemical symbol. Show a wall chart of the nuclides. Refer to Table 47–1 when discussing properties of nuclides.

B. Explain that one nucleon attracts another by means of the strong nuclear force and that this force is different from the electromagnetic force. It does not depend on electrical charge and is apparently the same for all pairs of nucleons. It is basically attractive; at short distances (a few fm) it is much stronger than the electrostatic force between protons, but it becomes very weak at larger distances. Two protons exert attractive strong forces on each other only at small separations but they exert repulsive electric forces at all separations. Because of the short range, a nucleon interacts only with its nearest neighbors via the strong force. Because the nucleus is small, the much stronger nuclear force dominates and both protons and neutrons can be bound in stable nuclei. Explain that the force is thought to a manifestation of the strong force that binds quarks together to form nucleons.

C. Show Fig. 47–4 and point out the $Z = N$ line and the stability zone. Explain why heavy nuclei have more neutrons than protons. Also explain that unstable nuclei are said to be radioactive and convert to more stable ones with the emission of one or more particles. Show Fig. 47–12 and point out the stable and unstable nuclei.

D. Explain that the surface of a nucleus is not sharply defined but nuclei can be characterized by their mean radii and these are given by $R = R_0 A^{1/3}$, where $R_0 \approx 1.2\,\text{fm}$ ($1\,\text{fm} = 10^{-15}\,\text{m}$). Stress how small this is compared to atomic radii. Show that this relationship between R and A leads to the conclusion that the mass densities of all nuclei are nearly the same. Show that the density of nuclear matter is about $2 \times 10^{17}\,\text{kg/m}^3$.

E. Explain that the mass of a nucleus is less than the sum of the masses of its constituent nucleons, well separated. The difference in mass is accounted for by the binding energy through $E_b = \Delta m\, c^2$, where Δm is the magnitude of the mass difference. The binding energy is the energy which must be supplied to separate the nucleus into well separated particles, at rest. Generalize this equation to the case of a nucleus with Z protons and N neutrons: $E_b = Zm_p c^2 + Nm_n c^2 - mc^2$. Show Fig. 47–6 and point out that there is a region of greatest stability, near iron. For heavier nuclei the binding energy per nucleon falls slowly but nevertheless does fall. For lighter nuclei the binding energy per nucleon rises rapidly with increasing mass number. Explain the terms fission and fusion, then remark that the high mass number region is important for fission processes, the low mass number region is important for fusion processes.

F. State that nuclear masses are difficult to measure with precision so binding energies are usually expressed in atomic mass units: $1\,\text{u} = 1.6605 \times 10^{-27}\,\text{kg}$. Also state that tables usually give atomic rather than nuclear masses and so include the mass of the atomic electrons. Show that the electron masses cancel in the expression for the binding energy. Give the mass-energy conversion factor: 931.5 MeV/u.

G. Explain that nuclei have discrete energy levels, with separations on the order of MeV. An excited nucleus can make a transition to a lower energy state with the emission of a photon, typically in the gamma ray region of the spectrum. Explain that a nucleus may have intrinsic angular momentum and a magnetic moment. Spins are on the order of $\hbar$, like atomic electrons, but moments are much less than electron moments because the mass of a nucleon is much greater than the mass of an electron.

II. Radioactive decay.

A. Explain that nuclei may be either stable or unstable and those which are unstable ultimately decay to stable nuclei. Decay occurs by spontaneous emission of an electron (e^-), a helium nucleus (α), a positron (e^+), or larger fragments. The resulting nucleus has a different complement of neutrons and protons than the original nucleus.

B. Explain that decay is energetically favorable if the total mass of the products is less than the original mass. Define a decay symbolically as $X \rightarrow Y + b$, where X is the original nucleus, Y is the daughter nucleus, and b is everything else. Point out that charge, number of nucleons, and energy are all conserved. Define the disintegration energy by $Q = (m_X - m_Y - m_b)c^2$. Note that an appropriate number of electron rest energies must be added or subtracted so that atomic masses may be used. Note also that Q must be positive for spontaneous decays and Q appears as the kinetic energy of the decay products or as an excitation energy if the daughter nucleus is left in an excited state.

C. Explain that each radioactive nucleus in a sample has the same chance of decaying and that the decay rate ($R = -\mathrm{d}N/\mathrm{d}t$) is proportional to the number of undecayed nuclei present at time t: $-\mathrm{d}N/\mathrm{d}t = \lambda N$. This has the solution $N = N_0 \exp(-\lambda t)$, so the decay rate is given by $R = R_0 \exp(-\lambda t)$. Define the term half-life and show that $\tau = (\ln 2)/\lambda$. Go over Sample Problems 47–4 and 5, show Fig. 47–8, and point out the half-life. Remark that R decreases by a factor of 2 in every half-life interval.

D. Discuss α decay. Write down Eq. 47–9 and explain that the daughter nucleus has 2 fewer neutrons and 2 fewer protons than the parent. Go over Sample Problem 47–6 to show that α decay is energetically favorable for ^{238}U. Show Fig. 47–9 and explain that the deep potential well is due to the strong attraction of the residual nucleus for the nucleons in the α particle, while the positive potential is due to Coulomb repulsion. The two forces form a barrier to decay. Explain that the α particle can tunnel through the barrier. Its wave function does not go to zero at the inside edge but has a finite amplitude in the barrier and on the outside. There is a non-zero probability of finding the α particle on the outside. High, wide barriers produce a small probability of tunneling and a long half-life while low, narrow barriers produce the opposite effect. Note the wide range of half-lives that occur in nature (Table 47–2).

E. Discuss β decay. Explain that a neutron can transform into a proton with the emission of an electron and a neutrino (strictly, an antineutrino) and that a proton can transform into a neutron with the emission of a positron and a neutrino. Mention the properties of a neutrino: massless, neutral, weakly interacting. Only protons bound in nuclei can undergo β decay but both free and bound neutrons can decay. These transformations lead to decays such as the ones given in Eqs. 47–10 and 11. Explain that the energy is shared by the decay products and that the electrons or positrons show a continuous spectrum of energy up to some maximum amount (see Fig. 47–10). Explain that neutron rich nuclides generally undergo β^- decay while proton rich nuclides generally undergo β^+ decay. This is a mechanism for bringing the nucleus closer to stability. Carefully discuss the inclusion of electron rest energies in the equation for Q so that atomic masses can be used. In particular, show that in β^- decay there is no excess electron mass but in β^+ decay there is an excess of 2 electron masses.

F. Define the units used to describe radioactivity and radiation dosage: curie, roentgen, rad, and rem.

SUPPLEMENTARY TOPICS

1. Radioactive dating. If time permits, cover this topic as an application of radioactive decay processes.
2. Nuclear models. This topic adds a little breadth to the nuclear physics section and helps students understand nuclear processes a little better.

SUGGESTIONS

1. Nuclear constitution is covered in problems 5, 8, and 11. Nuclear radius and density are covered in problems 4, 6, and 12.
2. Include questions 2, 3, 5, and 11 in the discussion of the strong interaction and its role in nuclear binding. Problems 9, 10, 13, 19, and 21 illustrate nuclear stability and binding. Be sure to include problem 10 if you intend to discuss fission (Chapter 48).
3. Problems 26 and 27 cover basic half-life calculations. Problems 32 through 45 involve half-life calculations drawn from many interesting applications. Assign some of them.
4. Following the discussion of α decay, students should be able to answer questions 13 and 20 in their own words. The disintegration energy and barrier height are covered in problems 46 and 50. Problem 48 asks students to take into account the recoil of the residual nucleus. Problem 49 shows why alphas are emitted rather than well separated nucleons.
5. After discussing β decay, ask questions 21, 22, and 23. Also assign one or more of problems 52, 53, and 55. Problem 56 shows that β particles do not exist inside nuclei before decay occurs. The β decay discussion can be broadened somewhat by including the recoil of the nucleus. See problem 61.
6. Films:
   a. *The Discovery of Radioactivity*, 16 mm or 3/4" video cassette, color, 15 min. International Film Bureau, 332 S. Michigan Ave., Chicago, IL 60604. The experiments and discoveries of Roentgen, Curie, Becquerel, Elster, Rutherford, and Geitel are featured.
   b. *The Rutherford Scattering of Alpha-Particles*; Films for the Humanities and Sciences, Inc., Box 2053, Princeton, NJ 08543. Experimental setup is explained and data taken. The analysis is left for the student.
   c. *The Determination of a Radioactive Half-Life*; Films for the Humanities and Sciences, Inc., Box 2053, Princeton, NJ 08543. Experimental setup is explained and data taken. The analysis is left for the student.
7. Demonstrations
   a. Geiger counter: Freier and Anderson MPa2.
   b. Radioactivity: Hilton A15, A16, A18.
8. Laboratory
   Many of the following experiments make use of a Geiger tube and scalar.
   a. BE Experiment 47: *The Characteristics of a Geiger Tube* describes how students can systematically investigate the plateau and resolving time of a Geiger tube. They also learn how to operate a scalar. Consider prefacing the other experiments either with this experiment or with a demonstration of the same material.
   b. BE Experiment 48: *The Nature of Radioactive Emission.* Statistical fluctuations in the counting rate for a long half-life source provide a demonstration of the statistical nature

of radioactivity. Students also study variations in the counting rate as the source-counter separation is increased.

c. MEOS Experiment 14–7: *Half-Life of Radioactive Sources.* A Geiger counter and scalar are used to measure the decay rate as a function of time for indium, cesium 137, and barium 137. For the first and last, the data is used to compute the half-life. Other sections explain how to use a microcomputer to collect data and make the calculation and how to use a emanation electroscope to collect data. A neutron howitzer or minigenerator is required to produce radioactive sources.

d. BE Experiment 50: *Measurement of Radioactive Half-Life.* Nearly the same as MEOS 14–7. The generation of sources with short half-lives is discussed.

e. MEOS Experiment 14–6: *Absorption of Gamma and Beta Rays.* The particles are incident on sheets of aluminum and the number which pass through per unit time is counted. Students make a logarithmic plot of the counting rate as a function of the thickness of the aluminum and determine the range of the particles.

f. BE Experiment 49: *Properties of Radioactive Radiation.* Essentially the same as MEOS 14–6 but cardboard and lead as well as aluminum absorbers are used. Students can compare the relative absorbing power of these materials.

Chapter 48 ENERGY FROM THE NUCLEUS

BASIC TOPICS

I. The fission process.

A. Refer back to the binding energy per nucleon vs. A curve (Fig. 47–6). It suggests that a massive nucleus might split into two or more fragments nearer to iron, thereby increasing the total binding energy. Each fragment is more stable than the original nucleus. This is the fission process.

B. Remark that many massive nuclei can be rendered fissionable by the absorption of a thermal neutron. Such nuclei are called fissile. Give the example $^{235}\text{U} + \text{n} \rightarrow {}^{236}\text{U}^* \rightarrow \text{X} + \text{Y} + b\text{n}$. Explain that a thermal neutron ($\approx 0.04\,\text{eV}$) is absorbed by a ^{235}U nucleus and together they form the intermediate fissionable $^{236}\text{U}^*$ nucleus. This nucleus splits into 2 fragments (X and Y) and several neutrons. The sequence of events is illustrated in Fig. 48–2. Point out ^{236}U on Fig. 47–6. The disintegration energy for one possible fission event is calculated in Sample Problem 48–1.

C. Explain that different fission events, starting with the same nucleus, might produce different fragments. The fraction of events that produce a fragment of a given mass number A is graphed in Fig. 48–1. Point out that fragments of equal mass occur only rarely. Explain that the parent nucleus is neutron rich, the original fragments are neutron rich, and that the original fragments expel neutrons to produce the fragments X and Y. These generally decay further by β emission and some may emit delayed neutrons following β decay.

D. Show Fig. 48–3 and explain that the parent nucleus starts in the energy well near $r = 0$. The incoming neutron must supply energy to start the fission process. The required energy is slightly less than E_b since tunneling can occur. Point out the energy Q released by the process. Point out Table 48–2 and explain that E_n is the actual energy supplied by an incoming thermal neutron. Point out nuclides in the table for which fission does not occur.

E. Write out several fission modes for ^{235}U and note that on average more than one neutron is emitted. Explain that some neutrons come promptly while others come from later decays (the delayed neutrons). Point out that the average mode yields $Q \approx 200\,\text{MeV}$, of which 190 MeV or so appears as the kinetic energy of the fission fragments and 10 MeV goes to the neutrons.

II. Fission reactors.

A. Note that to have a practical reactor the fission process must be self sustaining, once started. Also, there must be a way to control the rate of the process and to stop it, if desired.

B. To be self sustaining, a chain reaction must occur: neutrons from one fission event are used to trigger another. The neutrons emitted from a typical fission event share about 5 to 10 MeV energy and they must be slowed to thermal speeds to be useful. Some sort of moderator, often water, is used.

C. Explain that on average about 2.5 neutrons are produced per fission event. Describe in detail what happens to them. Some leak out of the system, some of the slowed neutrons are captured by ^{238}U, some are captured by fission fragments, and the rest start fission in ^{235}U. Fig. 48–4 gives some typical numbers.

D. Explain the terms critical, subcritical, and supercritical. Note that the control rods, which absorb slow neutrons, are used to achieve criticality. Point out that without the delayed neutrons, control would not be possible since time is needed to move the rods into or out of the reactor.

E. Define the multiplication factor k as the ratio of the number of neutrons present at one time that participate in fission to the number present in the previous generation. Remark that $k = 1$ for critical operation, $k < 1$ for subcritical operation, and $k > 1$ for supercritical operation. Explain that k is determined by the positions of the control rods. The rods are pulled out to increase k and thereby increase power output. They are pushed in to decrease k and thereby decrease power output. When the desired power level is obtained the rods are positioned so $k = 1$.

F. Use Fig. 48–5 to describe the essential features of a nuclear power plant. Apart from the fact that the fission process is used to heat water or generate steam, this schematic could apply to any power plant. Remark on the special problems attendant on nuclear plants.

III. Fusion.

A. Return to Fig. 47–6 and remark that if two low mass nuclei are combined to form a higher mass nucleus the binding energy is increased considerably. The energy is transformed to the kinetic energy of the resulting nucleus and any particles emitted. In order to carry out the fusion process, the nuclei must be given sufficient energy to overcome the electrostatic repulsion of their protons. They can then approach each other closely enough for the attraction of the strong force to bind them. For ^{3}He the height of the barrier is about 1 MeV. Since tunneling is possible, fusion can occur at somewhat smaller energies.

B. To achieve a large number of fusion events, hydrogen or helium gases must be raised to high temperatures. Even at the temperature of the sun only a small fraction of the nuclei have sufficient energy to overcome the Coulomb barrier. Go over Fig. 48–9.

C. Discuss fusion in the sun. Remark that the core of sun is 35% hydrogen and 65% helium by mass. Outline the principal proton-proton cycle: 2 protons fuse to form a deuteron, a positron, and a neutrino. A deuteron fuses with a proton to form ^{3}He and two ^{3}He nuclei fuse to form ^{4}He and two protons. Remark that 6 protons are consumed and two are produced for a net loss of 4. The two positrons are annihilated with electrons to produce photons. Note that the process can be simplified to $4p + 2e^- \rightarrow \alpha + 2\nu + 6\gamma$ and the Q value is computed from the mass difference between the alpha particle and the 4 protons.

D. Calculate the energy released. Show that $Q = 26.7\,\text{MeV}$ and note that the neutrinos take about 0.5 MeV with them when they leave the sun. Point out that the fusion process produces about 20 million times as much energy per kg of fuel as the burning of coal.

E. If time permits, discuss helium burning. Use the solar constant to calculate the rate at which the sun converts mass to energy. Speculate on the future of the sun. Also mention the carbon cycle, which is essentially the same as the proton-proton cycle. Carbon acts as a catalyst.

F. Discuss controlled thermonuclear fusion. Explain that deuteron-deuteron and deuteron-triton fusion events are being studied. Point out that high particle concentrations at high temperatures must be maintained for sufficiently long times in order to make the process work. Discuss some means for doing this: the tokamak for plasma confinement by magnetic fields, inertial confinement, and laser fusion. State that the right combination has not yet been achieved but work continues.

SUGGESTIONS

1. After explaining the basic fission process, test understanding with questions 4, 5, and 6. Use question 15 to discuss critical size. Also assign problems 4, 11, and 13.
2. Following the discussion of the fission reactor, ask some of questions 8, 9, 10, 11, 12, and 13. To help students understand the role of a moderator, assign problem 27. To illustrate the role of the control rods assign problems 23 and 24.
3. Following the discussion of the basic fusion process, assign problems 33 and 34. Also ask question 17.
4. To help students understand the fusion process as an energy source, assign problems 42 and 44. The carbon cycle is covered in problem 47.
5. Film
   Fusion: The Ultimate Fire, 16 mm, color, 14 min. BFA Educational Media, Division of Phoenix Films, 468 Park Avenue, New York, NY 10016. Various types of fusion research are presented. An explanation of the fundamental concepts of fusion is aided by animation. Includes visits to fusion labs.
6. Demonstrations
   Chain reaction: Freier and Anderson MPa1.

Chapter 49 QUARKS, LEPTONS, AND THE BIG BANG

BASIC TOPICS

I. The particle "zoo".

A. Show a list of particles already familiar to students. Include the electron, proton, neutron, and neutrino, then add the muon and pion. Explain that many other particles have been discovered in cosmic ray and accelerator experiments. To impress students with the vast array of particles and the enormous collection of data, make available to them a Review of Particle Properties paper, published roughly every 2 years in Reviews of Modern Physics.

B. Explain that many new particles are discovered by bombarding protons or neutrons with electrons or protons and show a picture of a detector, such as Fig. 49–1 or a bubble chamber picture. State that the picture shows tracks of charged particles in a strong magnetic field, hence the curvature. Remind students that the radius of curvature can be used to find the momentum of a particle if the charge is known. Indicate the collision point and emphasize that the new particles were not present before the collision: the original particles disappear and new particles appear. In most cases the total rest energy after the collision is much greater than the total rest energy before. Kinetic energy was converted to mass.

C. Mention that a few particles seem to be stable (electron, proton, neutrino) but most decay spontaneously to other particles. Point out decays on a bubble chamber picture. Explain the statistical nature of decays and remind students of the meaning of half-life. Examples: $n \rightarrow p + e^- + \nu$, $\pi^+ \rightarrow \mu^+ + \nu$.

D. Explain that for each particle there is an antiparticle with the same mass. A charged particle and its antiparticle have charge of the same magnitude but opposite sign. Their magnetic moments are also opposite. A particle and its antiparticle can annihilate each other, the energy (including rest energy) being carried by photons or other particles produced in the annihilation. Example: $e^+ + e^- \rightarrow \gamma + \gamma$. Antiparticles (except the positron) are denoted by a bar over the particle symbol. Some uncharged particles (such as the photon and π^0) are their own antiparticles. The universe seems to be made of particles, not antiparticles.

II. Particle properties.

A. Spin angular momentum. Remind students that many particles have intrinsic angular momentum. Explain that the magnitude is always an integer or half integer times $\hbar$. Remark that particles with half integer spins are called fermions while particles with integer spins are called bosons. Remind students of the Pauli exclusion principle and its significance, then state that fermions obey the principle while bosons do not. Give examples: electrons, protons, neutrons, and neutrinos are fermions; photons, pions, and muons are bosons. Remark that spin angular momentum is conserved in particle decays and interactions. An odd number of fermions, for example, cannot interact to yield bosons only.

B. Charge. Remind students of charge quantization and charge conservation. Even if the character and number of particles change in an interaction, the total charge before is the same as the total charge after. Example: $n \rightarrow p + e^- + \nu$.

C. Momentum and energy. Explain that energy and momentum are conserved in decays and interactions. Give masses and rest energies for the particles in the list of part I. Give the expressions for relativistic energy and momentum in terms of particle velocity.

D. Forces. Remark that all particles interact via the force of gravity and all charged particles interact via the electromagnetic force. The force of gravity is too weak to have observable influence at energies presently of interest. Remark that there are two additional forces, called strong and weak, respectively. Remind students of the role played by the strong force in holding a nucleus together and the role played by the weak force in beta decay. These topics were covered in Chapter 47. Note that lifetimes for strong decays are about 10^{-23} s, lifetimes for electromagnetic decays are about 10^{-14} to 10^{-20} s, and lifetimes for weak decays are about 10^{-8} to 10^{-13} s.

E. Leptons. State that particles that interact via the strong force (as well as the weak) are called hadrons and that particles that interact via the weak force but not the strong are called leptons. List the leptons (electron, muon, tauon, and their neutrinos) and explain that a different neutrino is associated with each of the leptons. Remark that the neutrino that appears following muon decay is not the same as the neutrino that appears following beta decay. Neutrinos are labelled with subscripts giving the associated lepton: ν_e, ν_μ, and ν_τ.

F. Baryons and mesons. Remark that some strongly interacting particles (proton, neutron) are fermions and are called baryons while others (pion, kaon) are bosons and are called mesons. Explain that a baryon number of +1 is assigned to each baryon particle, a baryon number of −1 is assigned to each baryon antiparticle, and a baryon number of 0 is assigned to each meson. Then baryon number is conserved in exactly the same way charge is conserved: the total baryon number before a collision or decay is the same as the total baryon number after. This conservation law (and conservation of energy) accounts for the stability of the proton, the baryon with the smallest mass. There is some speculation that baryon number is not strictly conserved and that protons may decay to other particles, but the half-life is much longer than the age of the universe. Some physicists are trying to observe proton decay.

G. Strangeness. Explain that another quantity, called strangeness, is conserved in strong interactions. Neutrons and protons have $S = 0$, K^- and Σ^+ have $S = -1$. A particle and its antiparticle have strangeness of opposite sign. Conservation of strangeness allows $\pi^+ + p \rightarrow K^+ + \Sigma^+$ but prohibits $\pi^+ + p \rightarrow \pi^+ + \Sigma^+$, for example.

III. Quarks and the eight-fold way.

A. Show the eight-fold way patterns (Fig. 48–3) and point out the oblique axes. Remark that these patterns are to fundamental particles as the periodic table of chemistry is to atoms and that they have provided clues to the existence of particles not previously observed.

B. Remark that the properties of strongly interacting particles can be explained if we assume they are made up of more fundamental particles (called quarks). List the u, d, and s quarks and their properties (Table 49–6). Particularly note the fractional charge and baryon number. Baryons are constructed of three quarks, antibaryons of three antiquarks, and mesons of a quark and antiquark. Show that uud has the charge, spin, and baryon number of a proton and udd has the charge, spin, and baryon number of a neutron. Give the quark content of the spin 1/2 baryons (Figs. 49–3*a* and 49–4*a*) and the quark content of the spin 0 mesons (Figs. 49–3*b* and 49–4*b*). Point out that the strange quark accounts for the strangeness quantum number. Mention the charm, bottom, and top quarks and point out they lead to other particles.

C. Explain that the existence of internal structure allows for excited states: there are other particles with exactly the same quark content as those in the figures but they are different particles because the quarks have different motions. The additional energy results in greater mass. Contrast this with the leptons, which have no internal structure. Quarks and leptons are believed to be truly fundamental.

D. Messenger particles. Explain that particles interact by exchanging other particles. Electromagnetic interactions proceed by exchange of photons, for example. Also explain that energy may not be conserved over short periods of time but this is consistent with the uncertainty principle. State that the strong interaction proceeds by the exchange of gluons by quarks and the weak interaction proceeds by the exchange of Z and W particles by quarks and leptons. The interaction that binds nucleons in a nucleus is the same as the interaction that binds quarks in a baryon or meson. In the former case gluons are exchanged between quarks of different nucleons, in the latter they are exchanged between quarks of the same baryon or meson.

E. Explain that quarks are conserved in strong interactions. Either the original quarks are rearranged to form new particles or quark-antiquark pairs are created, then both the original and the new quarks are rearranged. This accounts for conservation of strangeness. Example: $K^+ \rightarrow K^0 + \pi^+$ ($u\bar{s} \rightarrow d\bar{s} + u\bar{d}$). A $d\bar{d}$ pair is formed. The d quark couples to the $\bar{s}$ quark to form a K^0 and the $\bar{d}$ quark couples to the u quark to form a π^+. Contrast this with the weak interaction, which can change one type quark into another. Illustrate with beta decay, in which a d quark is converted to a u quark.

F. Explain that quarks have another property, called color. Color produces the gluon field, much as charge produces the electromagnetic field: baryons interact via the strong interaction because quarks have color. Be sure students understand that "color" in this context has nothing to do with the frequency of light. Mention that gluons carry color. The emission or absorption of a gluon changes the color of a quark. Contrast this with the electromagnetic interaction: a photon does not carry charge.

IV. The big bang and cosmology.

A. Remind students of the doppler shift for light and state that spectroscopic evidence convinces us that on a large scale matter in the universe is receding from us and we are led to conclude that the universe is expanding. Write down Hubble's law and give the Hubble parameter: 17×10^{-3} m/(s · ly). Show that this implies a minimum age for the universe of about 15×10^9 y.

B. State that the future expansion (or contraction) of the universe depends on its mass density and that the density of matter that radiates is too small to prevent expansion forever. Explain that there is evidence for the existence of matter that does not radiate (dark matter). Explain how the rotational period of a star in a galaxy, as a function of its distance from the galactic center, provides such evidence. The nature of the dark matter is not presently known.

C. Discuss the microwave background radiation and state that physicists believe it was generated about 300,000 years after the big bang, when the universe became tenuous enough to allow photons to exist without being quickly absorbed.

D. Remark that in the early universe the temperature was sufficiently high that the exotic particles now being discovered (and others) existed naturally. We need the results of high energy physics to understand the early universe.

E. Go over the chronological record given at the end of Section 14.

SUGGESTIONS

1. To test for understanding of the conservation laws and the stability of particles, ask questions 10, 11, 12, 13, 14, and 19. Problems 2, 8, 10 (or 11), 12, 13, 15, 16, 17, and 20 each deal with one or more of the conservation laws. Assign several.
2. To help clarify particle properties and classifications, ask questions 4, 8, 18, 27, 28, and 29.
3. Problems 23, 24, 26, 27, and 28 provide excellent illustrations of the quark model. Assign a few of them. Also ask questions 30, 31, 33, and 34.
4. Questions 15, 21, 22, and 24 deal with the fundamental forces. Ask one or two of them.
5. Include questions 36, 37, 39, and 40 in discussions of cosmology. Also assign problem 30. Assign problem 31 or 32 in connection with the red shift. If you discussed the relativistic Doppler shift in connection with Chapter 42, assign problem 33. Problems 35 and 39 deal with the cosmic background radiation. Dark matter and the future of the universe are the subjects of problem 38.
6. Demonstrations
   a. Show nuclear emulsion plates, available from Brookhaven National Laboratory, Fermilab, and other high energy laboratories.
   b. Elementary particles: Hilton A23.
7. Laboratory
   MEOS Experiment 14–8: *Nuclear and High Energy Particles.* A dry ice and alcohol cloud chamber is used to observe the tracks of alpha and beta particles as well as the tracks produced by cosmic rays. A magnet is used to make circular tracks.

SECTION THREE
BIBLIOGRAPHY

This section contains a bibliography of pedagogic articles, arranged by textbook chapter. You will find listed here articles that discuss teaching strategies, new or innovative ways of introducing certain topics, lab experiments, demonstrations, and other topics of interest to introductory physics teachers.

The symbols used to denote frequently cited journals are: AJP (American Journal of Physics), TPT (The Physics Teacher), and SA (Scientific American).

Here are some articles that are applicable to the whole course.

General

1. Arnold B. Arons; *A Guide to Introductory Physics Teaching*; John Wiley & Sons, New York (1990). 342 pages of useful observations and helpful suggestions from a master teacher and leader in pedagogic research. Must reading for every introductory physics teacher.
2. Robert J. Beichner; *Applications of Macintosh Microcomputers in Introductory Physics*; TPT **27**, 348 (May 1989). A review of computer uses and recommendations of software.
3. Donna Berry, ed.; *A Potpourri of Physics Teaching Ideas*; American Association of Physics Teachers; Publications Sales, AAPT Executive Office, 5112 Berwyn Road, College Park, MD 20740-4100. Reprints from The Physics Teacher.
4. Stephen G. Brush; *History of Physics*; American Association of Physics Teachers; Publications Sales, AAPT Executive Office, 5112 Berwyn Road, College Park, MD 20740-4100. 13 articles and comprehensive bibliography.
5. Marvin L. De Jong; *Computers in Introductory Physics*; Computers in Physics **5**, 12 (January/February 1991). A review of some of the ways computers are used in the introductory course, with emphasis on computation.
6. Denis Donnelly; *Equation-solving software packages: Uses in the undergraduate curriculum*; AJP **58**, 585 (June 1990). Descriptions of various commercial software packages and discussion of their uses.
7. Ronald Edge; *String and Sticky Tape Experiments*; American Association of Physics Teachers; Publications Sales, AAPT Executive Office, 5112 Berwyn Road, College Park, MD 20740–4100. Inexpensive demonstration and lab experiments.
8. Cliff Frohlich; *Resource letter PS–1: Physics of sports*; AJP **54**, 590 (July 1986). Bibliography of various sports from the physicists point of view. Use to broaden interest.
9. C. Frohlich. ed.; *Physics of Sports*; American Association of Physics Teachers; Publications Sales, AAPT Executive Office, 5112 Berwyn Road, College Park, MD 20740-4100. 13 articles and resource letter.
10. Robert G. Fuller; *Computers in Physics Education*; American Association of Physics Teachers, AAPT Executive Office, 5112 Berwyn Road, College Park, MD 20740–4100. Reprints of 21 journal articles.
11. Rick Guglielmino; *Using Spreadsheets in an Introductory Physics Lab*; TPT **27**, 175 (March 1989). How to use electronic spreadsheets to record and analyze data.
12. William G. Harter; *Nothing Going Nowhere Fast: Computer Graphics in Physics Courses*; Computers in Physics **5**, 466 (Sep/Oct 1991). Some good ideas for computer simulations

of collisions, waves, and relativistic reference frames. Discusses their use in class and their pedagogic value. Some hints for developing your own simulations.

13. R.B. Hicks and H. Laue; *A computer-assisted approach to learning physics concepts*; AJP **57**, 807 (September 1989). A study of the effectiveness of computer tutorials on learning physics.

14. Michael E. Krieger and James H. Stith; *Spreadsheets in the Physics Laboratory*; TPT **28**; 378 (May 1990). A description of spreadsheet use at West Point to analyze and plot experimental data.

15. Priscilla W. Laws; *Calculus-Based Physics without Lectures*; Physics Today **44**, 24 (December 1991). Description of an introductory course based on highly interactive lab experiments.

16. William M. MacDonald, Edward F. Redish, and Jack M. Wilson; *The M.U.P.P.E.T. Manifesto*; Computers in Physics **2**, 23 (July/August 1986). The computer-based instructional program at the University of Maryland.

17. Lillian C. McDermott; *Millikan Lecture 1990: What we teach and what is learned – Closing the gap*; AJP **59**, 301 (April 1991). An analysis of physics pedagogy and a review of pedagogical research, with many implications for teaching.

18. Melba Phillips, ed.; *Physics History from AAPT Journals*; American Association of Physics Teachers; Publications Sales, AAPT Executive Office, 5112 Berwyn Road, College Park, MD 20740-4100. 27 reprints.

19. Albert J. Read; *"Hands-on" exhibits in physics education*; AJP **57**, 393 (May 1989). A discussion of hallway exhibits.

20. John S. Risley; *Using Physics Courseware*; TPT **27**, 188 (March 1989). A discussion of the uses of computers in physics classes and labs.

21. Robert F. Tinker; *Computer-aided Student Investigations*; Computers in Physics **2**, 46 (January/February 1988). A wide range of computer-aided lab experiments.

22. Alan Van Heuvelen; *Learning to think like a physicist: A review of research-based instructional strategies*; AJP **59**, 891 (October 1991). Discussion of teaching strategies based on results of pedagogic research.

23. Alan Van Heuvelen; *Overview, Case Study Physics*; AJP **59**, 898 (October 1991). Use of case studies that force students to confront their misconceptions.

Chapter 1

1. E. Roger Cowley; *A classroom exercise to determine the Earth-Moon distance*; AJP **57**, 351 (April 1989). Compares the size of moon's image to the size of earth shadow during lunar eclipse. A good example of an indirect measurement.

2. J.L. Heilbron; *The politics of the meter stick*; AJP **57**, 988 (November 1989). History of the reform of weights and measures during the French Revolution.

3. Wayne E. McGovern; *The range of a data set: Its relationship to the standard deviation for various distributions*; AJP **60**, 943 (October 1992). Sometimes the difference between the greatest and the least values of a data set, suitably corrected, can be used as a measure of the uncertainty in the measurement.

4. Robert O'Keefe and Bahman Ghavimi-Alagha; *The World Trade Center and the distance to the world's center*; AJP **60**, 183 (February 1992). A nice example of an indirect measurement. The height of a tall building is measured from far away and the result is used to calculate the earth's radius.

5. Mark A. Peterson; *Error analysis by simulation*; AJP **59**, 355 (April 1991). A proposal to teach error analysis using computer generated data. Students compare actual data to simulated data.

Chapter 2

1. Stanislaw Bednarek; *Magnetic track for experiments in mechanics*; AJP **60**, 664 (July 1992). The description and construction details for an apparatus to replace the air track in introductory labs. Magnetic levitation is used to separate the carts from the track and thereby reduce friction.
2. M.G. Calkin; *The motion of an accelerating automobile*; AJP **58**, 573 (June 1990). Experiments indicate the square of the speed is linear in the time. You might use these results here or when rolling is studied.
3. Bill Crummett; *Measurements of Acceleration Due to Gravity*; TPT **28**, 291 (May 1990). A compendium of the techniques used for a precise measurement of g. You might use this material as a source for discussion when your students measure g in the lab.
4. Marvin L. Dejong; *Derivatives in Calculus-Based Physics*; TPT **24**, 412 (October 1986). The notation used in most introductory physics texts is different from that used in most introductory calculus texts and sometimes confuses students. It helps to be aware of the differences and point them out to students.
5. Bruce Denardo, Selmer Wong and Alpha Lo; *Errors due to average velocities*; AJP **57**, 528 (June 1989). Discussion of errors that arise when the average velocity is used to approximate the instantaneous velocity at the spatial or temporal midpoint of an interval. Application to measurement of velocity of a glider on an air track. Useful for lab presentations.
6. John H. Dodge; *Fluid Resistance and Terminal Velocity*; TPT **30**, 420 (October 1992). Describes a simple apparatus to demonstrate the influence of a fluid medium on a falling object.
7. B. Duchesne, C.W. Fischer, and C.G. Gray; *Inexpensive and accurate position tracking with an ultrasonic ranging module and a personal computer*; AJP **59**, 998 (November 1991). Apparatus for recording and analyzing position, velocity, and acceleration data. Schematic of interface is given.
8. A. Edgar; *A low-cost timer for free-fall experiments*; AJP **59**, 568 (June 1991). An inexpensive stop watch is used, modified with external circuitry that starts and stops the watch. About 10 ms accuracy is obtained.
9. Ian R. Gatland, Robert Kahlscheuer, and Hicham Menkara; *Experiments utilizing an ultrasonic range finder*; AJP **60**, 451 (May 1992). A list of experiments for the introductory course, including motion on an inclined plane, oscillations, and collisions.
10. Fred M. Goldberg and John H. Anderson: *Student Difficulties with Graphical Representations of Negative Values of Velocity*; TPT **27**, 254 (April 1989). Results of pedagogic research, with some recommendations for teaching.
11. Laurence I. Gould and Harry Workman; *Air track with a distributed infrared detector system*; AJP **56**, 739 (August 1988). Many permanently mounted emitter-detector pairs, interfaced to a computer. Article describes the data acquisition system and 8 basic experiments.
12. Edwin Kaiser; *Instantaneous Velocity: A Different Approach*; TPT **29**, 394 (September 1991). Uses spark-timer data to find the limit of the average velocity as the time interval becomes small.
13. Edward Kluk and John L. Lopez; *Don't Use Airtracks to Measure Gravity Acceleration*; TPT **30**, 48 (January 1992). An experimental scheme to obtain 1% or better accuracy with inexpensive apparatus.

14. Lillian C. McDermott, Mark L. Rosenquist, and Emily H. vanZee; *Student difficulties in connecting graphs and physics: Examples from kinematics*; AJP **55**, 503 (June 1987). Discussion of student difficulties and some instructional strategies that might be used to overcome the difficulties.

15. Mark L. Rosenquist and Lillian C. McDermott; *A conceptual approach to teaching kinematics*; AJP **55**, 407 (May 1987). A discussion of conceptual problems students have with velocity and acceleration. Implications for teaching.

16. Wolfgang Rueckner and Paul Titcomb; *An accurate determination of the acceleration of gravity for lecture hall demonstration*; AJP **55**, 324 (April 1987). An apparatus that measures the acceleration of a falling body with an accuracy of 0.022%. Can be used to find drag coefficients.

17. T.R. Sandin; *The Jerk*; TPT **28**, 36 (January 1990). Jerk is defined as the time rate of change of acceleration. Here is a discussion of how the concept can be used. Both one- and two-dimensional examples.

18. Fritz Schoch and Walter Winiger; *How to Measure g Easily with $\approx 10^{-4}$ Precision in the Beginners' Lab*; TPT **29**, 98 (February 1991). Free fall experiment is done with spheres of various densities and the results extrapolated to an infinite-density sphere (for which air resistance is negligible).

19. Ronald K. Thornton and David R. Sokoloff; *Learning motion concepts using real-time microcomputer-based laboratory tools*; AJP **58**, 858 (September 1990). Studies of the learning curves of students who use motion detectors to learn about velocity and acceleration.

20. Stuart M. Quick; *A computer-assisted free-fall experiment for the freshman laboratory*; AJP **57**, 814 (September 1989). Photogate signals are sent to a computer for analysis. Least square fits are used.

21. Ray G. Van Ausdal; *Structured Problem Solving in Kinematics*; TPT **26**, 518 (November 1988). A plan for teaching problem solving that takes into account many of the pitfalls encountered by students.

Chapter 3

1. Robert P. Bauman; *Physics that Textbook Writers Usually Get Wrong III. Forces and Vectors*; TPT **30**, 402 (October 1992). An analysis of the way vectors are described in some texts, with application to non-inertial forces.

2. Walter Hauser; *Vector products and pseudovectors*; AJP **54**, 168 (February 1986). Cross products do not behave the same as vectors upon reversal of the coordinate system. A way of teaching the distinction in an introductory course.

3. R. Ramirez-Bon; *An Interesting Problem Solved by Vectors*; TPT **28**, 594 (December 1990). Vector analysis is used to reconstruct a treasure map.

Chapter 4

1. B.A. Aničin; *The rattle in the cradle*; AJP **55**, 533 (June 1987). A plea to use Newton's original discussion of centripetal acceleration. Some of these ideas might help students understand the concept.

2. Peter J. Brancazio; *The Physics of Kicking a Football*; TPT **23**, 403 (October 1985). Projectile motion analysis applied to a football in flight, with data from actual games. Use as lecture material or as extra reading for sports-minded students.

3. Ronald A. Brown; *Maximizing the Range of a Projectile*; TPT 30, 344 (September 1992). A projectile is fired from a height h above the landing point. The author shows how to find the angle for maximum range without using the calculus.
4. Stillman Drake; *Galileo's Gravitational Units*; TPT **27**, 432 (September 1989). Discussion of the units used by Galileo in writing about projectile motion, pendula, and falling bodies. Use as background information or for extra reading by students.
5. Howard E. Evans II; *Raindrops Keep Falling on My Head ...*; TPT **29**, 120 (February 1991). A problem in relative velocity. How should you walk (or run) in the rain to get the least wet?
6. Carey S. Inouye and Eric W.T. Chong; *Maximum Range of a Projectile*; TPT **30**, 168 (March 1992). Studies of the range when the projectile is fired from a point above the landing point.
7. David Keeports; *Numerical Calculation of Model Rocket Trajectories*; TPT **28**, 274 (May 1990). Model rocket experiments are often used as applications of Newtonian mechanics. Here's how to analyze the flight to predict maximum altitude and range for comparison with experiment.
8. E. Keshishoglou and P. Seligmann; *Experiments in two dimensions using a video camera and microcomputer*; AJP **57**, 179 (February 1989). Standard air table experiments (uniform circular motion, projectile motion, collisions, etc) but displayed on a monitor using a video camera. A computer is used to carry out analysis.
9. William M. MacDonald; *The physics of the drive in golf*; AJP **59**, 213 (March 1991). A computer is used to generate the trajectory. Spin and drag effects are included. Use the data to illustrate "real" projectile motion.
10. Bengt Magnusson and Bruce Tiemann; *The Physics of Juggling*; TPT **27**, 584 (November 1989). Projectile motion analysis applied to juggling. Some parts require knowledge of rotational kinematics but discussion of these can be postponed.
11. Ernie McFarland; *How Olympic records depend on location*; AJP **54**, 513 (June 1986). An investigation into how changes in g and air density affect track and field events. Use to add interest to kinematics.
12. Andre Mirabelli; *A new projectile problem and the attribution of continuity*; AJP **54**, 278 (March 1986). Problem: find the time for which a projectile is farthest from the firing point. There is an interesting discontinuity in the time as a function of firing angle.
13. Miky Ronen and Aharon Lipman; *The V-Scope: An "Oscilloscope" for Motion*; TPT **29**, 298 (May 1991). Discusses a commercially available 3-D ranging apparatus for tracking and displaying the motions of several objects simultaneously. Several experiments are described. Also see: Harold A. Daw; *An Assessment of the PASCO V-Scope*; TPT **29**, 304 (May 1991).
14. Clifford Swartz; *Reference Frames and Relativity*; TPT **27**, 437 (September 1989). Discussion of the Galilean transformation with applications to inertial and non-inertial frames. Many good ideas for teaching this topic.
15. A. Tan and A.C. Giere; *Maxima problems in projectile motion*; AJP **55**, 750 (August 1987). Some problems dealing with maximum range, time of flight, trajectory length, and distance from plane for a projectile fired over an inclined plane.

Chapter 5

1. Edward A. Desloge; *The empirical foundation of classical dynamics*; AJP **57**, 704 (August 1989). An alternative viewpoint: the use of momentum conservation as the foundation of dynamics rather than force. Avoids the implicit use of Newton's third law in the definition of mass.

2. David L.D. Green and David T. Hartney; *Newton's Truck: Determining Mass and Finding g by Pushing a Truck across a Parking Lot*; TPT **26**, 448 (October 1988). Experimental verification of $\mathbf{F} = ma$.
3. David Hestenes, Malcolm Wells, and Gregg Swackhamer; *Force Concept Inventory*; TPT **30**, 141 (March 1992). Results of tests given to determine students perceptions of the concepts of force and motion. Test questions are included.
4. Joseph B. Keller; *Newton's second law*; AJP **55**, 1145 (December 1987). An interesting way to look at the second law. Author concludes it may be used to define force and mass, while retaining a great deal of empirical content.
5. Douglas A. Kurtze; *Teaching Newton's Second Law — A Better Way*; TPT **29**, 350 (September 1991). The author advocates writing the law $\mathbf{a} = \sum \mathbf{F}/m$ since many students interpret the right side of an equation as the cause of the left side.
6. David P. Maloney; *Forces As Interactions*; TPT **28**, 386 (September 1990). Results of pedagogic studies of student misconceptions with consequences for teaching about force.
7. Erwin Marquit; *A plea for a correct translation of Newton's law of inertia*; AJP **58**, 867 (September 1990). The interpretation of the first law given here makes it important for the definition of force.
8. John S. Rigden; *Editorial: High thoughts about Newton's First Law*; AJP **55**, 297 (April 1987). Some comments about the first law and our inability to communicate the wonder of nature to students.
9. Michael Svonavec; *Accelerated motion with a variable weight*; AJP 55, 753 (August 1987). Analysis of an Atwood machine taking into account the mass of the string.
10. Yvette A. Van Hise; *Student Misconceptions in Mechanics: An International Problem?*; TPT **26**, 498 (November 1988). Summary of some pedagogic problems in teaching mechanics, along with some suggestions.
11. R.E. Vermillion, G.O. Cook; *A particle sliding down a movable incline: An experiment*; AJP **56**, 438 (1988). A new twist on an old experiment.

Chapter 6

1. Albert A. Bartlett; *Physics and the Measurement of Automobile Performance*; TPT **26**, 433 (October 1988). Studies of air resistance and fuel efficiency as related to the motion of a car.
2. Albert A. Bartlett and J. Parker Lamb; *The Train Left the Track*; TPT **28**, 586 (December 1990). Analysis of a railroad accident on a curved track uses evidence at the scene to determine the speed of the train just before the accident. A nice application of the idea of centripetal acceleration.
3. Peter J. Brancazio; *Trajectory of a fly ball*; TPT **23**, 20 (January 1985). Projectile motion analysis, including the effects of air resistance, applied to baseball. Use for background or suggest to sports-minded students for extra reading.
4. Ronald A. Bryan, Robert Beck Clark, and Pat Sadberry; *Illustrating Newton's Second Law with the Automobile Coast-Down Test*; TPT **26**, 442 (October 1988). Uses what students already know about acceleration and deceleration of cars to teach the second law.
5. Uri Haber-Schaim and John H. Dodge; *There's More to It than Friction*; TPT **29**, 56 (January 1991). Examines the parlor trick in which a table cloth is pulled from under a drinking glass.
6. Se-yuen Mal; *Extreme value problems in mechanics without calculus*; AJP **55**, 929 (October 1987). Three extremum problems that can be solved using only the knowledge that the sine

or cosine function must have a value in the range −1 to +1 for real angles: minimum force to keep a block moving with constant velocity on a rough surface, maximum height of mud thrown from the rim of a rolling wheel, direction in which a jogger should run to avoid being hit by a truck.

7. Eugene E. Nalence; *Using Automobile Road Test Data*; TPT **26**, 278 (May 1988). Force of air resistance can be calculated from coast-down data given in automotive magazines.
8. Channon P. Price; *Teacup Physics: Centripetal Acceleration*; TPT **28**, 49 (January 1990). Analysis of the surface of a spinning liquid. Use for demonstration.
9. Robert R. Speers; *Physics and roller coasters – The Blue Streak at Cedar Point*; AJP **59**, 528 (June 1991). Track profile is used to predict forces acting on passenger. Results are compared to actual accelerometer measurements. May also be used in conjunction with Chapter 8.
10. C.W. Tompson and J.L. Wragg; *Terminal Velocity on an Air Track*; TPT **29**, 178 (March 1991). Magnetic braking of carts on an air track provides velocity-dependent force.
11. William S. Wagner; *Automobile deceleration force by the coast-down method*; AJP **54**, 1049 (November 1986). Calculate force of friction on tires and drag coefficient by measuring speed vs. time as auto coasts to rest. Experiment suitable for introductory lab.
12. William M. Wehrbein; *Frictional forces on an inclined plane*; AJP **60**, 57 (January 1992). Detailed analysis of the conditions for which sliding takes place if the box on the plane starts at rest.
13. Robert L. Wilde; *A correction for spring mass in the ubiquitous centripetal force experiment of freshman physics*; AJP **57**, 1098 (December 1989). Neglect of the spring mass may lead to errors in excess of 1%. Corrections are discussed.
14. Metin Yersel; *A Simple Demonstration of Terminal Velocity*; TPT **29**, 335 (September 1991). The author recommends the use of air bubbles because they reach terminal velocity quickly. The measuring apparatus is described.
15. Joseph M. Zayas; *Experimental determination of the coefficient of drag of a tennis ball*; AJP **54**, 622 (July 1986). Read for experimental technique or for results.

Chapter 7

1. Robert P. Bauman; *Physics that Textbook Writers Usually Get Wrong*; TPT **30**, 264 (May 1992). Clarification of the concepts of work and energy.
2. Stan Jakuba; *Effect of Exercise Expressed in Joules and Watts*; TPT **29**, 512 (November 1991). Some data you might find useful for your lectures.
3. Ronald A. Lawson and Lillian C. McDermott; *Student understanding of the work-energy and impulse-momentum theorems*; AJP **55**, 811 (September 1987). Tests to discover what students really understand and the implications of the results for teaching.
4. A. John Mallinckrodt and Harvey S. Leff; *All about work*; AJP **60**, 356 (April 1992). Identifies seven types of work that can be done on a system of particles that interact with each other and with their environment. Shows how the fundamental definition leads to the first law of thermodynamics.

Chapter 8

1. Dale R. Blaszczak; *The roller coaster experiment*; AJP **59**, 283 (March 1991). An experiment to demonstrate conservation of energy.

2. D. P. Shelton and M.E. Kettner; *Potential energy of interaction for two magnetic pucks*; AJP **56**, 51 (January 1988). Using an air table to investigate the interaction between two magnetic pucks. May be treated as an exercise in determining the force and potential energy of an interaction. Knowledge of magnetism is not necessary.

3. David G. Willey; *Conservation of Mechanical Energy Using a Pendulum*; TPT **29**, 567 (December 1991). A pendulum is released from a known height and passes through a photogate timer. Data is used to compute its speed at the lowest point. The final kinetic energy is compared to the initial potential energy.

Chapter 9

1. A.B. Arons; *Developing the Energy Concepts in Introductory Physics*; TPT 27, 506 (October 1989). A useful comparison of the "work-energy theorem" for center of mass motion and the first law of thermodynamics. Should be read by anyone teaching this chapter of the text. Also see B.A. Sherwood; *Pseudowork and Real Work*; AJP **51**, 597 (1983) and B.A. Sherwood and W.H. Bernard; *Work and Heat Transfer in the Presence of Sliding Friction*; AJP **52**, 1001 (1984).

2. Marie Baehr; *Center of Mass of a Can with a Varying Level of Liquid*; TPT **30**, 34 (January 1992). Some physical insight into a popular introductory problem.

3. George Barnes; *Conservation of momentum demonstration using a piece of sewer pipe*; AJP **54**, 741 (August 1986). A demonstration experiment in which a sheet of paper is pulled from under a pipe.

4. Robert R. Cadmus, Jr.; *A video technique to facilitate the visualization of physics phenomena*; AJP **58**, 397 (April 1990). A device for superposing video frames. Can view trajectory of center of mass and spin of an object, for example.

5. Margaret Stautberg Greenwood; *Conservation of Momentum and the Center of Mass of a Cart-Truck System*; TPT **25**, 370 (September 1987). A radio controlled toy truck runs on the top of a low friction cart. Demonstration can be used to show conservation of momentum and the constancy of the position of the center of mass.

6. M. Stautberg Greenwood, R. Bernett, M. Benavides, S. Granger, R. Plass, and S. Walters; *Using a Smart-pulley Atwood machine to study rocket motion*; AJP **57**, 943 (October 1989). One side is a funnel filled with sand while the other is a fixed mass.

7. Margaret Stautberg Greenwood; *Inclined Plane on a Frictionless Surface*; TPT **28**, 109 (February 1990). Solution to the problem using an accelerated reference frame.

8. F. Herrmann and M. Schubart; *Measuring momentum without the use of $p = mv$ in a demonstration experiment*; AJP **57**, 858 (September 1989). Inelastic collisions between bodies that carry 1 unit of momentum each.

9. Eugene Levin; *Energy in the center of mass*; AJP **55**, 909 (October 1987). Two carts on an air track are coupled by magnets. A falling weight exerts an impulsive force on one, breaking the bond. Find that only the energy in the center of mass frame is available to break the bond. The remainder of the energy is needed to conserve momentum. Can calculate the energy required to break a given bond.

10. J. Matolyak and G. Matous; *Simple Variable Mass Systems: Newton's Second Law*; TPT **28**, 328 (May 1990). Another approach to the problem, using Newton's second law rather than conservation of momentum.

11. T.A. McMath; *A Dynamics Cart Demonstration: Momentum, Kinetic Energy, and More*; TPT **24**, 282 (May 1986). A rod is attached to a compressed spring on a cart. When the spring is

released the rod pushes against a rigid wall and the cart moves away from the wall. Changes in momentum and energy can be found. Useful for demonstrating energy and momentum changes in collisions.

12. J. Sherfinski; *Acceleration from the Energy Function Derivative*; TPT **26**, 228 (April 1988). When the motion is one dimensional the acceleration can be obtained from energy considerations. Examples are given.
13. R. Stephenson; *Timely equations of rocket motion and the surprising power of rockets*; AJP **57**, 322 (April 1989). Analysis of rocket position as a function of time rather than rocket velocity and acceleration as functions of mass. Power is also calculated. Find that the power of a rocket can exceed the power of its engine.
14. K. Voyenli and E. Eriksen; *On the motion of an ice hockey puck*; AJP **53**, 1149 (1985). Analysis of the effect of friction on the motion of an object across a horizontal surface. If started with both translational and rotational motion, the motions stop simultaneously. Also see letter from J.M. Daniels and response in AJP **54**, 777 (September 1986).
15. Mu-Shiang Wu; *Note on a Conveyor-belt Problem*; TPT **24**, 220 (April 1986). Sand is dropped on a moving conveyor-belt. The power supplied to keep the belt moving at constant speed is exactly twice the rate of increase of kinetic energy of the sand. The rest of the energy is lost to internal energy. This note provides a detailed analysis. Also see Maurice Bruce Stewart; *The Conveyor Belt Problem and Newton's Third Law*; TPT **27**, 193 (March 1989); Frank S. Crawford; *The Famous Conveyor-Belt Problem*; TPT **27**, 547 (October 1989); Margaret Stautberg Greenwood; *Inclined Plane on a Frictionless Surface*; TPT **28**, 109 (February 1990); Harvey S. Leff; *Conveyor-Belt Work, Displacement, and Dissipation*; TPT **28**, 172 (March 1990).

Chapter 10

1. Howard Brody; *Models of baseball bats*; AJP **58**, 756 (August 1990). Experimental evidence that hand-held bats behave like free bodies in collisions with balls.
2. Nicholas E. Brown; *Impulsive thoughts on some elastic collisions*; TPT **23**, 421 (October 1985). Analysis of the often-used swinging balls demonstration in which one or more balls collide with a group of balls initially at rest. The outcome depends on whether or not the resting balls are in contact with each other.
3. D. Easton; *Can a Fly Stop a Train?*; TPT **25**, 374 (September 1987). A thought experiment to show that two bodies deform when undergoing a collision.
4. Shulamith G. Eckstein; *Verification of fundamental principles of mechanics in the computerized student laboratory*; AJP **58**, 909 (October 1990). Computer shows position, velocity, momentum, force, and kinetic energy as functions of time during a collision between two magnetically-interacting carts on an air track.
5. F. Herrmann; *Demonstration of a slow inelastic collision*; AJP **54**, 658 (July 1986). Two gliders on an air track are coupled by a string wound on a pulley attached to one glider. The collisions last for as long as it takes for the string to unwind. Mechanical energy is obviously not conserved but momentum is. Use to convince students the time of the interaction is not important. They can see momentum being transferred.
6. David T. Kagan; *The effects of coefficient of restitution variations on long fly balls*; AJP **58**, 151 (February 1990). Of value for the data. Legally allowed variations in the coefficient of restitution might result in ranges that differ by about 15 ft.

7. W. Klein and G. Nimitz; *Inelastic collision and the motion of the center of mass*; AJP **57**, 182 (February 1989). Air track experiment. Two gliders are coupled with a spring and are initially at rest. They are hit by third glider. A flag is placed at the center of mass and its motion is observed.

8. F.C. Peterson; *Air tables and the mystery of the lost momentum*; AJP **56**, 473 (May 1988). Collision points on some air table pucks are not along the line of their center of masses. The pucks scrape on the table during the collision and momentum is lost. A remedy is suggested.

9. Robert G. Watts and Steven Baroni; *Baseball — bat collisions and the resulting trajectories of spinning balls*; AJP **57**, 40 (January 1989). Detailed analysis of the trajectory of a spinning baseball.

Chapter 11

1. Peter J. Brancazio; *Rigid-body dynamics of a football*; AJP **55**, 415 (May 1987). Analysis is beyond the introductory course but many results can be used in discussing rigid-body motion.

2. Howard Brody; *The moment of inertia of a tennis racket*; TPT **23**, 213 (April 1985). Suggest as extra reading for sports-minded students.

3. H. Brody; *The sweet spot of a baseball bat*; AJP **54**, 640 (July 1986). Interesting mechanics problem.

4. Harold A. Daw; *Coriolis lecture demonstration*; AJP **55**, 1010 (November 1987). Apparatus to show motion in a rotating frame using a camera or TV camera fixed to the frame. Includes some ideas for its use.

5. Hans C. Ohanian; *Rotational motion and the law of the lever*; AJP **59**, 182 (February 1991). An alternate derivation of $\tau = I\alpha$ for rigid body rotation about a fixed axis with a critique of a flawed derivation presented in many texts.

6. Radoslaw Szmytkowski; *Simple method of calculation of moments of inertia*; AJP **56**, 754 (August 1988). Using the parallel axes and perpendicular axes theorems to avoid calculus.

7. W.F.D. Theron; *The "faster than gravity" demonstration revisited*; AJP **56**, 736 (August 1988). One end of an initially vertical rod is fixed, then the rod rotates to the ground. Acceleration and time to fall are examined.

Chapter 12

1. B. Bagchi and Paul Holody; *Study of Projectile Motion by Angular Momentum and Torque*; TPT **29**, 376 (September 1991). Shows how to use the ideas of rotational motion to study projectile trajectories.

2. Richard E. Berg; *Traction Force on Accelerated Rolling Bodies*; TPT **28**, 600 (December 1990). Demonstration that shows the direction of the force of friction acting at the point of contact between a rolling body and the surface on which it rolls. The force can be studied as a function of the rotational inertia of the rolling body.

3. Roger Blickensdefer; *The Wheel and the Galilean Transformation*; TPT **26**, 160 (March 1988). A demonstration to show the trajectory of a point on a rolling wheel.

4. A. Domenech, T. Domenech, and J. Cebrian; *Introduction to the study of rolling friction*; AJP **55**, 231 (March 1987). Experiments for the introductory lab.

5. Martin H. Edwards; *Zero angular momentum turns*; AJP **54**, 846 (September 1986). Conservation of angular momentum when a body turns in the absence of external torque. Application to a falling cat. Also see comments by Kenneth Laws and the response by Edwards in AJP **56**, 81 (January 1988)

6. J.E. Fredrickson; *The Tail-less Cat in Free-Fall*; TPT **27**, 620 (November 1989). Photographs showing how the cat twists and extends its legs to conserve angular momentum while landing on its feet.

7. T.M. Kaalotas and A.R. Lee; *A simple device to illustrate angular momentum conservation and instability*; AJP **58**, 80 (January 1990). A spring-loaded dumbbell. While it is spinning a trigger is pressed and mass originally at the ends moves to the center, thus decreasing the rotational inertia.

8. Sol Krasner; *Why Wheels Work: A Second Version*; TPT **30**, 212 (April 1992). Takes into account deformations of the wheel.

9. L. Lam and E. Lowry; *Static Friction of a Rolling Wheel*; TPT **25**, 504 (November 1987). A plausibility argument that the point on a wheel that is contact with the surface has zero velocity if the wheel is not slipping. Hence static, not kinetic, friction is present.

10. James A. Lock; *An alternative approach to the teaching of rotational dynamics*; AJP **57**, 428 (May 1989). Physical reasoning and elementary mathematics is used to discuss torque-free rotation, with applications to motions of a top.

11. S.Y. Mak and K.Y. Wong; *A qualitative demonstration of the conservation of angular momentum in a system of two noncoaxial rotating disks*; AJP **57**, 951 (October 1989). Broadens the usual conservation of angular momentum experiments. Useful as a demonstration.

12. Robert H. March; *Who will win the race?*; TPT **26**, 297 (May 1988). How to construct an object that minimizes the time to roll down an incline plane. Use this when demonstrating the rolling of spheres and cylinders.

13. Robert J. Reiland; *Two Fundamental Surprises*; TPT **27**, 326 (May 1989). Some thoughts about pulling a yo-yo.

14. John Sherfinski; *Rotational Dynamics — Two Fundamental Issues*; TPT **26**, 290 (May 1988). Another view of rolling without slipping and a plausibility argument that the frictional force does no work.

15. Qing-gong Song; *The requirement of a sphere rolling without slipping down a grooved track for the coefficient of static friction*; AJP **56**, 1145 (December 1988). For balls moving in a wide groove the coefficient of friction required for no slipping is much less that for balls moving on a plane. Easier to obtain condition of no slipping in experiments.

16. Peter L. Tea, Jr.; *Trouble on the loop-the-loop*; AJP **55**, 826 (September 1987). Analysis of a sphere rolling without slipping on a loop-the-loop. Conditions for negotiating the top are derived.

17. Peter L. Tea, Jr.; *On seeing instantaneous centers of velocity*; AJP **58**, 495 (May 1990). Still photographs of a body with many small lights are taken. Moving lights produce curved lines on the photo, lights at rest produce dots. Photo is used to identify the point that is instantaneously at rest.

18. Martin S. Tiersten; *Moments not to forget — The conditions for equating torque and rate of change of angular momentum around the instantaneous center*; AJP **59**, 733 (August 1991). The conditions are the well-known ones but the author shows how they may be used to solve many rotational motion problems.

19. Fredy R. Zypman; *Moments to remember — The conditions for equating torque and rate of change of angular momentum*; AJP **58**, 41 (January 1990). A reminder that the moments

must be taken about a point with zero acceleration, about the center of mass, or about a point with acceleration directed toward or away from the center of mass.

Chapter 13

1. D.L. Mathieson; *The Tensile Strength of Paper*; TPT **29**, 412 (September 1991). Describes apparatus for measuring the ultimate strength.

Chapter 14

1. L.H. Cadwell and E.R. Boyko; *Linearization of the simple pendulum*; AJP **59**, 979 (November 1991). Analytic and geometric approaches to obtaining leading corrections to the simple harmonic approximation for a simple pendulum.
2. John E. Carlson; *The Pendulum Clock*; TPT **29**, 8 (January 1991). Design of pendulums that are independent of temperature and keep time to about 2 seconds per month. A little history.
3. Michael T. Frank and Edward Kluk; *Equations of Motion on a Computer Spreadsheet: The Damped Harmonic Oscillator and More*; TPT **28**, 308 (May 1990). An algorithm is given for this sometimes tricky numerical problem.
4. Margaret Stautberg Greenberg, Frances Fazio, Marie Russotto, and Aaron Wilkosz; *Using videotapes to study damped harmonic motion and to measure terminal speeds: A laboratory project*; AJP **54**, 897 (October 1986). Designed for intermediate mechanics but can be modified for an introductory lab in kinematics.
5. Thomas B. Greenslade, Jr.; *"Atwood's" oscillator*; AJP **56**, 1151 (December 1988). An oscillating system that can be used as a demonstration. Qualitative explanation can be given but detailed analysis is probably above the level of the introductory course.
6. S. Eubank, W. Miner, T. Tajima, and J. Wiley: *Interactive computer simulation and analysis of Newtonian dynamics*; AJP **57**, 457 (May 1989). A computer program for numerical simulation of one-dimensional oscillators.
7. Robert A. Nelson and M.G. Olsson; *The pendulum — Rich physics from a simple system*; AJP **54**, 112 (February 1986). Using a pendulum to measure g. The many corrections are discussed. Probably beyond the level of an introductory course but the paper provides a useful list of corrections that might be discussed in lab.
8. R.D. Peters and J.A. Shepard; *A pendulum with adjustable trends in period*; AJP **57**, 535 (June 1989). A pendulum oscillating in a container with variable pressure. Can use a leaky vacuum system to hold the period constant to microseconds over 10 min intervals.
9. Antonio Soares de Castro; *Damped harmonic oscillator: A correction in some standard textbooks*; AJP **54**, 741 (August 1986). Strictly, the amplitude is NOT given by $Ae^{-\gamma t}$. Author computes expressions for the true turning points of the motion.
10. Robert W. Stanley; *Numerical methods in mechanics*; AJP **52**, 499 (June 1984). An analysis of various numerical integration schemes. Read this before attempting numerical integration at the introductory level. Also see Margaret Stautberg Greenwood; *Comment on "Numerical methods in mechanics" [AJP **52**, 499 (1984)]*; AJP **56**, 1040 (November 1988). This paper gives a numerical integration scheme for motion of a damped pendulum.
11. Keith Turvey; *An undergraduate experiment on the vibration of a cantilever and its application to the determination of Young's modulus*; AJP **58**, 483 (May 1990). Young's modulus is determined from the fundamental frequency of vibration as a function of length.

Chapter 15

1. Albert A. Bartlett; *The Slingshot Effect: Explanation and Analogies*; TPT **23**, 466 (November 1985). Details of spacecraft orbits designed to utilize interactions with planets to increase the speed. Use for background material or assign as extra reading for interested students.
2. R.R. Boedeker; *Gravitation as an Early Example of Correspondence*; TPT **29**, 569 (December 1991). An analysis of the error made in making the usual approximation for the acceleration due to gravity near the surface of the earth.
3. S.K. Bose; *Projectiles in Circular Orbits*; TPT **29**, 568 (December 1991). An analysis is made of the correspondence between the usual equations of projectile motion and the equations of an object in orbit. Students can learn when the orbit is very nearly parabolic and when it is circular (or elliptical).
4. Robert Garisto; *An error in Isaac Newton's determination of planetary properties*; AJP **59**, 42 (January 1991). A reconstruction of Newton's calculations of the mass, surface gravity, and density of Jupiter, Saturn, and the Earth. Traces an error made in using the geocentric elongation of an orbit rather than the heliocentric elongation.
5. G.T. Gillies; *Resource Letter MNG-1: Measurements of Newtonian gravitation*; AJP **58**, 525 (June 1990). Bibliography of measurements of G and tests of the inverse-square law.
6. Laurent Hodges; *Gravitational field strength inside the Earth*; AJP **59**, 954 (October 1991). Because the density of the Earth is not uniform the gravitational field strength is greater throughout most of its volume than at the surface.
7. Donald G. Ivey; *Gravity in the Real World*; TPT **30**, 242 (April 1992). The earth's gravitational field is complicated by variations in density. The author uses a simple two-layer model of the earth demonstrate the idea.
8. Michael Martin Nieto, Richard J. Hughes, T. Goldman; *Actually, Eotvos did publish his results in 1910, it's just that no one knows about it ...*; AJP **57**, 397 (May, 1989). History of his gravity research.
9. Michael S. Saulnier and David Frisch; *Measurement of the gravitational constant without torsion*; AJP **57**, 417 (May 1989). A modification of Cavendish balance that can be used in introductory lab.
10. Jean Sivardiere; *A simple look at the Kepler motion*; AJP **56**, 132 (February 1988). Kepler's laws and some analytic geometry of conic sections is used to derive many properties of the Kepler motion.

Chapter 16

1. Henry S. Bader and Costas E. Synolakis; *The Bernoulli — Poiseuille Equation*; TPT **27**, 598 (November 1989). A discussion of how to treat viscous flow, with examples suitable for the introductory course.
2. George Barnes; *A Flettner rotor ship demonstration*; AJP **55**, 1040 (November 1987). A revolving cylinder on an air track glider propels the glider by Bernoulli's principle.
3. Robert P. Bauman; *Archimedes' Bath*; TPT **25**, 162 (March 1987). Interesting conjectures on the discovery of the principle.
4. Daniel E. Beeker; *Depth Dependence of Pressure*; TPT **28**, 486 (October 1990). The volume of a gas-filled bag is measured as it is lowered to various depths in a fluid, then the ideal gas law is used to obtain the pressure.

5. Edward H. Carlson; *A microscopic picture of Reynolds number and Stokes' law*; AJP **56**, 1045 (November 1988). Use this paper to introduce the idea of Reynolds number and to show its relationship to the drag force.

6. Ronald M. Cosby and Douglas E. Petry; *Simple Buoyancy Demonstrations Using Saltwater*; TPT **27**, 550 (October 1989). Modified versions of the Cartesian diver. Preparation details are given.

7. Samuel Derman; *A Pointed Demonstration of Surface Tension*; TPT **29**, 414 (September 1991). Describes some demonstration experiments.

8. John N. Fox, Jerry K. Eddy, and Norman W. Gaggini; *A real-time demonstration of the depth dependence of pressure in a liquid*; AJP **56**, 620 (July 1988). The output of a pressure transducer is interfaced to a computer. Pressure vs. depth is displayed on the monitor.

9. Margaret Stautberg Greenberg, Frances Fazio, Marie Russotto, and Aaron Wilkosz; *Using the Atwood machine to study Stokes' law*; AJP **54**, 904 (October 1986). Designed for intermediate mechanics but can be modified for an introductory lab. Also see E. Rune Lindgren; *Comments on "Using the Atwood machine to study Stokes' law"* [AJP **54**, 904(1986)]; AJP **56**, 940 (October 1988). Some ways to improve the experiment.

10. Thomas B. Greenslade, Jr.; *Demonstrations with a vacuum: Old Demonstrations for New Vacuum Pumps*; TPT **27**, 332 (May 1989). Guinea and feather experiment as well as some demonstrations that depend on pressure differences.

11. Dean O. Kuethe; *Confusion about Pressure*; TPT **29**, 20 (January 1991). Three common misconceptions about pressure are examined. Implications for teaching are discussed.

12. Alan L. Lehman and Thomas A. Lehman; *An illustration of buoyancy in the horizontal plane*; AJP **56**, 1046 (November 1988). Gas filled balloons and a flame in uniform circular motion in air. Air pressure increases from the center of the circle toward its circumference and balloons respond to the buoyant force thus generated. Illustrated.

13. S.Y. Mak and K.Y. Wong; *The measurement of surface tension by the method of direct pull*; AJP **58**, 791 (August 1990). A discussion of sources of experimental errors.

14. John J. McPhee and Gordon C. Andrews; *Effect of sidespin and wind on projectile trajectory, with particular application to golf*; AJP **56**, 933 (October 1988). Computer simulations predict the results of hooking and slicing.

15. Robert H. Stinson; *Classroom demonstration of streamline and turbulent flow*; AJP **59**, 1051 (November 1991). An inexpensive demonstration using an overhead projector.

16. Costa Emmanuel Synolakis, Henry S. Badeer; *On combining the Bernoulli and Poiseuille equation — A plea to authors of college physics texts*; AJP **57**, 1013 (November 1989). Fluid dynamics of a viscous fluid.

17. A. Tan; *The shape of streamlined tap water flow*; TPT **23**, 494 (November 1985). An analysis of the diameter of a stream a water flowing from a tap, with a photograph.

18. R.E. Vermillion; *Derivations of Archimedes' principle*; AJP **59**, 761 (August 1991). The principle is derived by considering the gravitational potential energy of the fluid-object system as a function of the depth to which the object is immersed.

19. Robert G. Watts, Ricardo Ferrer; *The lateral force on a spinning sphere: Aerodynamics of a curveball*; AJP **55**, 40 (January 1987). Results of wind tunnel studies.

20. Klaus Weltner; *A comparison of explanations of the aerodynamic lifting force*; AJP **55**, 50 (January 1987). Explanations based on Bernoulli's equation and on the deflection of air are examined and are shown to be consistent. Second method is better since it can be used to explain the higher air velocity above the wing.

21. Klaus Weltner; *Aerodynamic Lifting Force*; TPT **28**, 78 (February 1990). Some experiments with simple apparatus.
22. Klaus Weltner; *Bernoulli's Law and Aerodynamic Lifting Force*; TPT **28**, 84 (February 1990). Interpretation of the lifting force as a reaction to the deflection of the airstream. Gives reasoning behind Bernoulli's law.

Chapter 17

1. K. Yusuf Billah and Robert H. Scanlan; *Resonance, Tacoma Narrows bridge failure, and undergraduate physics textbooks*; AJP **59**, 118 (February 1991). The bridge failure explained from an engineering point of view, a view that is substantially different from that presented in most physics texts.
2. W.N. Mathews, Jr.; *Superposition and energy conservation for small amplitude mechanical waves*; AJP **54**, 233 (March 1986). Shows that the creation of constructively interfering waves implies the creation of destructively interfering waves and this results in the conservation of total energy. Derivation may not be appropriate for an introductory course but the results can be incorporated into lectures. Also see Comment by Abbas Asgharian and Laieh Asgharian and the response by Mathews, AJP **56**, 183 (February 1988).
3. N. Gauthier; *Derivation of the one-dimensional wave equation*; AJP **55**, 477 (May 1987). Uses the condition that a wave moves without change in shape to obtain the wave equation. Ignores forces so cannot answer the question as to why shape does not change.
4. R.E. Vermillion, H.M. Simpson; *Spinning a pulley with a vibrating cord: A marvel*; AJP **57**, 540 (June 1989). A longitudinally vibrating string, over a pulley, causes the pulley to spin in one direction. Interesting physical phenomena. Analysis given. Might be a good project for student analysis.

Chapter 18

1. Michael Bretz, M.L. Shapiro, and M.R. Moldover; *Spherical acoustic resonators in the undergraduate laboratory*; AJP **57**, 129 (February 1989). Simple equipment used to measure the speed of sound and show its dependence on pressure and temperature.
2. Jan Paul Dabrowski; *Speed of Sound in a Parking Lot*; TPT **28**, 410 (September 1990). A simple, inexpensive experiment.
3. Charlotte Farrell; *A Sound-Wave Demonstrator — For about* $10; TPT **29**, 185 (March 1991). Uses index cards, fastened to a wood frame, with small magnets attached.
4. Michael T. Frank and Edward Kluk; *Velocity of Sound in Solids*; TPT **29**, 246 (April 1991). Metal rod is hit at one end. Microphone picks up sound at the other end. Oscilloscope is used to display intensity as a function of time.
5. Donald E. Hall; *Sacrificing a Cheap Guitar in the Name of Science*; TPT **27**, 673 (December 1989). Using a musical instrument to demonstrate standing waves and resonance.
6. Brian Holmes; *The Helium-Filled Organ Pipe*; TPT **27**, 218 (March 1989). Adding helium increases the speed of sound and the fundamental frequency. Use to demonstrate the influence of the speed of sound on the frequency.
7. G.B. Karshner; *Direct method for measuring the speed of sound*; AJP **57**, 920 (October 1989). Time of flight measurement using electronic gates and an oscilloscope.
8. Thomas D. Rossing; *Resource Letter MA-2: Musical acoustics*; AJP **55**, 589 (July 1987). A bibliography. The first resource letter in this subject was in AJP 43, 944 (1975).

9. Thomas D. Rossing, ed.; *Musical Acoustics*; American Association of Physics Teachers book; Publications Sales, AAPT Executive Office, 5112 Berwyn Road, College Park, MD 20740-4100. Reprints of journal articles.
10. Donald E. Shult; *Tone Holes and Frequency of Open Pipes*; TPT **29**, 16 (January 1991). Laboratory experiments to find the end correction and the influence of holes on resonance frequencies of open pipes.
11. C.T. Tindle; *Pressure and displacement in sound waves*; AJP **54**, 749 (August 1986). Argues that sound should be described in terms of pressure waves, not displacement waves. Also see Frank Munley; *Phase and displacement in sound waves*; AJP **56**, 1144 (December 1988).
12. Herbert T. Wood; *Mechanical Analogue of the Doppler Effect*; TPT **30**, 340 (September 1992). A demonstration to help students visualize the effect. A long string is marked every "wavelength" and is pulled from a spool at a constant rate. The observer, with a stopwatch, runs toward or away from the spool, counting the number of marks passed.
13. Junry Wu; *Are sound waves isothermal or adiabatic?*; AJP **58**, 694 (July 1990). A discussion of the various factors that enter into the compressibility of a sample carrying a sound wave. Also see the comment by Pieter B. Visscher in AJP **59**, 948 (October 1991).
14. An Zhong; *An acoustic Doppler shift experiment with the signal-receiving relay*; AJP **57**, 49 (January 1989). A sensitive experiment to measure or demonstrate acoustic Doppler shifts.

Chapter 19

1. Ralph Baierlein; *The Meaning of Temperature*; TPT **28**, 94 (February 1990). The relationship between temperature and kinetic energy, negative temperatures, and changes in temperature on adiabatic compression are discussed.
2. Richard A. Bartels; *Do darker objects really cool faster?*; AJP **58**, 244 (March 1990). An interesting experiment in radiative heat transfer.
3. Manfred Bucher and Hugh A. Williamson; *Conversion of Temperature Scales*; TPT **24**, 288 (May 1986). A diagram used to show the relationship between any two scales. Use to compare Fahrenheit and Celsius, Celsius and Kelvin, etc.
4. John N. Fox; *Measurement of thermal expansion coefficients using a strain gauge*; AJP **58**, 875 (September 1990). A computer is used to make a plot of displacement vs. temperature.
5. George D. Nickas; *A thermometer based on Archimedes' principle*; AJP **57**, 845 (September 1989). A commercial toy. As the temperature changes so does the density of the fluid and various spheres in the fluid rise or fall.
6. R. Mostert; *Classroom Experiments on Thermal Expansion of Solids*; TPT **30**, 15 (January 1992). Experiments that allow students to calculate the coefficient of thermal expansion. Some use liquid nitrogen.

Chapter 20

1. Robert P. Bauman; *Physics that Textbook Writers Usually Get Wrong II. Heat and Energy*; TPT **30**, 353 (September 1992). A careful exposition of the definition of heat.
2. F. Paul Inscho; *Mechanical Equivalent of Heat*; TPT **30**, 372 (September 1992). An improvement to the common apparatus in which a long tube containing lead shot is rotated. Instructions for constructing the tube are given.
3. W.P. Lonc; *Heat experiment with a microwave oven*; AJP **57**, 51 (January 1989). A calorimetry experiment for the introductory lab.

4. W.G. Rees and C. Viney; *On cooling tea and coffee*; AJP **56**, 434 (May 1988). An application of Newton's law of cooling and an experimental investigation of factors influencing cooling.
5. Aklo Saltoh; *A New Apparatus for Measuring the Mechanical Equivalent of Heat*; TPT **25**, 97 (February 1987). Easy to construct apparatus for demonstration or lab. A hollow cylinder containing lead grains is mounted on low friction bearings and is turned by means of a crank. The work done is computed and the change in temperature of the grains is measured.

Chapter 21

1. Jorge Berger; *Kinetic illustration for thermalization*; AJP **56**, 923 (October 1988). Particles of a gas make elastic collisions with a moving piston in contact with a thermal reservoir and thereby achieve the Boltzmann velocity distribution. Too advanced for introductory course but provides background material for lectures.
2. A. Compagner; *Thermodynamics as the continuum limit of statistical mechanics*; AJP **57**, 106 (February 1989). Thermodynamics is obtained from the continuum limit rather than the usual thermodynamic limit. The analysis is beyond the introductory level but a distilled version of the idea can be presented. Useful for background information.
3. Bernard H. Lavenda; *Brownian Motion*; SA **252**, no. 2, 70 (February 1985). Use of Brownian motion to measure the mass of an atom and as a model for random processes.
4. Robert Otani and Peter Siegel; *Determining Absolute Zero in the Kitchen Sink*; TPT **29**, 316 (May 1991). Students construct a constant pressure thermometer and use it, along with the ideal gas law, to find the absolute temperature.
5. R.D. Russell; *Demonstrating Adiabatic Temperature Changes*; TPT **25**, 450 (October 1987). Experiments with simple and inexpensive apparatus.

Chapter 22

1. Albert A. Bartlett; *Applications of Refrigerator Systems*; TPT **24**, 92 (February 1986). Discussion of the differences between refrigerators, heat pumps, dehumidifiers, and air conditioners. Use to supplement discussion of first and second laws of thermodynamics.
2. Charles H. Bennett; *Demons, Engines and the Second Law*; SA **257**, no.5, 108 (November 1987). The story of the Maxwell demon and why it cannot exist, even in principle.
3. William H. Cropper; *Carnot's function: Origins of the thermodynamic concept of temperature*; AJP **55**, 120 (February 1987). A history of the idea of thermodynamic temperature and related thermodynamic concepts.
4. Hans U. Fuchs; *Entropy in the teaching of introductory thermodynamics*; AJP **55**, 215 (March 1987). Author argues that entropy is more easily understood than heat. Treats entropy and temperature as fundamental and bases thermodynamics on them.
5. J.M. Gordon; *Maximum power point characteristics of heat engines as a general thermodynamic problem*; AJP **57**, 1136. The influence of irreversibilities, including the use of reservoirs with finite heat capacities, are analyzed and are shown to lead to a maximum power output that does not coincide with maximum efficiency.
6. Harvey S. Leff and Andrew F. Rex; *Resource Letter MD-1: Maxwell's demon*; AJP **58**, 201 (March 1990). A bibliography of ideas and history.
7. Thomas V. Marcella; *Entropy production and the second law of thermodynamics: An introduction to second law analysis*; AJP **60**, 888 (October 1992). The first law of thermodynamics is

written in terms of the entropy increase and used to analyze various processes. The analysis can be used to calculate the energy that becomes unavailable to do work.

8. Levi Tansjp; *Comment on the discovery of the Second Law*; AJP **56**, 179 (February 1988). A history of the discovery of the second law of thermodynamics.

Chapter 23

1. Gary Benoit and Mauri Gould; *A New Device for Studying Electric Fields*; TPT **29**, 182 (March 1991). An electronic circuit, featuring a field-effect transistor, replaces the usual gold-leaf electroscope.
2. Peter Heering; *On Coulomb's inverse square law*; AJP **60**, 988 (November 1992). A history of Coulomb's famous experiment and attempts to reproduce the results. The authors conclude that the validity of the data was in doubt and that formulation of the inverse square law of electrostatics was based more on theory than on experiment.
3. Leslie E. Mathes; *Where to Buy Electricity and Magnetism Software*; Computers in Physics **2**, 89 (November/December 1988). A compendium of software covering many topics in electricity and magnetism, with short descriptions and vendor addresses.
4. Walter Roy Mellen; *Inexpensive fun with Electrostatics*; TPT **27**, 86 (February 1989). Description of demonstration experiments.

Chapter 24

1. Ludwik Kowalski; *A Short History of the SI Units in Electricity*; TPT **24**, 97 (February 1986). Background information for curious students (and their instructors).
2. L. Kristjansson; *On the drawing of lines of force and equipotentials*; TPT **23**, 203 (April 1985). An examination of diagrams from textbooks with some thoughts on why students might be confused. Implications for teaching are discussed.
3. Ross L. Spencer; *Electric field lines near an oddly shaped conductor in a uniform electric field*; AJP **56**, 510 (June 1988). Numerical computation of potential and field. Plots by computer.

Chapter 26

1. Richard E. Berg; *Van de Graaff Generators: Theory, Maintenance, and Belt Fabrication*; TPT **28**, 281 (May 1990). If you use these machines in lab, you might find this article useful either as a source of discussion material or for its description of maintenance procedures.
2. Wayne A. Bowers; *Still more on the Coulomb potential*; AJP **57**, 375 (April 1989). A clever scaling technique is used to derive the coulomb potential without the use of calculus.
3. Douglas A. Kurtze; *A noncalculus derivation of the $1/r$ potential*; AJP **60**, 457 (May 1992). Dimensional analysis is used.
4. Robert D. Smith; *Electrostatic field plotting apparatus*; AJP **58**, 410 (April 1990). An easy to make, durable apparatus that replaces the apparatus using conductive paper.

Chapter 27

1. R.E. Benenson; *Direct observation of the force on a dielectric*; AJP **59**, 763 (August 1991). Description of the apparatus and results.

2. Edwin A. Karlow; *Let's Measure the Dielectric Constant of a Piece of Paper!*; TPT **29**, 23 (January 1991). The time to charge a parallel plate capacitor to a fixed potential difference is measured, with and without a piece of paper between its plates. The results are used to calculate the dielectric constant.
3. L. Kowalski; *A myth about Capacitors in Series*; TPT **26**, 286 (May 1988). Potential differences across capacitors, taking leakage resistance into account.
4. Helene F. Perry, Randall S. Jones, and Gregory N. Derry; *Capacitors in Parallel and Series*; TPT **29**, 348 (September 1991). An experiment to find the distribution of charge and electric potential for systems of capacitors.
5. Donald M. Trotter, Jr.; *Capacitors*; SA **259**, no. 1, 86 (July 1988). Construction and uses of modern capacitors.
6. David E. Wilson; *A direct laboratory approach to the study of capacitors*; AJP **57**, 630 (July 1989). Description of an experiment in which a constant current source is used to charge a capacitor. The charging is timed so the charge can be computed.
7. Constantino A. Utreras-Diaz; *Dielectric slab in a parallel-plate condenser*; AJP **56**, 700 (August 1988). Direct calculation of the force without recourse to the well-known energy technique.

Chapter 28

1. Samuel D. Harper; *The energy dissipated in a switch*; AJP **56**, 886 (October 1988). Energy dissipated is computed by taking the switch to be a resistor whose resistance varies from 0 to infinity as it is opened. Problem is worked for both resistive and capacitive loads.

Chapter 29

1. Robert P. Bauman and Saleh Adams; *Misunderstandings of Electric Current*; TPT **28**, 334 (May 1990). A discussion of some common misconceptions. It is worthwhile reading this article before teaching about electric current.
2. L.V. Hmurcik and J.P. Micinilio; *Contrasts between Maximum Power Transfer and Maximum Efficiency*; TPT **24**, 492 (November 1986). The standard problem of a battery with internal resistance coupled to a load resistor. Maximum power transfer and maximum efficiency do not occur for the same load resistance. A plea to discuss efficiency when discussing power transfer.
3. Lillian C. McDermott and Peter S. Shaffer; *Research as a guide for curriculum development: An example from introductory electricity. Part I: Investigation of student understanding*; AJP **60**, 994 (November 1992) and *Part II: Design of instructional strategies*; AJP **60**, 1003 (November 1992). Conceptual difficulties students have with electrical circuits and suggestions for teaching.
4. Melvin S. Steinberg; *Transient Lamp Lighting with High-Tech Capacitors*; TPT **25**, 95 (February 1987). Uses non-polar capacitors of up to 1 F to achieve long time constants in circuits with lamps. Can demonstrate, for example, that lamps near the battery do not turn on sooner than lamps far away.

Chapter 30

1. A.K.T. Assis and F.M. Peixoto; *On the Velocity in the Lorentz Force Law*; TPT **30**, 480 (November 1992). An argument that the velocity must be measured relative to the inertial frame in which the fields are **E** and **B** and not, for example, relative to the sources of the fields.

2. Rodney C. Cross; *Magnetic lines of force and rubber bands*; AJP **57**, 722 (August 1989). Detailed analysis of the validity of thinking of magnetic field lines as stretched strings, pulling and pushing on each other.

3. S.Y. Mak and K. Young; *Floating metal ring in an alternating magnetic field*; AJP **54**, 808 (September 1986). Analysis of a levitating ring in an alternating field. Net force is due to some rather small phase differences.

4. Joseph E. Price; *Electron trajectory in an e/m experiment*; AJP **55**, 18 (January 1987). Detailed calculation at the undergraduate level. Includes inhomogeneity of field and explains fanning out of the beam. References to other detailed calculations.

Chapter 31

1. C. Christodoulides; *Comparison of the Ampère and Biot-Savart magnetostatic force laws in their line-current-element forms*; AJP **56**, 357 (April 1988). Shows the two laws can produce different results and claims differences are due to oversimplified and unrealistic geometrical models for current loops. Also see Cynthia Kolb Whitney; *On the Ampère/Biot-Savart discussion*; AJP **56**, 871 (October 1988) and C. Christodoulides; *On the equivalence of the Ampère and Biot-Savart magnetostatic force laws*; AJP **57**, 680 (August 1989).

2. W.G. Delinger; *Magnetic Field Inside a Hole in a Conductor*; TPT **28**, 234 (March 1990). A nice way to work the problem using the vector nature of current density.

3. Herman Erlichson; *The magnetic field of a circular turn*; AJP **57**, 607 (July 1989). Numerical solution for the field anywhere. Program can be run by introductory students. Useful for lab.

4. Lawrence B. Golden, James R. Klein, and Luisito Tongson; *An introductory low-cost magnetic field experiment*; AJP **56**, 846 (September 1988). Students investigate the field of a solenoid with a Hall effect probe.

5. H.G. Gnanatilaka and P.C.B. Fernando; *An investigation of the magnetic field in the plane of a circular current loop*; AJP **55**, 341 (April 1987). Experimental details. Making graphs of the field as a function of position gives students a qualitative understanding of the shape of the field.

6. Peter Heller; *Analog demonstrations of Ampere's law and magnetic flux*; AJP **60**, 17 (January 1992). An apparatus for measuring the magnetic field and details of the data can be used to verify Ampere's law. Also see the erratum that appears in AJP **60**, 274 (March 1992).

7. Arthur Hovey; *On the magnetic field generated by a short segment of current*; AJP **57**, 613 (July 1989). A series of simple experiments to convince students of the validity of the Biot-Savart law.

8. Edward M. Purcell; *Helmholtz coils revisited*; AJP **57**, 18 (January 1989). Four coils can be positioned so constant and 32-pole fields dominate. Analysis is beyond the introductory course but the paper might be used as a guide in setting up some interesting experimental situations.

9. C.E. Zaspel; *An inexpensive Ampère's law experiment*; AJP **56**, 859 (September 1988). Uses a magnetic compass and knowledge of the earth's field to measure the magnetic field of a long straight wire.

Chapter 32

1. Christopher C. Jones; *Faraday's law apparatus for the freshman laboratory*; AJP **55**, 1148 (December 1987). A search coil with output to an oscilloscope is used to investigate the emf

induced by a time varying current in a field coil. Several exercises are described. Can be used as a laboratory exercise or a demonstration.

2. David T. Kagan; *Measuring the Earth's Magnetic Field in an Introductory Laboratory with a Spinning Coil*; TPT **24**, 423 (October 1986). Describes apparatus that can replace the usual flip coil and avoids the measurement of charge.
3. R.C. Nicklin; *Faraday's law — Quantitative experiments*; AJP **54**, 422 (May 1986). Computer is used to calculate induced emf. Several experiments are described.
4. Joseph Priest and Bryant Wade; *A Lenz Law Experiment*; TPT **30**, 106 (February 1992). The popular demonstration of a cylindrical magnet falling through a metal tube is made quantitative.
5. Thomas D. Rossing and John R. Hull; *Magnetic Levitation*; TPT **29**, 552 (December 1991). An important application of magnetic fields and induced currents. Use it to help motivate students or as a source for a term paper.
6. W.M. Saslow; *Electromechanical implications of Faraday's law: A problem collection*; AJP **55**, 986 (November 1987). A collection of problems dealing with Faraday's law, some suitable for the introductory course.
7. P. Seligmann; *The Earth's magnetic field — A new technique*; AJP **55**, 379 (April 1987). Standard Faraday's law experiment, but the ballistic galvanometer is replaced by an inexpensive integration circuit. Circuit is described.
8. L. Pearce Williams; *Why Ampère did not discover electromagnetic induction*; AJP **54**, 306 (April 1986). Interesting history.

Chapter 33

1. Frank S. Crawford; *Mutual inductance $M_{12} = M_{21}$: An elementary derivation*; AJP **60**, 186 (February 1992). Energy considerations are applied to two coupled circuits. This paper provides the derivation omitted from the text.
2. An Zhong; *Determine the Magnetic Induction of a Coil with a Hall Element*; TPT **28**, 123 (February 1990). Experimental apparatus is described.

Chapter 34

1. Alfred S. Goldhaber; *Resource Letter MM-1: Magnetic monopoles*; AJP **58**, 429 (May 1990). A bibliography of the search and implications of the results.
2. Kenneth A. Hoffman; *Ancient Magnetic Reversals: Clues to the Geodynamo*; SA **258**, no. 5, 76 (May 1988). Reversals of the earth's magnetic field, how they are studied, and what they tell us about the source of the field.
3. Edward W. Holmes, Jr.; *The Earth's Magnetotail*; SA **254**, no.3, 40 (March 1986). Solar wind deforms earth's magnetic field, giving it a tail.
4. S.L. O'Dell and R.K.P. Zia; *Classical and semiclassical diamagnetism: A critique of treatment in elementary texts*; AJP **54**, 32 (January 1986). Classical physics predicts diamagnetism does not exist but semiclassical arguments can be used to discuss it.
5. Yunchi Meng, Zhujian Liang; *Improvements in the demonstration of the hysteresis loops of ferromagnetic materials*; AJP **55**, 933. Hysteresis loops can be displayed on an oscilloscope. Use for lecture demonstration.

6. S.K. Runcorn; *The Moon's Ancient Magnetism*; SA **257**, no. 6, 60 (December 1987). Evidence for a magnetic field of the moon.

Chapter 38

1. V.M. Babović, D.M. Davidović, and B.A. Aničin; *The Doppler interpretation of Rømer's method*; AJP **59**, 515 (June 1991). Using the ideas of the Doppler effect to derive Rømer's expression for the speed of light from observations of the moon's of Jupiter.
2. Harry E. Bates; *Resource Letter RMSL–1: Recent measurements of the speed of light and the redefinition of the meter*; AJP **56**, 682 (August 1988). Bibliography.
3. Craig F. Bohren; *Multiple scattering of light and some of its observable consequences*; AJP **55**, 524 (June 1987). Analysis is beyond introductory course but listing and qualitatively discussing some of the consequences of multiple scattering might liven a lecture.
4. E.F. Carr and J.P. McClymer; *A laboratory experiment on interference of polarized light using a liquid crystal*; AJP **59**, 366. Can be used to demonstrate polarized light, properties of liquid crystals, or interference.
5. G.R. Davies; *Polarized Light Corridor Demonstrations*; TPT **28**, 464 (October 1990). A dozen demonstrations that can be setup in a hallway; includes various polarizing mechanisms, the Faraday effect, interference, and double refraction.
6. Werner B. Schneider; *A surprising optical property of Plexiglas rods — An unusual approach to birefringence*; AJP **59**, 1086 (December 1991). Paper shows how to use Plexiglas to demonstrate elliptical polarization and to construct quarter-wave and half-wave plates.
7. Joseph L. Spradley; *Hertz and the Discovery of Radio Waves and the Photoelectric Effect*; TPT **26**, 492 (November 1988). A short history.
8. Herman Winick; *Synchrotron Radiation*; SA **257**, no. 5, 88 (November 1987). Generation and uses of electromagnetic radiation from particle accelerators.

Chapter 39

1. Sue Allen; *The Gaussian formula and the elusive fourth principal ray*; AJP **60**, 160 (February 1992). A nice derivation of the lens and mirror equation using symmetry arguments rather than the law of refraction or reflection.
2. Richard Atneosen and Richard Feinberg; *Learning optics with optical design software*; AJP **59**, 242. Software used to design lenses can be used to teach about reflection, refraction, spherical mirrors, lenses, spherical and chromatic aberrations.
3. Craig F. Bohren and Allistair B. Fraser; *Newton's zero-order rainbow: Unobservable or nonexistent?*; AJP **59**, 325 (April 1991). Newton predicted a zero-order rainbow, resulting from two refractions and no reflection, but it has never been observed. Reasons are discussed. Paper can be used as part of general discussion of rainbows.
4. John W.W. Burrows; *Derivation of the mirror equation*; AJP **54**, 432 (May 1986). An alternative derivation that indicates the mirror equation is applicable to some mirrors that are not circular in cross section.
5. Paul Chagnon; *Animated Displays II: Multiple Reflections*; TPT **30**, 488 (November 1992). A hall display consisting of multiple mirrors that rotate and objects that move in front of multiple-mirror systems. Construction details are given.

6. Samuel Derman; *Ray Tracing with Spherical and Parabolic Reflectors*; TPT **28**, 590 (December 1990). Ray tracing exercises to show that off-axis rays are defocused by a spherical mirror and that all paraxial rays are focused by a parabolic mirror. You might have students try these exercises in conjunction with a laboratory experiment.

7. Harry D. Downing and Walter L. Trikosko; *Ray Tracing for a Few £*; TPT **29**, 369 (September 1991). Demonstration apparatus for ray tracing. Moving parts allow you to show how the image position and size change as the object is moved.

8. Igal Galili, Fred Goldberg, and Sharon Bendall; *Some Reflections on Plane Mirrors and Images*; TPT **29**, 471 (October 1991). Using demonstrations and ray diagrams to explore common misconceptions about images.

9. Fred M. Goldberg and Lillian C. McDermott; *Student Difficulties in Understanding Image Formation by a Plane Mirror*; TPT **24**, 472 (November 1986). Summary of pedagogic research results, with implications for teaching.

10. Fred M. Goldberg and Lillian C. McDermott; *An investigation of student understanding of the real image formed by a converging lens or concave mirror*; AJP **55**, 108 (February 1987). Results of a study of student misconceptions about elementary optics with implications for instruction.

11. Fred Goldberg, Sharon Bendall, and Igal Galili; *Lenses, Pinholes, Screens, and the Eye*; TPT **29**, 221 (April 1991). Some demonstrations to illustrate the roles of the eye and screen in observing images.

12. Ludwik Kowalski; *On Field Lenses*; TPT **30**, 366 (September 1992). A suggestion on how to draw ray diagrams for multi-lens optical systems. The implication that a ray bends at an intermediate image should be avoided.

13. A.F. Leung; *Wavelength of light in water*; AJP **54**, 956 (October 1986). A demonstration experiment.

14. Alfred F. Leung, Simon George, and Robert Doebler; *Refractive Index of a Liquid Measured with a He-Ne Laser*; TPT **29**, 226 (April 1991). Describes apparatus for measuring angles of incidence and refraction.

15. Alfred F. Leung and Simon George; *Simple Homemade Container for Measuring Refractive Index*; TPT **30**, 438 (October 1992). A container made of microscope slides holds the material whose index is to be found. The bending of a laser beam by the material is measured and the result is used to calculate the refractive index.

16. B.F. Melton; *A Surprising Demonstration of Total Internal Reflection*; TPT **29**, 539 (November 1991). Marbles in a Florence flask, covered by sufficient water, cannot be seen from above. Use as a hallway demonstration.

17. Malcolm R. Howells, Janos Kirz, and David Sayre; *X-ray Microscopes*; SA **264**, no. 2, 80 (February 1991). Use for special reading assignments. Students can compare and optical microscopes.

18. William S. Heaps; *How not to focus a small source with a single lens*; AJP **55**, 888 (October 1987). The chief message is that the image is not in focus when it has the smallest diameter.

19. Stuart Leinoff; *Ray Tracing with Virtual Objects*; TPT **29**, 275 (May 1991). Discusses some of the pitfalls of ray tracing when the image of one lens is behind a second lens.

20. Gordon P. Ramsey; *Reflective Properties of a Parabolic Mirror*; TPT **29**, 240 (April 1991). A simple mathematical proof that light incident along the optic axis is reflected to the focal point and light through the focal point is reflected parallel to the axis.

21. Kenneth N. Taylor; *Measuring Focal Lengths and Principal Planes of Complex Lens Systems*; TPT **25**, 484 (November 1987). Experiments using what is known as the magnification method are described.

Chapter 40

1. Ralph Baierlein and Vacek Miglus; *Illustrating double-slit interference: Yet another way*; AJP **59**, 857 (September 1991). An easily built model featuring representations of wave trains painted on long wooden slats.
2. Göran Rämme; *Colors on Soap Films — An Interference Phenomenon*; TPT **28**, 479 (October 1990). Experimental setup to observe interference patterns of both reflected and transmitted light.

Chapter 41

1. B.A. Aničin, V.M. Babovic, and D.M. Davidovic; *Fresnel lenses*; AJP **57**, 312 (April 1989). A history of their invention.
2. Andrew DePino, Jr.; *Unusual Diffraction Patterns*; TPT **25**, 219 (April 1987). Showing the diffraction patterns produced by various common articles (screws, capacitors, pin, etc.), illuminated by a He-Ne laser.
3. Simon George and Robert Doebler; *Index of Refraction Using a Diffraction Pattern in a Fishtank*; TPT **29**, 462 (October 1991). The radius of the first minimum of the diffraction pattern of fine powder particles, measured in air and in a liquid, is used to find the index of refraction of the liquid.
4. James E. Kettler; *The compact disk as a diffraction grating*; AJP **59**, 367 (April 1991). Students measure the groove spacing using light with a known wavelength or else measure the wavelength and assume the groove spacing has the value stated by the manufacturer.
5. A.F. Leung and Simon George; *Diffraction intensity of a phase grating submerged in different liquids*; AJP **57**, 854 (September 1989). The intensity of diffracted light is strongly dependent on the index of refraction of the liquid surrounding the grating. Can be made into a demonstration experiment.
6. Christian Nöldeke; *Compact Disc Diffraction*; TPT **28**, 484 (October 1990). An experiment to use diffraction to measure the separation of tracks on a compact disc.
7. Philip Sadler; *Projecting Spectra for Classroom Demonstrations*; TPT **29**, 423 (October 1991). How to build and use an inexpensive spectrum projector.
8. Mu-Shiang Wu, Shu-Ming Yang; *Dispersion and resolving power of a grating*; AJP **54**, 735 (August 1986). Explains distinction between these two concepts.

Chapter 42

1. Ralph Baierlein; *Teaching $E = mc^2$*; AJP **57**, 391 (May 1989). Emphasizes that energy and mass change in parallel. One is not converted to the other. Also see: Ralph Baierlein; *Teaching $E = mc^2$: An Exploration of Some Issues*; TPT **29**, 170 (March 1991). Use these papers here or when covering the nuclear physics chapters.
2. S.P. Boughn; *The case of the identically accelerated twin*; AJP **57**, 791 (September 1989). Variation of twin paradox: both twins accelerate identically over identical times, but they age

differently. Use to dispel the notion that the acceleration of one twin accounts for the different aging in the usual statement of the paradox. Also see: Edward A. Desloge and R.J. Philpott; *Comment on "The case of the identically accelerated twins," by S.P Boughn [Am.J.Phys. 57, 791-793 (1989)]*; AJP **59**, 280 (March 1991).

3. G. Cook and T. Lesoing; A simple derivation of the Doppler effect for sound; AJP **59**, 218 (March 1991). The Lorentz transformation is used.
4. Edward A. Desloge and R.J. Philpott; *Uniformly accelerated reference frames in special relativity*; AJP **55**, 252 (March 1987). Deals with twin paradox and some phenomena of general relativity.
5. W.L. Fadner; *Did Einstein really discover "$E = mc^2$"?*; AJP **56**, 114 (February 1988). A history of the relationship between mass and energy.
6. Robert W. Flynn; *The Relativistic Velocity Addition Formula*; TPT **29**, 524 (November 1991). A graphical interpretation.
7. N. David Merman; *The amazing many colored relativity engine*; AJP **56**, 601 (July 1988). Description of a computer program that helps students explore special relativity. No knowledge of physics, algebra, or geometry is required to use the program.
8. Asher Peres; *Relativistic telemetry*; AJP **55**, 516 (June 1987). Some well-known results of special relativity (addition of velocities, time dilation, etc.) derived directly from the postulates without use of the Lorentz transformation.
9. P.C. Peters; *An alternate derivation of relativistic momentum*; AJP **54**, 804 (September 1986). Uses a non-head-on elastic collision rather than the usual one dimensional collision.
10. Fritz Rohrlich; *An elementary derivation of $E = mc^2$*; AJP **58**, 348 (April 1990). Uses non-relativistic definitions of momentum and kinetic energy and the Doppler shift for light to analyze an experiment in which two photons are emitted by a mass at rest and produces the famous Einstein result. Also see the comment by Lawrence Ruby and Robert E. Reynolds; AJP **59**, 756 (August 1991) and the response by F. Rohrlich; AJP **59**, 757 (August 1991).
11. T.R. Sandin; *In defense of relativistic mass*; AJP **59**, 1032 (November 1991). The author advocates use of relativistic mass as a unifying and simplifying concept.
12. G.P. Sastry; *Is length contraction really paradoxical?*; AJP **55**, 943 (October 1987). A new paradox is described and explained. Appendix contains a bibliography of length contraction paradoxes.
13. Daniel D. Skwire and Lawrence J. Badar; *The Life and Legacy of Edward Williams Morley*; TPT **25**, 556 (December 1987). Biography. Use to augment information about Michelson.
14. James M. Supplee; *Relativistic buoyancy*; AJP **57**, 75 (January 1989). The Lorentz contraction means that the buoyant force is velocity dependent. Some ramifications are discussed.
15. Edwin F. Taylor; *Space–time software: Computer graphics utilities in special relativity*; AJP **57**, 508 (June 1989). Description of the programs and how to use them.
16. Edwin F. Taylor; *Why does nothing move faster than light? Because ahead is ahead!*; AJP **58**, 889 (September 1990). An argument you might use in lecture.

Chapter 43

1. Richard Kidd, James Ardini, and Anatol Anton; *Evolution of the modern photon*; AJP **57**, 27 (January 1989). Examines four distinct models of the photon, in historical context.
2. Jesusa Valdez Kinderman; *Investigating the Compton Effect with a Spreadsheet*; TPT **30**, 426 (October 1992). Spreadsheet-generated pie graphs are used to display the results of a Compton scattering event. The random number generator is used to select the scattering angle.

3. Abner Shimony; *The Reality of the Quantum World*; SA **258**, no. 1, 46 (January 1988). Experiments that demonstrate wave-particle duality for photons.
4. Daniel Wilkins; *General Compton effect via a Lorentz transformation*; AJP **56**, 1044 (November 1988). Uses Lorentz transformation to discuss Compton effect when target electron is not initially at rest.

Chapter 44

1. François Bardou; *Transition between particle behavior and wave behavior*; AJP **59**, 458 (May 1991). Double slit experiment with particles. Use photographic plate to detect particles. The distance from the slits to the plate determines the ability to decide which slit the particle came through.
2. Robert Deltete and Reed Guy; *Einstein's opposition to the quantum theory*; AJP **58**, 673 (July 1990). Point out this paper to students who are interested in the ideas of quantum theory.
3. G. Matteucci; *Electron wavelike behavior: A historical and experimental introduction*; AJP **58**, 1143 (December 1990). Includes photographs of electron diffraction patterns.
4. A. Tonomura, J. Endo, T. Matsuda, T. Kawasaki, and H. Ezawa; *Demonstration of single-electron buildup of an interference pattern*; AJP **57**, 117 (February 1989). Two-slit diffraction with electrons from an electron microscope. Cannot be done in introductory lab for lack of equipment but should be referenced to show it can be done.

Chapter 45

1. R. Gupta; *Resource Letter LS-1: Laser spectroscopy*; AJP **59**, 874 (October 1991). A bibliography of the techniques of laser spectroscopy. Some papers are elementary and might be used by introductory students as the basis for term papers.
2. P.W. Milonni; *Why $\ell(\ell+1)$ instead of ℓ^2?*; AJP **58**, 1012 (October 1990). A retelling of Feynmann's proof, at a level for introductory students.

Chapter 46

1. E.H. Brandt; *Rigid levitation and suspension of high-temperature superconductors by magnets*; AJP **58**, 43 (January 1990). Use for background material if you perform the demonstration.
2. Judith Bransky; *Superconductivity — A New Demonstration*; TPT **28**, 392 (September 1990). A superconducting ring encloses a transformer core and is wired into the coil. When it is cooled below the transition temperature it excludes flux and the output voltage of the transformer drops to zero.
3. E.A. Early, C.L. Seaman, K.N. Yang, M.B. Maple; *Demonstrating superconductivity at liquid nitrogen temperatures*; AJP **56**, 617 (July 1988). Some demonstration experiments.
4. Yoshihiro Hamakawa; *Photovoltaic Power*; SA **256**, no. 4, 86 (April 1987). Power plants based on conversion of power from the sun.
5. W. Klein; *In memoriam J. Jaumann: A direct demonstration of the drift velocity in metals*; AJP **55**, 22 (January 1987). Move sample so the electrons are at rest relative to the magnetic field in a Hall effect experiment. Hall voltage then vanishes.
6. Alan M. Wolsky, Robert F. Giese, and Edward J. Daniels; *The New Superconductors: Prospects for Applications*; SA **260**, no. 2, 60 (February 1989). Possible technological applications of high transition temperature superconductors.

Chapter 47

1. Harry Manos; *Using linear accelerators with high-school students*; AJP **57**, 139 (February 1989). Experiments students can do if a small (4 MeV) linear accelerator is available.
2. Michael K. Moe and Simon Peter Rosen; *Double-Beta Decay*; SA **261**, no. 5, 48 (November 1989). Descriptions of the experiments and what we can learn from them.
3. M.F. Rayner-Canham and G.W. Rayner-Canham; *Pioneer women in nuclear science*; AJP **58**, 1036 (November 1990). Background material.
4. Joseph L. Spradley; *Yukawa and the birth of meson theory*; TPT **23**, 283 (May 1985). A history of the beginnings of the meson theory of nuclear forces. Use for background material.
5. Joseph L. Spradley; *Women and the Elements*; TPT **27**, 656 (December 1989). Short biographies of 6 women who played important roles in nuclear physics.

Chapter 48

1. Derek A. Boyd; *Taking the Temperature of a Tokamak*; TPT **25**, 22 (January 1987). A low-level discussion of the light emitted by the plasma of a tokamak and how it is used to find the temperature.
2. R. Stephen Craxton, Robert L. McCrory, and John M. Soures; *Progress in Laser Fusion*; SA **255**, no. 2, 68 (August 1986). Discussion of latest techniques.
3. J.A. Fillo and P. Lindenfeld; *Introduction to Nuclear Fusion Power and the Design of Fusion Reactors*; American Association of Physics Teachers book; Publications Sales, AAPT Executive Office, 5112 Berwyn Road, College Park, MD 20740-4100.
4. Melvin M. Levine; *Fission Reactors*; American Association of Physics Teachers book; Publications Sales, AAPT Executive Office, 5112 Berwyn Road, College Park, MD 20740-4100.

Chapter 49

1. *Fundamental Particles and Interactions, A Wall Chart of Modern Physics*; TPT **26**, 556 (December 1988). Hang on wall of hallway or lecture hall.
2. Gordon Aubrecht, ed.; *Quarks, Quasars, and Quandries*; American Association of Physics Teachers book; Publications Sales, AAPT Executive Office, 5112 Berwyn Road, College Park, MD 20740-4100. Summaries of advances in particle physics and cosmology to 1986.
3. Steven Detweiler, ed.;*Black Holes*; American Association of Physics Teachers book; Publications Sales, AAPT Executive Office, 5112 Berwyn Road, College Park, MD 20740-4100. Reprints of 16 journal articles.
4. Gary J. Feldman and Jack Steinberger; *The Number of Families of Matter*; SA **264**, no. 2, 70 (February 1991). Description of electron-positron collision experiments at CERN and SLAC and the conclusions drawn from them.
5. Samuel Gulkis, Philip M. Lubin, Stephen S. Meyer, and Robert F. Silverberg; *The Cosmic Background Explorer*; SA **262**, no. 1, 132 (January 1990). Description of a satellite to detect the microwave background radiation.
6. Donald E. Hall; *The Hazards of Encountering a Black Hole*; TPT **23**, 540 (December 1985) and TPT **24**, 29 (January 1986). Discussion of black holes at the introductory level. Good for background material or as extra reading for curious students.

7. Wendell G. Holladay; *Hadronic numerics with quarks of f flavors*; AJP **60**, 711 (August 1992). A listing of the various possibilities for baryons and mesons, without resort to group theory. This might help to make the hadron multiples more understandable to your students.
8. John Horgan; *Universal Truths*; SA **263**, no. 4, 108 (October 1990). Discussion of cosmological issues, particularly the origin of the universe.
9. Goronwy Tudor Jones; *The Physical Principles of Particle Detectors*; TPT **29**, 578 (December 1991). A description of how detectors are used to measure physical quantities such as the mass, energy, momentum, and charge. Some parts may be above the level of an introductory course but you can use most of the article for lecture material.
10. Edward W. Kolb and Chris Quigg; *Probing the Structure of the Universe from Quarks to Cosmology*; TPT **24**, 528 (December 1986). How research into fundamental particles sheds light on cosmological problems. A long review of the work in both fields and a discussion of how the fields are related.
11. Helge Kragh; *The negative proton: Its earliest history*; AJP **57**, 1034 (November 1989).
12. Lawrence M. Krauss; *Dark Matter in the Universe*; SA **255**, no. 6, 58 (December 1986). Motions of galaxies indicate the existence of more matter than is seen. Particle physics and astrophysics give clues as to what it is.
13. David Lindley, Edward W. Kolb, and David N. Schramm; *Resource Letter CPP-1: Cosmology and particle physics*; AJP **56**, 492 (June 1988). Bibliography.
14. Chris Quigg; *Elementary Particles and Forces*; SA **252**, no. 4, 84 (April 1985). Describes coherent view of matter and forces in terms of fundamental particles.
15. Thomas A. Roman; *General relativity, black holes, and cosmology: A course for nonscientists*; AJP **54**, 144 (February 1986). Good ideas for teaching the concepts without advanced mathematics.
16. David N. Schramm and Gary Steigman; *Particle Accelerators Test Cosmological Theories*; SA **258**, no. 6, 66 (June 1988). The close connection between cosmology and particle physics.
17. Roger H. Stuewer; *The naming of the deuteron*; AJP **54**, 206 (March 1986). Interesting history. Tidbits for lectures.
18. Massimo Tinto; *The search for gravitational waves*; AJP **56**, 1066 (December 1988). An introduction to sources of gravitational waves and their detection.
19. Charles H. Townes and Reinhard Genzel; *What is Happening at the Center of the Galaxy?*; SA **262**, no. 4, 46 (April 1990). Evidence that the center is massive black hole.

SECTION FOUR
ANSWERS TO SELECTED QUESTIONS

In this section of the manual, answers are given to many of the end-of-chapter questions suggested in the Lecture Notes. Roughly a fourth of the questions in the extended version of *Fundamentals of Physics* are answered. Questions that contain references to discussions elsewhere are not answered here and usually only one or two of a set of similar questions are answered. The answers given cover the main points; detailed answers are not given where these are long and involved.

Some of the questions highlight common misconceptions of students. It is often worthwhile to assign some of these questions prior to or along with problems. In some instances, questions and problems are linked in the SUGGESTIONS sections of the Lecture Notes.

Many of the questions concern applications of the theory to areas not discussed in the text. These are an excellent source of lecture and recitation section material and might form the basis for student projects and term papers. In some cases, reference is made in the question to material in the literature and the references, in turn, give bibliographies or suggest a series of experiments.

Some instructors, in an attempt to have students think qualitatively about physical phenomena, ask questions similar to those in the text as examination questions. In some cases, both qualitative and quantitative aspects can be combined as separate parts of the same question.

Chapter 1

1. If a great many measurements show the same systematic variation, the variation might be attributed to the standard and a new standard sought. Remark on the variation of the solar day as measured by a cesium clock. When the current mean solar day was used as a standard, systematic variations on the order of 10^{-8} s occurred in time measurements. The standard was changed to a particular mean solar day, then to a cesium clock.

2. Ideally, comparison with secondary standards should be easy to carry out with high precision. The standard should be indestructible, easy to store, safe to use, and convenient in size.

7. The bar expands and contracts with changes in temperature. If two length measurements of the same object at the same temperature are made using the standard at two different temperatures, different results are obtained. To avoid this, the bar is taken as the standard only when it is at a certain temperature. No inconsistencies occur as long as the measurement of temperature is well defined.

8. Both possibilities were open to them. The definition they accepted does not change length and speed measurements made prior to the new definition.

10 a. Simultaneously at two points on the Earth's surface find the angle between the vertical and the line to a star. Use these angles and the arc length to calculate the radius of the Earth. For the case shown $\alpha = \theta_1 + \theta_2$ and $r = s/\alpha$, where α is in radians.

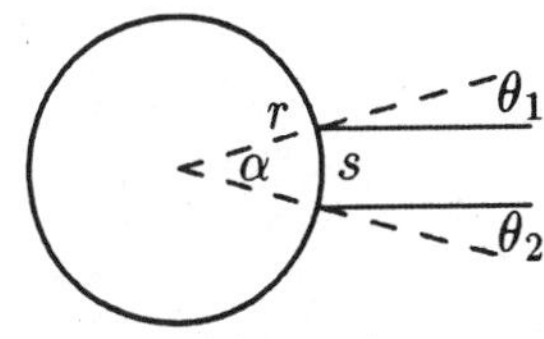

b. When the sun is overhead at A, sight to the center of the sun from B and use $d = r \sin\theta / \sin(\theta - \alpha)$. The radius r is known from part (a) and $\alpha = s/r$ is in radians.

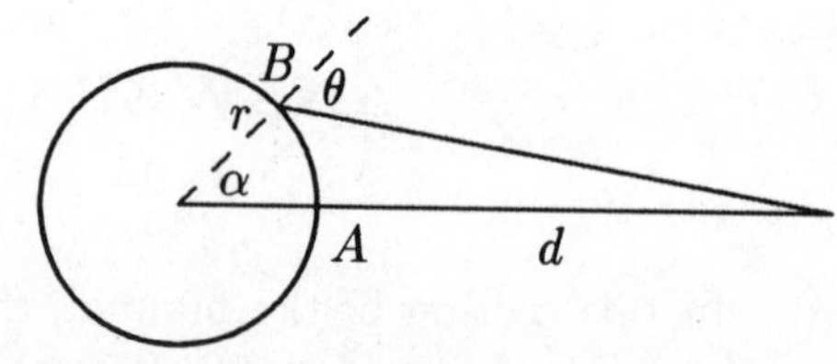

c. Measure the angular separation of the sun's edge across a diameter, then use the known distance to the sun to calculate its radius.

11 a. Measure the total thickness of 100 sheets, then divide by 100.

b. Measure the volume of water used to blow a bubble (or 100 bubbles and divide by 100). Also measure the radius of the bubble. Use $V = 4\pi r^2 \,\Delta r$.

c. Measure the mass of one atom using a mass spectrometer, measure the mass of a sample of the material, and take the ratio to find the number of atoms in the sample. Measure the volume of the sample, compute the volume per atom, and take the cube root. This assumes the atoms are closely packed. If the atomic configuration is known (from x-ray analysis, say), geometry can be used to find the interatomic spacing.

13. If such a definition were used, time intervals that we believe to be regular would actually vary. Time intervals arising from diverse physical phenomena would show similar variations, indicating that the fault lies with the irregularity of the pulse rather than with the interval being measured. We might define the second as the pulse beat of a particular person at a particular time, say noon on January 1, 1980. It would be difficult to make comparison measurements with such a clock.

18. An atomic standard is certainly more accessible, less easily destroyed, less variable, and more reproducible than the present standard kilogram. It is difficult, however, to make high precision comparisons of atomic and macroscopic masses. Techniques are improving with technology.

Chapter 2

4. The meanings are different. Consider a round trip and point out that the magnitude of the average velocity is zero but the distance divided by the time is not.

7. No. When $v =$ constant, $x = x_0 + vt$ for any interval of duration t. The average velocity is $(x - x_0)/t = v$, the same as the instantaneous velocity.

8. No in general. Graph v(t) for a variable acceleration. Pick an interval (a,b) and draw the straight line through the end points. The velocity for the variable acceleration motion is greater at every instant in the interval so the average velocity is greater than for the constant acceleration case. Other situations can be considered.

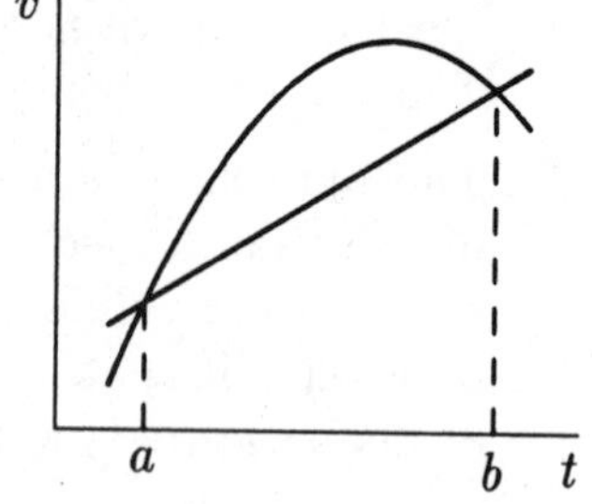

9 a. Yes. Point out that an object thrown straight up has zero velocity at its highest point but that it has a non-zero acceleration. If it did not it would remain at the highest point.

b. No. If the velocity is constant then its magnitude must also be constant.

10. Yes. A ball thrown upward has constant acceleration, if air resistance can be neglected. At the top of its trajectory it stops, then starts down.

11. Yes. If the velocity and acceleration remain in the same direction, the speed increases regardless of changes in the magnitude of the acceleration.

12. The negative root does not necessarily have physical meaning. The motion of the particle before $t = 0$ is not specified and the acceleration may not have been the given value a then. If the particle had the same constant acceleration a before $t = 0$ then the negative root does have physical meaning. The particle is in fact at position x two different times.

15. In the absence of air resistance and near the surface of the earth the acceleration of the apple is $9.8\,\mathrm{m/s^2}$, downward. The acceleration of the apple, once released, does not depend on the acceleration of the balloon. At release the speed of the apple is $2\,\mathrm{m/s}$. Point out that if the apple is released without applying a force to it, its initial velocity is the velocity of the balloon at the time of release.

Chapter 3

1. The beginning and end points are the same, so the displacements are the same. Emphasize that the path between is not relevant.

2. No. Draw one vector and a circle, centered at its head, with a radius equal to the magnitude of the second vector. Point out that the vectors from the tail of the first to points on the circle give all possible resultants as the orientation of the second vector changes. None of these is zero. It is possible to find many sets of three vectors which add to zero. Vectors of such a set form the sides of a triangle.

6. Explain that 3 vectors that sum to zero form a triangle and therefore define a plane.

8. Explain that units are attached to the components, not to the unit vectors. This allows the use of the same unit vectors for quantities with different units. Also explain that if a unit vector is defined by $\mathbf{A}/A$, the units cancel.

13. If $\mathbf{a} \cdot \mathbf{b} = 0$ then either the vectors are perpendicular to each other, $\mathbf{a} = 0$, or $\mathbf{b} = 0$. If $\mathbf{a} \cdot \mathbf{b} = \mathbf{a} \cdot \mathbf{c}$ then the components of $\mathbf{b}$ and $\mathbf{c}$ along the direction of $\mathbf{a}$ are the same. Components along directions perpendicular to $\mathbf{a}$ are not necessarily the same, so $\mathbf{b}$ may be different from $\mathbf{c}$. This question is useful in connection with the discussion of work in Chapter 7.

14. If $\mathbf{a} \times \mathbf{b} = 0$ then either $\mathbf{a}$ is parallel to $\mathbf{b}$, $\mathbf{a} = 0$, or $\mathbf{b} = 0$. If $\mathbf{a}$ and $\mathbf{b}$ are non-zero vectors that are parallel to each other then $\mathbf{a} \times \mathbf{b} = 0$ because the sine of the angle between them is 0.

Chapter 4

2. For a given initial speed the length of the jump depends on the angle of take-off and so does the height of the jump. So different heights mean different initial angles and different lengths of jump.

3. The minimum speed occurs at the highest point on the trajectory, where the vertical component of the velocity vanishes. Remind students that the horizontal component is constant. If the projectile remains above the firing point, it has its maximum speed at firing. If it goes below the firing point it has its maximum speed at the end of its flight.

10. No. The initial velocity of the projectile and the acceleration due to gravity determine a plane. The projectile remains in that plane throughout its flight. If air resistance is significant and a wind is blowing perpendicularly to the plane of $\mathbf{v}_0$ and $\mathbf{g}$, then the projectile will move out of the plane.

11. If the direction of the velocity changes with time the acceleration does not vanish, regardless of whether the magnitude of the velocity changes or not. In rounding a curve the direction of the velocity changes so the acceleration cannot vanish. In addition, the acceleration must

have a non-vanishing component that is perpendicular to the velocity. An object with constant acceleration moves on a curved path if the acceleration and velocity are not parallel. Point out that a projectile moves on a curved path.

16. No. It is not constant in a frame that has a changing acceleration relative to the given frame. Differentiate Eq. 26 three times.

14. If the train does not accelerate the ball falls into his hands. The horizontal components of the velocities of the ball and train are the same. If the acceleration of the train is forward, the ball falls behind the boy. If the train goes around a curve the ball falls to the side and in front of the boy.

Chapter 5

1. When the bus decelerates, the upper part of the body continues forward if the force exerted on it (by muscles of the lower back) is not great enough to give it the same deceleration as the bus. The lower body is restrained by the frictional force of the floor, unless the force required is too great. If it is, you slide. The situation when the bus accelerates is similar. The force of the floor pulls your feet along with the bus but the lower body does not provide sufficient force to accelerate the upper body and you bend backward. It is easiest for a standing passenger to maintain balance if the line through his feet is parallel to the acceleration of the bus.

4 a. The body moves at constant speed if the resultant force changes with time such that it is always perpendicular to the velocity. If the resultant force is constant in direction, the object cannot move with constant speed unless the two forces sum to zero.

b. The two forces must sum to zero.

c. The velocity can be instantaneously zero, even though the acceleration is not. An example is a ball thrown straight up, at the top of its trajectory.

d. The velocity must be zero initially and the sum of the forces must be zero.

9. The horizontal component of the resultant force on the horse is the net result of the wagon pulling back and the ground pushing forward. The horse can accelerate if he pushes back on the ground with a force greater than that of the wagon on him. The ground then pushes forward with sufficient force. The horizontal component of the resultant force on the wagon is due to the horse and to the ground. If this does not vanish, the wagon accelerates. Point out that the two forces in an action-reaction pair act on different bodies.

10 a. Yes.

b. Yes. The propeller pushes air back and the air pushes the propeller forward.

c. Yes.

d. Yes.

e. No. Three different bodies are involved.

f. No. The pull of the earth on the cart is gravitational, the push of the earth is a normal force, essentially electric in nature. If the cart moved into a deep mud hole, the force of gravity would not change but the normal force would and the cart would accelerate downward for a time.

11 a. Not true. Mass is the inertial property of an object. It is a measure of the acceleration of the object when it is subjected to a 1 N force. Weight is the force of gravity on the object.

b. True.

c. True only if the gravitational field acting on the body remains the same as the mass is changed.

d. Not true. Variations in weight come about because the local gravitational field varies. The mass remains the same no matter what the local field.

18. Yes. The resultant of a vertical and horizontal force can never vanish, so the mass accelerates.

20. In most cases, none. The acceleration of the body is in the direction of the total force acting on it. The velocity, on the other hand, can be in any direction whatsoever. Some forces, such as drag and magnetic forces are velocity dependent, however. The force of air resistance is opposite the direction of the velocity of the object relative to the air and a magnetic force is perpendicular to the velocity of the object.
28. The tension is the greatest when the elevator is descending with decreasing speed. It is then greater than the weight of the object. When the elevator is at rest or moving with constant speed the tension in the chord equals the weight of the object. The tension is the least when the elevator is descending with increasing speed. The magnitude of the tension is then less than the weight of the object.
29. The forces on the woman are those due to gravity and to the spring balance. The spring balance reads a maximum when the elevator and woman are accelerating upward. It reads less if the elevator and woman are at rest or moving with constant velocity, still less if they are accelerating downward, and least (0) if they are in free fall.
30. If the pulley is frictionless, the elevator is in free fall.

Chapter 6

1. At first polishing removes some microscopic roughness from the surfaces. There are fewer "bumps" where the surfaces can catch on each other or where welds can form. When the surfaces are highly polished a large fraction of the atoms of one surface are close to atoms of the other surface and the surfaces tend to adhere. The materials tend to form a single sample with the atoms of one attracting atoms of the other.
3. Yes, by pushing with a force that has an upward vertical component as well as a horizontal component. The magnitude of the force that starts the crate is then less than $\mu_s W_c$, where W_c is the weight of the crate. The crate pushes downward as well as horizontally on you so the normal force of the floor on you is greater than your weight. The magnitude of the force you can generate without slipping is greater than $\mu_s W$, where W is your weight.
4. An external force must act on her. Such a force cannot be exerted by the ice so she cannot walk or roll off the ice. Throwing her arm out causes the rest of her body to move for a short time but this is of limited value since she must soon pull back on her arm and this stops the motion. She can, however, throw a shoe. She exerts a force on the shoe and it exerts an equal and opposite force on her. Once started, she would continue to move were it not for air resistance. She could have been lowered from a stationary helicopter, pushed backward off a moving sled with just the right force to stop her, or been moving across the ice and hit a boulder with mass equal to hers, resting on the ice.
5. The normal force of the road on the tires is less when the car is going up or down a hill than when it is on level road. Thus the maximum force of static friction is less and the tires will skid at a smaller acceleration.
7. Assume the drag force is proportional to the speed: $D = bv$ in magnitude. Then the terminal speed is given by $v_T = mg/b$. It is proportional to the mass and inversely proportional to the drag coefficient. We expect the drag coefficient to be proportional to the cross-sectional area of the drop. The mass, of course, is proportional to the volume of the drop. This means m/b is proportional to the radius of the drop. We conclude that larger drops fall faster than small drops. A similar argument holds if D is proportional to v^2 or any other positive power of v.
10. At first the drag force of the air, acting upward, is greater than the force of gravity, acting downward. The object slows and the drag force on it decreases until it reaches terminal speed. Then the forces balance and it continues with constant velocity.

13. The force of the air on the wings is nearly normal to them. When the plane is banked and turning this force has a component toward the center of the circular path. The banking angle is adjusted so this component equals mv^2/r.

15. The force of friction supplies the centripetal force required for the coin to go around with the record. As the record speeds up, the force required becomes larger. When the upper limit for static friction is reached, friction can no longer supply a sufficiently large force and the coin flies off.

Chapter 7

3. Generally there are forces of friction acting in all machines and some of the work done by the operator changes the internal energy of the moving parts. For some machines (the lever and multiple pulleys, for example) some work acts to change the potential energy of the machine parts. This energy is recoverable but, in practice, it is rarely recovered in a useful way. Machines are used because a small force applied at the input produces a large force at the output. Nevertheless the work in always equals or exceeds the work obtained.

4. Suppose the winning team pulls the rope through its hands and does not move. It does positive work on the rope and the rope does negative work on the team. Suppose the losing team holds onto the rope. It does negative work on the rope and the rope does an equal amount of positive work on it. If the losing team drags its feet, the earth does negative work on them. Look at the system composed of the two teams and the rope. The net work done by external forces is negative. But the center of mass is accelerating toward the winning team and kinetic energy is increasing. The net work done at the hands of the winning team is internal work and changes internal energy to translational kinetic energy.

7. You have done positive work mgh and the earth has done negative work $-mgh$. It is important to specify the force being considered when work is discussed.

12. Suppose the whole spring is stretched by x. Then each half is stretched by $x/2$. Since the spring is massless the applied force is simply transmitted without change in value and the force on each half is the same as the applied force. Thus the spring constant for each half must be 2k. You might derive the equation $1/k = 1/k_1 + 1/k_2$ for the effective spring constant of two springs in series.

13. Since $W = \frac{1}{2}kx^2$ the greater work is done on the spring with the greater spring constant when they are stretched through the same distance. Since $W = \frac{1}{2}F^2/k$ the greater work is done on the spring with the smaller spring constant when they are stretched by the same force.

15. While the ball is moving upward, the force of gravity does negative work on it and if this is the only force acting the ball loses kinetic energy and slows down. The ball stops when the magnitude of the work equals the initial kinetic energy. On the way down the force of gravity does positive work. If this is the only force acting the ball gains kinetic energy. Since the magnitude of the work done when the ball falls equals the magnitude of the work done when it rises, the ball returns to the throwing point with speed equal to its initial speed. If air resistance is present it does negative work throughout the trip. The ball does not rise as far and it returns with less kinetic energy and a slower speed than it had initially.

Chapter 8

1. It is transformed almost wholly to internal energy in the brake while wheel rotation is being stopped by the road, while the tires skid. Once the wheels stop rotating no more work is

done in the brakes. The brakes, tires, and road become warmer than their surroundings and eventually dissipate energy as heat to the surroundings.

3. If it bounces to one half its original height mechanical energy was lost in the collision with the floor and to the air. If it bounces to 1.5 times its original height energy from somewhere increased the mechanical energy of the object. For example, the object may originally have had stored elastic potential energy (like a compressed spring) that was released on impact or an explosion may have occurred.
4. It appears as potential energy of the counterweight, internal energy in the hydraulic fluid, and electric energy in the lines behind the motor.
6. The same amount of energy must be transferred, with or without an air bag. With an air bag the transfer occurs over a larger distance than without, so the average force is less.
7. Both are right. The kinetic energy of an object depends on the reference frame used to measure the speed. In your frame the first duck is moving, in the frame of the second duck it is not. In both frames the mechanical energy of the duck is conserved if no forces do work on it.
17. The potential energy stored in the spring appears as kinetic energy of the molecules freed in the reaction that dissolved the spring.

Chapter 9

1. Suppose a cone is generated by positioning a large number of thin triangular prisms so their bases are diameters of a circle. The center of mass of each prism is one third of the way up but the mass density of the cone they generate is greater near the apex than near the base. Mass must be moved down to achieve a uniform density and this lowers the center of mass.
3. The center of mass of the earth's atmosphere is quite close to the center of the earth.
4. Yes, if his body bends around the bar.
9. After being thrown, the external forces acting on the firecracker are the force of gravity and the normal force of the ice on each piece after it lands. After the toss but before the first piece lands, the firecracker's center of mass follows the parabolic trajectory of a projectile. As each piece lands, the normal force of the ice causes the vertical component of the center of mass velocity to decrease until, when all pieces have landed, that component vanishes. Meanwhile the horizontal component remains constant throughout. After all the pieces have landed, the firecracker's center of mass moves across the ice. For the man-firecracker system, the net force is the same as for the firecracker alone. After throwing the firecracker, the force of gravity on the man is canceled by the normal force of the ice on him. The center of mass is initially projected straight upward by the act of throwing since the ice can exert only a normal force. It rises, then falls along a vertical line with acceleration g. When the fragments hit, the center of mass velocity decreases until it vanishes when the last fragment has landed. It then remains zero. The center of mass of the man-firecracker system is then stationary.

Chapter 10

2. If the direction of the force is reversed so the impulse during one interval is equal in magnitude and opposite in direction to the impulse during a second interval, then the total impulse over the two intervals is zero.
6. The impulse delivered is the same whether the collision takes place with padding or without. Padding (gloves, snow, tree branches, etc.) lengthen the time of the interaction and reduce the

average force. When hitting the side of a ravine the change is momentum is less so the impulse and average force are less than when landing on horizontal ground from the same height.

9. The terms *elastic* and *inelastic* describe collisions in which kinetic energy is conserved and not conserved, respectively. In a collision between two helium atoms momentum is conserved whether or not kinetic energy is conserved.
11. The incident particle passes through the target particle. No interaction takes place. Clearly momentum and kinetic energy are conserved.
14. There is a change in energy associated with the permanent deformation of the balls. This is chiefly a potential energy due to the changed configuration of the atoms. In addition, the motions of the atoms change on collision and the internal energy increases. The total momentum was zero before the collision and remains zero throughout. Each ball exerts a force of the same magnitude but opposite direction on the other. So the changes in their momenta have the same magnitude and opposite directions. The momentum changes sum to zero.
16. The velocity of B after the collision is given by $v_{Bf} = 2m_Av_{Ai}/(m_A + m_B)$. For the greatest speed, $m_A/(m_A+m_B)$ should be as large as possible or $m_A \gg m_B$. For the greatest momentum $m_Bm_A/(m_A+m_B)$ should be as great as possible or $m_A \ll m_B$. For the greatest kinetic energy, $m_Bm_A^2/(m_A + m_B)^2$ should be as large as possible. This function has its maximum value for $m_A = m_B$. If $m_A \gg m_B$, then $v_{Bf} \approx 2v_{Ai}$ and A continues on with little decrease in speed. If $m_A \ll m_B$, then A rebounds backward with $v_{Af} \approx -v_{Ai}$. In the first case B acquires the greatest speed it can, whereas in the second case it acquires the greatest momentum. Initially all the kinetic energy is vested in A. This is the most that B can acquire and it does so only if A stops. For this to happen m_B must be the same as m_A

Chapter 11

1. The units in $\omega = \omega_0 + \alpha t$ and $\phi = \omega_0 t + \frac{1}{2}\alpha t^2$ are consistent if ϕ is in degrees, ω is in deg/s, and α is in deg/s^2. Remind students that $s = r\phi$, $v = r\omega$, and $a_T = r\alpha$ are valid only if ϕ is in radians, ω is in rad/s, and α is in rad/s^2. The latter relations follow directly from the definition of the radian. If ϕ is given in degrees, then $s = (\pi/180)r\phi$, for example.
13. If $\alpha = 0$, the point has radial acceleration $r\omega^2$ but $a_t = 0$. If $\alpha \neq 0$, then the point has both radial and tangential acceleration: $a_r = r\omega^2$, $a_t = r\alpha$. For α constant, $\omega = \omega_0 + \alpha t$ and the radial component (but not the tangential component) changes with time, so both the magnitude and direction of **a** change with time.
15. The linear speeds of points on the rims of the two gears are the same, so $R_1\omega_1 = R_2\omega_2$.
18. The disk with the greater density has the smaller radius. Since $I = \frac{1}{2}MR^2$ it also has the smaller rotational inertia. The disk with the smaller density has the larger rotational inertia.
19. Apply a known torque about the chosen axis, measure the resulting angular acceleration, and calculate I using $\tau = I\alpha$. In practice you must correct for frictional torque in the bearings. It is usually convenient to use a constant known torque, measure the angular displacement for a given time, and use $\theta = \frac{1}{2}\alpha t^2$ to find α.
20. The hoop has the largest rotational inertia since all its mass is at a large distance from the axis. On the other hand, most of the mass comprising the prism is relatively close to the axis, so this object has the smallest rotational inertia.
21. The rotational inertia about O is greater for (a), when the heavy steel portion is far from the axis of rotation, than for (b), when the steel portion is near the axis, so the stick has a greater angular acceleration for (b) than for (a).

Chapter 12

4. If both cylinders are uniform they have the same rotational inertia and reach the bottom at the same time. After the hole is drilled, the brass cylinder has the larger rotational inertia and takes longer than the wooden cylinder to reach the bottom.

5. For either bike $v_{cm} = R\omega$, where R is the tire radius. Since both are going at the same speed the wheel with the larger radius (Ruth's) has the smaller angular speed. The speed of the point at the top of a rolling wheel is twice the speed of the center of mass, so points at the tops of the two wheels have the same speed.

8 a. Conservation of energy, applied at the bottom of the ramp, yields $\frac{1}{2}v_1^2(m + I/r^2) = \frac{1}{2}v_2^2(m + I/R^2)$, where r is the radius of the rod and R is the radius of a disk. Since $R > r$, $v_2 > v_1$.

b. The rotational inertia parameter β is given by $\beta = \frac{1}{2}(mr^2 + 2MR^2)/(m + 2M)r^2$, where r is the radius of the rod and m is its mass, R is the radius of a disk and M is its mass. If $M \gg m$ and $R \gg r$ then $\beta \approx R^2/2r^2$. It is much larger than the parameter for a hoop, which is 1. A large value of β means a low speed at the bottom, so the hoop wins.

c. See above.

11. Viewed from the rear of the car, the torque exerted by the drive train on the rear wheels is to the left. The wheels exert an oppositely directed torque on the car body, causing the car to "nose up". The same thing happens for a front-wheel-drive car.

15. The angular momentum of an object about a fixed axis is given by $\tau = I\omega$. It does not change if the net external torque vanishes. If $\tau = 0$ and the rotational inertia decreases then the angular velocity increases. Example: a spinning skater who drops her arms. If torque is applied in the direction of the angular velocity and the rotational inertia does not change, then the angular velocity increases. Example: a wheel with a tangential force applied at its rim.

16. The student exerts no torque on the dumbbells and they exert no torque on him, so his angular velocity does not change. The dumbbells become projectiles with velocities that are initially horizontal. Their angular momentum just after release is same as it was just before release. Once released, however, the force of gravity is no longer balanced by the force of the student's hands and there is a net torque on each dumbbell. Because the dumbbells are identical and are released from diametrically opposite points the total torque about a point on the axis of rotation vanishes.

Chapter 13

1. Yes. At the bottom of its swing, the net force and net torque are both zero.

2. If the body does not spin no matter what its orientation when it is thrown, the center of mass and center of gravity must coincide. Otherwise, when the two centers are not on a vertical line gravity would exert a torque about the center of mass and the angular acceleration would not vanish.

3. Individual particles are not in equilibrium but because Newton's third law holds, the wheel as a whole meets the criteria for equilibrium.

4. Turn a bicycle over and apply forces to the axle and rim of the front wheel so that the center of mass does not move. Turn on a phonograph and watch it come up to speed. Push down on the accelerator when the car is on ice. Explain the case when two tangential forces, equal in magnitude but opposite in direction act at the rim of a disk. The disk then has an angular acceleration but its center of mass does not accelerate.

5. One that is stretched tightly. In equilibrium, the sum of the vertical components of the forces of the ropes equals the weight of the hammock and its contents and is the same whether the hammock is stretched tightly or sags. If it is stretched tightly, however, the tension in the ropes must be great to achieve the same vertical component.

6. Neglect the force of friction at the wall and suppose the ladder makes the angle θ with the horizontal. Then if the person is a distance x up the ladder the force of friction at the ground must be $f = (wx + \frac{1}{2}W\ell)/\ell \tan\theta$ for the ladder to remain in equilibrium. Here $w =$ the weight of the person and $W =$ the weight of the ladder. As x increases f must increase to hold the ladder, so the danger of slipping increases as the person climbs.

9. When you lift up on your toes, a torque about your center of mass is associated with the normal force of the floor, then through your toes. You tend to turn about a horizontal axis so the upper part of your body moves backward away from the door.

Chapter 14

2. Because both springs stretch by the same amount we know that $m_1/k_A = m_2/k_B$. Since $m_2 < m_2$, $k_B < k_A$. Since the mechanical energy is given by $E = \frac{1}{2}kx_m^2$ it follows that $E_B < E_A$.

4. The extension of the spring is $\Delta y = mg/k$, so $k/m = g/\Delta y$. This is the square of the natural angular frequency so the period is $T = 2\pi/\omega = 2\pi\sqrt{\Delta y/g}$. The natural frequency depends on the ratio k/m and not on k and m separately.

6. The period and spring constant do not depend on the amplitude. The period depends on k/m and the spring constant is a property of the spring. The total mechanical energy is given by $\frac{1}{2}kx_m^2$, the potential energy when the displacement is a maximum and the velocity vanishes. It clearly increases by a factor of 4 if the amplitude is doubled. The total energy can also be written $\frac{1}{2}mv_m^2$, the kinetic energy when the displacement is zero. So $v_m = \sqrt{k/m}x_m$ and is proportional to x_m. The maximum acceleration is given by kx_m and is proportional to x_m. Both these quantities double if the amplitude is doubled.

7. Since $v_m = x_m\omega$, you could double the amplitude (by doubling both the initial displacement and initial speed) or double the frequency (without changing the amplitude). In the later case, k/m should be made 4 times as large. This means changing the spring or mass and might not be allowed by the wording of the question.

10. For every angular displacement θ, $\sin\theta < \theta$ so we expect the true acceleration to be less than the value given by the small angle approximation. This makes the period longer.

11. If the upward direction is taken to be positive the tangential acceleration of a simple pendulum in an accelerating elevator is given by given by $a_t = -(g + a)\sin\theta$, where a is the acceleration of the elevator and θ is the angle of the pendulum from the vertical. For small angle oscillations the period is given by $T = 2\pi\sqrt{\ell/(g + a)}$, where ℓ is the length of the pendulum. The period is not affected by the velocity of the elevator. In detail the answers are: (*a*) not affected; (*b*) not affected; (*c*) T increases; (*d*) T decreases; (*e*) T increases; (*f*) T increases if $g < |a| < 2g$ and decreases if $|a| > 2g$. (*g*) For the conditions of (*f*) the pendulum swings upside down. a_t is then positive and the pendulum tends to swing toward the highest point rather than the lowest. To obtain the usual pendulum equation when a is negative and has magnitude greater than g replace θ with $\theta \pm \pi$.

Chapter 15

1. A gravitational force is proportional to the gravitational mass of the object on which it acts. The acceleration of the object is given by the force divided by the inertial mass. Since the gravitational and inertial masses are equal the acceleration is independent of the mass and light objects have the same acceleration as heavy objects.
4. For the same reason that the earth bulges: the effective force is reduced by the centrifugal force (in the frame of the earth) as the river flows south.
6. The car going toward the west, opposite to the velocity of the earth's surface under it, presses down harder on the road. Its centripetal acceleration is less so the normal force of the road on it is greater.
8. In (b) the two end masses are closer together than in (a) so the potential energy has a larger negative value in (b) than in (a).
11. Viewed on a large scale, the moon travels around the sun in a nearly elliptical path, pulled by the gravitational force of the sun. Its orbit is nearly the same as that of the Earth and even if there were no gravitational attraction between the moon and the Earth they would travel close together. Superposed on this motion is its motion around the Earth. This motion is on a much smaller scale.
16. No. The centripetal force acting on one of the objects is due to gravity and the normal force of the earth's surface. The force of gravity is too strong to provide the right centripetal force for the period of rotation. If the period were about 84 min the normal force would be zero and the object would be in orbit.
22. Rockets can be launched to the east over water, taking advantage of the rotational motion of the Earth and keeping away from populated areas.
24. The period is proportional to $r^{3/2}$, where r is the orbit radius. Saturn's orbit has the larger radius so Saturn has the greater period. The orbital speed is given by $2\pi r/T$ and is proportional to $r^{-1/2}$ so Mars has the greater orbital speed. The angular speed is given by v/r and is proportional to $r^{-3/2}$ so Mars has the greater angular speed.

Chapter 16

4. The water also exerts a downward force on the sides of the first vessel and an upward force on the sides of the second vessel. The net force (including the forces on the sides and bottom) in each case is equal in magnitude to the weight of the water. This is different for each of the three cases shown.
9. No. It floats at the same height. The difference in pressure across the block is less so the buoyant force is less but it is less by exactly the right amount so the acceleration of the block is the same as the acceleration of the water. The same statement is true whether the elevator is accelerating upward or downward.
10. They weigh the same. The block displaces a volume of water that has the same weight as the block.
12. The ice displaces a volume of water with weight equal to its own weight. When it melts, it produces an amount of liquid with the same weight and hence with a volume equal to the volume of water originally displaced. The water level does not change. When ice containing sand floats, it displaces a volume of water with weight equal to the weight of the ice and sand. When it melts the ice portion produces a volume of liquid with weight equal to its own weight. But the sand sinks showing that it alone displaces a volume of water with weight less than its

weight. In all, less water is displaced after melting than before so the water level falls. If the ice contains air, the level rises when it melts.

13. As the air pressure increases the ball rises slightly. It drops slightly as the air is pumped out. To simplify the problem, consider a cylinder of length ℓ, floating upright with length h submerged. If p_a is the air pressure on its top, then $p_a + \rho_a g(\ell - h)$ is the air pressure at the water level, and $p_a + \rho_a g(\ell - h) + \rho_w gh$ is the water pressure on its bottom. The net force on the cylinder is $mg + p_a A - [p_a + \rho_a g(\ell - h) + \rho_w gh]A$. This vanishes so $h/\ell = (\rho - \rho_a)/(\rho_w - \rho_a)$, where ρ is the density of the cylinder. As the air pressure increases, ρ_a increases and h decreases. The derivative $dh/d\rho_a$ is negative for $\rho < \rho_w$. Since ρ_a is so much smaller than either ρ or ρ_w, the effect is very slight.

16. The denser salt water of the North Sea could provide sufficient buoyant force to keep the ship afloat. The less dense fresh water of the estuary could not.

22. If the upright log tips to the right the center of buoyancy will be to the left of the center of mass and the torque about the center of mass will continue tipping the log in the same direction. The log thus floats horizontally. If sufficient iron is added to the lower end the center of gravity will be to the left of the center of buoyancy when the log tips to the right. There is now a restoring torque.

27. Consider the view from the rest frame of the boats, as shown. The water speed v far from the boats is the same as the boat speed in the frame of the water. The region between the boats narrows from point 1 to point 2. Since vA remains constant, the water speed at 2 exceeds v. Therefore the pressure drops and the boats are sucked together. A similar effect, due to air, occurs between cars. It is most noticeable when a large truck passes slowly by your car with both car and truck going at high speed.

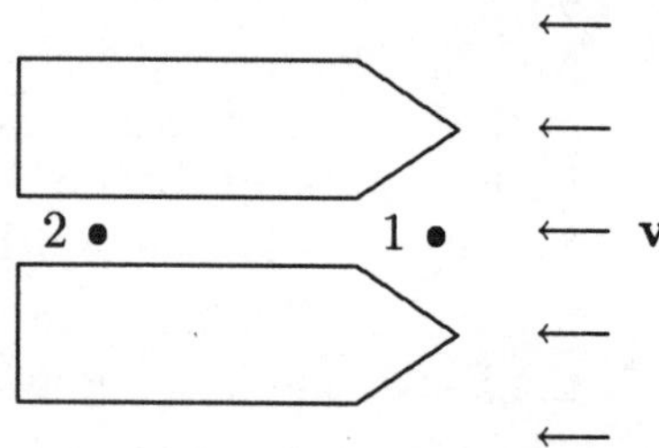

28. As the moving air enters the region between the paper and the wide portion of the funnel its speed increases greatly (continuity equation) and the pressure drops to well below atmospheric pressure (Bernoulli's equation). The force of the air in the central part holds the paper inside the funnel.

29. When thrown against the wind the lift is greater and the discus remains in the air longer. Drag is also greater but drag is small to begin with.

Chapter 17

1. Tie an object to one end of a taut string and create a wave by jiggling the other end. Before the wave reaches the object it is motionless but once the wave gets there it moves. You have done work on the string and the object acquires kinetic energy. Energy was transferred from your hand to the object via the wave on the string.

2. When energy is transferred by particles, as in a collision, the transfer takes place over an extremely short time. For a reasonably long wave train the transfer of energy takes place over a longer time. For short wave trains the two types of energy transfer become harder to distinguish. Interference can be used to distinguish particles and waves. If the energy reaching a point can be diminished by the addition of another source, the mechanism is a wave. If it cannot the mechanism is a classical particle. The distinction is blurred at the quantum level.

4. For any given value of t all these functions increase indefinitely with x. Such disturbances are impossible to produce since they require that the source have infinite amplitude at a time infinitely far in the past.

8. No. In the small amplitude approximation each wave continues to travel at the same speed, in the same direction, without change in shape. When amplitudes are large, one wave might alter the elastic properties of the medium or its density and thereby change the speed, wavelength, or shape of a second wave.

9. No. Energy is redistributed but none is lost. No energy reaches points where there is complete destructive interference but more energy reaches points where the interference is constructive.

10. Even when the string is straight it is still moving and has kinetic energy. At maximum displacement the string is not moving and all the energy is potential energy, stored as elastic energy in the stretched string. At intermediate displacements the energy is partially kinetic and partially potential.

15 *a*. Use $\ell = n\lambda/2$ for a fixed string and $d = \lambda/2$ for the distance between nodes to show that $d = \ell/n$, where ℓ is the string length. A node occurs at $\ell/3$ for any harmonic with a harmonic number that is a multiple of 3.

b. For the fifth harmonic the distance between nodes is $d = \ell/5$ and the point at $2\ell/5$ is a node.

c. For the tenth harmonic the distance between nodes is $d = \ell/10$ and the point at $2\ell/5 = 4\ell/10$ is a node.

16. Since $\ell = n\lambda/2$, $f = v/\lambda = nv/2\ell$. Since $v_B > v_A$, n_B must be less than n_A for the two strings to have the same frequency. This occurs only for the situation shown in (*d*).

Chapter 18

3. Highly accurate timing devices can be used to measure the speed of waves with various wavelengths. High and low notes played by an orchestra reach the back of a concert hall at times that are the same to even well-trained ears.

5. It's present value permits negative values of the sound level. They are, however, below the normal threshold of hearing. Negative values would be associated with audible sound if the reference intensity were increased.

6. The valves and slide change the length of the air column in the instrument and thus change the resonant frequencies of the instrument.

7. With no valves the bugler is limited to the natural resonant frequencies of the instrument. Notes are produced by blowing air through the lips into the instrument. The lips are closed by the low pressure behind the moving air and are pulled open by the pressure wave reflected at the far end of the bugle. Different notes can be sounded by changing the response of the lips to the column of vibrating air. This is done chiefly by changing the tension in the lips, pursing and relaxing them.

9. Sound waves generated by the struck prong set the second prong in motion. Emphasize that the two prongs have the same natural frequency so the sound wave that strikes the second prong is in resonance with it.

12. Pitch, loudness, and tone quality are all subjective characteristics of sound. Pitch roughly corresponds to frequency but the relationship also depends on the intensity. Two notes, near each other in frequency, might be ordered with either one at the higher pitch, depending on the relative intensity. Loudness corresponds roughly to intensity but, as a look at Fig. 18–8 reveals, there are low frequency notes which are inaudible and higher frequency notes, played at the same intensity, which can be heard and so are louder. Tone quality is determined chiefly by

the mixture of overtones present, their relative amplitudes, when they reach peak amplitude, and when they die out (relative to the fundamental).

17. No they are not. This is revealed by the Doppler effect equation: the velocity factor is $(v+v_D)/v$ when she runs toward him and is $v/(v-v_S)$ when he runs toward her. These are not the same. When he is stationary the distance between successive maxima is $\lambda = v/f$ but she sees them approach at a speed of $v + v_D$ and so hears maxima at intervals of $(v/f)/(v + v_D)$. This corresponds to a frequency of $f(v + v_D)/v$. The source emits maxima with an interval of $T = 1/f$. If it is moving this corresponds to a distance of $\lambda' = T(v - v_S) = (v - v_S)/f$ between successive maxima. They move toward the observer with speed v so the observer hears maximum at an interval of $\lambda'/v = (v - v_S)/vf$. This corresponds to a frequency of $fv/(v - v_S)$.

18. No. Suppose the medium is moving from the source toward the observer. In the frame for which the medium is at rest the source is moving away from the observer and the observer is moving toward the source. The Doppler effect equation becomes $f' = f(v + v_D)/(v + v_S)$. Since $v_D = v_S$ this yields $f' = f$.

Chapter 19

3. Energy is transferred from the thermometer to the ice. Since the enclosure is evacuated the transfer takes place via electromagnetic radiation. At first the thermometer radiates more energy to the ice than the ice radiates to the thermometer.

11 a. The sun's temperature can be found by examining the radiation from it. One method is to measure the radiancy and apply Stefan's law. Another is to find the wavelength for which the spectral radiancy is a maximum and calculate the temperature of a blackbody which gives the same peak.

b. The temperature of the upper atmosphere can be inferred from the drag on a satellite. Atmospheric force on the satellite depends on the density of the atmosphere which, in turn, depends on the temperature. It is also possible to measure the distribution of the speeds of charged particles in the upper atmosphere. The distribution is temperature dependent.

c. The thermometer for this application must be quite small. Thermocouples are usually used. These are made of two fine wires of different metals, joined together at each end. There is a potential difference, end to end, which depends on the temperature difference between the ends. This instrument can be made in the form of a hypodermic needle and inserted into the insect.

d. Most of the light from the moon is directly reflected sunlight. Some sunlight is absorbed, however, and reradiated in the infrared. This is the portion which heats the moon and analysis of the spectrum gives the temperature.

e. The property measured should be insensitive to pressure and show wide variation in the range from $-15°$ C to $+15°$ C. The resistivities of metals meet the requirements. The resistance of a platinum sample is usually measured.

f. The resistance of a semiconductor (usually GaAs) is measured. This shows great variation with small changes in temperature in the region around 4 K. It does not change phase or undergo structural changes from room temperature down.

12. Many gases give the same temperature reading (the ideal gas temperature) in the limit of small concentration. They differ in convenience of measurement and range of usefulness, however. The gas cannot undergo a phase change in the pressure and temperature range of interest, so a low boiling point is desirable. Since the temperature limit as $p_3 \longrightarrow 0$ must be evaluated, it is desirable that the curve shown in Fig. 19–6 have as small a slope as possible. Generally

helium is picked as the gas which best satisfies these requirements. In addition, the gas and the container material must be picked to minimize diffusion of the gas into the container walls. If platinum is used for the container, helium also satisfies this requirement.

14. The glass is heated first and expands, causing a drop in the mercury level. The mercury is then heated and its level rises.
17. The ball expands so it does not fit through the hole. If the ring is heated, the hole becomes larger and the ball passes through with more room to spare than originally.
18. When heated, the two strips expand by different amounts and curve into an arc. The strip with the larger coefficient of expansion is on the outside, where the arc length is greater.

Chapter 20

2. Any adiabatic process in which work is done on or by the system results in a temperature change. For example, use a piston to compress a gas in a well insulated container.
5. Heat added to ice at the melting point causes the formation of liquid water at the same temperature. In the definition of heat the temperature difference referred to is between the environment and the system. The environment must be at a higher temperature to add heat to the system but the temperature of the system need not change while heat is being added.
9. If heat with magnitude Q is transferred the temperature of the hotter body drops by $\Delta T_H = -Q/m_H c_H$, where m_H is its mass and c_H is its specific heat. The temperature of the colder body increases by $\Delta T_C = Q/m_C c_C$, where m_C is its mass and c_C is its specific heat. The two temperature changes are not equal unless $m_H c_H = m_C c_C$. In general, the object with the larger heat capacity experiences the smaller temperature change.
10. Because the heat conductivities are different, energy is transferred to the hand touching the objects at a different rates. It is chiefly the rate of energy transfer that is sensed. They will feel equally hot or cold when they are both at body temperature.
19. The outside air is much colder than the inside air and heat flows from the air inside to the wall, through the wall, to the air outside. The rate of heat flow per unit temperature difference is greater for flow through the wall than for heat transfer at either surface. If the temperature difference between the inside air and the inside wall surface is small or zero, the energy leaving the surface through the wall is replaced at a lower rate than it leaves and the temperature of the surface decreases. The decrease reduces the rate at which energy leaves the surface and increases the rate at which energy is transferred from the air to the wall. Steady state is reached when the two rates are the same. There is then a temperature difference. At the outside surface the temperature of the wall increases until the rate at which energy passes through the wall equals the rate at which energy is transferred to the air. There is then a temperature difference between the surface and the air. Mention that the thermal conductivity of air is quite small and energy gets to the wall on the inside and is removed from the wall on the outside chiefly by convection. The convection rate is to be compared with the conduction rate through the wall.
24. The internal energy is increased if heat flows from the environment to the system or if the environment does positive work on the system. If a gas in a well insulated container is compressed, its internal energy increases but no heat flows. If the gas expands adiabatically against a piston, it does work and its internal energy decreases. No heat flows.
25. No. The internal energy is the sum of the kinetic energies of the moving molecules and the potential energies of their interactions. Both heat and work may change these energies but once changed they retain no memory of how the change was brought about.

32. The double walls cut down on convection and serves to retain the vacuum. The vacuum cuts down on both convection and conduction. Silvering cuts down on radiation losses.

Chapter 21

1. If the concentration is reduced by removing molecules there will be a point where the assumption of equal numbers of molecules moving in all directions fails. Temperature and pressure are then not well defined and we must consider the motion of individual molecules to calculate the force on the container walls, for example. On the other hand, as the concentration is increased the gas becomes non-ideal because the sum of the molecular volumes is too large a fraction of the container volume and because the potential energy of molecular interaction is too large a fraction of the total energy. In general, real gases do not meet the requirements to be ideal. However, for many gases, particularly those with small, weakly interacting molecules, there is a fairly broad range of concentrations for which kinetic theory works and for which the gas is nearly ideal. For some gases, say a plasma of charged particles (which interact strongly over large distances), such a concentration range may not exist.

2. If the walls are at the same temperature as the gas there is no net transfer of energy. In some collisions molecules give up energy to the walls; in others they receive energy from the walls. Over many collisions just as much energy is given up as is received.

5. The pressure is the same in the two rooms since the doorway between them is open. The volumes are the same. Use $n = pV/kT$ to show there are fewer molecules in the room with the higher temperature.

8. We consider equal time intervals which are so long that during any one of them a large number of molecules strike the walls. The number of collisions with the walls is not precisely the same for each interval but it nearly the same and as the interval becomes large the fractional deviation from the average becomes small. Except for these small fluctuations the force exerted by the molecules on the walls is constant. It is important that the interval be large compared to the mean time between collisions of a single molecule. It is if the gas concentration is sufficiently high.

10. If the speed of the cold ball as a whole is roughly the same as the root-mean-square speed of molecules in the hot ball, the total kinetic energy of the two balls is about the same. Only the kinetic energy as measured in the center of mass frame, however, influences the temperature. The temperature of the cold ball does not increases by virtue of its motion as a whole. Interactions with air molecules may result in increases in its internal energy and temperature, of course.

27. The parameter d can be defined so $\pi d^2/4$ gives the scattering cross section for the molecules. Monatomic molecules act most nearly like spheres.

29. A sound disturbance must be passed from molecule to molecule via collisions. A density fluctuation, for example, is passed along when molecules from one region compress molecules into a neighboring region. This cannot occur in an orchestrated way over distances that are shorter than a few mean free paths.

30. Think of each different kind of molecule as forming a system that exchanges energy with the other systems of molecules. The systems reach thermal equilibrium with each other and with the walls of the container (or thermal reservoir). Each has the same temperature and hence the same speed distribution it would have if the other systems of molecules were not present.

33. As the temperature increases more molecules will have high speeds. Since the total number of molecules remains the same, fewer have speeds in a given range near the average.

40. The gas does work on its environment and as a result its internal energy decreases. This means a decrease in temperature. For an ideal gas pV^γ is constant and thus so is $nRTV^{\gamma-1}$. Since $\gamma > 1$ an increase in V means a decrease in T.

Chapter 22

6. When one surface is close to the another, bonds are formed by the attraction of molecules to each other across the boundary. When the bodies move relative to each other, work is done to break the bonds. This energy is supplied by the agent pushing the bodies past each other. The molecules in the same body also exert forces on each other and when one molecule near the surface is pulled out of position to form a bond, other molecules, deep inside the material, are also pulled away from their normal positions. When the bond is broken, the energy is distributed among the various out-of-place molecules and eventually spreads to a great many, if not all, the molecules of the body. The temperature of the material increases and if the temperature becomes greater than that of the environment, energy is transferred as heat to the environment. The reverse process does not occur since that would require the energy to return to a relatively few molecules near the surface, where it would be used to move surface molecules and finally to provide the kinetic energy of the other body.
10. Work cannot be calculated using a p–V diagram if the process is irreversible. In fact, pressure is undefined. Work is done, however, if the gas expands or contracts against a piston. Molecules strike the moving piston and exert a force on it.
14. For any heat engine, the greater the ratio of work obtained to heat added, the greater the efficiency. If all the heat added were converted to work, with no heat rejected or retained as internal energy, the engine would be 100% efficient. This is prohibited by the second law of thermodynamics unless the temperature of the cold reservoir is 0 K. The third law asserts that it is impossible to achieve exactly 0 K.
15. All reversible heat engines have the same efficiency and irreversible engines have lower efficiencies, so reduction of efficiency occurs when there are departures from reversibility: compression and expansion strokes are not quasi-static or heat transfer is accomplished by contact with an environment at a greatly different temperature than that of the working substance. There is also heat loss: not all of the heat leaving the boiler enters the working substance.
16. Use $e = 1 - T_C/T_H$ to show that $de/dT_C = -1/T_H$ and $de/dT_H = T_C/T_H^2$. Since $T_C < T_H$, the efficiency increases more rapidly with a decrease in T_C than with an increase in T_H.
21. Put a hot and a cold body in thermal contact. As they come to thermal equilibrium the entropy of the hot body decreases. The increase in entropy of the cold body is at least as great so the entropy of the system consisting of the two bodies either increases or remains the same. Melt ice by putting it in warm water. The entropy of the water decreases but the entropy of the ice increases at least as much. The entropy of any object might decrease if the object exchanges energy with its environment.
28. For the Earth $\Delta S_e = \Delta Q/T_e$ and for the sun $\Delta S_s = -\Delta Q/T_s$, where ΔQ is the heat absorbed by the earth. Since $T_s > T_e$, it follows that $|\Delta S_e| > |\Delta S_s|$ and $\Delta S = \Delta S_e + \Delta S_s > 0$.

Chapter 23

1. Start with neutral spheres and, while they are touching each other, bring the charged rod close to one of them. The positive charge on the rod attracts negative charge to the surface near the rod, some of it coming from the other sphere where it leaves positive charge behind. Now

separate the spheres and finally remove the rod. The sphere which was nearest the rod is negatively charged while the other sphere is positively charged. Since charge is conserved, the magnitudes of the charges are the same. The spheres need not be the same size.

2. Use two identical spheres and charge one of them (by touching it to the glass rod or by induction). Then remove the rod and touch the spheres together. Since they are identical, they will carry identical charge. The spheres must be identical for the final charges to be the same.

3. The rod polarizes the cork dust so that the surface of a dust particle nearest the rod is charged oppositely to the rod and the surface away from the rod is charged with the same sign charge as the rod. Since the electric force decreases with distance, the near side of the dust is attracted more strongly than the far side is repelled and dust jumps to the rod. When in contact, charge flows and the dust obtains a net charge of the same sign as the charge on the rod. The dust may then be repelled.

5 a. Not charged.

b. Charged negatively.

11. The electrons in the metal exert forces on each other. Electrons already at the end tend to repel other electrons, for example. Flow stops when the net force on each electron, due to charge on the insulator and to other electrons in the metal, vanishes.

14 a. No. Charge separation occurs in the object, with negative charge closer to the rod than positive charge. Since the force between two charges decreases with their separation, the result is attraction, even if the object is neutral.

b. Yes. Charge separation in a neutral rod would cause attraction. For repulsion the net charge must be positive.

15. The electrons are held in the material by the attractive forces of positive ions near the surface. Some fast electrons may escape, to be replaced by other electrons from the environment. More escape if the metal is heated.

18. Each force is proportional to the product of the charges so the two particles exert forces of equal magnitude on each other.

Chapter 24

3. Wherever there is an electric field its direction is not ambiguous. It can be found, for example, by evaluating the vector sum of the fields due to all charges. On the other hand, if two field lines cross, each would give a different direction for the field at the point of crossing. We conclude that field lines do not cross.

4. At a point that is far from the charge distribution compared to the distance between charges, the field must become like the field of a single charge, equal to the net charge in the distribution. For net positive charge, this field is radially outward from the charge.

5. At any instant the acceleration of a charge is along the field line through its position at that instant. Stress that it is the field line due to all other charges. Since the particle starts from rest, it starts moving along the field line through its initial position. If the line is straight, it continues along that line. If the line bends, the particle does not follow it once the particle acquires a non-vanishing velocity.

6. Yes. The motion of the charge does not enter into the determination of the electric force on it. Remark that the situation is different for magnetic fields.

10. The field lines shown are for the total field, created by both charges. The total field is large near either one of the charges because the field of that charge dominates there. The force on a charge, however, is due to the field of all <u>other</u> charges, not its own field. To find the force on the lower charge in the figure only the field due to the upper charge should be used.

12 a. From the positive charge toward the negative charge.

b. Away from the positive charge.

c. Toward the negative charge.

d. Perpendicular to the median plane, pointing from the region closest to the positive charge toward the region closest to the negative charge.

14 a. The dipoles attract each other. Note that the two charges closest to each other, on different dipoles, have opposite signs.

b. They repel each other. Note that the two charge closest to each other, on different dipoles, have the same sign.

19. Yes, if the field varies along the direction of the dipole moment. Then the forces on the two charges do not have the same magnitude and their sum is not zero.

Chapter 25

4 a. *All* the charges.

b. Equal. Charges q_3 and q_4 do not contribute to the net flux through the surface.

5 a. No change occurs since the net charge inside the two surfaces is the same. The shape of the surface is immaterial.

b. No change. The size of the surface is immaterial as long as the charge remains inside.

c. No change. The position of the charge is immaterial as long as it remains inside.

d. The net flux is now zero. There is zero charge inside the surface.

e. No change from the original situation. The charge inside contributes the same flux as before while the charge outside contributes zero flux.

f. The flux is now that due to the two charges and is $(q_1 + q_2)/\epsilon_0$.

7. If the surface encloses zero net charge, the integral for the flux vanishes but $\mathbf{E}$ need not be zero anywhere on the surface. Think of the surface divided into many small elements $\Delta\mathbf{A}$. For some elements $\mathbf{E}\cdot\Delta\mathbf{A}$ is positive while for others $\mathbf{E}\cdot\Delta\mathbf{A}$ is negative. The sum may vanish even though none of the individual terms are zero. Consider a closed surface with a single charge outside. $\mathbf{E}$ does not vanish anywhere but the integral is zero. The converse is true, however. If $\mathbf{E} = 0$ everywhere on the surface then the flux through the surface vanishes and the net charge enclosed must be zero.

10. $\mathbf{E} = 0$ inside a spherical balloon with a spherically symmetric charge distribution on its surface, provided no other charge is present. For a sausage shaped balloon with a uniform charge distribution on its surface, $\mathbf{E} \neq 0$ inside. Note that the flux through any closed surface wholly inside either balloon vanishes. Also note that the proof that $\mathbf{E} = 0$ inside a uniform distribution depends on the symmetry of the situation.

13. Zero. When the touch the ball and hollow conductor form a single conductor and any excess charge moves to the outer surface, the outer surface of the hollow conductor. No excess charge remains on the ball.

16. Gauss' law holds (as it does in every situation) but there is not enough symmetry to solve it for the electric field.

17. Charge $-q$ appears on the inner surface and, if the spherical shell is neutral, charge $+q$ appears on the outer surface. If the shell has total charge Q then $Q+q$ is on the outer surface. A nearby

metal object will not change the charge on the shell surfaces but it will cause a redistribution of the charge on each surface. Since it experiences an electric field, charge separation occurs on its surface and it generates a field of its own. Charge shifts on the shell surfaces to make the total field vanish inside the shell.

19. The charge is distributed along the rod so for any field point some is far away and contributes little to the total field. The charge nearest the field point contributes the most to the field and this charge is finite.

22. Imagine an electric field with a tangential component. Rotate the line by 180° around the dotted line. The field rotates to the direction shown in the second diagram. The charge distribution, however, is the same before and after the rotation, so the field must be the same. Only a radial field, not a field with a tangential component, remains the same after the rotation. A similar argument can be used to show that the magnitude of **E** is the same at every point on a cylinder centered on the line.

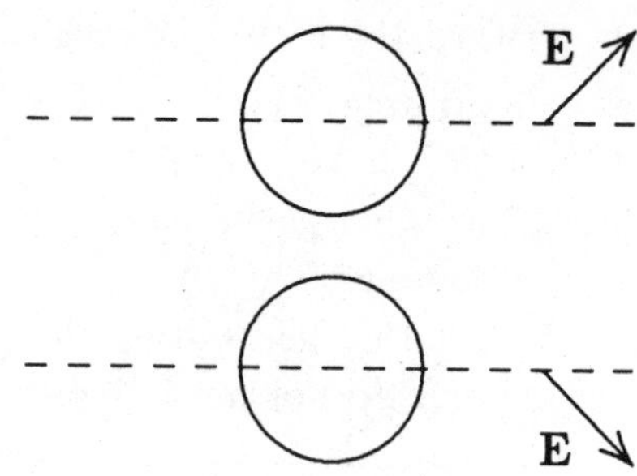

Chapter 26

1. The potential at any point (not occupied by a point charge) can be given any value whatsoever. If a point is said to be at +100 V instead of 0, the potential at all points would be given values +100 V higher than previously and the potential difference between any two points would remain the same. Only potential *differences* have physical meaning.

5. Electric fields point from regions of higher potential toward regions of lower potential. Electrons are accelerated in the direction opposite the field, so they tend to go to regions of high electric potential. Protons, on the other hand, are accelerated in the direction of the field and tend to go to regions of low electric potential.

9. The electric potential has a definite value at each point in space. Two different intersecting equipotential surfaces would indicate two different values for the potential at the point of intersection. We conclude that equipotential surfaces do not intersect.

12. No. The electrostatic field is zero inside a conductor but the potential has the same value as at the surface. If the conductor is charged and the zero of potential is at infinity, for example, then $\mathbf{E} = 0$ but $V \neq 0$ at points inside.

13. No. **E** must be known for all points along some path joining the point where $V = 0$ to the point of interest. Then the path integral that defines the potential difference can be evaluated.

14. E can be estimated using $|\Delta V/\Delta s|$, where Δs is the distance between two equipotential surfaces whose potentials differ by ΔV. The surfaces are closer together on the left side of the figure so the field has a larger magnitude there.

18. Charge is not necessarily distributed uniformly over the surface of a conductor. In fact, the charge per unit area is large where the radius of curvature is small and vice versa. If, on the other hand, the magnitude of the field is uniform over the surface then the charge distribution is uniform, as shown by $E = \sigma/\epsilon_0$.

21. $\mathbf{E} = 0$ in that region. Note that V must be constant throughout a volume, not just along a line or on a surface, for **E** to vanish.

22. The electric field must be zero in that region. Remove all charge from the region and completely surround it with a good conductor. The field is zero, not only in the conductor, but in the cavity.

23. In one possible arrangement the charges are at the vertices of an equilateral triangle and $q_1 = q_2 = -2q_3$.

25. In each case the charge originally on the metal object moves to the outer surface of the shell. In (a) the shell remains negatively charged but the magnitude of the charge is less. In (b) the shell becomes neutral. In (c) the shell becomes positively charged.

Chapter 27

1. Charge is conserved. All charge removed from one plate goes to the other. Neither the battery nor the wire collect charge. The plates were originally neutral, so they end up with charge of equal magnitude and opposite sign, regardless of their size or shape.

2. Yes. As a simple example, consider two spherical shells, one inside the other. The outer shell is $(q/4\pi\epsilon_0)(1/b - 1/a)$ higher in potential than the inner shell. Here a is the radius of the inner shell and b is the radius of the outer shell.

6 a. The capacitance increases. Use $C = \epsilon_0 A/d$ to see this.

b. The new capacitance is $C = \epsilon_0 A/(d - t)$, where d is the plate separation and t is the slab thickness. The effect is small if $t \ll d$.

c. The capacitance is doubled.

d. The charge on the larger plate spreads out into an area slightly larger than that of the smaller plate. For the same charge, the field and the potential difference are smaller (the field lines are further apart) and the capacitance is larger. This is a small effect since most of the charge on the larger plate is opposite charge on the smaller plate.

e. The capacitance is reduced by slightly more than 50%.

f. There is no effect. The charge also doubles.

g. The capacitance increases. Think of the situation as a collection of capacitors, wired in parallel, each succeeding one with a slightly smaller separation than the one before.

9. When fringing is taken into account the field lines are further apart so the field is smaller in magnitude for the same charge on the plates. The potential difference is less so the capacitance is greater.

14. In a uniform field the polarization is also uniform so the dielectric acts like a body with charge distributed only on its surface. The polarization charge has the same magnitude and opposite sign on opposite surfaces. Since the field is the same at each surface, the net force is zero.

17. The charge remains the same. It cannot change once the battery is removed. Because the dielectric is polarized with positive polarization charge on its surface nearest the negative plate and negative polarization charge on its surface near the positive plate, the field between the plates is reduced. The potential difference, which is the path integral of the field, is also reduced. A smaller potential difference for the same charge means a larger capacitance. The energy stored is reduced since $U = \frac{1}{2}q^2/C$. Negative work was done by the agent that moved the dielectric into place (the field pulled the slab in, the agent had to pull back on the slab to keep it from accelerating).

18. The potential difference and electric field remain the same since they are maintained by the battery. The sum of the charge on either plate and the polarization charge on the nearby surface of the dielectric is the same as the charge on the plate before the slab was inserted. These charges create the field and the field is the same. Since the polarization charge and the plate

charge have opposite signs, the charge on the plate must increase. An increase in the charge for the same potential difference means a greater capacitance. Since $U = \frac{1}{2}CV^2$, the energy stored increases by $\frac{1}{2}(\kappa - 1)C_0V^2$. The battery does work $(\Delta q)V = (\Delta C)V^2 = (\kappa - 1)C_0V^2$, so the agent does work $-\frac{1}{2}(\kappa - 1)C_0V^2$. The agent had to hold back on the slab to keep it from accelerating.

Chapter 28

1. All steady currents are in closed loops so any current that enters the closed surface must leave. If $d\mathbf{A}$ is taken to point outward then a current leaving the surface contributes a positive quantity to the integral and a current entering contributes a negative quantity. As a result, the value of the integral is 0.

2. Whatever the convention, it must describe all possible situations. In particular, if positive and negative charge move in opposite directions the current densities must be in the same direction and the currents must have the same sign. If they move in the same direction, the current densities must be in opposite directions and the currents must have opposite signs. This stems from an important property of nature. Positive charge moving in one direction and negative charge moving in the opposite direction are both brought about by the same electric field or emf and they produce the same effects (magnetic field, for example). For most applications (excepting the Hall effect, for example), the current may be considered to be composed of positive charge moving in one direction or negative charge moving in the opposite direction. With this limitation, it is possible to take the electron to be either positive or negative and, in a separate convention, to take the current to be in the direction of electron flow or in the opposite direction. The laws of physics must then be written to conform to the convention and describe nature.

4. The situation here is not static, in contrast to the situations discussed in Chapter 25. Here charges are kept in motion by a battery or other seat of emf and in steady state there is a non-equilibrium distribution of charge within the conductor. This charge creates a field in the interior. The charge, of course, moves in response to the field but it is replaced by action of the battery. When the battery is removed the charge reaches an equilibrium distribution, for which the electric field vanishes in the interior.

5a. In the steady state, no charge builds up anywhere and the current into the one corner of the cube must be the same as the current passing through any plane completely through the cube, regardless of the position and orientation of the plane as long as the two terminals are on opposite sides.

b. Near the sides of the cube the current density is parallel to the sides, while near the diagonal from one lead to the other it is parallel to the diagonal. The current density changes direction and increases in magnitude from a side to the diagonal. The plane cannot be oriented so the current density is uniform on it.

c. Yes.

d. Yes.

6a. Doubling V doubles the electric field in the wire and so doubles the acceleration of the electrons between collisions. The drift speed is therefore doubled.

b. If L is doubled, with the same potential difference, the electric field is halved and so is the drift speed.

c. There is no effect as far as the drift speed is concerned. The electric field is the same if the potential difference between the ends is the same. The current, however, increases by a factor of 4 since the cross sectional area increases by that factor and the current density remains the same.

19. The apparent paradox is resolved by specifying what is held fixed while R changes. If V is fixed, P decreases as R increases according to $P = V^2/R$. If i is fixed, P increases as R increases according to $P = i^2 R$.

Chapter 29

2. To measure the emf, place a high resistance voltmeter across the terminals, with no other circuit attached. To find the internal resistance, first place a small resistance across the terminals and measure both the current and the potential difference across the terminals. Then use $V = \mathcal{E} - ir$ to calculate r.

4. If R is the resistance of the light bulb, r is the internal resistance of the battery, and $\mathcal{E}$ is the emf, then the current is $i = \mathcal{E}/(R + r)$. If the internal resistance is large then $i = \mathcal{E}/r$, independently of the bulb resistance. The power dissipated by the bulb is $i^2 R = \mathcal{E}^2 R/r^2$, which is higher for the high resistance bulb (25 W). If $r = R$ and $\mathcal{E} = 120\,\text{V}$, then the bulb glows with normal brightness. The low resistance bulb (500 W), however, glows more dimly than normal. The situation is reversed if the internal resistance is low. Then the potential difference across the battery is nearly $\mathcal{E}$, independently of the bulb resistance. The power dissipated by the bulb is $\mathcal{E}^2/R$, which is lower for the high resistance bulb. Both bulbs glow "normally".

5. When the current is opposite to the emf, which occurs when the battery is being charged.

8. Connect batteries in parallel when you expect to draw a large current using a small emf. The current through each battery is less then the total. Connect them in series to obtain a greater emf than any of the individual emfs. The current is now the same in all of them.

11. 1 farad = 1 coulomb/volt and 1 ohm = 1 volt · second/coulomb, so 1 farad × 1 ohm = 1 second.

12. Along with the capacitance, it controls the charging time. A large resistance means a small current at any given instant during charging. This, in turn, means charge is accumulating on the plates at a slower rate.

16. Connect a known capacitor and unknown resistor in series with an emf. Measure the current as a function of time and use the measured function to find the time constant for the circuit, then compute $R = \tau/C$.

Chapter 30

2. The magnetic field would then be different for a test charge moving in different directions. When it is defined as usual, we may think of the field as existing independently of the test charge.

5. No. The electron might be traveling parallel or antiparallel to the field. To detect a magnetic field a test charge must be made to move in at least two different directions (in separate experiments).

6. No. An electric field may be pointing in the direction perpendicular to the electron's velocity. For example, the electric field of a positive point charge might cause an electron to travel uniformly around a circle just as a uniform magnetic field does. The test for a magnetic field must start with a test for an electric field, using a stationary test charge. After the electric force is subtracted from the total force, what remains is the magnetic force.

8. The magnetic field acting on the charged particle is produced by other moving charges, somewhere. The charged particle also creates a magnetic field and exerts a magnetic force on these other charges.

9 a. For typical magnetic fields ($\approx 1\,\mathrm{T}$), the dominant force is the force of the magnetic field on the electrons and the interaction between electrons can be ignored. At first they repel each other and speed up but once they are separated even slightly this effect can be neglected. Look down on the room. Each electron travels in a horizontal circle, through a point very near the point of release and tangent to its initial velocity. Both travel in a clockwise direction so the circles are on opposite sides of the line along the initial velocities.

b. The positron travels around its orbit in a counterclockwise direction while the electron travels in a clockwise direction as before. The circles coincide. The charges do not move in circles but actually spiral inward as they lose speed in collisions.

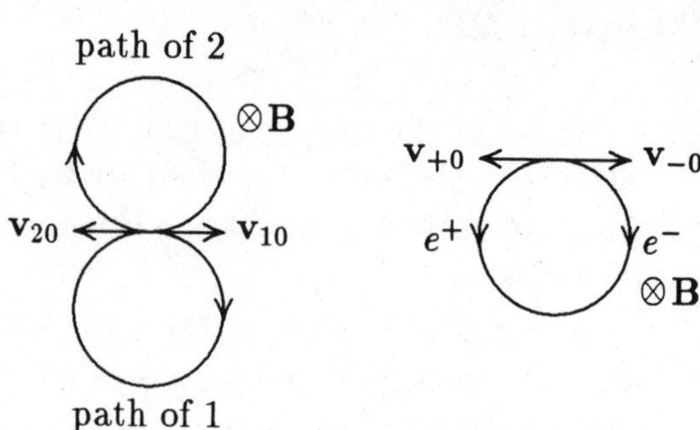

14. Yes. The force on the free electrons must be transmitted to the wire as a whole; that is, to the ions. This comes about through collisions between electrons and ions. If there were no interaction between electrons and ions, electrons would pass out of the wire and the wire would not move.

17. If electrons form the current, they are moving and experience magnetic forces. The protons are at rest (macroscopically) and do not experience magnetic forces.

21. No. If U_i is the initial potential energy, U_f is the final potential energy, θ_i is the initial angle between the dipole moment and the field, and θ_f is the final angle, then the work required is given by $W = U_f - U_i = -\mu B\cos\theta_f + \mu B\cos\theta_i = -\mu B\cos(\theta_i + \pi) + \mu B\cos\theta_i = 2\mu B\cos\theta_i$. Remark that if the moment is initially perpendicular to the field the net work done in turning it end-for-end is 0. Positive work is done during part of the rotation and negative work is done during another part.

22. The torque on the wire is zero since the dipole moment is parallel or antiparallel to the field. For stable equilibrium the dipole moment must be in the same direction as the field. This means the current must be counterclockwise as viewed from above.

Chapter 31

3. Not necessarily. If **B** is constant in magnitude the same number of field lines pass through each unit area perpendicular to the lines. The lines do not spread or come together. If the lines are more dense in the neighborhood of one portion of a line than in the neighborhood of another portion, the magnitude of **B** is not constant along that line. The magnitude of **B** is constant along any field line produced by a long straight wire; it is not constant along a field line produced by two parallel wires.

5. If the electrons are moving to the right in the laboratory frame, then an observer moving with the electrons sees the positive ions moving to the left. If the wire is neutral the two currents are the same, as are the magnetic fields they produce. These statements neglect relativistic effects.

6 a. The charges repel each other electrically and attract each other magnetically.

b. The charges attract each other electrically and repel each other magnetically.

7. Each of the 3 other wires attracts the left-hand wire with a force that is along the line joining the wires. The net force is to the right.

8. Near each wire the lines are circular since the field of the nearest wire dominates. In the region between the wires the individual fields tend to cancel and there are few lines. In the region beyond either wire the individual fields tend to reenforce each other and the density of lines is greater. Far away from both wires the field is like that of a single wire carrying current $2i$ and the lines are circular again.

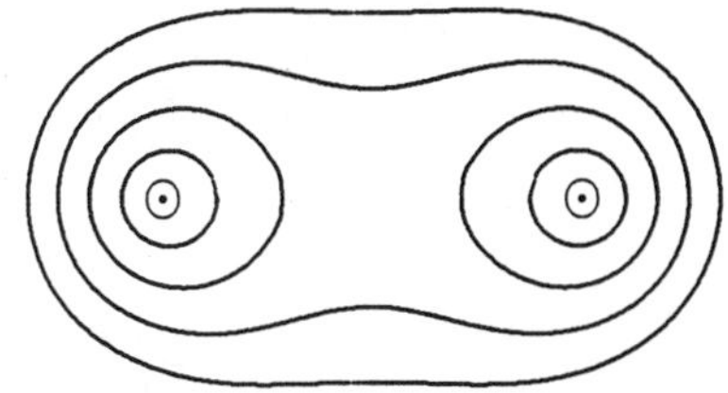

10. They tend to rotate about the point of intersection, remaining in the same plane. The angle between them decreases. Once they pass each other, the torque reverses sign, so they slow down, stop, and begin rotating in the opposite direction.

11. The magnetic fields are in opposite directions in regions II and IV. There are points in these regions when the resultant field is zero.

13. I: $\oint \mathbf{B} \cdot d\mathbf{s} = -\mu_0 i$.
II: $\oint \mathbf{B} \cdot d\mathbf{s} = 0$.
III: $\oint \mathbf{B} \cdot d\mathbf{s} = +\mu_0 i$.

Chapter 32

2. In most cases the stronger the magnet, the greater the rate of change of flux through the coil and the greater the induced emf. Consider a magnetic dipole moving in the direction of its moment. The field at a point a distance r from the dipole is proportional to μ/r^3, where μ is the dipole moment. The rate of change of the field a distance r from the dipole is proportional to $\mu v/r^4$. Magnets with larger dipole moments produce larger emfs.

5. As it falls the magnet induces eddy currents in the tube. The field of the eddy currents pushes upward on the magnet and retards its fall. It soon reaches a terminal velocity. Gravitational potential energy is converted to thermal energy in the tube.

10 a. The magnetic field due to the current in the larger loop is left to right through the smaller loop as seen in the diagram. Since the field is increasing in magnitude, the field of the smaller loop must point right to left through that loop. The current must be in the counterclockwise direction.

b. The coils behave like two magnets with like poles facing each other. They repel each other.

15. Current in the left loop is counterclockwise and is decreasing with time. The field through the right loop is into the page and is decreasing in magnitude. The current induced in the right loop creates a field which is into the page inside the loop and, to create this field, the current must be clockwise.

20. The induced current produces a magnetic field that is out of the page in the interior of the loop. The external field must be into the page since its flux increases as the rod moves. This increase is then opposed by the induced field.

22. It can be translated, without rotation, in any direction. Since the field is uniform the flux through the loop does not change. It can be rotated about any axis parallel to the field, through the center of the loop or elsewhere. Rotation about any other axis will produce a changing flux and an induced emf.

24. Suppose a strip without slots starts as shown and swings into the field. The flux through it increases at first and currents are induced. The induced field must oppose the increase in flux, so the induced currents are counterclockwise. Current flows up the right portion of the strip (in the field) and down the left portion (outside the field). The magnetic force on the current in the right portion retards the motion of the strip. Similarly, as the strip leaves the field on the other side, the magnetic force tends to pull it back. The slots force the current loops to close inside the field, for the most part, and damping is reduced.

26 a. No. Induced electric fields and emfs exist whether charge flows or not.

b. Cylindrical symmetry does not then exist and we cannot expect the electric field to be tangent to the circle. It is not. The emf would have a different value.

c. Yes. Electric fields and emfs exist outside the cylindrical region containing the magnetic field, as predicted by Faraday's law.

28. More electrons are at the tip of the wing on a passenger's right side. Consider the magnetic force on an electron moving forward with the plane.

Chapter 33

2. Use $\mathcal{E} = -L\,di/dt$. Assume a changing current and calculate the flux and emf for each loop of the coil, using a consistent sign convention. Add the emfs, then solve for L.

4. Wind the coil in layers with the current in opposite directions in successive layers.

7. After the switch is thrown current continues to flow for a long time and charge collects at the switch blades. The potential across the blades increases (it must match $L\,di/dt$ across the inductor) and when it is sufficient to ionize the air, an arc jumps. Point out that the open switch acts like a capacitor.

11. The derivative of a function can be large at a point where the function vanishes. The function passes through zero with a large slope. In this case, di/dt is large but i itself is zero. At the instant the switch is closed the electrons start to experience an electric field and they accelerate. But at that instant the drift speed is zero.

13. Yes. When current passes through the coil the potential difference is $iR + L\,di/dt$ just as for two separate elements. The current obeys $\mathcal{E} = iR + L\,di/dt$ and this equation has the solution given in the text. The time constant is L/R. The point here is that the resistance and inductance pictured in Fig. 33–5 may actually be in the same physical circuit element, a coil for example. When the time constant is calculated, one must use the total inductance and total resistance in the circuit and take all sources into account. The circuit itself forms a loop and so has a nonvanishing inductance. For completeness it should be mentioned that the resistance of many materials depends on the magnetic field so it might matter whether or not the resistance and inductance are physically close. This is not usually an important practical consideration and, in any event, it is accounted for when those properties are measured.

15 a. Let r be the radius of the wire and ℓ be the length of the solenoid. Then the number of turns is given by $N = \ell/2r$. The inductance is given by $L = \mu_0 N^2 A/\ell$, where A is the cross-sectional area of the solenoid. Thus $L = \mu_0 \ell A/4r^2$. The solenoid with the finer wire (A) has the greater inductance.

b. If a is the radius of the solenoid the total length of wire is $2\pi a N = 2\pi a\ell/2r$ and the resistance of the solenoid is $R = \rho a\ell/r^3$. The wire with the finer wire (A) has the greater resistance. The inductive time constant is $\tau_L = L/R = \mu_0 Ar/\rho a$. The wire with the thicker wire (B) has the greater inductive time constant.

19. **B** changes direction but u, which depends on B^2, does not change.

22. Yes. Magnetic field lines due to one coil pass through the coil that produces them as well as through the other coil.

Chapter 34

1. Lay one bar on a table. With their long axes at right angles to each other, pass the other bar along the length of the first. If the force of attraction is strong when the second bar is at either end of the first and weaker when it is at the midpoint, the stationary bar is magnetized. If the force of attraction is fairly uniform, the moving bar is magnetized. This technique works because the magnetic field of a magnetized bar is strong at the ends and weak in the middle.

2. One of the bars is weakly magnetized or unmagnetized. If both were strongly magnetized, repulsion would occur for some orientations.

8 a. Place the needle in a strong magnetic field, near a strong permanent magnet or inside a solenoid, say.

b. The dipole moment of the needle is in the direction of the applied field. The end where the field exits the needle is painted.

c. A north magnetic pole.

9. About the same. The dipole moments of all magnetic atoms are a few Bohr magnetons and the number of atoms per unit volume is the same order of magnitude for all materials.

10. In a paramagnetic substance, the atoms have dipole moments that tend to align with the field. Collisions disorient the atoms and reduce the magnetization. The effect of collisions increases as the temperature increases and the number of dipoles aligned with the field is therefore temperature dependent. In a diamagnetic substance, the part of the dipole moment of interest comes about through a distortion of the electron orbit by the field. A collision does not change the distortion much, so the magnetization is relatively insensitive to temperature.

13. The potential energy of the dipoles in the external field. Some become aligned with the field and their potential energy is reduced.

15. Wrap the wire into a coil and connect the batteries of one flashlight. Carefully note the direction of the current and determine the direction of the magnetic dipole moment. Place the compass on the axis of the coil above the north magnetic pole of the coil, say. The end of the needle that points toward the coil is a south magnetic pole. The other flashlight is used to see.

Chapter 35

1. When the capacitor is completely discharged ($q = 0$) current is flowing and the capacitor immediately begins to collect charge again. The situation is analogous to a mass on the end of a spring as it passes the equilibrium point. Its displacement is zero but its velocity is not, so it continues on by.

4 a. The product LC: $\omega = 1/\sqrt{LC}$.

b. The initial charge on the capacitor and the initial current, as well as ω: $Q^2 = q_0^2 + (i_0/\omega)^2$. This can be obtained from $q_0 = Q\cos\phi$ and $i_0 = -\omega Q \sin\phi$.

7. Take q to be positive if the upper plate has positive charge on it; take the current to be positive in the clockwise direction, and use $q_0 = Q\cos\phi$, $i_0 = -\omega Q\sin\phi$.

a. $q_0 = Q$, $i_0 = 0$ so $\phi = 0$.

b. $0 < q_0 < Q$, $i_0 < 0$ so $0 < \phi < \pi/2$.

c. $q_0 = 0$, $i_0 < 0$ so $\phi = \pi/2$.

d. $-Q < q_0 < 0$, $i_0 < 0$ so $\pi/2 < \phi < \pi$.
e. $q_0 = -Q$, $i_0 = 0$ so $\phi = \pi$.
f. $-Q < q_0 < 0$, $i_0 > 0$ so $\pi < \phi < 3\pi/2$.
g. $q_0 = 0$, $i_0 > 0$ so $\phi = 3\pi/2$.
h. $0 < q_0 < Q$, $i_0 > 0$ so $3\pi/2 < \phi < 2\pi$.

10. The potential difference across the capacitor is given by $V = q/C$. Replace q with x and C with $1/k$ to obtain kx. The potential difference corresponds to the force of the spring on the mass.

12. The system shown in Fig. 35–9 obeys $m\,d^2x/dt^2 = -k_1k_2x/(k_1+k_2)$ and the charge on the capacitor of an LC circuit obeys $L\,d^2q/dt^2 = -q/C$. Capacitance corresponds to the reciprocal of the spring constant. If we replace k_1 with $1/C_1$ and k_2 with $1/C_2$, $k_1k_2/(k_1+k_2)$ becomes $1/(C_1+C_2)$. Thus we want an effective capacitance of $C = C_1 + C_2$. This means the two capacitors are in parallel. q corresponds to the displacement of the mass from its equilibrium position and L corresponds to the mass m.

Chapter 36

1 a. The emf drives the current. If they did not have the same frequency, the phase difference between $\mathcal{E}$ and i would change with time. Kirchhoff's loop rule would not be satisfied for all values of the time.

b. If the voltages across various elements of the circuit summed to zero at one time, they would not at a later time. Kirchhoff's loop rule would not be satisfied for all times and energy would not be conserved.

2. Potential differences, currents, and emfs are not vectors and, for example, do not add as vectors. The direction of a phasor does not represent the direction of any physical quantity. Its projection on an axis, however, has the same time dependence and phase as the physical quantity associated with it. A phasor diagram shows the phase relationships of the various physical quantities represented. Furthermore, when two quantities with the same frequency but different phases are added, the result is the projection of the resultant phasor, the "vector sum" of the two phasors. So a phasor diagram provides a useful way to add such quantities.

4. At very low frequency $\tan\phi \approx -1/\omega RC$, a large negative number. ϕ is nearly $-90°$. At very high frequency $\tan\phi \approx \omega L/R$, a large positive number. ϕ is nearly $+90°$. $\phi = 0$ at resonance ($\omega = 1/\sqrt{LC}$). The phase constant varies continuously from $-90°$ to $+90°$.

7. If the phase of the voltage is between 0 and 180° greater than the phase of the current, the voltage is said to lead the current. Plot both the voltage and current as functions of time and look at a peak of one function and the nearest (in time) peak of the other. The peak of the voltage occurs at an earlier time than the peak of the current. If the phase of the voltage is between 0 and 180° less (or between 180° and 360° more) than that of the current, the voltage is said to lag the current. It reaches its peak later than the nearest current peak.

8 a. $X_L > X_C$ means $\omega^2 LC > 1$. For fixed ω, the product LC should be relatively large.

b. For fixed L and C, ω should be relatively large.

11. The loop and junction rules hold for each instant of time.

13. Write $\overline{P} = \mathcal{E}_{\text{rms}}^2 R/Z^2$ and differentiate with respect to R, C, or L, holding $\mathcal{E}_{\text{rms}}$ constant, then evaluate the derivative for the given values of R, C, L, and ω. To find how the power factor changes, differentiate R/Z. For the circuit of Sample Problem 36–4:

a. The sign of $d\overline{P}/dR$ is negative so $\overline{P}$ decreases with increasing R. The power factor increases and, since $\phi = -29.3°$ for the circuit, ϕ increases.

b. The sign of $d\overline{P}/dC$ is positive so $\overline{P}$ increases with increasing C. The power factor increases and ϕ increases.

c. The sign of $d\overline{P}/dL$ is positive so $\overline{P}$ increases with increasing L. The power factor increases and ϕ increases.

16. Since the emf leads the current, $X_L > X_C$ and $\omega L > 1/\omega C$. If the frequency is decreased slightly $\omega L - 1/\omega C$ decreases and the impedance decreases with it. Also explain what happens if $X_C > X_L$.

17. No. The current lags the applied emf if ϕ is positive and leads the applied emf if ϕ is negative. You cannot tell the sign of ϕ from the value of $\cos\phi$ since $\cos(-\phi) = \cos\phi$.

Chapter 37

3. Even if a laboratory electric field is made to change at an extremely fast rate, only a small magnetic field is produced. Consider a uniform field that changes at 10^{10} V/m·s and is normal to a circle with a 1 cm radius. The magnetic field at the rim is about 5×10^{-10} T and is not easily measurable. On the other hand, if a uniform magnetic field in the same circle varies at 1 T/s, the electric field at the rim is about 5×10^{-3} V/m and is easily measurable.

5. Take **A** to be into the page. In the integral $\oint \mathbf{B} \cdot d\mathbf{s}$, $d\mathbf{s}$ is in the clockwise direction. The question can be answered by finding the sign of $d\Phi_E/dt$ and hence of $\mathbf{B} \cdot d\mathbf{s}$. It can also be answered by finding the direction of the displacement current and observing that the lines of **B** around i_d are the same as the lines around a true current.

a. $\Phi_E > 0$ and $d\Phi_E/dt < 0$ so the integral is negative and **B** points in the counterclockwise direction. Here i_d is opposite to **E**.

b. $\Phi_E > 0$ and $d\Phi_E/dt < 0$ so the integral is negative and **B** points in the counterclockwise direction. Here i_d is along **E**.

c. $\Phi_E < 0$ and $d\Phi_E/dt > 0$ so the integral is positive and **B** points in the clockwise direction. Here i_d is into the figure.

d. $d\Phi_E/dt = 0$ and, since the electric field has cylindrical symmetry, $\mathbf{B} = 0$. Here $i_d = 0$.

6. The electric field is zero only at the instant shown, not at any time just before or just after. That is, the derivative of the field is not zero, so there is a non-vanishing displacement current.

7. The displacement current density is in the direction of $d\mathbf{E}/dt$. Since the line of **E** does not change, it is in the direction of **E** if E is increasing and in the opposite direction if E is decreasing. For the situation shown, i_d is into the page (left diagram) or left to right (right diagram). The lines of **B** form circles around the direction of the displacement current. Point the right thumb in the direction of i_d (or $d\mathbf{E}/dt$), then the fingers curl in the direction of **B**.

Chapter 38

4. The statement is essentially true. A displacement current is associated with the time-varying electric field and is responsible for the spatial variation of the magnetic field.

7. Faraday's law describes the mechanism by which the changing magnetic field induces the electric field. The displacement current term in the Ampere-Maxwell law is needed to complete the cycle: it describes the mechanism by which the changing electric field induces the magnetic field.

8. No. Light carries momentum and, when it is absorbed, the conservation law demands that the momentum be transferred to the absorbing medium.

11. The sources of radio waves are usually fixed dipole antennas. These radiate linearly polarized waves with the direction of polarization always the same. The sources of visible light are atoms. While each wave train from any atom may be linearly polarized, trains from different atoms or trains from the same atom at different times have different polarization directions. Thus light from extended sources is unpolarized. The direction of its electric field fluctuates rapidly.

13. If you wish to rotate the plane through the angle θ, stack N polarizing sheets with the polarizing direction of the first sheet at the angle θ/N to the direction of polarization of the incident light and each successive sheet rotated by θ/N in the same direction relative to the previous sheet. The transmitted light is polarized in the desired direction and has an intensity that is proportional to $\cos^{2N}(\theta/N)$. The more sheets used the greater the intensity and the less the loss.

16. First use two sheets with their polarization axes perpendicular to each other. Light transmitted through the first sheet is polarized along its axis and is not transmitted through the second. Now add the middle sheet with its axis in any direction not parallel to that of either of the other sheets. Since the polarization direction of light transmitted by the first sheet is not perpendicular to the axis of the middle sheet, some light is transmitted through it. The polarization direction of the light transmitted by the middle sheet is along the axis of that sheet and is not perpendicular to the axis of the last sheet. Hence some light is transmitted through the last sheet.

Chapter 39

6. Stand in front of a mirror and look at your image. The "left" hand of the image is not a left hand at all but the image of a right hand. The symmetry of the human body helps give you the mistaken impression. Suppose the right arms of all humans were shorter than the left arms. Then the image arm in front of your short right arm would not be a long left arm and it would be clear that left and right were not reversed.

9. No. Light rays diverge so that the light appears to come from the position of the image but no light actually passes through that position.

10. Assume the source is real so $p > 0$.
   a. The image is real if $i > 0$ or $2/r > 1/p$. For a convex mirror ($r < 0$) the image is always virtual. For a concave mirror ($r > 0$) the image is virtual for $p < r/2$ and real for $p > r/2$.
   b. Since $m = -i/p$ is positive for erect images and negative for inverted images, all virtual images are erect and all real images are inverted.
   c. Combining the mirror equation and the expression for magnification yields $m = -r/(2p - r)$. The image is larger than the object if $|2p - r| < |r|$. This occurs for concave mirrors if $p < r$. The image is smaller if $p > r$. For convex mirrors the image is always smaller than the object.

11. The mirror is convex so $|i| < |p|$. The image is closer to the mirror than the object but the angle of vision is wider than it would be with a plane mirror. Differentiate $1/p + 1/i = 2/r$ with respect to time to show that $|di/dt| = (i^2/p^2)\,|dp/dt|$. Since $|i| < |p|$ the image moves more slowly than the object (toward or away from the mirror).

13. No. When the lens is reversed r_1 becomes $-r_2$ and r_2 becomes $-r_1$ in the lensmaker's equation. The focal length is the same, so the image does not change.

14. Converging lens: If $p > 2f$ the image is real, inverted, and smaller than the object. If $f < p < 2f$ the image is real, inverted, and larger than the object. If $p < f$ the image is virtual, erect, and larger than the object. Diverging lens: the image is always virtual, erect, and smaller than the object.

16. The magnification is given by $m = -i/p$, where p is the object distance and i is the image distance. If $m = -1$, $i = p$ and $(1/p) + (1/i) = (1/f)$ yields $p = 2f$. The magnification is -1 if the object distance is twice the focal length. If $m = +1$, $i = -p$ and the focal length is infinite. Such a "lens" is really a flat plate of glass and doesn't bring light to a focus.
18. Yes. In fact, $1/f = (n/n' - 1)(1/r_1 - 1/r_2)$, where n' is the index of refraction for the medium and n is the index for the lens. A lens for which r_1 is positive and r_2 is negative is converging if $n > n'$ and is diverging if $n < n'$.

Chapter 40

1. In classical theory the oscillating electric field in the wave causes charges in the material to vibrate and to emit electromagnetic radiation. The vibrational frequency of the charges is the same as the frequency of the incoming wave and so is the frequency of the reradiated light. The quantum mechanical description also leads to absorption and reradiation at the same frequency. The speed of the wave is different in the material and, since the frequency of the same, the wavelength is different. For completeness, you might mention that for some materials a small portion of the reradiated light has a frequency that is a multiple of the original frequency.
2. No. The color seen is associated with the wavelength of the light in the fluid of the eye. This is the same regardless of the refractive index of the medium outside the eye.
5. Maxima still occur for $d \sin\theta = m\lambda$ and minima still occur for $d \sin\theta = (m + \frac{1}{2})\lambda$ but now λ is the wavelength of the light in water and is somewhat shorter than the wavelength in air. The pattern is tighter, with the minima closer together, for example.
7. The double-slit interference pattern washes out but red and blue single-slit diffraction patterns remain, with their intensities superposed. This may happen for two reasons. The red and blue parts of the spectrum come from different sets of atoms or from the same set of atoms at different times. They are incoherent. Even if they are coherent in the sense that the phase constants do not change, the interference pattern fluctuates at the beat frequency. To see this, picture two phasors rotating with different periods. The observer sees the time average of the pattern and this does not show the double-slit interference effect.
10. The pattern would wash out. The vector amplitudes of the two waves at any point on the screen would be perpendicular to each other and the square of the resultant field would be the sum of the squares of the two fields. Thus the intensity, which is proportional to the average over a cycle, would be proportional to the sum of the intensities of the individual waves. These are independent of the phase difference.
15. It is transmitted. Point out that the condition for a minimum in reflection is the same as the condition for a maximum in transmission.
16. Light waves from the two sources are not coherent. The phase difference is rapidly changing. The eye perceives the time average intensity, which is nearly the same at every illuminated point.
17. In both cases the wave reflected from the outside or first surface suffers a phase change of π. In the case of the soap film the wave reflected from the second surface does not suffer a phase change on reflection since the refractive index for air is less than that for water. In the case of the oil film the wave reflected from the second surface suffers a phase change of π since the index of refraction of water is greater than that of oil. This means the thickness that produces fully constructive interference with a soap film produces fully destructive interference with an oil film and vice versa.

Chapter 41

7a. When the wavelength is increased without changing the slit width the pattern spreads out. Any two adjacent minima, for example, have a larger angular separation and the central bright band is wider.

b. When the slit width is increased without changing the wavelength the pattern becomes more compact. Any two adjacent minima have a smaller angular separation and the central bright band is narrower.

8. There are no minima. The entire forward region is within the central maximum. As λ increases beyond a the variation in intensity with angle becomes less and the illumination in the forward direction is more nearly uniform.

12. The resultant of wavelets r_1 and r_3 is out of phase with the resultant of wavelets r_2 and r_4. The order of adding the wavelets is immaterial.

18. The greater the number of slits the more sensitive the intensity is to the phase difference associated with adjacent slits. Imagine a large number of parallel phasors. This is the condition for the formation of a diffraction line. Now change the angle between adjacent phasors by some small amount and note the decrease in total intensity that occurs because the phase difference between the first and last phasor is large. Repeat for 2 or 3 phasors. For the same change in phase the decrease in intensity is significantly less. Thus a small change in θ results in a large change in intensity if there are many slits but only a small change if there are few slits. The diffraction lines are sharper for a large number of slits than for a small number. Since the phase difference for waves from adjacent slits is given by $(2\pi d/\lambda)\sin\theta$, increasing d or decreasing λ makes the intensity pattern more sensitive to changes in angle and so results in sharper lines.

23. The order is limited by the condition that θ be less than 90°. Use $d\sin\theta = m\lambda$ to show that the highest order is given by the largest integer value of m that is less than d/λ.

24. Use $R = Nm$ and $m = (d/\lambda)\sin\theta$ to show that $R = (Nd/\lambda)\sin\theta = (L/\lambda)\sin\theta$.

Chapter 42

2. Ocean currents, air currents, and ballistic missile trajectories all show significant effects. Rotation of the earth must be taken into account in launching space craft. On a smaller scale, precise measurements of weight also show effects. They must be taken into account when the measurements are used (by geologists, for example) to find the local density of the Earth.

5. The speed is c in both cases.

9. As the equations in Table 42–3 reveal, if Δx and Δt both vanish then both $\Delta x'$ and $\Delta t'$ also vanish, regardless of the relative velocity of the two frames. Thus the events occur at the same time and at the same place in all frames.

10. The events recorded by the two observers are different so different results should not seem paradoxical. Each observer marks the positions of the ends of the other meter stick at the same time, according to his clocks. These events are not simultaneous according to clocks of the other observer, in the rest frame of the stick being measured. Note that the processes are different because different meter sticks at rest in different frames are involved. Also note the symmetry. Point out that an observer in S', watching an observer in S measure the stick in S', agrees that the result is less than 1 m. S' sees S mark one end first, then the other end. During the interval the first mark moves toward the position of the second so, according to S', the marks are closer than 1 m at the time the second mark is made. If the observer in S' did not know about relativity he would claim that S did not carry out a length measurement since the making of the marks was not simultaneous.

14. Consider a particle moving with speed c along a line in the $x'y'$ plane that makes the angle θ with the x' axis. The components of its velocity, in S', are $v'_x = c\cos\theta$, $v'_y = c\sin\theta$. Frame S' moves along the x axis of frame S with velocity u. In frame S, the components are given by $v_x = (v'_x + u)/(1 + v'_x u/c^2) = c(c\cos\theta + u)/(c + u\cos\theta)$ and $v_y = v'_y/\gamma(1 + v'_x u/c^2) = c^2\sin\theta/\gamma(c + u\cos\theta)$. The first equation is an application of Eq. 42–23 while the second can be derived by substituting the Lorentz transformation equations into $v_y = \Delta y/\Delta t$. A little straightforward algebra and use of $\cos^2\theta + \sin^2\theta = 1$ gives $(v)^2 = v_x^2 + v_y^2 = c^2$, independently of u. Thus the speed of the particle is c in every inertial frame.
According to Eq. 42–34 the momentum of a massless particle moving at a speed different from c vanishes. Since $E = pc$ so does its energy. No change can occur in any interaction so such a particle cannot be observed.
16. The rest mass of a collection of particles is given by $m = E/c^2$, where E is the total energy of the system as measured in a frame for which the total momentum vanishes. The total energy includes the kinetic energies of the individual particles and is greater than the sum of the rest energies of the particles. As the material cools the total kinetic energy decreases and so does the rest mass of the collection. A sufficiently sensitive scale would indicate the change.

Chapter 43

2. The energy of the photon source is reduced by hf when a photon is emitted. When the source is receding its loss in energy is reduced according to relativity theory. The reduction is the same as the reduction in photon energy and the conservation of energy principle is preserved.
4. For a particle with mass m and momentum p, $E^2 = (mc^2)^2 + (pc)^2$. For a photon $m = 0$ but $p \neq 0$ and $E = pc$. $E = mc^2$ relates rest energy to mass. A photon has energy but no rest energy.
9. The existence of a cutoff frequency means that the energy transferred to an electron depends on the frequency of the light. When the frequency is too low, not enough energy is transferred to drive the electron from the material. The fact that the cutoff frequency is independent of the light intensity means that the energy transferred is not proportional to the square of the amplitude as it would be if the classical wave theory were correct. In the photon theory the effect is explained by postulating that energy can be transferred only in units of hf. These are the photons.
12. All of them can result in the emission of electrons. All that is required is that electrons near the surface be given an additional energy equal to at least the work function.
15. The photon reverses its direction of travel and so picks up momentum opposite to the direction of incidence. Since the electron is initially at rest its momentum after the scattering event must be in the direction of incidence for the total momentum of the photon-electron system to be conserved.
16. The change in wavelength occurs when light is scattered from free or nearly free electrons. Materials differ in the concentration of these electrons and in their energies. The energy transferred in Compton scattering is large compared to the original electron energy so, to a good approximation, the scattering is nearly as if the electrons were at rest. The concentration of electrons does not influence $\Delta\lambda$; it does influence the number of scattering events which occur, not the result of any single event. So $\Delta\lambda$ is nearly the same for all materials.
17. The maximum change in wavelength that can occur is about 0.0024 nm and is independent of the wavelength of the incident light. This change is extremely small compared to the wavelength of visible light (about 500 nm). It occurs only in the sixth significant figure and is difficult to detect.

23. The potential energy is taken to be zero when the electron and proton are infinitely far apart. A negative potential energy means their separation is finite. A negative total energy means the electron does not have sufficient kinetic energy to escape from the proton. The same zero of potential energy can be chosen for a classical system. Negative total energy implies a bound system classically as well as quantum mechanically.

27. Yes. The electron and proton are then no longer bound. The atom is ionized. Since the energy of a free electron is not quantized, any photon with energy greater than the binding energy can be absorbed.

28. Cavity radiation: oscillators in the cavity walls have energies which are integer multiples of hf, where f is the natural frequency; they emit or absorb radiation energy in units of hf; the number of oscillators with a given energy is given by the usual law of statistical mechanics.
Photoelectric effect: light energy comes in integer multiples of hf, where f is the frequency of the light; in the interaction with an electron, energy hf is transferred to the electron, which then gives up some of its energy to the material and escapes with the remaining energy in the form of kinetic energy.
Compton effect: light behaves like a collection of particles each having energy hf and momentum $hf/c = h/\lambda$; the interaction between a photon and an electron can be analyzed as any other 2 particle collision in which energy and momentum are conserved.
Structure of hydrogen: the electron can have only certain discrete energies. When it jumps from a state with energy E_i to a state with energy E_f it emits a photon with energy given by $hf = E_f - E_i$.

Chapter 44

1. Both wave and particle properties are associated with any physical object. The wavelength of the wave is related to the particle momentum; the frequency is related to the particle energy. Some experiments, like diffraction experiments, are sensitive to wave properties and some, like the Compton effect experiment, to particle properties.

9. The de Broglie wavelength can be smaller than the particle size. A proton with a high momentum, for example, has an extremely short de Broglie wavelength. You might mention that protons and neutrons are really composite particles and that the dimensions of more fundamental particles, such as quarks and electrons, are too small to be measured and may be zero. If they are, the wavelengths obviously cannot be smaller. The de Broglie wavelength can also be larger than the particle size. This happens if the momentum is small.

13. Quantum mechanics is capable, in principle, of giving exact predictions for the probabilities of the occurrence of events. It is not capable of giving an exact prediction of the outcome of any single event. Quantum mechanics can be used to predict the average outcome and the distribution of results for a large number of identical experiments.

14. Look at the $n = 2$ graph of Fig. 44–10. The probability of finding the particle within Δx of the left wall is given by $\int_0^{\Delta x} |\psi|^2 \, dx$. It is very small but it is not zero since the volume element extends away from the wall and the probability density is slightly greater than zero a small distance from the wall.

15. $|\psi|^2$ is an oscillating function of x with 101 zeros (counting the end points). As n becomes large the wavelength becomes small. If the wavelength is much smaller than the detector, the probability of detection is the average of P_n over many cycles times the width of the detector. This the same as the classical value for the probability of detection.

21. The probability amplitude wave is diffracted by each slit and so spread out beyond the geometric shadow of the slit. Placing a slitted barrier in the beam reveals the wave nature of matter rather than the particle nature.

23. Since the probability density is zero at two points, the particle will never be found at those points while it is in that state. It might be found on either side of the points. In addition, it is not possible to find the particle first on one side of a zero and then on the other side, without altering the wave function. Performance of the first position measurement alters the wave function. After the first measurement it might be placed in the original state and the position measured again. This time it might be found on the other side of the zero. To get it into the original state, we had to give it a certain well defined energy and momentum magnitude and we lost control of its position.

24. In the Bohr theory the electron is assigned a definite orbit, a circle of definite radius, for example. Since the value of r is not uncertain, the radial component of the electron's momentum has large (infinite) uncertainty. If this momentum component is measured many times, with the electron initially placed in the same orbit, the standard deviation of the results is huge. This is not substantiated by experiment.

25 a. Problem 44–47 provides an excellent example. Here a photon is used to detect an electron in a hydrogen orbit. The wavelength must be much smaller than the orbit radius. This means the energy transferred in a head-on Compton event is much greater than that required to ionize the atom. The text also gives an example of the use of a single slit to localize an electron. The narrower the slit the greater are the diffraction effects.

b. The disturbance can be taken into account only insofar as the new wave function can be computed if details of the measurement are known. The momentum, energy, and position of the particle cannot be predicted with certainty.

Chapter 45

1. Wave mechanics predicts the probabilities of finding the electron in various regions around the proton, information that is necessary, for example, to understand the interaction of the atom with neighboring atoms. Bohr theory erroneously ascribes a definite orbit to the electron and is therefore not useful for the understanding of chemical properties. In addition, Bohr theory does not correctly predict the correct allowed values of the angular momentum and magnetic dipole moment. Wave mechanics does. Wave mechanics also predicts the probability that the electron makes a transition to another state when it is subjected to an external force.

6. The magnetic dipole moment of the electron is directed opposite to its spin angular momentum because the electron is negatively charged.

9. The angular momentum does not have a definite direction in space. Suppose it has a definite magnitude and a definite z component. Then in the vector model the angular momentum vector precesses about the z axis and its x and y components vary with time. Quantum mechanically the result of a measurement of either the x or y component is one of the values $m_\ell \hbar$ and in a series of measurements a distribution of values is obtained. The x and y components cannot be specified.

13. The probability density function gives the probability per unit volume that the particle is in an infinitesimal volume in the neighborhood of $\mathbf{r}$ at time t. The wave function $\psi(\mathbf{r})$ is a function such that the probability density is given by $|\psi|^2$. It satisfies a differential equation (the Schrodinger equation) and it is complex, while $|\psi|^2$ is real and positive. Since ψ can be constructed as a linear combination of waves; it can exhibit interference and diffraction effects. The radial probability density, defined by $P(r) = 4\pi r^2 |\psi|^2$, gives the probability per

unit radial distance that the particle is in a spherical shell of infinitesimal thickness dr. The radial probability density is closely related to the average of the probability density over angles.

17. The torque is produced by the magnetic field acting on the magnetic moment of the atom: $\boldsymbol{\tau} = \boldsymbol{\mu} \times \mathbf{B}$. In the field (assumed to be along the z axis), states with different values of m_ℓ (and hence different values of μ_z) have different energy. The atoms are distributed among these states, so different atoms may have different values of μ_z. One may think of the angular momentum and dipole moment as precessing about the direction of the field, the precession being produced by the torque. The force on the atom is produced by the field gradient: $F_z = \mu_z \partial B/\partial z$. This causes atoms with dipole moments in different directions to follow different trajectories as they cross the field.

20. The lengths of the periods in the periodic table support the need for a fourth quantum number. For example, the chemical properties of the inert gases are similar and these atoms occur for $Z = 2, 10, 18, 36, 54$, and 86. The numbers of electrons in the various orbitals can be found using these numbers. When the rules for specifying ℓ and m_ℓ are applied, it is found that each shell contains only half as many states as are needed to produce the periodic table. An additional quantum number, with two possible values, is needed. If there were no physical manifestation of the fourth quantum number, perhaps its need could be eliminated by revising the Pauli exclusion principle. However, a magnetic dipole moment is associated with spin and the exclusion principle is supported by relativity theory.

22. In the ground state, the atoms of the lanthanide series all have two electrons in $n = 6$, $\ell = 0$ states and have empty $n = 6$, $\ell = 1$ states. The wave functions of the $n = 6$, $\ell = 0$ electrons extend the farthest distance from the nucleus of any of wave functions for electrons in a lanthanide atom. It is these electrons that interact with neighboring atoms and determine the chemical properties of the elements, so the chemical properties are similar. Atoms of the series differ in the number of $n = 4$, $\ell = 3$ electrons. These have wave functions which are more confined than the $n = 6$, $\ell = 0$ functions and they do not participate in chemical interactions. The atoms of the series also differ in the number of protons, so the energies of the less shielded inner electrons are reduced from atom to atom through the series. This affects the characteristic x-ray spectrum and the atoms fit on the Mosley plot in their predicted places.

24. The cutoff wavelength is the wavelength of a photon with energy equal to the original kinetic energy of the electron. It is a clue to the photon nature of light because it ties the wavelength (and hence the frequency) of the emitted light to the energy lost by an electron. The electron loses all its kinetic energy in the creation of a photon with the same energy.

26. A characteristic x ray is emitted when a higher energy atomic electron falls to a lower vacant state. The lower state was vacated when its original occupant was knocked out of the atom by an electron incident from outside the atom. The frequencies of characteristic x rays are proportional to the energy differences of the atomic states involved and these, in turn, depend on the atomic number (the nuclear charge). A plot of $\sqrt{f}$ vs. Z is nearly linear.

28. Atomic hydrogen does not emit characteristic x rays. If the single electron in a hydrogen atom were removed, the atom could pick up a free electron and one or more photons would be emitted as the electron dropped to the ground state. The energy of the most energetic photon possible is 13.6 eV, too small by a factor of 50 to be an x ray.

37. Without population inversion the intensity of the beam would be small or non-existent. When the population of the upper state is less than the population of the lower state the system is closer to thermodynamic equilibrium and the radiation energy absorbed is more nearly equal to the radiation energy emitted in any time interval. It is only when the populations are inverted that significant light amplification takes place.

Chapter 46

5. Without the Pauli exclusion principle all electrons would be in low lying states, tightly bound to atoms, and the conductivity would be extremely small. With the exclusion principle, one or more electrons per atom are in states with sufficiently high energy that the electrons are loosely bound and can accelerate in an electric field to produce a large current.

6. The density of states function $n(E)$, when multiplied by dE, gives the number of quantum mechanical states per unit volume with energy between E and $E + dE$. The states need not be occupied and, in fact, the function is not influenced by whether they are or not. This function is different for different systems since, in general, different systems have different sets of electron states, with different energies. The Fermi-Dirac probability function $p(E)$ gives the probability that a state with energy E is occupied at temperature T. It ranges from 0 to 1. It is completely independent of the states of the system, and the same function is valid for all large collections of electrons, regardless of the density of states function. In fact, a particular system may not have a state with the energy E for which the function is evaluated. The density of occupied states function $n_0(E)$, when multiplied by dE, gives the thermodynamic average number of electrons per unit volume with energy between E and $E + dE$. This must be the product of the number of states per unit volume in the energy range and the probability of occupation: $n_0(E) = n(E)p(E)$. As an example, suppose in some small energy range there are 5 states per unit volume and each has a 25% chance of being occupied. We then expect 1.25 electrons per unit volume to have energy in that range, on average.

12. In each case look at the distribution of electrons among the states at $T = 0\,\mathrm{K}$ and examine the most energetic electrons. If there are a large number of vacant states nearby in energy, the material is a conductor. If the vacant states are separated in energy from the occupied states but a significant number of electrons are promoted thermally from occupied states as the temperature increases to room temperature, the material is a semiconductor. This may come about because the gap between the valence and conduction bands is small or because there are impurity states in the gap. If a significant number of electrons are not thermally promoted, the material is an insulator.

13. Band theory can be used to predict whether a given material is a conductor, insulator, or semiconductor. It gives us the density of states function and so is used to predict the number of electrons that contribute to the current when an electric field is turned on. Band theory is used, in conjunction with a known mean free time, to compute the electron drift speed.

21. Collisions with atoms control the temperature dependence of the conductivity of metals. They enter the simple theory through the mean free time. As the temperature increases so do the number of collisions; the mean free time decreases and so does the conductivity. For intrinsic semiconductors at room temperature and above, the number of electrons in the conduction band and the number of holes in the valence band control the temperature dependence of the conductivity. As the temperature increases the number of electrons thermally promoted across the gap increases and so does the conductivity. The number of collisions also increases, as for metals, but thermal promotion is far more important at these temperatures.

22. Germanium has a higher density of charge carriers at room temperature since electrons have a higher probability of being thermally excited across a narrow gap than across a wide gap. At $T = 0\,\mathrm{K}$ neither has any charge carriers because the conduction bands of both materials are empty and the valence bands are completely filled.

24. As far as electrical conduction is concerned, the net effect of all electrons in a band with one empty state is the same as that of a single positive charge with momentum equal to the total momentum of the electrons. The positive particle is called a hole. Normally holes in the valence

band, rather than the conduction band, enter the description of a semiconductor because there are relatively few holes in this band and all of them have nearly the same energy.

Chapter 47

2. The strong force is independent of the electrical charge on the particles, the electrostatic force is not. The strong force has a short range and nearly vanishes when the separation of the particles is on the order of 2 fm or more, while the electrostatic force is long range and dies out much more slowly with separation. At separations where both exist, the strong force is much stronger than the electrostatic force. Except at very small separations the strong force is attractive, while the electrostatic force can be attractive or repulsive, depending on the signs of the charges carried by the particles involved.

5. All nucleons attract neighboring nucleons via the strong force but the force is weak for nucleons with greater separations. On the other hand, protons repel all other protons that are beyond the range of the strong force. The nucleus becomes more stable if there are more neutrons than protons because protons can then be separated from other protons and held in the nucleus by nearby neutrons.

11. The shape of the binding energy curve is due chiefly to saturation of the strong force and to the reduction in binding energy because of the mutual electrostatic repulsion of protons. For small A the strong interaction is not saturated so each new nucleon interacts with all nucleons already there and the binding energy per nucleon increases strongly with A. At large A each new nucleon interacts via the strong force with about the same number of nucleons as the nucleons already there. If no other forces were present the curve showing binding energy per nucleon vs. A would level off. In fact, it decreases as A increases. As A increases so does Z and the increase in electrostatic interactions between protons tends to decrease the binding energy.

17. Carefully measure the number of ^{238}U nuclei in a sample and the decay rate of the sample. Use $dN/dt = -\lambda N$ to calculate λ and $\lambda\tau = \ln 2$ to calculate τ.

20. Look at the potential barrier diagramed in Fig. 47–9. The larger the energy of the α particle the narrower the barrier at the energy of the α. This means the probability of tunneling is greater and the half-life is shorter. Once the α is outside and moving away, its potential energy drops toward zero. The energy of the α while inside becomes part of the disintegration energy and is shared with the recoil nucleus as the particles repel each other. Thus the larger the energy of the α, the greater the disintegration energy.

22. The energy lost by the nucleus in β decay can be measured and this energy defines the upper limit to the β spectrum. For most β decays, however, the β has less energy and physicists must account for the missing energy or declare the principle of energy conservation to be invalid. It was postulated that another particle, a neutrino, is emitted with the β and that the neutrino carries the previously missing energy. Later experiments, in which interactions involving incident neutrinos were observed, substantiated the existence of these particles.

23. Neutrinos have spin with m_s either $+\frac{1}{2}\hbar$ or $-\frac{1}{2}\hbar$, like electrons. They obey the Pauli exclusion principle and the Fermi-Dirac probability function is valid for them. Photons, on the other hand, have $m_s = -1$, 0, or $+1$ and do not obey the Pauli exclusion principle. It is possible to have any number of photons in any given state. The number of photons is not conserved in interactions.

Chapter 48

3. Strictly speaking, Q depends on the masses of the fragments. However, for the overwhelming majority of fission events X and Y are medium mass nuclei. These have nearly the same binding energy per nucleon, so Q does not deviate much from 200 MeV, which is characteristic.
4. The curve is approximately symmetric. For each heavy fragment there is a companion light fragment. The mass numbers of the fragments sum to 236, less the number of emitted neutrons. The heavier fragments have a greater ratio of neutrons to protons and the number of neutrons emitted varies. This destroys the symmetry slightly.
5. The decaying fragments are all neutron rich and become more stable by conversion of a neutron to a proton, with the emission of a β^- particle. β^+ decay would make the fragments even more neutron rich.
13. While the power level is being decreased, the rods are pushed in and k decreases. When the desired level of neutron flux is achieved the rods are adjusted so $k = 1$ again. For long term sustained operation $k = 1$ no matter what the power level.
14. The two isotopes are nearly identical in their chemical properties and they react chemically with other elements in identical ways. They differ in mass, however, and this difference is exploited when diffusion is used to separate them. A mass spectrometer can also be used.
17. No. The number of fusion events per unit volume is given by the *product* $n(K)p(K)$. Thus the greatest number of events occur for the kinetic energy at which the product of the two curves is a maximum.

Chapter 49

2. Waves associated with high energy particles have greater frequencies and shorter wavelengths than those associated with low energy particles. They can therefore be used to probe smaller regions. Of equal importance, high energy is needed to produce massive particles from the constituents of ordinary matter.
5. To leave tracks in a detecting chamber a particle must interact with the material in the chamber. Neutrinos have such a low probability of interaction that they leave no tracks.
6. Although all neutrinos travel at the speed of light they may have different momenta. The relativistic expression $p = mv/\sqrt{1 - v^2/c^2}$ becomes indeterminate as m approaches 0 and v approaches c. A zero mass particle can have non-vanishing momentum only if it travels at the speed of light. Since $E = pc$ for a zero mass particle, a different momentum leads to a different energy.
8. Photons are bosons (spin $\hbar$) while neutrinos are fermions (spin $\hbar/2$). Neutrinos participate in the weak interaction while photons are the messenger particles for the electromagnetic interaction. Neutrinos are produced in weak decays, such as beta decays. Photons are produced in electromagnetic transitions of other particles, such as the change in the state of an electron in an atom. Both particles can be detected by observing the process that is the inverse to the producing process: the absorption of photons by atoms and the change of a proton to a neutron when a neutrino is absorbed.
11. Of all the charged particles, the electron (and the positron) has the smallest mass. Its decay, if it occurs, must be to a particle with equal or less mass and so would not conserve charge.
13. Spin angular momentum cannot be conserved by such a decay. The z component of the spin angular momentum of the products must be either $-\hbar$, 0, or $+\hbar$ while the z component of the spin angular momentum of the neutron must be either $-\hbar/2$ or $+\hbar/2$.

28. Baryons interact via the strong interaction, leptons do not. Baryon number (+1 for baryons, −1 for antibaryons) seems to be conserved in all interactions. Baryons are composed of more fundamental particles (quarks), while leptons are not.

29. A meson has a spin angular momentum that is a multiple of $h/2\pi$, while a baryon has a spin angular momentum that is an odd multiple of $h/4\pi$. This means mesons are bosons and obey Bose-Einstein statistics while baryons are fermions and obey Fermi-Dirac statistics. Baryons obey the Pauli exclusion principle; mesons do not. In addition, a conservation principle (the conservation of baryon number) is associated with baryons. There is none associated with mesons.

33. To decay, one of the quarks in a charged pion must change flavor (i.e. u must change to d or $\bar{d}$ must change to $\bar{u}$). This occurs as a result of a weak interaction and requires a relatively long time.

40. To take a simple example, suppose a galaxy is spherical, with each part rotating about the center under the gravitational pull of the other parts. If the mass density is uniform all parts will have the same angular speed. On the other hand, if the density decreases with increasing radius, the angular speed of the outer parts will be less than that of the inner parts. For real galaxies, the density of luminous matter does decrease from the center to the rim but the angular speed goes not decrease in proportion. We conclude that there must be more mass. The argument can be put on a quantitative basis and generalized to non-spherical galaxies. The angular speed as a function of distance from the center can be used to map the mass distribution.

41. Yes. No matter what we look at we see the object as it was when the light we see left it, not as it is at the time of observation. The direction in which we look does not matter.

SECTION FIVE
COMPARISON OF PROBLEMS WITH THIRD EDITION

In the table below the left column (in bold type) of each group gives the numbers of problems and exercises in the current (fourth) edition of *Fundamentals of Physics* and for each entry the right column gives the number of the same problem or exercise in the third edition. Any problem that did not appear in that edition is labeled "new". Nearly every problem that appeared in the third edition has been changed, often by adjusting the number of significant figures in the data or by rewording the problem statement to clarify its meaning. Only more significant changes are noted here. If parts have been dropped or if the meaning of the problem statement has changed the entry is labeled "revised". If one or more new parts have been added the entry is labeled "supplemented". If the numerical values of the given data have been changed (other than by inserting zeros to change the number of significant figures) the entry is labeled "data changed". Data in some problems in the third edition were given in both SI and British units. If either has been dropped the label reads "SI units only" or "British units only", as appropriate. Sometimes a hint has been added or dropped. The labels reflect this.

Chapter 1

1E	1E
2E	2E
3E	3E
4E	4E
5E	5E
6E	6E
7E	7E
8E	8E
9P	9P
10P	10P
11P	11P
12P	12P
13P	13P
14P	14P
14P	14P
15P	15P
16P	16P
17P	17
18P*	18P*
19P*	19P*
20E	20E
21E	21E
22E	22E
23E	23E
24E	24E
25E	25E
26E	26E revised
27P	27P
28P	28P
29P	29P
30P	30P
31P	31P
32P*	32P*
33E	33E
34E	34E
35E	35E
36P	37P
37P	38P
38P	39P
39P	40P
40P	41P revised

Chapter 2

1E	4E
2E	new
3E	new
4E	3E
5E	5E
6E	6E
7E	7E
8E	8E revised, supplemented
9P	9P
10P	13P
11P	10P supplemented
12P	11P supplemented
13P	27P supplemented
14P	12P
15P	14P
16E	new
17E	new
18E	new
19P	25P
20E	15E
21E	16E
22E	17E
23E	18E
24E	19E
25E	20E revised
26E	21E
27E	22E
28E	23E supplemented
29E	new
30P	24P
31P	26P supplemented
32P	28P
33P	29P

34E	32E
35E	35E
36E	41E
37E	30E
38E	31E
39E	33E
40E	34E
41E	36E
42E	37E
43E	38E
44E	39E
45E	40E supplemented
46P	42P
47P	43P supplemented
48P	44P data changed
49P	45P supplemented
50P	46P supplemented
51P	47P
52P	48P
53P	49P data changed to SI, supplemented
54P	50P
55P	51P
56P	52P supplemented
57P	53P supplemented
58P	new
59E	55E supplemented
60E	56E supplemented
61E	57E
62E	54E
63E	75P
64E	58E
65E	59E
66E	60E
67P	61P revised
68P	62P
69P	63P
70P	64P
71P	65P
72P	67P
73P	68P revised
74P	69P hint added
75P	70P
76P	71P
77P	72P data changed
78P	73P
79P	74P
80P	76P
81P	77P
82P	78P
83P	79P
84P	80P
85P	82P
86P	83P
87P	81P
88P	84P*
89P	85P*
90	new
91	new
92	new
93	new
94	new
95	new

Chapter 3

1E	1E
2E	2E
3E	3E
4E	4E
5E	5E
6P	6P
7P	7P
8P	8P
9E	9E
10E	10E
11E	11E
12E	12E
13E	13E
14E	14E
15P	15P
16P	16P
17P	17P
18P	18P
19P	19P*
20E	20E
21E	21E
22E	22E
23E	23E
24E	24E
25E	25E supplemented
26P	new
27P	27P
28P	28P
29P	29P
30P	30P
31P	31P
32P	32P
33P	33P
34P	34P
35P*	35P*
36E	36E
37E	37E revised
38E	38E revised
39E	39E revised
40E	40E revised
41E	41E
42E	42E
43E	43E revised
44E	44E
45E	45E
46E	46E
47P	47P
48P	48P
49P	49P
50P	51P
51P	50P
52P	52P
53P	53P
54P	54P
55P	55P
56P	57P
57P	58P
58P	60P
59P	59P
60	new
61	new
62	new
63	new

Chapter 4

1E	new
2E	new
3E	new
4E	new
5E	1E
6E	2E
7E	3E
8E	new
9E	new
10E	new
11E	4E
12E	5E supplemented
13E	6E

14P 7P
15P 5P
16P 9P
17E 12E
18E 13E
19E 11E supplemented
20E 10E
21E 14E
22E 15E
23E 16E
24E 17E
25E 18E
26E 19E
27E 20E
28E 21E
29E 22E
30E 23E data changed to SI
31P 24P
32P 25P
33P 26P
34P 27P
35P 28P
36P 29P
37P 30P
38P 31P
39P 32P
40P 33P
41P 34P
42P 35P
43P 36P
44P 37P revised
45P 39P revised
46P 40P
47P 41P
48P 42P
49P 43P
50P 44P
51P 45P
52P 46P
53P 47P
54P 48P
55P* 49P revised
56P* 50P
57P* 51P*
58E 52E
59E 53E
60E 54E
61E 55E
62E 56E
63E 57E
64E 58E
65E 59E
66E 60E revised
67E 61E
68E 62E
69P 64P
70P 63P
71P 66P
72P 68P
73E 69E
74E 70E
75E 71E
76E new
77P 72P
78E new
79E 75E supplemented
80P 76E
81E 77E
82P 78P
83P 79P
84P 80P
85P 81P
86P 82P
87P 83P
88P 84P
89P* 86P
90P* 85P
91E 73E
92P 74P
93 new
94 new
95 new
96 new
97 new
98 new
99 new
100 new

Chapter 5

1E new
2E new
3P new
4E new
5E new
6E new
7E new
8E new
9P new
10P new
11E 14E
12E 15E
13E 16E
14E 17E data changed
15E new
16E new
17E new
18E new
19E new
20E 20E
21E 5E
22E 6E data changed
23E 7E
24E 8E
25E 21E
26E 22E data changed
27E 23E revised
28E 25E
29E 26E
30E 27E
31E 28E
32E 29E
33E 30E
34E 31E revised
35E 33E
36E 34E
37P 35P
38P 9P
39P 11P
40P 13P
41P 36P data changed
42P new
43P new
44P 38P
45P 39P
46P 40P
47P 41P
48P 42P
49P 43P
50P new
51P 44P data changed
52P 45P
53P 46P data changed
54P 47P
55P 49P
56P 50P data changed,

supplemented
57P 51P
58P 52P
59P 53P supplemented
60P 54P
61P 55P SI units only
62P 56P
63P 57P
64P 58P
65P 59P revised
66P 60P
67P 61P
68P 62P
69P 63P
70P 64P
71P 65P
72P 66P
73P 67P supplemented
74 new
75 new
76 new
77 new
78 new
79 new
80 new
81 new
82 new

Chapter 6

1E 1E
2E 2E
3E 3E
4E 4E
5E 6E
6E new
7E 9E
8E 10E
9E 11E
10E 12E SI units only
11E 13E
12E 5E data changed, supplemented
13E 14E
14E 15E supplemented
15E 16E
16E 17E
17P 18P
18P 19P
19P 20P data changed
20P 21P
21P 22P
22P 23P data in SI units
23P 24P
24P 25P
25P 26P
26P 27P
27P 28P
28P 30P
29P 31P
30P 32P
31P 33P data changed to SI
32P 34P
33P 35P
34P 36P
35P 37P
36P 38P
37P 39P
38P 40P
39P 41P
40P 42P
41P new
42P* 43P*
43E 44E
44E 45E
45E 46E
46P 47P
47E 48E
48E 49E
49E 50E
50E 51E SI units only
51E 52E
52E 53E
53E 55E
54E 56E
55E 57E
56E 58E
57E 59E
58P 61P
59P 62P
60P 63P
61P 64P
62P 65P
63P 66P
64P 67P
65P 68P
66P 69P
67P 70P
68P 71P
69P 72P
70P 73P
71P 74P
72 new
73 new
74 new
75 new
76 new
77 new
78 new
79 new

Chapter 7

1E new
2E 1E data changed
3E 2E data changed
4E 3E data changed
5E 4E SI units only
6E new
7E new
8E new
9 6P
10 7P SI units only
11 11P
12 9P
13E 11E
14E 12E
15P 13P
16P 14P
17P new
18E 15E
19E new
20P 16P
21E 26E revised
22E 17E
23E 18E
24E 19E
25E 20E
26E 23E
27E 25E
28E 27E SI units only
29E 28E
30E new
31E 36P revised
32P 29P

33P 30P
34P 32P
35P 33P SI units only
36E 5P supplemented
37P new
38P 38P
39E 42E data changed
40E 43E British units only
41E 48E
42E new
43P 52P revised
44P 53P revised
45P 54P revised
46P 56P supplemented
47P 57P
48P 58P
49P 60P
50E 66E revised
51E 67E
52P 68P
53 new
54 new
55 new
56 new
57 new

Chapter 8

1E 1E
2E 2E
3E 7E
4E 8E
5E 9E
6E 10E
7E 11E revised
8E 12E
9E 13E hint added
10E 16E
11E 17E
12E 18E
13E 19E
14E 20E
15P 21E
16P 24P
17P 25P
18P 7–35P
19P 26P supplemented
20P 27P
21P 28P
22P 29P revised
23P 30P
24P 32P
25P 33P
26P 34P
27P 35P
28P 14E
29P 15E supplemented
30P 36P
31P 37P data changed to SI
32P 38P
33P 39P
34P 40P
35P* 31P
36P* 41P*
37P* 42P*
38E 43E
39P 6E supplemented
40P 44P
41P 45P supplemented
42E 46E
43E 3E
44E 4E
45E 5E
46E 22P
47E 7–24E
48E 7–21E British units only
49E 7–59P
50E 7–40E revised
51E 7–41E SI units only
52E 7–44E
53E 7–45E
54E 7–46E
55E 7–47E
56E 7–49E data changed to SI
57E 7–50E
58E 7–51P
59E 7–22E
60E 47E
61E 48E
62E 49E
63E 7–31P
64E 50E
65E 51E
66P 52E
67P 7–37P data changed
68P 53E revised
69P 54P
70P 55P supplemented
71P 56P
72P 57P
73P 58P
74P 59P
75P 7–34P SI units only
76P 7–55P
77P new
78P 60P
79P 61P
80P 62P supplemented
81P 63P no numerical data
82P 64P
83P 65P
84P 66P*
85P 7–61P
86P 7–62P
87P 7–63P
88P 7–64P
89P* 7-65P*
90P* new
91P* 9–53P*
92E 67E
93E 68E
94E 70E
95P 71P
96P 72P
97P 73P
98P 74P data changed
99P 75P
100E 76E
101P 77P
102 new
103 new
104 new
105 new
106 new
107 new
108 new
109 new
110 new

Chapter 9

1E 1E supplemented
2E 2E hint added
3E 3E supplemented

4E	4E	
5E	5E	
6P	6P	
7P	7P	
8P	8P	
9P	9P	
10P	new	
11P*	new	
12P*	10P	
13E	11E	
14E	12E	
15E	13E	
16E	14E	
17E	15E	
18P	16P	
19P	17P	
20P	18P	
21P	19P	supplemented
22P	20P	
23E	21E	
24E	22E	SI units only
25E	24E	revised
26E	25E	
27E	26E	
28P	27P	supplemented
29P	new	
30P	29P	data changed
31P	30P	
32P	31P	
33E	32E	
34E	33E	
35E	34E	data changed
36E	35E	
37E	36E	
38P	40P	
39P	41P	
40P	42P	
41P	43P	data changed, supplemented
42P	44P	revised
43P	45P	
44P	49P	
45P	new	
46P	52P	
47E	54E	
48E	55E	
49E	56E	
50E	57E	
51E	59E	
52E	58E	
53P	60P	
54P	62P	
55P	61P	hint added
56P	63P	
57P	64P	SI units only
58E	65E	SI units only, revised
59E	66E	
60E	67E	
61P	68P	revised
62P	69P	
63	new	
64	new	
65	new	
66	new	

Chapter 10

1E	1E	
2E	2E	
3E	3E	
4E	new	
5E	4E	
6E	6E	
7E	new	
8E	new	
9E	7E	data changed
10E	8E	
11E	9E	
12P	10E	supplemented
13P	12P	
14P	9–46P	
15P	9–47P	
16P	31P	
17P	9–48	
18P	9–51P	supplemented
19P	5E	
20P	13P	
21P	14P	
22P	15P	
23P	16P	
24P	17P	
25P	18P	
26P	19P	
27P	20P	
28P	11E	
29E	21E	supplemented
30E	23E	
31E	24E	
32E	25E	
33E	26E	
34E	28E	
35P	29E	
36P	30P	
37P	32P	
38P	new	
39P	33P	
40P*	35P	hint added
41E	36E	
42E	37E	
43E	38E	
44E	9–37E	
45E	39E	
46E	40E	supplemented
47E	41E	
48P	42P	
49P	43P	
50P	44P	
51P	45P	
52P	46P	
53P	47P	
54P	48P	
55P	49P	
56P	34P	
57P	50P	
58P	52P	
59P*	51P	
60E	53E	
61E	54E	
62E	55E	
63E	56E	
64E	57E	
65E	58E	
66P	59P	
67P	60P	
68P	61P	
69P	62P	
70P	63P	
71P	64P	
72P	65P	
73P	66P	
74E	67E	
75E	68E	
76P*	69P	
77P*	70P	
78	new	

79 new
80 new
81 new
82 new
83 new
84 new
85 new

Chapter 11

1E 1E revised
2E 3E
3E 4E
4E 5E
5E 6E
6P 7P
7P 8P
8P 9P
9P 10P
10P 11P
11E 12E
12E 13E data changed
13E 14E
14E 15E
15E 16E
16E 17E
17E 18E
18P 19P
19P 20P
20P 21P
21P 22P
22P new
23P 23P
24P 24P
25P 25P
26E 26E
27E 27E
28E 28E
29E 29E
30E 30E
31E 31E
32E 32E
33E 33E SI units only
34E 34E
35P 35P
36P 36P
37P 37P
38P 38P SI units only
39P 39P
40P 40P
41P 41P
42P 42P
43P 44P
44P 43P
45E 45E
46P 46P
47E 47E
48E 48E
49E 49E
50E 50E
51E 51E
52E 52E
53E 53E
54E 57P hint added
55P 54P
56P new
57P 55P hint added
58P 56P
59E 58P
60E 59E
61E 61E
62E 60E SI units only
63P 63P
64P 62P
65E 64E
66E 65E
67E 66E
68E 67E data changed
69E 68E
70P 69E
71P 70P
72P 71P
73P 72P
74P 73P
75P 74P
76P 75P
77P 76P revised
78E 77E supplemented
79E 78E
80E 79E
81E 80E
82P 81P
83P 82P
84P 83P supplemented
85P 84P
86P 85P
87P 86P
88P* 87P*
89 new
90 new
91 new
92 new
93 new
94 new
95 new
96 new

Chapter 12

1E 1E
2E 3E
3E 4E
4E 5E
5E 6E
6E 2E
7E 7E
8E 31E
9P new
10P 9P
11P 10P
12P 11P supplemented
13P 12P
14P 13P
15P* 14P
16E 15E
17P 16P
18E 17E hint deleted
19E 18P
20E new
21E new
22P new
23P new
24P new
25P new
26E 19E
27E 20E
28E 21E
29E 22E hint deleted
30E 23E
31P 24P
32P 25P
33P 26P
34P 27P
35P 28P
36E 30E

37E	new
38E	new
39E	new
40P	new
41P	33P
42E	29E
43E	32E
44E	37E
45E	38E
46P	39P
47P	34P
48P*	35P*
49P*	36P*
50E	40E
51E	41E supplemented
52E	42E
53E	43E
54E	44E
55E	45E
56E	46P
57P	47P
58P	48P data changed, supplemented
59P	49P
60P	50P
61P	51P
62P	52P
63P	53P
64P*	54P
65P*	55P
66E	56E
67P	57P
68	new
69	new
70	new
71	new
72	new
73	new

Chapter 13

1E	1E
2E	2E
3E	3E
4E	4E
5E	5E
6E	6E
7E	7E
8E	8E
9E	9E
10E	10E
11E	11E SI units only
12E	12E
13E	13E supplemented
14E	14E
15E	15E
16E	16E
17E	17E revised
18E	18E
19P	19P
20P	20P
21P	21P
22P	22P
23P	23P
24P	24P
25P	25P
26P	new
27P	26P
28P	27P
29P	28P
30P	29P revised
31P	30P
32P	31P SI units only
33P	32P
34P	33P
35P	34P
36P	35P
37P	36P
38P	37P
39P	38P
40P	39P
41P	40P
42P	41P
43P	42P hint added
44P	43P hint added
45P*	44P*
46E	45E supplemented
47E	46E
48E	47E
49E	48E
50E	49E
51E	50E
52P	51P
53P	52P
54P	new
55P	new
56P	new
57	new
58	new
59	new
60	new
61	new

Chapter 14

1E	1E
2E	2E
3E	3E
4E	4E
5E	5E
6E	6E
7E	7E
8E	8E
9E	9E
10E	10E
11E	11E
12E	12E
13E	13E
14E	14E
15E	15E
16E	16E
17E	17E
18P	19P
19P	18P
20P	20P
21P	21P
22P	22P
23P	23P
24P	24P
25P	25P
26P	26P
27P	27P
28P	28P
29P	29P
30P	30P
31P	31P
32P	32P
33P	33P
34P	34P
35P	35P
36P	36P
37P	37P supplemented
38P	38P
39E	39E
40E	40E

41E	41E
42E	42E
43E	43E
44E	44E
45E	45E
46P	46P
47P	47P
48P	48P
49P	50P
50P*	51P*
51E	52E
52P	55P
53P	53E
54P	54P
55E	56E
56E	57E supplemented
57E	58E
58E	59R
59E	60E
60E	61E
61E	62E
62E	63E supplemented
63E	64E
64E	65E
65E	66E revised
66E	67E
67E	69P
68E	70P
69P	71P revised
70P	72P
71P	73P
72P	74P
73P	68P
74P	75P
75P	76P
76P	77P
77P	78P
78P	79P
79P*	new
80E	80E
81E	81E
82E	82E
83P	83P
84P	84P
85P	85P
86E	86E
87P	87P
88	new
89	new
90	new
91	new

Chapter 15

1E	2E
2E	3E
3E	4E
4E	8E
5P	new
6E	5P
7E	6P
8P	7P
9P	new
10P	new
11P	new
12P	new
13P	35E
14P*	new
15P	15P
16E	18E
17E	19E
18E	20E
19E	21E
20E	22E
21E	23E
22E	24E revised
23P	25P
24P	26P
25P	27P
26P	28P
27P	new
28P	32P
29P	33P
30P*	31P data changed
31E	9E
32E	10E
33E	11E
34P	13P
35P	14P
36P	new
37E	new
38E	new
39E	new
40E	36E
41E	37E
42E	38E
43E	39E
44E	new
45E	new
46P	new
47P	44P hint deleted
48P	45P
49P	46P
50P	47P
51P	48P
52P	50P revised
53P*	49P
54P*	51P
55E	53E
56E	54E
57E	55E
58E	40E
59E	57E
60E	58E
61E	59E
62E	60E
63E	61E
64E	62E
65E	new
66E	70E
67E	69E
68P	84P
69P	63P
70P	64P
71P	65P
72P	66P
73P	67P*
74P*	68P*
75P*	83P
76E	71E
77E	72E
78P	73E revised
79P	74E
80P	75E
81P	77P
82P	78P
83P	79P
84P	81P
85P	82P
86	new
87	new
88	new
89	new
90	new
91	new

Chapter 16

1E	1E
2E	2E
3E	3E SI units only
4E	4E
5E	5E
6E	6E
7P	7P
8P	9P
9E	10E
10E	11E SI units only
11E	12E
12E	13E
13E	14E
14E	15E
15E	16E
16E	17E
17E	18E
18P	19P
19P	20P
20P	21P
21P	22P
22P	23P
23P	24P
24P	25P
25P	26P
26E	27E
27E	new
28P	28P
29E	29E
30E	30E
31E	31E
32E	32E
33E	33E
34E	34E
35E	35E
36E	36E
37E	37E
38E	38E
39E	39E
40E	40E
41P	41P
42P	42P
43P	43P
44P	44P
45P	45P
46P	46P
47P	47P
48P	48P
49P	49P
50P	new
51P	50P
52P*	51P*
53E	52E
54E	53E
55E	54E
56P	55P
57P	56P
58E	57E
59E	58E
60E	59E
61E	60E
62E	61E
63E	62E
64E	65E
65E	64E
66E	65E
67E	66E
68E	67E
69E	68E
70E	69E
71P	70P
72P	71P
73P	72P
74P	73P
75P	74P
76P	75P
77P	76P
78P	77P
79P	78P
80P	79P
81P	80P
82P	81P
83P	82P
84P*	83P*
85P*	8P
86	new
87	new
88	new
89	new
90	new

Chapter 17

1E	1E
2E	2E
3E	3E
4E	4E
5E	5E
6E	6E
7E	7E
8E	8E
9E	9E revised
10E	10E
11E	11P
12E	12P
13P	13P
14P	14P
15P	17P
16P	18P
17E	19E SI units only
18E	20E
19E	21E
20E	22E
21E	23E
22E	24E
23E	25E
24E	26E
25P	27P
26P	16P
27P	28P
28P	29P
29P	30P
30P	31P
31P	32P
32P*	33P*
33E	new
34E	35E
35P	36P
36E	37E
37E	38E
38P	39P
39P	41P
40P	42P
41P*	40P
42E	new
43E	new
44E	54P revised
45E	44E
46E	45E
47E	46E
48E	48E
49E	49E
50E	50E
51E	51E

52E	52E
53E	new
54P	43E
55P	53P
56P	55P
57P	56P
58P	57P
59P	58P
60P	59P
61P	60P
62P	61P SI units only
63P	62P
64P	63P
65	new
66	new
67	new

Chapter 18

1E	5E
2E	6E
3E	7E
4E	8E
5E	9E
6E	10E
7E	11E
8E	12E
9E	13E
10E	14E
11E	15E
12E	new
13E	1E
14E	2E
15E	3E note deleted
16E	4E
17P	15P
18P	16E
19P	17P
20P	18P
21P	19P
22P	36P
23P	new
24P	31P
25P	43P
26P*	41P
27E	21E
28E	22E
29E	23E
30E	24E
31E	25E
32E	26E
33E	27E
34E	28E
35E	29E
36P	32P
37P	33P
38P	34P
39P	35P
40P	37P
41P	38P
42P	39P
43P	40P
44P	42P
45P	30P
46P*	44P*
47P*	45P*
48E	46E
49E	47E
50E	48E
51E	17-47E
52E	49E supplemented
53E	50E revised
54E	52E data changed
55E	53E
56E	51E supplemented
57P	54P supplemented
58P	55P
59P	56P
60E	57P revised
61P	58P
62P	59P
63P	60P
64E	61E
65E	62E
66E	63E
67P	64P
68P	65P
69E	66E
70E	67E
71E	68E
72E	69E
73E	70E
74E	71E
75E	72E
76E	73E
77E	74E
78E	75E
79P	76P
80P	77P
81P	78P
82P	79P
83P	80P
84P	81P
85P	82P
86P	83P
87P	84P data changed
88P	85P supplemented
89P	86P
90P	87P
91P	88P revised
92P	89P
93P	90P
94P	91P
95E	92E
96E	93E
97E	94E
98E	95E
99P	96P
100P	97P
101P	98P

Chapter 19

1E	1E
2E	2E
3E	3E
4E	4E
5E	5E
6P	6P
7P	7P
8P	8P
9E	13E
10E	14E
11E	14E
12E	new
13E	16E
14E	17E
15E	new
16P	9P
17P	10P
18P	11P
19P	12P
20E	18E
21E	19E
22E	20E

23E	21E
24E	22E
25E	23E
26E	24E
27E	25E data changed
28E	26E
29E	27E SI units only
30E	28E
31E	29E
32E	30E
33E	31E
34E	32E
35E	33E
36P	34P
37P	35P
38P	36P
39P	37P
40P	38P
41P	39P
42P	40P
43P	41P
44P	42P
45P	43P
46P	44P
47P	45P
48P	46P
49P	47P
50P	48P
51P	49P
52P*	51P
53P*	52P
54P*	53P

Chapter 20

1E	1E
2E	2E data changed
3E	3E
4E	4E
5E	5E
6E	6E data changed
7E	7E
8E	8E
9E	9E
10E	10E
11E	11E revised
12E	12E
13E	13E
14E	14E
15E	15E
16E	16E
17E	17E
18P	18P
19P	19P data changed
20P	20P
21P	21P
22P	22P
23P	23P
24P	24P
25P	25P
26P	26P
27P	27P
28P	28P
29P	29P
30P	30P
31P	31P
32P	32P
33P	33P
34P*	34P
35E	35E
36E	36E
37E	37E
38E	38E
39E	39E revised
40E	40E
41P	41P
42P	42P
43P*	43P*
44E	44E
45E	45E
46E	46E
47E	47E
48E	48E
49E	49E
50E	50E
51E	51E
52P	52P data changed
53P	53P SI units only
54P	54P
55P	55P
56P	56P
57P	57P
58P	58P
59P	59P hint omitted
60P*	60P
61	new
62	new
63	new

Chapter 21

1E	1E
2E	2E
3P	3P
4P	4P
5P	5P
6E	6E
7E	7E
8E	8E
9E	9E
10E	10E
11E	11E
12E	12E
13P	13P
14P	14P
15P	15P SI units only
16P	16P
17P	17P
18P	18P
19P	19P
20P	20P
21P	21P
22P	22P
23P	23P
24E	24E
25E	25E
26E	26E
27E	27E
28E	28E
29E	29E
30P	30P
31P	31P
32E	32E
33E	33E
34E	34E
35E	35E
36P	36P
37P	37P
38P	38P
39E	39E
40E	40E
41E	41E
42E	42E
43P	43P
44P	44P

45P	45P
46P	46P
47P	47P
48E	48E
49E	49E
50E	50E
51E	51E
52E	52E
53P	53P
54P	54P
55P	55P
56P	56P
57P	57P
58E	58E
59E	59E
60E	60E
61P	61P
62P	62P
63P	63P
64P	64P
65E	new
66E	65E hint added
67E	66E hint added
68P	67P
69P	68P
70E	70E
71E	71E
72E	72E
73E	73E
74E	74E
75E	75E data changed
76E	76E
77P	77P
78P	78P
79P	79P
80P	80P
81P	81P
82P	82P
83P	83P
84P	84P
85	new
86	new
87	new
88	new

Chapter 22

1E	1E
2E	2E
3E	3E
4P	4P
5P	5P
6E	6E
7E	7E
8E	8E
9E	9E
10E	10E
11E	11E
12E	12E
13E	13E
14E	14E
15E	15E
16E	16E
17E	17E
18E	18E
19E	19E
20P	20P
21P	21P
22P	22P
23P	23P
24P	24P
25P	25P
26P	26P
27P	27P
28P	28P
29P	29P
30P	30P
31P	31P
32P*	32P*
33E	33E
34E	34E
35E	35E
36P	36P
37P	37P
38E	38E
39E	39E
40E	40E
41E	41E
42E	42E
43E	43E
44P	44P
45P	45P
46P	46P
47P	47P
48P	48P
49P	49P
50P	50P
51P	51P
52P	52P
53P	53P
54	new
55	new
56	new
57	new
58	new

Chapter 23

1E	1E
2E	2E
3E	3E
4E	4E
5E	5E
6E	6E
7E	7E
8P	8P
9P	9P
10P	10P
11P	11P
12P	12P
13P	13P
14P	14P
15P	15P
16P	16P
17P	17P
18P	18P
19P	19P
20P	20P
21P	21P
22E	22E
23E	23E
24E	24E
25E	25E
26E	26E
27E	27E
28E	28E
29E	29E
30E	30E
31P	31P
32P	32P
33P	33P
34P	34P
35P	35P
36P	36P
37E	37E

38E 38E
39E 39E
40E 40E
41E 41E
42P 42P

Chapter 24

1E 7E
2E 8E
3E 9E
4E 10E
5E 11E
6P 12P revised
7E 14E
8E 15E
9E 16E
10E 17E
11E 18E
12E 19R revised
13E new
14P 20P
15P 21P
16P 22P
17P 23P
18P new
19P 24P
20P 25P
21P 26P
22P 27P
23E 28E
24E 29E
25E 30E
26P 31P
27P* 32P*
28E 23E
29P 29E
30P 35P
31P new
32P 36P
33P 37P
34P 38P
35P* 39P*
36E 40E
37E 41E
38P 42P
39E 1E
40E 2E
41E 3E
42E 4E data changed
43E 5E
44E 6E
45E 43E
46E 44E
47E 45E
48E 46E
49E 47E
50P 48P
51P 49P
52P 50P
53P 51P
54P 52P
55P 53P
56P 54P
57E 55E
58E 56E
59P 57P
60P 58P
61E
62E
63E
64E
65E
66E
67E
68E
69E
70E
71E
72E
73E
74E
75E
76E
77E
78E
79E
80E
81E
82E
83E
84E
85E
86E
87E
88E
89E
90E
91E
92E
93E
94E
95E
96E
97E
98E
99E

Chapter 25

1E 1E
2E 2E
3E 3E
4P 4P
5E 5E
6E 6E
7E 7E
8E 8E
9E 9E
10E 10E
11P 11P
12P 12P
13P 13P
14P 14P
15E new
16E 15E
17E 16E
18E 17E
19P 18P
20P 19P
21E 20E
22E 21E
23P 23P
24P 24P
25P new
26P 25P
27P 26P
28P 27P
29P 28P
30P 29P
31E 30E
32E 31E revised
33E 32E
34P 33P
35P 34P

36P	35P revised
37P	36P
38P	37P supplemented
39P*	38P*
40E	39E
41E	40E
42E	41E
43E	42E
44E	42E
45E	44E
46E	46P
47P	45P
48P	47P
49P	48P
50P	49P
51P	50P
52P	51P
53P	52P hint added
54P*	53P*
55P*	54P*
56	new
57	new
58	new
59	new
60	new
61	new

Chapter 26

1E	1E
2E	2E
3P	3P
4E	4E
5E	5E
6E	6E
7E	7E
8E	8E
9E	9E
10P	10P
11P	11P
12P	12P
13P*	13P* supplemented
14P*	29P* hint added
15E	14E
16E	15E
17E	16E
18E	17E
19E	18E
20E	19E hint deleted
21E	20E
22E	21P revised
23P	22P
24P	23P
25P	24P
26P	25P
27P	27P
28E	new
29E	new
30E	30E
31P	26P
32P	28P
33P	31P
34P	new
35P	new
36E	new
37E	new
38P	new
39P	new
40P	new
41P	39P revised
42E	33E
43E	34E
44E	35E
45E	36E
46E	37E
47E	38E
48P	32P
49P	39P
50P	40P
51E	41E
52E	42E
53E	43E
54E	44E
55E	45E
56E	46E
57E	47E
58P	48P
59P	49P supplemented
60P	new
61P	50P
62P	51P
63P	52P
64P	53P
65P	54P
66P	55P data changed
67P	new
68P	57P
69P	58P
70P	59P
71P	60P hint added
72P	61P
73E	62E
74E	63E
75E	64E
76E	65E supplemented
77P	66P
78P	67P
79P	68P
80P	69P
81P	70P
82E	71E
83E	72E
84E	73E
85P	74E
86P	75P
87	new
88	new
89	new
90	new
91	new

Chapter 27

1E	1E
2E	2E
3E	3E
4E	4E
5E	5E
6E	6E
7E	7E
8E	8E
9E	9E
10E	10E
11E	15P
12P	13P
13P	14P
14P	16P
15E	18E
16E	19E
17E	20E
18E	21E
19E	22E
20E	23E
21P	24E
22P	25P

23P 26P
24P 27P
25P 29P
26P 30P
27P 31P
28P 32P
29P 33P
30P 34P
31P 35P
32E 36E
33E 37E
34E 38E
35E 39E
36E 40E
37E 41E
38E 42E
39E 43E
40P 44P
41P 45P
42P 46P
43P 47P
44P 48P
45P 28P
46P 49P
47P 50P
48P 51P
49P 11P
50P 12P
51P* 17P*
52E 52E
53E 53E
54E 54E
55E 55E
56E 56E
57P 57P
58P 58P
59P 59P
60P 60P
61P 61P
62P 62P
63P 63P
64P 64P
65P 65P
66E 66E
67E 67E
68P 68P
69P 69P
70P 70P
71P 71P
72 new
73 new
74 new

Chapter 28

1E 1E
2E 2E
3E 3E
4E 4E
5E 5E
6E 6E
7E 7E
8E 8E
9E 9E
10E 10E
11P 11P
12P 12P
13P 13P
14P 14P
15P 15P
16E 16E
17E 17E
18E 18E
19E 19E
20E 20E
21E 21E
22E 22E
23E 23E
24E 24E
25E 25E
26E 26E
27E 27E
28E 28E
29P 29P
30P 30P
31P 31P
32P 32P
33P 34P
34P 35P
35P 36P
36P 37P
37P 38P
38P 39P
39P 40P
40P 41P
41P 42P
42P 43P
43P 44P
44E 45E
45E 46E
46E 47E
47E 48E
48E 49E
49E 50E
50E 51E
51E 52E
52E 53E
53P 33P
54P 54P
55P 55P
56P 56P
57P 57P
58P 58P
59P 59P
60P 60P
61P 61P
62P 62P
63P 63P

Chapter 29

1E 1E
2E 2E
3E 3E
4P 4P
5E 5E
6E 6E
7E 7E
8E 8E
9E 9E
10E 10E
11E 11E
12E 12E
13E 13E
14E 14E
15P 15P
16P 16P
17P 17P
18P 18P
19P 19P
20P 20P
21P 21P
22P 22P
23P 23P
24P 24P

25P 25P
26E 26E
27E 27E
28E 28E
29E 29E
30E 30E
31E 31E
32E 32E
33E 33E
34E 34E
35E 35E
36E 36E
37E 37E
38P 39P
39P 40P
40P 41P
41P 43P
42P 44P
43P 45P supplemented
44P 46P
45P 49P
46P 50P
47P 51P supplemented
48P 52P
49P 53P
50P 54P
51P 55P
52P 56P
53E 57E
54E 38E
55P 42P
56P 47P
57P 48P
58P 58P
59P 59P
60P 60P
61P 61P
62P 62P
63P 63P
64P 64P
65E 65E
66E 66E
67E new
68E 68E
69P 69P
70P 70P
71P 71P
72P 72P
73P 73P
74P 74P
75P 75P
76P 76P
77P 77P
78P 78P
79 new

Chapter 30

1E 1E revised
2E 2E
3E 3E
4E 4E
5P 5P
6P 6P
7P 7P
8P* 8P*
9E 9E
10E 10E
11E 11E
12P 12P
13P 13P
14E 14E
15E 15E
16P 16P
17P 17P
18P 18P
19E 19E
20E 20E
21E 21E
22E 22E
23E 23E
24E 24E
25E 25E
26E 26E
27E 27E
28P 28P
29P 29P
30P 30P
31P 31P
32P 32P
33P 33P
34P 34P
35P 35P
36P 36P
37P 37P
38P 38P
39P 39P
40P 40P
41P 41P
42P 42P
43E 43E
44E 44E
45E 45E
46P 46P
47P 47P
48P 48P
49P 49P supplemented
50P 50P
51P 51P
52P 52P
53E 53E
54E 54E
55E 55E
56P 56P
57P 57P
58P 58P
59P 59P
60P 60P
61P 61P
62E 62E
63E 63E
64E 64E
65E 65E
66E 66E
67P 67P
68P 68P
69 new
70 new
71 new
72 new

Chapter 31

1E 1E
2E 2E
3E 3E supplement
4E 4E
5E 5E
6E 6E
7E 7E
8E 8E
9E 9E
10E 10E
11P 16P
12P 11E

13P	12P
14P	13P
15P	14P
16P	15P
17P	17P
18P	18P
19P	19P
20P	20P
21P	21P
22P	22P
23P	23P
24P	24P
25P	25P
26P	26P
27E	27E
28E	28E
29E	29E
30E	30E
31E	31E
32E	32E
33P	33P
34P	34P
35P	35P
36P	36P
37P	37P
38P	38P
39P	39P
40E	40E
41E	41E
42E	42E
43E	43E
44P	44P
45P	45P
46P	46P
47P	47P
48P	48P
49P	49P
50P	50P
51P	51P
52P*	52P*
53E	53E
54E	54E
55E	55E
56E	56E
57E	57E
58P	58P
59P	59P
60P	60P
61P	61P
62P	62P
63E	63E
64E	64E
65E	65E
66E	66E
67E	67E
68P	68P
69P	69P
70P	70P
71P	71P
72P	72P
73P	73P

Chapter 32

1E	1E
2E	2E
3E	3E
4E	4E
5E	5E
6E	6E
7E	7E
8E	8E
9P	9P
10P	10P
11P	11P
12P	12P
13P	13P
14P	14P
15P	15P
16P	16P
17P	17P
18P	18P
19P	19P
20P	20P
21P*	21P*
22E	24E
23E	25E
24E	26E
25E	27E
26P	28P
27P	29P
28P	30P
29P	31P
30P	32P
31P	33P
32P	34P
33P	35P
34P	36P
35P	37P
36P	38P
37P	39P
38P*	41P
39P*	42P*
40E	43E
41E	44E
42P	45P
43P	46P
44P	47P
45E	48E
46E	49E
47P	50P
48	new
49	new
50	new
51	new

Chapter 33

1E	1E
2E	2E
3E	3E
4P	4P
5P	5P
6P	6P
7P	7P
8P	9P
9E	9E
10E	10E
11E	11E
12E	12E
13P	13P revised
14E	14E
15E	15E
16E	16E
17E	17E
18E	18E
19E	19E
20E	20E
21P	21P
22P	22P
23P	23P
24P	24P
25P	25P
26P	26P
27P	27P

28P*	28P*
29E	29E
30E	31E
31E	32E
32E	33E
33P	34P
34P	35P
35P	36P
36P	38P
37P	39P
38E	40E
39E	30E
40E	41E
41E	42E
42E	43E
43E	44E
44P	37P
45P	45P
46P	46P
47E	47E
48E	48E
49P	49P
50P	50P
51P	51P
52P	52P
53P	53P
54	new
55	new
56	new
57	new

Chapter 34

1E	1E
2P	2P
3P	3P
4E	4E supplemented
5P	5P
6E	6E
7E	7E
8P	8P
9P*	9P*
10E	10E
11E	11E
12E	12E
13P	13P
14P	14P
15P	15P
16P	16P
17P	17P
18E	18E
19E	19E
20E	20E
21P	21P
22P	22P
23P	23P
24P	24P
25P	25P
26E	26E
27E	new
28E	28E
29E	29E
30E	30E
31E	31E
32P	32P
33P	33P
34P	34P
35	new
36	new
37	new
38	new

Chapter 35

1E	1E
2E	2E
3E	3E
4E	4E
5E	5E
6P	new
7E	6E
8P	7P
9E	8E
10E	9E
11E	10E
12E	11E
13E	12E
14E	13E
15P	14P
16P	15P
17P	16P
18P	17P
19P	18P
20P	19P
21P	20P
22P	21P
23P	22P
24P	23P
25P	24P
26P	25P
27P	26P
28P	27P
29P	28P
30P*	29P*
31E	30E
32E	31E
33P	32P
34P	33P
35P	34P
36P*	35P*
37E	new
38E	new
39P	new
40P	new
41	new
42	new
43	new

Chapter 36

1E	1E
2E	2E
3E	3E
4E	4E
5E	5E
6E	6E
7E	7E
8P	8P
9P	9P
10P	10P
11P	11P
12P	12P
13E	13E
14E	14E
15E	15E
16E	16E
17P	17P
18P	18P
19P	19P
20P	20P
21P	21P
22P	22P
23P	23P
24P	24P

25P	25P
26P	26P
27P	27P
28P*	28P*
29E	29E
30E	30E
31E	31E
32E	32E
33E	33E
34E	34E
35E	35E
36P	36P
37P	37P
38P	38P
39P	39P
40P	40P
41P	41P
42P	42P
43P	43P
44E	44E
45E	45E
46E	46E
47P	47P
48P	48P
49	new
50	new
51	new
52	new

Chapter 37

1E	1E
2E	2E
3E	3E
4P	4P
5P	5P
6E	6E
7E	7E
8E	8E
9E	9E
10P	10P
11P	11P
12P	12P
13P	13P
14P	14P
15P	15P
16E	16E
17P	17P
18P	18P
19P	19P
20P*	20P
21	new
22	new
23	new

Chapter 38

1E	2E
2E	2E
3E	3E
4E	4E
5E	5E
6E	6E
7P	7P
8E	new
9E	9E
10P	10P
11E	11E
12E	12E
13P	13P
14P	14P
15E	15E
16E	16E
17E	17E
18E	18E
19E	19E
20E	20E
21E	21E
22E	22E
23E	23E
24E	24E
25P	25P
26P	26P
27P	27P
28P	28P
29P	29P
30P	30P
31P	31P
32P	32P
33P	33P supplemented
34E	new
35E	37E
36E	38E
37E	39E
38E	40E
39P	41P
40P	42P
41P	43P
42P	44P
43P	45P
44P	46P
45P	47P
46P	48P
47P	49P note deleted
48E	50E
49E	51E
50E	52E
51E	53E
52E	new
53E	new
54E	new
55E	new
56P	56P
57P	57P
58P	58P
59P	59P
60P	60P
61	new
62	new
63	new
64	new

Chapter 39

1E	1E
2E	2E
3E	7E
4E	8E
5P	9P
6P	10P
7P	12P
8P	13P
9P	14P
10P	16P hint added
11P	15P
12P	17P
13P	18P
14P	19P
15P	20P
16E	21E
17E	22E
18E	23E
19E	24E
20E	25E

21P 26P
22P 27P
23P new
24P new
25P 30P note added
26P 31P
27P 32P
28E 35E
29E 36E
30E 37E
31P 38P
32E new
33E 42E
34E 39E revised
35E 40E
36E 46P
37E new
38E new
39P new
40P new
41E 44P
42E 48P
43P 49P
44P 50P
45P* 52P*
46E 53E
47E 54E
48P 55P
49P 56P
50P 57P
51P 58P
52P 59P
53P 60P
54E 61E
55E 62E
56E 63E
57E 64E
58E 65E
59E 66E
60E 67E
61E 68E
62E 69E
63P 70P
64P 71P
65P 72P
66P 73P
67P 74P
68P 75P
69P 76P
70P 77P supplemented
71P 78P
72P 79P
73P 80P
74P 81P
75P 82P
76E 83P
77E 84P
78P 85P
79P 86P
80P 87P
81P 88P
82P 89P
83P 90P
84 new
85 new
86 new
87 new
88 new

Chapter 40

1E 1E
2E 2E
3E 3E
4E 39–3E
5E 39–4E
6E 39–5E
7E 39–6E
8P 5P
9P 11P
10P new
11P new
12P new
13P new
14P new
15E 6E
16E 7E
17E 8E
18E 9E
19E 10E
20E 11E
21E 12E
22E 13E
23E 14E
24P 15P
25P 16P
26P 17P
27P 18P
28P 19P
29P 20P
30P 22P revised
31P 23P
32P 24P
33P 25P
34P 26P
35P 27P
36E 29E
37E 30E
38E 31E
39P 32E
40P 33P hint added
41P 34P
42P 35P
43P* 36P*
44E new
45E new
46E new
47E 41E
48E new
49E 42E
50E new
51E new
52E 43E
53E 38E
54P new
55P new
56P 40E
57P 45P
58P 46P
59P 47P
60P 48P
61P 49P
62P 50P
63P 51P revised
64P 52P
65P 53P
66P 54P
67P 55P
68P 56P
69P 57P
70P 58P
71P 59P
72P 60P
73P 61P
74P 62P
75P 63P

76P	64P
77P	21P revised
78E	65E
79E	66E
80P	67P
81P	68P
82	new
83	new
84	new
85	new

Chapter 41

1E	1E
2E	2E
3E	3E
4E	4E
5E	5E
6E	6E revised
7E	7E
8P	8P
9P	9P
10P	10P
11E	11E
12E	12E
13P	13P
14P	14P
15P	15P supplemented
16P	16P
17P*	17P*
18E	18E
19E	19E
20E	20E
21E	21E
22E	22E
23E	23E
24E	24E
25E	25E
26P	26P
27P	new
28P	28P
29P	29P
30P	30P
31P	31P
32P	32P
33P	33P revised
34P	34P
35E	35E
36E	36E
37P	37P
38P	38P
39P	39P
40P	40P
41P	41P
42P	42P
43E	43E
44E	44E
45E	45E
46E	46E
47E	47E
48E	48E
49P	49P
50P	50P
51P	51P
52P	52P
53P	53P
54P	54P
55P	55P
56P	56P
57P	57P
58P	58P
59P	59P
60E	60E
61E	61E
62E	62E
63E	63E
64E	64E
65E	65E
66E	66E
67P	67P
68P	68P
69P	69P
70E	70E
71E	71E
72E	72E
73E	73E
74E	74E data changed
75E	75E
76P	76P
77P	77P
78P	78P
79P	79P
80P	80P

Chapter 42

1E	1E
2E	2E
3P	3P data changed
4E	4E
5E	5E
6P	6P
7P	7P
8P	8P
9E	9E
10E	10E
11E	11E
12E	12E
13E	13E
14P	14P
15P	15P
16P	16P
17E	17E
18E	18E
19E	19E
20E	20E
21E	21E
22P	22P
23P	23P
24P	24P
25E	25E
26E	26E
27E	27E
28E	28E
29E	29E
30P	30P
31P	31P
32P	32P
33E	33E
34E	34E
35E	35E
36P	36P
37P	37P
38E	38E
39E	39E
40E	40E
41E	41E
42E	42E
43E	43E
44E	44E
45E	45E
46P	46P
47P	47P

48P	48P
49P	49P revised
50P	50P
51P	51P
52P	52P
53P	53P revised
54P	54P
55P	55P
56P	56P
57P	57P
58P	58P
59P	59P
60P	60P
61P	61P
62	new
63	new
64	new
65	new

Chapter 43

1E	1E
2E	2E
3E	3E
4E	4E
5E	5E
6P	6P
7P	7P
8P	8P
9P	9P
10P	10P
11P	11P
12P	12P
13P	13P
14E	14E
15E	15E
16E	16E
17E	17E
18E	18E
19E	19E
20E	20E
21E	21E
22E	22E
23P	23P
24P	24P
25P	25P
26P	26P
27P	27P
28P*	28P*
29E	29E
30E	30E
31E	31E
32E	32E
33E	33E
34E	34E
35E	35E
36P	36P
37P	37P
38P	38P
39P	39P
40P	40P
41P	41P
42E	42E
43E	43E
44E	44E
45E	45E
46E	46E
47E	47E
48P	48P
49P	49P
50P	50P
51P	51P
52P	52P
53P	53P
54P	54P
55E	55E
56E	56E
57E	57E
58E	58E
59E	59E
60E	60E
61E	61E
62E	62E
63E	63E
64P	64P
65P	65P
66P	66P
67P	67P
68P	68P
69P	69P
70P	70P
71P	71P
72P	72P
73E	73E
74E	74E
75E	75E
76P	76P
77P	77P
78P	78P
79	new
80	new
81	new
82	new
83	new
84	new

Chapter 44

1E	1E
2E	2E
3E	3E
4E	4E hint deleted
5E	5E hint deleted
6E	6E
7E	7E
8E	8E
9P	9P
10P	10P
11P	11P
12P	12P
13P	13P
14P	14P
15P	15P
16P	16P
17P	17P
18P	18P
19E	19E
20P	20P
21E	21E
22E	22E
23E	23E
24E	24E
25E	25E
26P	26P
27P	27P
28P	28P
29P	29P
30E	30E
31E	31E
32E	32E
33E	33E
34E	34E
35P	35P
36P	36P
37E	37E

38P 38P
39P 39P
40P 40P
41E 41E
42E 42E
43E 43E
44E 44E
45E 45E
46P 46P
47P 47P
48 new
49 new
50 new
51 new

Chapter 45

1E 1E
2E 2E
3E 3E
4E 4E
5E 5E
6E 6E
7E 7E
8E 8E
9E 9E
10P 10P
11P 11P
12P 12P
13P 13P
14P 14P
15P 15P
16P* 16P*
17E 17E
18E 18E
19E 19E
20E 20E
21E 21E
22E 22E
23E 23E
24P 24P
25P 25P
26P 26P
27P 27P
28P 28P
29E 29E
30E 30E
31E 31E
32P 32P
33E 33E
34E 34E
35E 35E
36E 36E
37E 37E
38E 38E
39E 39E
40P 40P
41P 41P
42P 42P
43E 43E
44E 44E
45E 45E
46P 46P
47P 47P
48P 48P
49E 49E
50E 50E
51E 51E data changed
52E 52E
53E 53E
54P 54P
55P 55P
56P 56P
57P 57P data changed
58P 58P
59P 59P
60E 60E
61E 61E
62E 62E
63E 63E
64E 64E
65E 65E
66E 66E
67E 67E
68P 68P
69P 69P
70P 70P
71P 71P
72P 72P

Chapter 46

1E 1E
2E 2E
3P 3P
4P 4P
5E 5E
6E 6E
7E 7E
8E 8E
9E 9E
10E 10E
11E 11E
12E 12E
13E 13E
14E 14E
15E 15E
16P 16P
17P 17P
18P 18P
19P 19P
20P 20P
21P 21P
22P 22P
23P 23P
24P 24P
25P 25P
26P 26P
27P 27P
28P 28P
29P 29P
30E 30E
31E 31E
32E 32E
33P 33P
34P 34P
35P 35P
36E 36E
37P 37P
38E 38E
39E 39E
40E 40E
41P 41P

Chapter 47

1E 1E
2E 2E
3P 3P
4E 4E
5E 5E
6E 6E
7E 7E
8E 8E

9E	9E supplemented
10E	10E
11E	11E
12E	12E
13E	13E
14E	14E
15E	15E
16E	16E
17E	17E
18E	18E
19P	19P
20P	20P
21P	21P
22P	22P
23P	23P
24P	24P
25P	25P
26E	26E
27E	27E
28E	28E
29E	29E
30E	30E
31E	31E
32E	32E
33E	33E
34P	34P
35P	35P
36P	36P
37P	37P
38P	38P
39P	39P
40P	40P
41P	41P
42P	42P
43P	43P
44P	44P
45P	45P
46E	46E
47E	47E
48P	48P
49P	49P
50P	50P
51P	51P
52E	52E
53E	53E
54E	54E
55E	55E
56E	56E
57P	57P
58P	58P
59P	59P
60P	60P
61P*	61P*
62E	62E
63E	63E
64P	64P
65P	65P
66E	66E
67E	67E
68E	68E
69E	69E
70P	70P
71P	71P
72E	72E
73E	73E
74E	74E
75P	75P
76P	76P
77P	77P
78P	78P
79P	79P
80	new
81	new

Chapter 48

1E	1E
2E	2E
3E	3E
4E	4E
5E	5E
6E	6E
7E	7E
8E	8E
9E	9E
10P	10P
11P	11P
12P	12P
13P	13P
14P	14P
15E	15E
16E	16E
17E	17E
18P	18P
19P	19P
20P	20P
21P	21P
22P	22P
23P	23P
24P	24P
25P	25P
26P	26P
27P	27P
28E	28E
29E	29E
30P	30P
31P	31P
32E	32E
33E	33E
34E	34E
35E	35E
36P	36P
37P	37P
38P	38P
39E	39E
40E	40E
41E	41E
42P	42P
43P	43P
44P	44P
45P	45P
46P	46P
47P	47P
48P	48P
49P	49P
50P	50P
51P	51P
52E	52E
53P	53P
54P	54P
55E	55E
56E	56E
57P	57P
58	new
59	new
60	new
61	new

Chapter 49

1E	1E
2E	2E
3E	3E
4E	4E
5E	5E

6P	6P	**18E**	18E	**30E**	30E
7P	7P	**19E**	19E	**31E**	31E
8P	8P	**20P**	20P	**32E**	32E
9P	9P	**21P**	21P	**33P**	33P
10P	10P	**22P**	22P	**34P**	34P
11P	11P	**23E**	23E	**35P**	35P
12E	12E	**24E**	24E	**36E**	36E
13E	13E	**25E**	25E	**37P**	37P
14P	14P	**26E**	26E	**38P**	38P
15E	15E	**27E**	27E	**39E**	39E
16E	16E	**28P**	28E	**40E**	40E
17E	17E	**29P**	29P		

SECTION SIX
ANSWERS TO PROBLEMS

Chapter 1

2. (*a*) \$632.31; (*b*) \$41,667; (*c*) 3.93×10^6
3. (*a*) 186 mi; (*b*) 3.0×10^8 mm
4. 5 ft 10 in. = 1.78 m, etc.
5. (*a*) 10^9; (*b*) 10^{-4}; (*c*) 9.1×10^5
6. (*a*) 4.00×10^4 km; (*b*) 5.10×10^8 km^2; (*c*) 1.08×10^{21} m^3
7. 32.2 km
8. (*a*) 1 ft^2 = 0.111 yd^2; (*b*) 1 in.^{2} = 6.45 cm^2; (*c*) 1 mi^2 = 2.59 km^2; (*d*) 1 m^3 = 10^6 cm^3
9. 0.020 km^3
10. 0.276 cords
11. (*a*) 250 ft^2; (*b*) 23.3 m^2; (*c*) 3060 ft^3; (*d*) 86.6 m^3
12. 2.0×10^{22} cm^3
13. 8×10^2 km
14. 1070 acre-feet
15. (*a*) 11 m^2/L; (*b*) 1.13×10^4 m^{-1}; (*c*) 2.17×10^{-3} gal/ft^2
16. (*a*) 4.85×10^{-6} pc, 1.58×10^{-5} ly; (*b*) 5.87×10^{12} mi, 1.91×10^{13} mi
17. (*a*) $d_{\text{sun}}/d_{\text{moon}} = 400$; (*b*) $V_{\text{sun}}/V_{\text{moon}} = 6.4 \times 10^7$; (*c*) 3.5×10^3 km
19. (*a*) 31 m; (*b*) 21 m; (*c*) Lake Ontario
20. (*a*) 0.98 ft/ns; (*b*) 0.30 mm/ps
21. 52.6 min, 5.2%
22. 3.156×10^7 s
23. 720 days
24. (*a*) 5.79×10^{12} days
25. (*a*) yes; (*b*) 8.6 s
26. (*a*) 0.013; (*b*) 0.54; (*c*) 10.3; (*d*) 31 m/s
27. 0.12 AU/min
28. 15°
29. 3.3 ft
30. C, D, A, B, E (best to worst); the important criterion is the constancy of the daily variation, not its magnitude
31. 2.1 h
32. 2 days 5 hours
33. 6.0×10^{26}
34. (*a*) 2.99×10^{-26} kg; (*b*) 4.68×10^{46}
35. 9.0×10^{49}
36. 1.32×10^9 kg
37. (*a*) 10^3 km/m^3; (*b*) 158 kg/s
38. 3.8 mg/s
39. 0.260 kg
40. (*a*) 1.18×10^{-29} m^3; (*b*) 0.282 nm

Chapter 2

1. (*a*) Lewis: 10.0 m/s, Rodgers: 5.41 m/s; (*b*) 1 h 10 min
2. 13 m
3. 310 ft
4. 414 ms
5. 2 cm/y
6. 1 h 13 min (1 h 14 min)
7. 6.71×10^8 mi/h, 9.84×10^8 ft/s, 1.00 ly/y
8. (*a*) +40 km/h; (*b*) 40 km/h
9. (*a*) 5.7 ft/s; (*b*) 7.0 ft/s
10. 48 km/h
11. (*a*) 45 mi/h (72 km/h); (*b*) 43 mi/h (69 km/h); (*c*) 44 mi/h (71 km/h); (*d*) 0
12. (*a*) 0, −2, 0, 12 m; (*b*) +12 m; (*c*) +7 m/s
13. (*a*) 28.5 cm/s; (*b*) 18.0 cm/s; (*c*) 40.5 cm/s; (*d*) 28.1 cm/s; (*e*) 30.3 cm/s
14. 1.29 s
15. (*a*) mathematically, an infinite number; (*b*) 60 km
16. (*a*) −6 m/s; (*b*) negative x direction; (*c*) 6 m/s; (*d*) first smaller, then zero, and then larger; (*e*) yes ($t = 2$ s); (*f*) no
17. (*a*) $4\text{ s} > t > 2\text{ s}$; (*b*) $3\text{ s} > t > 0$; (*c*) $6\text{ s} > t > 3\text{ s}$; (*d*) $t = 3$ s
19. 100 m
20. 2.56 m/s^2
21. 20 m/s^2, in the direction opposite to its initial velocity
23. (*a*) The signs of v and a are: AB: +, 0; BC: +, −; CD: 0, 0; DE: −, +; (*b*) no; (*c*) no
24. (*a*) The signs of v and a are: AB: +, −; BC: 0, 0; CD: +, +; DE: +, 0; (*b*) no;

(*c*) no
26. (*e*) situations (*a*), (*b*), and (*d*)
27. (*a*) no; (*b*) m^2/s^2, m/s^2
28. (*a*) 80 m/s; (*b*) 110 m/s; (*c*) 20 m/s^2
29. (*a*) $t = 1.2$ s; (*b*) $t = 0$; (*c*) $t > 0$, $t < 0$
30. (*a*) 1.10 m/s, 0.367 m/s^2; (*b*) 1.47 m/s, 0.367 m/s^2
31. (*a*) $\bar{v} = 14$ m/s, $\bar{a} = 18\ m/s^2$; (*b*) $v(2) = 24$ m/s, $v(1) = 6$ m/s, $a(2) = 24\ m/s^2$, $a(1) = 12\ m/s^2$
32. (*a*) 2.00 s; (*b*) 12 cm from left edge of screen; (*c*) 9.00 cm/s^2, to the left; (*d*) to the right; (*e*) to the left; (*f*) 3.46 s
33. (*a*) L/T^2, ft/s^2, L/T^3, ft/s^3; (*b*) 2.0 s; (*c*) 24 ft; (*d*) −16 ft; (*e*) 3, 0 , −9, −24 ft/s; (*f*) 0, −6, −12, −18 ft/s^2
34. 0.556 s
35. (*a*) 1.6 m/s; (*b*) 18 m/s
36. each, 0.28 m/s^2
37. (*a*) 3.1×10^6 s; (*b*) 4.6×10^{13} m
38. 2.8 m/s^2
39. 0.10 m
40. $1.62 \times 10^{15}\ m/s^2$
41. (*a*) 8.3 m/s^2, 0.85g; (*b*) 3.2 s, $\approx 8T$
42. 21g
43. (*a*) 5.00 s; (*b*) 61.5 m
44. (*a*) 25g; (*b*) 400 m
45. 2.6 s
46. (*a*) 3.1 m/s^2; (*b*) 45 m; (*c*) 13 s
47. (*a*) 5.0 m/s^2; (*b*) 4.0 s; (*c*) 6.0 s; (*d*) 90 m
48. (*a*) 3.56 m/s^2; (*b*) 8.43 m/s
49. (*a*) 5.00 m/s; (*b*) 1.67 m/s^2; (*c*) 7.50 m
50. (*a*) 60.6 s; (*b*) 36.3 m/s
51. (*a*) 0.74 s; (*b*) −20 ft/s^2
52. (*a*) 0.75 s; (*b*) 50 m
53. (*a*) 32.9 m/s; (*b*) 49.1 s; (*c*) 11.7 m/s
54. (*a*) 82 m; (*b*) 19 m/s
55. (*a*) 34.7 ft; (*b*) 41.6 s
56. (*a*) 3.26 ft/s^2
57. collide
59. (*a*) 29.4 m; (*b*) 2.45 s
60. (*a*) 31 m/s; (*b*) 6.4 s
61. 183 m/s
62. (*a*) 48.5 m/s; (*b*) 4.95 s; (*c*) 34.3 m/s; (*d*) 3.50 s
63. (*a*) 1.54 s; (*b*) 27.1 m/s
64. (*a*) 5.44 s; (*b*) 53.3 m/s; (*d*) 5.80 m
66. (*a*) 3.2 s; (*b*) 1.3 s
67. (*a*) 3.70 m/s; (*b*) 1.74 m/s; (*c*) 0.154 m
68. (*a*) 101 m; (*b*) 13.0 s
69. 4.0 m/s
70. (*a*) 350 ms; (*b*) 82 ms (each is for ascent and descent through the 15 cm)
72. 857 m/s^2, upward
73. (*a*) $v = \sqrt{v_0^2 + 2gh}$, downward; (*b*) $t = [\sqrt{v_0^2 + 2gh} - v_0]/g$; (*c*) same; (*d*) $t = [\sqrt{v_0^2 + 2gh} + v_0]/g$, more
74. (*a*) 1.23 cm; (*b*) 4 times, 9 times, 16 times, 25 times
75. 4 times as high
76. (*a*) 8.85 m/s; (*b*) 1.00 m
77. $+1.65 \times 10^3\ m/s^2$, upward
78. 22 cm and 89 cm below the nozzle
79. (*a*) 38.1 m; (*b*) 9.02 m/s; (*c*) 14.5 m/s, up
80. (*a*) 3.41 s; (*b*) 57 m
81. 96g
82. (*a*) 40.0 ft/s
83. (*a*) 17 s; (*b*) 290 m
84. 1.5 s
85. ≈ 0.3 s
86. (*a*) 5.4 s; (*b*) 41 m/s
87. (*a*) 76 m above the ground; (*b*) 4.2 s
88. 20.4 m
89. 2.34 m
90. (*a*) 30 s; (*b*) 300 m
91. 90 m
92. (*a*) 15 m, 5.5 m, 2.0 m; (*b*) −13 m/s; (*c*) −30 m/s, −11 m/s, −4.1 m/s; (*d*) −11 m/s
93. 22 m/s
94. 4 m/s^2
95. (*a*) 2.5 m/s; (*b*) 8.0 m/s; (*c*) 1.0 m/s^2; (*d*) 0
96. (*a*) 15.7 m/s; (*b*) 12.5 m; (*c*) 82.3 m
97. (*a*) 1.0 cm/s; (*b*) 1.6 cm/s, 1.1 cm/s, 0; (*c*) −0.79 cm/s^2; (*d*) 0, −0.87 cm/s^2, −1.2 cm/s^2; (*e*) −0.84 cm/s^2
98. 3.7 ms
99. 5.9 m

Chapter 3

1. The displacements should be (*a*) parallel, (*b*) antiparallel, (*c*) perpendicular
2. (*a*) the vectors are parallel; (*b*) **b** = 0; (*c*) the vectors are perpendicular
3. (*a*) 370 m, 36° north of east; (*b*) displacement magnitude = 370 m. distance walked = 425 m

4. (*b*) 3.2 km, 41° south of west
5. 81 km, 40° north of east
6. **a** + **b**: 4.2, 40° east of north; **b** − **a**: 8.3, 70° north of west
7. (*a*) 38 units at 320°; (*b*) 130 units at 1.2°; (*c*) 62 units at 130°
8. Walpole (the state prison)
9. $a_x = -2.5$, $a_y = -6.9$
10. (*a*) 47.2 units; (*b*) 122°
11. $r_x = 13$ m, $r_y = 7.5$ m
12. (*a*) 4.28 m; (*b*) 11.7 m
13. (*a*) 14 cm, 45° left of straight down; (*b*) 20 cm, vertically up; (*c*) zero
14. 156 km, 39.8° west of north
15. 4.74 km
16. (*a*) 27.8 m; (*b*) 13.4 m
17. 168 cm, 32.5° above the floor
19. (*a*) 21.0 ft; (*b*) no, yes, yes; (*c*) 14**i** + 12**j** + 10**k**, a possible answer; (*d*) 26.1 ft
20. (*a*) 0.349 rad, 0.873 rad, 1.75 rad; (*b*) 18.9°, 120°, 441°
21. $r_x = 12$, $r_y = -5.8$, $r_z = -2.8$
22. (*a*) −9**i** + 10**j**; (*b*) 13 units at 132° relative to **i**
23. (*a*) 8**i** + 2**j**, 8.2, 14°; (*b*) 2**i** − 6**j**, 6.3, −72° relative to **i**
24. (*a*) 3**i** − 2**j** + 5**k**; (*b*) 5**i** − 4**j** − 3**k**; (*c*) −5**i** + 4**j** + 3**k**
25. (*a*) 5.0, −37°; (*b*) 10, 53°; (*c*) 11, 27°; (*d*) 11, 80°; (*e*) 11, 260°; the angles are relative to $+x$, the last two vectors are in opposite directions
26. **a** = 9**i** + 12**j**, **b** = 3**i** + 4**j**
27. (*a*) $r_x = 1.59$, $r_y = 12.1$; (*b*) 12.2; (*c*) 82.5°
28. 6.0 ft, 21° east of north
29. 3390 ft, horizontally
30. (*a*) Take the x axis to be east, the y axis to be north, and the z axis to be up. Then the displacement in meters is 1000**i** + 2000**j** − 500**k**; (*b*) zero
31. (*a*) −2.83 m, −2.83 m, +5.00 m, 0 m, 3.00 m, 5.20 m; (*b*) 5.17 m, 2.37 m; (*c*) 5.69 m, 24.6° north of east; (*d*) 5.69 m, 24.6° south of west
34. (*a*) Put axes along cube edges. Diagonals are $a\mathbf{i}+a\mathbf{j}+a\mathbf{k}$, $a\mathbf{i}+a\mathbf{j}-a\mathbf{k}$, $a\mathbf{i}-a\mathbf{j}-a\mathbf{k}$, $a\mathbf{i}-a\mathbf{j}+a\mathbf{k}$; (*b*) 54.7°; (*c*) $\sqrt{3}a$
35. (*b*) 11,200 km
36. (*a*) $a_x = 9.51$ m, $a_y = 14.1$ m; (*b*) $a'_x = 13.4$ m, $a'_y = 10.5$ m
37. (*a*) 10 m, north; (*b*) 7.5 m, north
38. (*a*) $+y$; (*b*) $-y$; (*c*) 0; (*d*) 0; (*e*) $+z$; (*f*) $-z$; (*g*) ab, both; (*h*) ab/d, $+z$
39. no
40. yes
42. (*a*) up, unit magnitude; (*b*) zero; (*c*) south, unit magnitude; (*d*) 1.00; (*e*) 0
43. (*a*) 30; (*b*) 52
44. (*a*) −18.8; (*b*) 26.9, $+z$ direction
45. (*a*) 0; (*b*) −16; (*c*) −9
46. (*a*) 12, out of page; (*b*) 12, into page; (*c*) 12, out of page
48. 22°
49. (*a*) 11**i** + 5**j** − 7**k**; (*b*) 120°
51. (*a*) 2**k**; (*b*) 26; (*c*) 46
52. (*a*) 57°; (*b*) $c_x = \pm 2.2$, $c_y = \mp 4.5$
53. (*a*) 2.97; (*b*) 1.51**i** + 2.67**j** − 1.36**k**; (*c*) 48
54. (*a*) −21; (*b*) −9; (*c*) 5**i** − 11**j** − 9**k**
55. 70.5°
57. (*b*) $a^2 b \sin\phi$
58. (*a*) $a_x = 3.0$, $a_y = 0$, $b_x = 3.46$, $b_y = 2.00$, $c_x = -5.00$, $c_y = 8.66$; (*b*) $p = -6.67$ $q = 4.33$
60. 2.6 km
61. 3.6
62. 4.1
63. 3.2

Chapter 4

1. (*a*) (−5.0**i** + 8.0**j**) m; (*b*) 9.4 m, 122° from $+x$
2. (*a*) 6.2 m
3. (*a*) (−7.0**i** + 12**j**) m; (*b*) xy plane
4. (−2.0**i** + 6.0**j** − 10**k**) m
5. (*a*) 671 mi, 63.4° south of east; (*b*) 298 mi/h, 63.4° south of east; (*c*) 400 mi/h
6. 7.59 km/h, 22.5° east of north
7. (*a*) 6.79 km/h; (*b*) 6.96°
8. (−0.70**i** + 1.4**j** − 0.40**k**) m/s
9. (*a*) (3**i** − 8t**j**) m/s; (*b*) (3**i** − 16**j**) m/s; (*c*) 16 m/s, −79° to $+x$
10. (*a*) (−1.5**i** + 0.50**k**) m/s^2; (*b*) 1.6 m/s^2, in xy plane, 162° from $+x$
11. (*a*) (8t**j** + **k**) m/s; (*b*) 8**j** m/s^2
12. (*a*) (6.00**i** − 106**j**) m; (*b*) (19.0**i** − 224**j**) m/s; (*c*) (24.0**i** − 336**j**) m/s^2; (*d*) −85.2° to $+x$

13. $(-2.10\mathbf{i} + 2.81\mathbf{j})\,\mathrm{m/s^2}$
14. 60.0°
15. (*a*) $-1.5\mathbf{j}$ m/s; (*b*) $(4.5\mathbf{i} - 2.25\mathbf{j})$ m
16. (*a*) $-18\mathbf{i}\,\mathrm{m/s^2}$; (*b*) 0.75 s; (*c*) never; (*d*) 2.2 s
17. (*a*) 18 cm; (*b*) 1.9 m
18. (*a*) 63 ms; (*b*) 1.6×10^3 ft/s
19. (*a*) 5.4×10^{-13} m; (*b*) decreases
20. (*a*) 2.0 ns; (*b*) 2.0 mm; (*c*) $(1.0 \times 10^9\,\mathbf{i} - 2.0 \times 10^8\,\mathbf{j})$ cm/s
21. (*a*) 0.50 s; (*b*) 10 ft/s
22. (*a*) 3.03 s; (*b*) 758 m; (*c*) 29.7 m/s
23. (*a*) 0.21 s, 0.21 s; (*b*) 8.1 in.; (*c*) 24 in.
24. (*a*) 16 m/s, 23° above the horizontal; (*b*) 27 m/s, 57° below the horizontal
25. (*a*) 16.9 m, 8.21 m; (*b*) 27.6 m, 7.26 m; (*c*) 40.2 m, 0
26. (*a*) 32.4 m; (*b*) −37.7 m
27. (*a*) 1.15 s; (*b*) 12.0 m; (*c*) 19.2 m/s, 4.80 m/s; (*d*) no
28. (*b*) 76°
29. (*b*) 27°
30. (*a*) 51.8 m; (*b*) 27.4 m/s; (*c*) 67.5 m
31. (*a*) 194 m/s; (*b*) 38°
32. 33 cm
33. 1.9 in.
35. not accidental, horizontal launch speed about 20% of world-class sprint speed
37. (*a*) 11 m; (*b*) 23 m; (*c*) 17 m/s, 63° below horizontal
38. His record would have been longer by about 1 cm.
39. (*a*) 73 ft; (*b*) 7.6°; (*c*) 1.0 s
40. (*a*) 260 m/s; (*b*) 45 s; (*c*) larger
41. 23 ft/s
42. 78 ft/s at 65° above the horizontal
43. ≈ 40 m
44. 6.29°, 83.7°
45. 30 m above the release point
46. 19 ft/s
47. the third
48. (*a*) 2900 ft; (*b*) 10 s
49. (*a*) 202 ft/s; (*b*) 806 m; (*c*) 161 m/s, −171 m/s
50. 7.0 m/s
51. 78.5°
52. (*a*) 20 cm; (*b*) no, the ball hits the net only 4.4 cm above the ground
53. 25 m
54. yes, its center passes about 4.1 ft above the fence
55. between the angles 31° and 63° above the horizontal
56. 5.7 s
57. (*a*) 310 s; (*b*) 105 km; (*c*) 139 km
58. $9.00 \times 10^{22}\,\mathrm{m/s^2}$, toward the center
59. (*a*) $4.0\,\mathrm{m/s^2}$; (*b*) toward the center of the circle
60. 6.7×10^6 m/s
61. (*a*) 22 m; (*b*) 15 s
62. (*a*) 7.49 km/s; (*b*) $8.00\,\mathrm{m/s^2}$
63. (*a*) 4.6×10^{12} m; (*b*) 2.8 d
64. (*a*) 0.94 m; (*b*) 19 m/s; (*c*) $2400\,\mathrm{m/s^2}$, toward center
65. (*a*) 7.3 km; (*b*) less than 80 km/h
66. (*a*) 1.3×10^5 m/s; (*b*) $7.9 \times 10^5\,\mathrm{m/s^2}$ or 8.0×10^4 g-units, toward the center; (*c*) both answers increase
67. (*a*) 19 m/s; (*b*) 35 rev/min
68. (*a*) $4.1\,\mathrm{m/s^2}$, down; (*b*) $4.1\,\mathrm{m/s^2}$, up
69. (*a*) $0.034\,\mathrm{m/s^2}$; (*b*) 84 min
70. $2.58\,\mathrm{cm/s^2}$
71. (*a*) 4.2 m, 45°; 5.5 m, 68°; 6.0 m, 90°; (*b*) 4.2 m, 135°; (*c*) 0.85 m/s, 135°; (*d*) 0.94 m/s, 90°; 0.94 m/s, 180°; (*e*) $0.30\,\mathrm{m/s^2}$, 180°; $0.30\,\mathrm{m/s^2}$, 270°. Angles measured counterclockwise from $+x$
72. $160\,\mathrm{m/s^2}$
73. (*a*) 5 km/h, upstream; (*b*) 1 km/h, downstream
74. 36 s, no
75. wind blows from the west at 53 mi/h
76. $0.018\,\mathrm{mi/s^2}$ from either frame
77. 48 s
78. 130°
79. (*a*) $(80\mathbf{i} - 60\mathbf{j})$ km//h; (*b*) **v** happens to be along the line of sight; (*c*) answers do not change
80. 60°
81. $(0.96\mathbf{j})$ m/s
82. (*a*) 5.8 m/s; (*b*) 16.7 m; (*c*) 67°
83. 80 m/s
84. (*a*) 38 knots, 1.5° east of north; (*b*) 4.2 h; (*c*) 1.5° west of south
85. 185 km/h, 22° south of west
86. (*a*) from 75° east of south; (*b*) 30° east of north; substitute west for east to get second solution

87. 93° from the direction of motion of the car
88. (*a*) 30° upstream; (*b*) 69 min; (*c*) 80 min; (*d*) 80 min; (*e*) perpendicular to the current, the least time is 60 min
89. (*a*) 47° east of north; (*b*) 6 min 35 s
90. (*a*) head the boat 25.3° upstream; (*b*) 12.6 min
91. 0.83 *c*
92. (*a*) 0.35 *c*; (*b*) 0.62 *c*
93. (*a*) 44 m; (*b*) 13 m; (*c*) 8.9 m
94. (*a*) 0, 0; 2.0 m, 1.4 m; 4.0 m, 2.0 m; 6.0 m, 1.4 m; 8.0 m, 0; (*b*) 2.0 m/s, 1.1 m/s; 2.0 m/s, 0; 2.0 m/s, −1.1 m/s; (*c*) 0, $-0.87\ \text{m/s}^2$; 0, $-1.2\ \text{m/s}^2$; 0, $-0.87\ \text{m/s}^2$
95. (*a*) 7.2 m/s, 16° west of north; (*b*) 29 s
96. (*a*) 20 m/s, 36 m/s; (*b*) 74 m
97. (*a*) 45 m; (*b*) 22 m/s
98. (*a*) 48 m, west of center; (*b*) 48 m, west of center
99. (*a*) 11 m; (*b*) 45 m/s
100. 240 km/h

Chapter 5

1. (*a*) $F_x = 1.88$ N, $F_y = 0.684$ N; (*b*) $(1.88\,\mathbf{i} + 0.684\,\mathbf{j})$ N
2. (*a*) $(1.0\,\mathbf{i} - 2.0\,\mathbf{j})$ N; (*b*) 2.2 N, −63° relative to $+x$; (*c*) $2.2\ \text{m/s}^2$, −63° relative to $+x$
3. (*a*) $(-6.26\,\mathbf{i} - 3.23\,\mathbf{j})$ N; (*b*) 7.0 N, 207° relative to $+x$
4. $(-2\,\mathbf{i} + +6\,\mathbf{j})$ N
5. $(3\,\mathbf{i} - 11\,\mathbf{j} + 4\,\mathbf{k})$ N
6. (*a*) 0; (*b*) +20 N; (*c*) −20 N; (*d*) −40 N; (*e*) −60 N
7. (*a*) $(-32\,\mathbf{i} - 21\,\mathbf{j})$ N; (*b*) 38 N, 213° from $+x$
8. (*a*) $(1\,\mathbf{i} - 1.3\,\mathbf{j})\ \text{m/s}^2$; (*b*) $1.6\ \text{m/s}^2$ at −50° from $+x$
9. (*a*) $(0.86\,\mathbf{i} - 0.16\,\mathbf{j})\ \text{m/s}^2$; (*b*) $0.88\ \text{m/s}^2$, −11° relative to $+x$
10. (*a*) $\mathbf{F}_2$ and $\mathbf{F}_3$ are in the $-x$ direction, $\mathbf{a} = 0$; (*b*) $\mathbf{F}_2$ and $\mathbf{F}_3$ are in the $-x$ direction, $\mathbf{a}$ is on the x axis, $a = 0.83\ \text{m/s}^2$; (*c*) $\mathbf{F}_2$ and $\mathbf{F}_3$ are at 34° from $-x$ direction; $\mathbf{a} = 0$
11. (*a*) mass = 44 slug, weight = 1400 lb; (*b*) mass = 421 kg, weight = 4100 N
12. (*a*) 22 N, 2.3 kg; (*b*) 1100 N, 110 kg; (*c*) 1.6×10^4 N, 1.6×10^3 kg
13. (*a*) 740 N; (*b*) 290 N; (*c*) 0; (*d*) 75 kg at each location
14. (*a*) 11 N, 2.2 kg; (*b*) 0, 2.2 kg
15. (*a*) 147 N, downward; (*b*) 147 N, upward; (*c*) 147 N
16. (*a*) 44 N; (*b*) 78 N
17. (*a*) 54 N; (*b*) 152 N
18. (*a*) 108 N; (*b*) 108 N; (*c*) 108 N
19. 1.18×10^4 N
20. 1400 lb
21. 1.2×10^5 N
22. 6.8×10^3 N
23. 16 N
24. (*a*) 0.24 μm; (*b*) 31 μm
25. $8.0\ \text{cm/s}^2$
26. (*a*) $13\ \text{ft/s}^2$; (*b*) 170 lb
27. (*a*) the 4 kg mass; (*b*) $6.5\ \text{m/s}^2$; (*c*) 13 N
28. (*a*) 42 N; (*b*) 72 N; (*c*) $4.9\ \text{m/s}^2$
29. 1.2×10^6 N
30. (*a*) $0.02\ \text{m/s}^2$; (*b*) 8×10^4 km; (*c*) 2×10^3 m/s
31. 69 lb
32. (*a*) 1.1×10^{-15} N; (*b*) 8.9×10^{-30} N
33. 1.5 mm
34. (*a*) 5500 N; (*b*) 2.7 s; (*c*) 4 times as far; (*d*) twice the time
35. $10\ \text{m/s}^2$
36. (*a*) 4.9×10^5 N; (*b*) 1.5×10^6 N
37. (*a*) 110 lb, up; (*b*) 110 lb, down
38. (*a*) 2.2×10^{-3} N; (*b*) 3.7×10^{-3} N
39. (*a*) $0.62\ \text{m/s}^2$; (*b*) $0.13\ \text{m/s}^2$; (*c*) 2.6 m
40. (*a*) 1.1 N; (*b*) 2.1 N
41. (*a*) $0.74\ \text{m/s}^2$; (*b*) $7.3\ \text{m/s}^2$
42. (*a*) $\cos\theta$; (*b*) $\cos\theta$
43. (*a*) $(5\,\mathbf{i} + 4.3\,\mathbf{j})$ m/s; (*b*) $(15\,\mathbf{i} + 6.4\,\mathbf{j})$ m
44. 1.8×10^4 N
45. (*a*) 65 N; (*b*) 49 N
46. (*a*) 4.6×10^3 N; (*b*) 5.8×10^3 N
47. (*a*) 220 kN; (*b*) 50 kN
48. (*a*) $250\ \text{m/s}^2$; (*b*) 2.0×10^4 N
49. (*a*) $0.970\ \text{m/s}^2$; (*b*) $T_1 = 11.6$ N, $T_2 = 34.9$ N
50. 23 kg
51. (*a*) 7020 lb; (*b*) 5460 lb
52. (*a*) 620 N; (*b*) 580 N
53. (*a*) 5.1 m/s
54. 1.9×10^5 lb
55. (*a*) 3260 N; (*b*) 2.7×10^3 kg; (*c*) $-1.2\ \text{m/s}^2$
56. (*a*) rope breaks; (*b*) $1.6\ \text{m/s}^2$
57. (*a*) 1.23, 2.46, 3.69, 4.92 N; (*b*) 6.15 N; (*c*) 0.25 N

58. (*a*) 0.735 m/s^2; (*b*) downward; (*c*) 20.8 N
59. (*a*) allow a downward acceleration with magnitude $\geq$ 4.2 ft/s^2; (*b*) 13 ft/s or greater
60. (*a*) 1.18 m; (*b*) 0.674 s; (*c*) 3.50 m/s
61. (*a*) 7.3 kg; (*b*) 89 N
62. (*a*) 566 N; (*b*) 1130 N
63. (*a*) 4.9 m/s^2; (*b*) 2.0 m/s^2, upward; (*c*) 120 N
64. 18,000 N
65. (*a*) 120 m/s^2; (*b*) $12g$; (*c*) 1.4×10^8 N; (*d*) 4.2 y
66. (*a*) 1.4×10^4 N; (*b*) 1.1×10^4 N; (*c*) 2700 N, toward the counterweight
67. (*a*) 2.18 m/s^2; (*b*) 116 N; (*c*) 21.0 m/s^2
68. 6800 N, at 21° to the line of motion of the barge
69. (*a*) 4.6 m/s^2; (*b*) 2.6 m/s^2
70. $2Ma/(a+g)$
71. (*a*) 9.4 km; (*b*) 61 km
72. (*b*) $F/(m+M)$; (*c*) $MF/(m+M)$; (*d*) $F(m+2M)/2(m+M)$
73. (*a*) −466 N; (*b*) −527 N; (*c*) −931 N, −1050 N; (*d*) first two cases: 931 N, third case: −1860 N, fourth case: −2100 N
74. (*a*) 180 N; (*b*) 640 N
75. 2.0 N, down
76. (*a*) 2.6 N; (*b*) 17°
77. 12 N
78. 300 N, up
79. (*a*) 2.8 N, due west; (*b*) 2.2 N, 22° west of south
80. 4.6 N
81. 2.9 m/s^2
82. (*a*) 3.0 N; (*b*) 0.34 kg

Chapter 6

1. (*a*) 200 N; (*b*) 120 N
2. 0.61
3. (*a*) 110 N; (*b*) 130 N; (*c*) no; (*d*) 46 N; (*e*) 17 N
4. 2°
5. (*a*) (*i*) 245 N, 100 N; (*ii*) 195 N, 86.6 N; (*iii*) 158 N, 50.0 N; (*b*) (*i*) at rest; (*ii*) slides; (*iii*) at rest
6. 440 N
7. 9.3 m/s^2
8. (*a*) 190 N; (*b*) 0.56 m/s^2
9. (*a*) 90 N; (*b*) 70 N; (*c*) 0.89 m/s^2
10. (*a*) no, 222 N; (*b*) no, 334 N; (*c*) yes, 310 N; (*d*) yes, 310 N
11. (*a*) no; (*b*) $(-12\mathbf{i} + 5\mathbf{j})$ N
12. (*a*) 240 N; (*b*) 0.60
13. 20°
14. (*a*) 7.5 m/s^2 down the slope; (*b*) 9.5 m/s^2 down the slope
15. (*a*) 0.13 N; (*b*) 0.12
16. $\mu_s = 0.58$, $\mu_k = 0.54$
18. (*a*) 0.11 m/s^2, 0.23 m/s^2; (*b*) 0.041, 0.029
19. (*a*) 56 N; (*b*) 59 N; (*c*) 1100 N
20. 36 m
21. (*a*) $v_0^2/(4g \sin\theta)$; (*b*) no
22. (*a*) 300 N; (*b*) 1.3 m/s^2
23. 0.53
24. (*a*) 66 N; (*b*) 2.3 m/s^2
25. (*a*) 11 N, rightward; (*b*) 0.14 m/s^2
26. (*a*) $\mu_k mg/(\sin\theta - \mu_k \cos\theta)$; (*b*) $\theta_0 = \tan^{-1}\mu_s$
27. (*a*) 2.0 m/s^2 down the plane; (*b*) 4.0 m; (*c*) it stays there
28. (*b*) 3.0×10^7 N
29. (*a*) 8.6 N; (*b*) 46 N; (*c*) 39 N
30. 100 N
31. (*a*) 0; (*b*) 3.9 m/s^2 down the incline; (*c*) 1.0 m/s^2 down the incline
32. 3.3 kg
33. (*a*) 13 N; (*b*) 1.6 m/s^2
34. (*a*) 11 ft/s^2; (*b*) 0.56 lb; (*c*) blocks move independently
35. (*a*) 1.05 N, in tension; (*b*) 3.62 m/s^2 down the plane; (*c*) answers are the same except that the rod is under compression
36. (*a*) 27 N; (*b*) 3.0 m/s^2
37. (*a*) 6.1 m/s^2, rightward; (*b*) 0.98 m/s^2, rightward
38. 490 N
39. $g(\sin\theta - \sqrt{2}\mu_k \cos\theta)$
40. (*a*) 3.0×10^5 N; (*b*) 1.2°
41. (*a*) 19°; (*b*) 3300 N
42. 9.9 s
43. 6200 N
44. 3.75
45. 2.3
46. 12 cm
47. 11 m/s
48. 9.6
49. 68 ft
50. (*a*) 3210 N; (*b*) yes

51. (*a*) 11°; (*b*) 0.19
52. 0.078
53. (*a*) 0.96 m/s; (*b*) 0.021
54. (*a*) 0.72 m/s; (*b*) 2.1 m/s^2; (*c*) 0.50 N
55. (*a*) 2.2×10^6 m/s; (*b*) 9.1×10^{22} m/s^2, toward the nucleus; (*c*) 8.3×10^{-8} N
56. $\sqrt{Mgr/m}$
57. 178 km/h
58. (*a*) 30 cm/s; (*b*) 180 cm/s^2, radially inward; (*c*) 3.6×10^{-3} N, radially inward; (*d*) 0.37
59. 0.12, 0.23
60. (*a*) 275 N; (*b*) 877 N
61. 874 N
62. (*a*) 175 lb; (*b*) 50 lb
63. $\sqrt{gR}$
64. (*a*) at the bottom of the circle; (*b*) 31 ft/s
65. 2.2 km
66. (*a*) 9.5 m/s; (*b*) 20 m
67. (*a*) 5.1 m/s^2, radially inward; (*b*) 4.8 N; (*c*) 10 N
68. 13°
69. (*a*) 0.0338 N; (*b*) 9.77 N
70. (*b*) 8.74 N; (*c*) 37.9 N, radially inward; (*d*) 6.45 m/s
71. (*a*) 5.8′; (*b*) 0; (*c*) 0
72. (*a*) 17 N, up the incline; (*b*) 20 N, up the incline; (*c*) 15 N, up the incline
73. 3.4 m/s^2
74. (*a*) 3.7 kN, up; (*b*) 2.3 kN, down
75. 7.6 N
76. (*a*) 6.0 N, to the left; (*b*) 3.6 N, to the left; (*c*) 3.1 N, to the left
77. 0.56
78. (*a*) 6.4 m/s; (*b*) 19 m/s
79. 4.6 N

Chapter 7

1. (*a*) 2.7×10^7 ft·lb; (*b*) 150 ft·lb
2. (*a*) 590 J; (*b*) 0; (*c*) 0; (*d*) 590 J
3. (*a*) 314 J; (*b*) −155 J; (*c*) 0; (*d*) 158 J
4. (*a*) 200 N; (*b*) 700 m; (*c*) -1.4×10^5 J; (*d*) 400 N, 350 m, -1.4×10^5 J
5. (*a*) 270 N; (*b*) −400 J; (*c*) 400 J; (*d*) 0; (*e*) 0
6. 5000 J
7. (*a*) $c = 4$ m; (*b*) $c < 4$ m; (*c*) $c > 4$ m
8. (*a*) 98 N; (*b*) 4.0 cm; (*c*) 3.9 J; (*d*) −3.9 J
9. (*a*) 215 lb; (*b*) 10,000 ft·lb; (*c*) 48 ft; (*d*) 10,300 ft·lb
10. 660 J
11. (*a*) 2200 J; (*b*) −1500 J
12. (*a*) 30.1 J; (*b*) 0.22
13. 25 J
14. 800 J
16. 0, by both methods
17. −6 J
18. (*a*) −0.043 J; (*b*) −0.13 J
19. 1250 J
20. (*a*) 23 mm; (*b*) 45 N
21. 1.8×10^{13} J
22. 1.2×10^6 m/s
23. (*a*) 3610 J; (*b*) 1900 J; (*c*) 1.1×10^{10} J
24. (*a*) 2.9×10^7 m/s; (*b*) 1.3 MeV
25. 47 keV
26. AB: +, BC: 0, CD: −, DE: +
27. 7.9 J
28. 31 m
29. (*a*) 48 km/h; (*b*) 8.9×10^4 J
30. (*a*) 5×10^{14} J; (*b*) 0.1 megaton TNT; (*c*) 8 bombs
31. (*a*) 1×10^5 megatons TNT; (*b*) 1×10^7 bombs
32. father, 2.4 m/s; son, 4.8 m/s
33. 530 J
34. 2.7×10^{33} J
35. (*a*) 1.2×10^4 J; (*b*) -1.1×10^4 J; (*c*) 1000 J; (*d*) 5.3 m/s
36. (*a*) $-3Mgd/4$; (*b*) Mgd; (*c*) $Mgd/4$; (*d*) $\sqrt{gd/2}$
37. (*a*) 797 N; (*b*) 0; (*c*) −1550 J; (*d*) 0; (*e*) 1550 J; (*f*) F varies during displacement
38. (*a*) 0.29 J; (*b*) −1.8 J; (*c*) 3.5 m/s; (*d*) 23 cm
39. 270 kW
40. 230 hp
41. 235 kW
42. (*a*) 28 W; (*b*) (6**j**) m/s
43. 17 kW
44. 490 W
45. (*a*) 1.8×10^5 ft·lb; (*b*) 0.55 hp
46. (*a*) 100 J; (*b*) 67 W; (*c*) 33 W
47. (*a*) 0.83, 2.5, 4.2 J; (*b*) 5.0 W
48. 0.99 hp
49. 90 hp
50. (*a*) 0.68 c; (*b*) 1.9×10^5 eV; (*c*) low by 37%
51. (*a*) 79.37 keV; (*b*) 3.12 MeV; (*c*) 10.9 MeV
52. (*a*) 10.9 MeV; (*b*) 43%
53. (*a*) 6.6 m/s; (*b*) 4.7 m
54. (*a*), (*b*) 13 J
55. (*a*) 12 J; (*b*) 4.0 m; (*c*) 18 J

56. -37 J
57. (*a*) 0; (*b*) -350 W

Chapter 8

1. 89 N/cm
2. 1.1×10^8 N/m
3. (*a*) 200 J; (*b*) 170 J; (*c*) 13 m/s
4. 8400 ft·lb
5. (*a*) 4.0×10^4 J; (*b*) 4.0×10^4 J
6. 2.1 m/s
7. (*a*) v_0; (*b*) $\sqrt{v_0^2 + gh}$; (*c*) $\sqrt{v_0^2 + 2gh}$; (*d*) $(v_0^2/2g) + h$
8. 830 ft
9. 56 m/s
10. (*a*) 27.0 kJ; (*b*) 2940 J; (*c*) 159 m/s
11. (*a*) 7.84 N/cm; (*b*) 62.7 J; (*c*) 80.0 cm
12. (*a*) 0.98 J; (*b*) 3.1 N/cm
13. (*a*) mgL; (*b*) $\sqrt{2gL}$
14. 2.29 m/s
15. (*a*) 2.8 m/s; (*b*) 2.7 m/s
16. 4.00 m
17. (*a*) 35 cm; (*b*) 1.7 m/s
18. (*a*) 54 m/s; (*b*) 52 m/s; (*c*) 76 m, below
19. (*a*) 1.2 J; (*b*) 11 m/s; (*c*) no; (*d*) no
20. (*a*) 39 ft/s; (*b*) 3.4 in.
21. (*a*) 25 kJ; (*b*) 7.8 kJ; (*c*) 160 m
22. (*a*) 5.0 m/s; (*b*) 79°; (*c*) 64 J
23. (*a*) 4.8 m/s; (*b*) 2.4 m/s
25. 10 cm
26. (*a*) $\sqrt{v_0^2 + 2gL(1 - \cos\theta_0)}$; (*b*) $\sqrt{2gL\cos\theta_0}$; (*c*) $\sqrt{gL(3 + 2\cos\theta_0)}$
27. 1.25 cm
28. (*a*) $U(x) = -Gm_1m_2/x$; (*b*) $Gm_1m_2d/x_1(x_1 + d)$
29. (*a*) 19 J; (*b*) 6.4 m/s; (*c*) 11 J, 6.4 m/s
30. (*a*) $8mg$ to the left and mg vertically down; (*b*) $\frac{5}{2}R$
31. it comes close to breaking but does not break
33. (*a*) $2\sqrt{gL}$: (*b*) $5mg$; (*c*) 71°
34. (*a*) 9.8 kW; (*b*) 3.1 kJ
35. $mgL/32$
36. (*a*) 39.6 cm; (*b*) 3.64 cm
39. (*a*) $1.12(A/B)^{1/6}$; (*b*) repulsive; (*c*) attractive
40. (*a*) 4.7 N; (*b*) between $x = 1.2$ m and $x = 14$ m; (*c*) 3.6 m/s
41. (*a*) turning point on left, none on right; (*b*) turning points on both left and right; (*c*) -1.2×10^{-19} J; (*d*) 2.2×10^{-19} J; (*e*) $\approx 1 \times 10^{-9}$ N on each, directed toward the other; (*f*) $r < 0.2$ nm; (*g*) $r > 0.2$ nm; (*h*) $r = 0.2$ nm
42. (*a*) 2.6×10^{12} J; (*b*) 3.8×10^8 J
43. (*a*) 7.9×10^4 J; (*b*) 1.8 W
44. (*a*) 7.8 MJ; (*b*) 6.2
45. (*a*) 2700 MJ; (*b*) 2700 MW; (*c*) 240 M$
46. 3.1×10^{11} W
47. (*a*) -0.74 J; (*b*) -0.53 J
48. -20 ft·lb
49. (*a*) 0.77 mi; (*b*) 71 kW
50. (*a*) 3000 J; (*b*) 300 W
51. 690 W
52. (*a*) 8.8 m/s; (*b*) 2600 J; (*c*) 1.6 kW
53. 5.5×10^6 N
54. 23.7 hp
55. 24 W
56. (*a*) 3.3 kW; (*b*) 1.8 kW
57. (*b*) 3.4
58. 880 MW
59. (*a*) -3800 J; (*b*) 3.1×10^4 N
60. 39 kW
61. 54%
62. 740 m
63. -12 J
64. (*a*) 2900 J; (*b*) 390 J; (*c*) 210 N
65. (*a*) 1.5 MJ; (*b*) 0.51 MJ; (*c*) 1.0 MJ; (*d*) 63 m/s
66. (*a*) 67 J; (*b*) 67 J; (*c*) 46 cm
67. 0.191
68. (*a*) -0.90 J; (*b*) 0.46 J; (*c*) 1.0 m/s
69. (*a*) 44 m/s; (*b*) 0.036
70. (*a*) 18 ft/s; (*b*) 18 ft
72. 4.3 m
73. (*a*) 5.0 in.; (*b*) 8.7 ft/s
74. (*a*) 31.0 J; (*b*) 5.35 m/s; (*c*) conservative
75. (*a*) 560 J; (*b*) 150 J; (*c*) 5.5 m/s
76. (*a*) 3.0×10^5 J; (*b*) 10 kW; (*c*) 20 kW
77. 1.2 m
78. (*a*) $\frac{1}{2}ke^2\left[\frac{1}{r_2} - \frac{1}{r_1}\right]$; (*b*) $-ke^2\left[\frac{1}{r_2} - \frac{1}{r_1}\right]$ (*c*) $-\frac{1}{2}ke^2\left[\frac{1}{r_2} - \frac{1}{r_1}\right]$
80. (*a*) 10 m; (*b*) 50 N; (*c*) 4.1 m; (*d*) 120 N
81. in the center of the flat part

82. (*a*) $\sqrt{2gL}$; (*b*) $5mg$; (*c*) $-mgL$; (*d*) $-2mgL$
83. (*a*) 24 ft/s; (*b*) 3.0 ft; (*c*) 9.0 ft; (*d*) 49 ft
84. (*a*) 1070 N; (*b*) 36 kW; (*c*) 6.4%
85. 180 W
86. 45 mi/h
87. (*a*) 2.1×10^6 kg; (*b*) $\sqrt{100 + 1.5t}$ m/s; (*c*) $(1.5 \times 10^6)/\sqrt{100+1.5t}$ N; (*d*) 6.7 km
88. 69 hp
89. (*a*) 110 rev/min; (*b*) 19 W
90. $t = (3d/2)^{2/3}(m/2P)^{1/3}$
91. (*a*) 216 J; (*b*) 1180 N; (*c*) 432 J
92. (*a*) 9.17×10^{15} J; (*b*) 2.91×10^5 y
93. (*a*) 1.1×10^{17} J; (*b*) 1.2 kg
94. 1560 MeV
95. 1.10 kg
96. 92.5 kg
97. 270 times the equatorial circumference of the earth
98. 8300 kg
99. 2×10^5 kg
100. 1.8 eV
101. (*a*) $2.46 \times 10^{15}\ \mathrm{s}^{-1}$; (*b*) emitted
102. (*a*) $K = 0.75$ J, $\Delta U_g = -1.0$ J, $\Delta U_e = 0.25$ J; (*b*) $K = 1.0$ J, $\Delta U_g = -2.0$ J, $\Delta U_e = 1.0$ J; (*c*) $K = 0.75$ J, $\Delta U_g = -3.0$ J, $\Delta U_e = 2.25$ J; (*d*) $K = 0$, $\Delta U_g = -4.0$ J, $\Delta U_e = 4.0$ J
103. 11 kJ
104. 15 J
105. (*a*) 7.0 J; (*b*) 22 J
106. 75 J
107. −320 J
108. 3.7 J
109. (*a*) 2.7 J; (*b*) 1.8 J; (*c*) 0.39 m
110. (*a*) 5.6 J; (*b*) 12 J; (*c*) 13 J

Chapter 9

1. (*a*) 4600 km; (*b*) $0.73R_e$
2. 6.46×10^{-11} m, along the line joining the atoms
3. (*a*) $x_{cm} = 1.1$ m, $y_{cm} = 1.3$ m; (*b*) shifts toward topmost particle
4. $0.2L$ from the heavy rod, along symmetry axis
5. $x_{cm} = -0.25$ m, $y_{cm} = 0$
7. in the iron, at midheight and midwidth, 2.7 cm from midlength
8. 6.8×10^{-12} m toward plane of the hydrogens, along axis of symmetry
9. $x_{cm} = y_{cm} = 20$ cm, $z_{cm} = 16$ cm
10. (*a*) (3.5 cm, 0.67 cm); (*b*) (3.5 cm, 0.67 cm)
11. 36.8 m
12. (*a*) $H/2$; (*b*) $H/2$; (*c*) descends to lowest point and then ascends to $H/2$; (*d*) $\frac{HM}{m}\sqrt{1+\frac{m}{M}} - 1$
13. 6.2 m
14. 72 km/h
15. (*a*) down, $mv/(m+M)$; (*b*) balloon again stationary
16. (*a*) center of mass does not move; (*b*) 0.75 m
17. (*a*) L; (*b*) 0
18. (*a*) 28 cm below the release point; (*b*) 2.3 m/s
19. 58 kg
20. 53 m
21. (*a*) 25 mm from each body; (*b*) 26 mm from lighter body, along an interconnecting line; (*c*) down; (*d*) $-1.6 \times 10^{-2}\ \mathrm{m/s^2}$
22. 13.6 ft
23. 8100 slug·ft/s, in the direction of motion
24. 24 km/h
25. (*a*) 52.0 km/h; (*b*) 28.8 km/h
26. 1.92×10^{-21} kg·m/s
27. a proton
28. (*a*) $(-4.0 \times 10^4\,\mathbf{i})$ kg·m/s; (*b*) west; (*c*) 0
29. (*a*) 30°; (*b*) $-0.572\mathbf{j}$ kg · m/s
30. (*a*) 7.5×10^4 J; (*b*) 3.8×10^4 kg·m/s, 30° south of east
31. (*a*) 6.4 J; (*b*) $P_i = 0.80$ kg·m/s, 30° above horizontal; $P_f = 0.80$ kg·m/s, 30° below horizontal
32. $0.707c$
33. 9.8×10^{-3} ft/s, backward
34. 0.57 m/s, toward center of mass
35. 4400 km/h
36. it increases by 4.4 m/s
37. $wv_{rel}/(W + w)$
38. (*a*) rocket case: 7290 m/s, payload: 8200 m/s; (*b*) before: 1.271×10^{10} J, after: 1.275×10^{10} J
39. 14 m/s, 135° from the other pieces
40. (*a*) 1.4×10^{-22} kg·m/s, 150° from the electron's velocity and 120° from the neutrino's velocity; (*b*) 1.0 eV
41. (*a*) 0.54 m/s; (*b*) 0; (*c*) 1.1 m/s

42. 190 m/s
43. (*a*) 721 m/s; (*b*) 937 m/s
44. one chunk stops, the other moves ahead with a speed of 4 m/s
45. (*a*) 0.200 v_{rel}; (*b*) 0.210 v_{rel}; (*c*) 0.209 v_{rel}
46. (*a*) 540 m/s; (*b*) 40.4°
47. (*a*) 8.0×10^4 N; (*b*) 27 kg/s
48. 108 m/s
49. (*a*) 1.57×10^6 N; (*b*) 1.35×10^5 kg; (*c*) 2.08 km/s
50. (*a*) 2.72; (*b*) 7.39
51. 2.2×10^{-2}
52. 28.8 N
54. (*a*) 50 kg/s; (*b*) 160 kg/s
55. 6.1 s
56. fast barge: 46 N more, slow barge: no change
57. (*a*) 2.3×10^4 N; (*b*) 4.2×10^6 W
58. (*a*) 100 m
59. 2.7 m/s
60. (*a*) 860 N; (*b*) 2.4 m/s
61. (*a*) −500 J; (*b*) 1700 N
62. (*a*) 1600 J; (*b*) 12.9 N; (*c*) 0.507 m/s^2; (*d*) 12.9 N
63. 4.9 kg · m/s
64. 4.8 m/s
65. 3.5 m/s
66. (*a*) 22 m; (*b*) 9..3 m/s

Chapter 10

1. 400 N·s
2. (*a*) 750 N; (*b*) 6.0 m/s
3. 2.5 m/s
4. 6.2×10^4 N
5. (*a*) $2mv/\Delta t$; (*b*) 580 N
6. 3000 N (= 660 lb)
7. 6400 lb
8. 1.1 m
9. 67 m/s
10. (*a*) 42 N·s; (*b*) 2100 N
11. (*a*) 2.3 N·s, in initial direction of flight; (*b*) 2.3 N·s, opposite initial direction of flight; (*c*) 1400 N, in initial direction of flight; (*d*) 58 J
12. (*a*) $(7.4 \times 10^3\,\mathbf{i} - 7.4 \times 10^4\,\mathbf{j})$ N·s; (*b*) $(-7.4 \times 10^4\,\mathbf{i})$ N·s; (*c*) 2.3×10^4 N; (*d*) 2.1×10^4 N; (*e*) −45°
13. 10 m/s
14. (*a*) 1.0 kg·m/s; (*b*) 250 J; (*c*) 10 N; (*d*) 1700 N
15. 216
16. 5 N
17. 29
18. (*a*) 0.48 g; (*b*) 7200 N
19. $2\mu v$
20. 1.50 N
21. 990 N
22. (*a*) 3.7 m/s; (*b*) 1.3 N·s; (*c*) 180 N
23. (*a*) 1.8 N·s, to the left; (*b*) 180 N, to the right
24. (*a*) 1.95×10^5 kg·m/s for each direction of thrust; (*b*) backward: −50.9 MJ, forward: +66.1 MJ, sideways: +7.61 MJ
26. 41.7 cm/s
27. 8 m/s
28. 38 km/s
29. (*a*) 1.9 m/s, to the right; (*b*) yes; (*c*) no, total kinetic energy would have increased
30. 4.2 m/s
31. 0.22%
32. 7.8%
33. (*a*) 99 g; (*b*) 1.9 m/s
34. 1.2 kg
35. (*a*) 2.47 m/s; (*b*) 1.23 m/s
36. 7.8 kg
37. 100 g
38. (*a*) 1/3; (*b*) $4h$
39. $m_1/3$
40. (*a*) 23 ft above top of shaft; (*b*) 23 ft below top of shaft
41. ≈ 2 mm/y
42. 3.0 m/s
43. 1.81 m/s
44. (*a*) $(10\,\mathbf{i} + 15\,\mathbf{j})$ m/s; (*b*) 500 J lost
45. 310 m/s
46. (*a*) +2.0 m/s; (*b*) −1.3 J; (*c*) +40 J; (*d*) got energy from some source, such as a small explosion
47. (*a*) 2.7 m/s; (*b*) 1400 m/s
48. (*a*) A: 4.6 m/s, B: 3.9 m/s; (*b*) 7.5 m/s
49. 190 tons
50. 20 J for the heavy particle, 40 J for the light particle
51. $mv^2/6$
52. 5.06 kg
53. 13 tons
54. (*a*) $mv_i/(m+M)$; (*b*) $M/(m+M)$
55. 25 cm

56. 0.975 m/s, 0.841 m/s
57. (*a*) 62.5 km/h; (*b*) 0.75
58. (*a*) 4.1 ft/s; (*b*) 1.8 ft·lb; (*c*) $v_{24} = 5.3$ ft/s, $v_{32} = 3.3$ ft/s
59. $\sqrt{2E(M+m)/mM}$
60. (*a*) 4.15×10^5 m/s; (*b*) 4.84×10^5 m/s
61. (*a*) 30° from the incoming proton's direction; (*b*) 250 m/s and 430 m/s
62. (*a*) $(-1.00\mathbf{i} + 0.668\mathbf{j}) \times 10^{-19}$ kg·m/s; (*b*) 1.19×10^{-12} J
63. (*a*) 41°; (*b*) 4.76 m/s; (*c*) no
64. 38 mi/h, 63° south of west
65. $v = V/4$
66. (*a*) 1010 m/s, 9.48° clockwise from the $+x$ axis; (*b*) 3.21 MJ
67. (*a*) 117° from the final direction of B; (*b*) no
69. 120°
70. $dm_1^2/(m_1+m_2)^2$
71. (*a*) 1.9 m/s, 30° to initial direction; (*b*) no
72. $\mathbf{v}_1$ and $\mathbf{v}_2$ will be at 30° to $\mathbf{v}_0$ and will have magnitude 6.9 m/s. $\mathbf{v}_1$ will be in the opposite direction to $\mathbf{v}_0$ and will have magnitude 2.0 m/s
73. (*a*) 3.4 m/s, deflected by 17° to the right; (*b*) 0.96 MJ
74. 8.12 MeV
75. (*a*) 117 MeV; (*b*) equal and opposite momenta; (*c*) π^-
76. (*a*) 2.08 MeV; (*b*) −1.18 MeV
77. (*a*) 4.94 MeV; (*b*) 0; (*c*) 4.85 MeV; (*d*) 0.09 MeV
78. (*a*) 4.4 m/s, toward the right; (*b*) 38 J
79. (*a*) 9.0 kg · m/s; (*b*) 3000 N; (*c*) 4500 N; (*d*) 20 m/s
80. 0.33 m
81. 2.6 m
82. (*a*) 2.5 m/s; (*b*) 42 J
83. 2.9×10^3 N
84. 5.0 kg
85. 8.1 m/s, 38° south of east

Chapter 11

1. (*a*) 1.50 rad; (*b*) 85.9°; (*c*) 1.49 m
2. (*a*) $a + 3bt^2 - 4ct^3$; (*b*) $6bt - 12ct^2$
3. (*a*) 5.5×10^{15} s; (*b*) 26
4. (*a*) $\omega(2) = 4.0$ rad/s, $\omega(4) = 28$ rad/s; (*b*) 12 rad/s²; (*c*) $\alpha(2) = 6.0$ rad/s², $\alpha(4) = 18$ rad/s²
5. (*a*) 2 rad; (*b*) 0; (*c*) 130 rad/s; (*d*) 32 rad/s²; (*e*) no
6. (*a*) $\omega_0 + at^4 - bt^3$; (*b*) $\theta_0 + \omega_0 t + at^5/5 - bt^4/4$
7. (*a*) 0.105 rad/s; (*b*) 1.75×10^{-3} rad/s; (*c*) 1.45×10^{-4} rad/s
8. 14 rev
9. 11 rad/s
10. (*a*) 4.0 m/s: (*b*) no
11. (*a*) 30 s; (*b*) 1800 rad
12. (*a*) −67 rev/min²; (*b*) 8.3 rev
13. (*a*) 9000 rev/min²; (*b*) 420 rev
14. 200 rev/min
15. (*a*) −1.25 rad/s²; (*b*) 250 rad; (*c*) 39.8 rev
16. (*a*) 2.0 rad/s²; (*b*) 5.0 rad/s; (*c*) 10 rad/s; (*d*) 75 rad
17. (*a*) 140 rad; (*b*) 14 s
18. (*a*) 2.0 rev/s; (*b*) 3.8 s
19. 8.0 s
20. (*a*) 13.5 s; (*b*) 27.0 rad/s
21. (*a*) 340 s; (*b*) -4.5×10^{-3} rad/s²; (*c*) 98 s
22. (*a*) 44 rad; (*b*) 5.5 s; (*c*) −2.1 s, 40 s
23. (*a*) 1.0 rev/s²; (*b*) 4.8 s; (*c*) 9.6 s; (*d*) 48 rev
24. 1500 rad
25. (*b*) -2.3×10^{-9} rad/s²; (*c*) 2600 y; (*d*) 24 ms
26. 6.1 ft/s² (1.8 m/s²), toward the center
27. (*a*) 3.5 rad/s; (*b*) 21 in./s; (*c*) 10 in./s
28. 0.13 rad/s
29. (*a*) 20.9 rad/s; (*b*) 12.5 m/s; (*c*) 800 rev/min²; (*d*) 600 rev
30. 5.6 rad/s²
31. (*a*) 2.0×10^{-7} rad/s; (*b*) 30 km/s; (*c*) 5.9 mm/s², toward the sun
32. (*a*) 3.0 rad/s; (*b*) 30 m/s; (*c*) 6.0 m/s²; (*d*) 90 m/s²
33. (*a*) 2.50×10^{-3} rad/s; (*b*) 20.2 m/s²; (*c*) 0
34. (*a*) $\omega_0 = \sqrt{\mu_s g/R}$
35. (*a*) 7.3×10^{-5} rad/s; (*b*) 350 m/s; (*c*) 7.3×10^{-5} rad/s, 460 m/s
36. (*a*) −1.1 rev/min²; (*b*) 9900 rev; (*c*) −0.99 mm/s²; (*d*) 31 m/s²
37. (*a*) 40.2 cm/s²; (*b*) 2.36×10^3 m/s²; (*c*) 83.2 m
38. (*a*) 310 m/s; (*b*) 340 m/s
39. (*a*) 3.8×10^3 rad/s; (*b*) 190 m/s
40. (*a*) 1.94 m/s²; (*b*) 75.1°, toward the center of the track
41. 16 s
42. (*a*) 150 cm/s; (*b*) 15 rad/s; (*c*) 15 rad/s;

(d) 75 cm/s; (e) 3.0 rad/s
43. (a) 73 cm/s^2; (b) 0.075; (c) 0.11
44. (a) 52.0 rad/s; (b) 22.4 rad/s; (c) no; (d) 5.38 km; (e) 1.15 h
45. 12.3 kg·m^2
46. 6.75×10^{12} rad/s
47. first cylinder: 1100 J, second cylinder: 9700 J
48. (a) 2.3×10^{-47} kg·m^2; (b) 1.3 meV
49. (a) 1300 g·cm^2; (b) 550 g·cm^2; (c) 1900 g·cm^2; (d) $A + B$
50. (a) 221 kg·m^2; (b) 1.10×10^4 J
51. (a) $5m\ell^2 + \frac{8}{3}M\ell^2$; ($b$) $(\frac{5}{2}m + \frac{4}{3}M)\ell^2\omega^2$
52. (a) 6490 kg·m^2; (b) 4.36 MJ
53. (a) 9.71×10^{37} kg·m^2; (b) 2.57×10^{29} J; (c) 1.9×10^9 y
54. 0.097 kg·m^2
57. $\frac{1}{3}M(a^2 + b^2)$
59. (a) 49 MJ; (b) 100 min
60. 4.6 N·m
61. 140 N·m
62. (a) 8.4 N·m; (b) 17 N·m; (d) 0
63. (a) $r_1F_1 \sin\theta_1 - r_2F_2 \sin\theta_2$; ($b$) −3.8 N·m
64. 12 N·m
65. 1.28 kg·m^2
66. (a) 28.2 rad/s^2; (b) 338 N·m
67. 9.7 rad/s^2, counterclockwise
68. (a) 0.791 kg·m^2; (b) 1.79×10^{-2} N·m
69. (a) 155 kg·m^2; (b) 64.4 kg
70. (a) 420 rad/s^2; (b) 500 rad/s
71. 130 N
72. small sphere: (a) 0.689 N·m; (b) 3.05 N large sphere: (a) 9.84 N·m; (b) 11.5 N
73. (a) 6.00 cm/s^2; (b) 4.87 N; (c) 4.54 N; (d) 1.20 rad/s^2; (e) 0.0138 kg·m^2
74. 1.73 m/s^2; 6.92 m/s^2
75. (a) $2\theta/t^2$; (b) $2R\theta/t^2$; (c) $T_1 = M(g - 2R\theta/t^2)$, $T_2 = Mg - (2\theta/t^2)(MR + I/R)$
76. (a) 1.9×10^{12} J/s; (b) -2.7×10^{-22} rad/s^2; (c) 4.1×10^9 N
77. (a) $3g(1 - \cos\theta)$; (b) $\frac{3}{2}g \sin\theta$; (c) 41.8°
78. (a) 1.4 m/s; (b) 1.4 m/s
79. 292 ft·lb (396 N·m)
80. (a) 19.8 kJ; (b) 1.32 kW
81. (a) $k\ell^2\omega^2/6$; (b) $\ell^2\omega^2/6g$
82. 5.42 m/s
83. $\sqrt{9g/4\ell}$
84. (a) 8.2×10^{28} N·m; (b) 2.6×10^{29} J; (c) 3.0×10^{21} kW
85. (a) 4.8×10^5 N; (b) 1.1×10^4 N·m; (c) 1.3×10^6 J
86. $\sqrt{2gh/(1 + 2M/3m + I/mr^2)}$
87. (a) −7.66 rad/s^2; (b) −11.7 N·m; (c) -4.60×10^4 J; (d) 624 rev; (e) the work done by friction, -4.60×10^4 J
88. (a) 42.1 km/h; (b) 3.09 rad/s^2; (c) 7.57 kW
89. (a) 2.0 kg · m^2; (b) 6.0 kg · m^2; (c) 2.0 kg · m^2
90. (a) 6.4 cm/s^2; (b) 2.6 cm/s^2
91. 0.054 kg · m^2
92. 18 rad
93. 3.2 kg · m^2
94. 2.6 J
95. (a) 5.6 rad/s^2; (b) 3.1 rad/s
96. (a) 0.15 kg · m^2; (b) 11 rad/s

Chapter 12

1. 1.00
2. −3.15 J
3. (a) 59.3 rad/s; (b) −9.31 rad/s^2; (c) 70.7 m
4. 1/50
5. (a) 990 J; (b) 3000 J; (c) 1.1×10^5 J
6. (a) 8.1°; (b) $0.14g$
7. (a) 44.8 ft·lb; (b) 11.2 ft; (c) no
8. (a) −4.11 m/s^2; (b) −16.4 rad/s^2; (c) −2.54 N·m
9. (a) 0, 0; (b) 22 m/s, 1500 m/s^2; (c) −22 m/s, 1500 m/s^2; (d) center: 22 m/s, 0; top: 44 m/s, 1500 m/s^2; bottom: 0, 1500 m/s^2
10. (a) $\frac{1}{2}mR^2$; (b) a solid circular cylinder
11. 48 m
12. (a) $mg(R - r)$; (b) 2/7; (c) $(17/7)mg$
13. (a) $2.7R$; (b) $(50/7)mg$
14. (a) 63 rad/s; (b) 4.0 m
15. (a) 1.13 s; (b) 13.6 m
16. (a) 13 cm/s^2; (b) 4.4 s; (c) 55 cm/s; (d) 1.8×10^{-2} J; (e) 1.4 J; (f) 27 rev/s
17. 70 rev/s
20. (a) 24 N·m, in $+y$ direction; (b) 24 N·m, $-y$; (c) 12 N·m, $+y$; (d) 12 N·m, $-y$
21. (a) 10 N·m, parallel to yz plane, at 53° to $+y$; (b) 22 N·m, $-x$
22. (a) $(-1.5\,\mathbf{i} - 4.0\,\mathbf{j} - \mathbf{k})$ N·m; (b) $(-1.5\,\mathbf{i} - 4.0\,\mathbf{j} - \mathbf{k})$ N·m
23. (a) $(6.0\,\mathbf{i} - 3.0\,\mathbf{j} - 6.0\,\mathbf{k})$ N·m; (b) $(26\,\mathbf{i} + 3.0\,\mathbf{j} - 18\,\mathbf{k})$ N·m; (c) $(32\,\mathbf{i} - 24\,\mathbf{k})$ N·m; (d) 0

24. $-2.0\,\mathbf{i}$ N·m
25. (*a*) $50\,\mathbf{k}$ N·m; (*b*) 90°
26. 1.3×10^8 kg·m²/s
27. (*a*) 9.8 kg·m²/s
28. (*a*) 12 kg·m²/s, out of page; (*b*) 3.0 N·m, out of page
30. (*a*) 0; (*b*) $(8.0\,\mathbf{i} + 8.0\,\mathbf{k})$ N·m
31. 2.5×10^{11} kg·m²/s
33. mvd, about any origin
34. (*a*) $600\,\mathbf{k}$ kg·m²/s; (*b*) $720\,\mathbf{k}$ kg·m²/s
35. (*a*) 3.15×10^{43} kg·m²/s; (*b*) 0.616
36. (*a*) $-170\,\mathbf{k}$ kg·m²/s; (*b*) $+56\,\mathbf{k}$ N·m; (*c*) $+56\,\mathbf{k}$ kg·m²/s²
37. 4.5 N·m, parallel to xy plane at −63° from $+x$
38. (*a*) 0; (*b*) $8t$ N·m, in $-z$ direction; (*c*) $2/\sqrt{t}$ N·m, $-z$; (*d*) $8/t^3$ N·m, $+z$
39. (*a*) 0; (*b*) 0; (*c*) $30t^3$ kg·m²/s, $90t^2$ N·m, both in $-z$ direction; (*d*) $30t^3$ kg·m²/s, $90t^2$ N·m, both in $+z$ direction
40. (*a*) $24(1 - t^4)\,\mathbf{k}$ kg·m²/s; (*b*) $-96t^3\,\mathbf{k}$ N·m; (*c*) $12(3 + t^4)\,\mathbf{k}$ kg·m²/s, $48t^3\,\mathbf{k}$ N·m
41. (*a*) $\frac{1}{2}mgt^2 v_0 \cos\theta_0$; (*b*) $mgtv_0 \cos\theta_0$; (*c*) $mgtv_0 \cos\theta_0$
42. (*a*) 0.53 kg·m²/s; (*b*) 4200 rev/min
43. (*a*) −1.47 N·m; (*b*) 20.4 rad; (*c*) −29.9 J; (*d*) 19.9 W
44. (*a*) $14m\ell^2$; (*b*) $4m\ell^2\omega$; (*c*) $14m\ell^2\omega$
45. (*a*) 12.2 kg·m²; (*b*) 308 kg·m²/s, down
46. (*a*) 1/3; (*b*) 1/9
48. $\omega_0 \left[\frac{R_1 I_2}{R_2 I_1} + \frac{R_2}{R_1}\right]^{-1}$
49. (*a*) 1.2 s; (*b*) 8.6 s; (*c*) 5.2 rev; (*d*) 6.1 m/s; (*e*) no
50. 500 rev
51. (*a*) 3.6 rev/s; (*b*) 3.0; (*c*) work done by man in moving weights inward
52. (*a*) 750 rev/min; (*b*) 450 rev/min, in the original direction of the second disk
53. (*a*) 267 rev/min; (*b*) 2/3
54. 3
55. 3.0 min
56. (*a*) 149 kg·m²; (*b*) 158 kg·m²/s; (*c*) 0.746 rad/s
57. 12.7 rad/s, clockwise as seen from above
58. (*a*) they revolve in a circle of 1.5-m radius at 0.93 rad/s; (*b*) 8.4 rad/s; (*c*) $K_a = 98$ J, $K_b = 880$ J; (*d*) from the work done in pulling inward
59. (*a*) $(7/12)ML^2$; (*b*) $(7/12)ML^2\omega_0$, down; (*c*) $(14/5)\omega_0$; (*d*) $(21/40)mL^2\omega_0^2$
60. $\frac{m}{M+m}\left(\frac{v}{R}\right)$
61. (*a*) $(mRv - I\omega_0)/(I + mR^2)$; (*b*) no, energy transferred to internal energy of cockroach
62. (*a*) $mvR/(I + MR^2)$; (*b*) $mvR^2/(I + MR^2)$
63. the day would be longer by about 0.8 s
64. $\theta = \cos^{-1}\left[1 - \frac{6m^2 h}{\ell(2m+M)(3m+M)}\right]$
65. (*a*) 0.148 rad/s; (*b*) 0.0123; (*c*) 181°
66. (*a*) 0.33 rev/s, clockwise as seen from above
67. 0.43 rev/min
68. 1300 m/s
69. 2.6 rad/s
70. (*a*) 4.0 N, to the left; (*b*) 0.60 kg · m²
71. $7.4\,\mathbf{k}$ kg · m²/s
72. 3.4 rad/s
73. (*a*) 7.4 J; (*b*) 2.9 m/s; (*c*) 5.3 J; (*d*) 1.5 m/s

Chapter 13

1. (*a*) two; (*b*) seven
2. (*a*) yes; (*b*) no; (*c*) no
3. 8.7 N
4. (*a*) 2.5 m; (*b*) 7.3°
5. (*a*) $(-27\,\mathbf{i} + 2\,\mathbf{j})$ N; (*b*) 176° counterclockwise from $+x$ direction
6. 120°
7. 7920 N
8. 1900 lb
9. 0.29
10. (*a*) $(W/L)\sqrt{L^2 + r^2}$; (*b*) Wr/L
11. (*a*) 2770 N; (*b*) 3890 N
12. (*a*) 340 lb; (*b*) 420 lb
13. (*a*) 1160 N, down; (*b*) 1740 N, up; (*c*) left, stretched; (*d*) right, compressed
14. 74 g
15. one-fourth the beam length from the free end
16. (*a*) 280 N; (*b*) 880 N, 71° above the horizontal
17. (*a*) $3W$; (*b*) $4W$
18. 0.536 m
19. (*a*) bottom: $2W$, sides: W; (*b*) $\sqrt{2}W$
20. (*a*) 1800 lb; (*b*) 822 lb; (*c*) 1270 lb
21. (*a*) 49 N; (*b*) 28 N; (*c*) 57 N; (*d*) 29°
22. 16 lb

24. (*a*) 1900 N, up; (*b*) 2100 N, down
25. (*a*) 408 N; (*b*) $F_h = 245$ N (right); (*c*) $F_v =$ 163 N (up)
26. (*a*) 340 N; (*b*) 0.88 m; (*c*) increases, decreases
27. $W\dfrac{\sqrt{2rh - h^2}}{r - h}$
28. (*a*) 5.0 N; (*b*) 30 N; (*c*) 1.3 m
29. (*a*) 42 N; (*b*) 66 N
30. (*a*) $L/2$; (*b*) $L/4$; (*c*) $L/6$; (*d*) $L/8$; (*e*) $25L/24$
31. (*a*) 43.3 lb; (*b*) 21.7 lb; (*c*) 12.5 lb
32. (*a*) 130 N; (*b*) 80 N, away from door at top, toward door at bottom
33. (*a*) 6630 N; (*b*) $F_h = 5740$ N; (*c*) $F_v = 5960$ N
34. 2.20 m
35. (*a*) $Wx/(L \sin\theta)$; (*b*) $Wx/(L \tan\theta)$; (*c*) $W(1 - x/L)$
36. (*a*) 1.50 m; (*b*) 433 N; (*c*) 250 N
37. (*a*) −180 lb, 60 lb; (*b*) 180 lb, 60 lb; (*c*) 180 lb, 210 lb; (*d*) −180 lb, −60 lb
38. (*a*) $a_1 = L/2$, $a_2 = 5L/8$, $h = 9L/8$; (*b*) $b_1 = 2L/3$, $b_2 = L/2$, $h = 7L/6$
39. 0.34
40. (*a*) 47 lb; (*b*) 120 lb; (*c*) 72 lb
41. bars BC, CD, and DA are in tension due to forces T, diagonals AC and BD are in compression due to forces $\sqrt{2}T$
42. (*a*) $\mu < L/2h$; (*b*) $\mu > L/2h$
43. (*a*) 445 N; (*b*) 0.50; (*c*) 315 N
44. (*a*) slides at 31°; (*b*) tips at 34°
45. (*a*) 3.9 m/s^2; (*b*) 2000 N on each rear wheel, 3500 N on each front wheel; (*c*) 790 N on each rear wheel, 1410 N on each front wheel
46. (*a*) 7.5×10^{10} N/m^2; (*b*) 2.9×10^8 N/m^2
47. (*a*) 1.9×10^{-3}; (*b*) 1.3×10^7 N/m^2; (*c*) 6.9×10^9 N/m^2
48. (*a*) 0.80 mm; (*b*) 2.4 cm
49. 3.1 cm
50. (*a*) 6.5×10^6 N/m^2; (*b*) 1.1×10^{-5} m
51. 2.4×10^9 N/m^2
52. (*a*) 1.4×10^9 N; (*b*) 75
53. (*a*) 1.8×10^7 N; (*b*) 1.4×10^7 N; (*c*) 16
54. (*a*) 4/5; (*b*) 1/5; (*c*) 1/4
55. (*a*) 867 N; (*b*) 143 N; (*c*) 0.165
56. (*a*) $RFr_A/(r_A^2 + r_B^2)$; (*b*) $RFr_B/(r_A^2 + r_B^2)$
57. (*a*) 670 N; (*b*) 400 N to the right, 670 N up
58. (*a*) 840 N; (*b*) 530 N
59. (*a*) 88 N; (*b*) 30 N to the right, 97 N up
60. (*a*) $(35\,\mathbf{i} + 200\,\mathbf{j})$ N; (*b*) $(-45\,\mathbf{i} + 200\,\mathbf{j})$ N; (*c*) 190 N
61. (*a*) 15 N; (*b*) 29 N

Chapter 14

1. (*a*) 0.50 s (*b*) 2.0 Hz; (*c*) 18 cm
2. (*a*) 0.75 s; (*b*) 1.3 Hz; (*c*) 8.4 rad/s
3. (*a*) 245 N/m; (*b*) 0.284 s
4. (*a*) 0.50 s; (*b*) 2.00 Hz; (*c*) 12.6 rad/s; (*d*) 79.0 N/m; (*e*) 4.40 m/s; (*f*) 27.6 N
5. 708 N/m
6. 37.8 m/s^2
7. $f > 500$ Hz
8. (*a*) 96.0 lb/ft; (*b*) 19.5 lb
9. (*a*) 100 N/m; (*b*) 0.45 s
10. (*a*) 7900 dyne/cm; (*b*) 1.19 cm; (*c*) 2.00 Hz
11. (*a*) 6.28×10^5 rad/s; (*b*) 1.59 mm
12. (*a*) 10 N; (*b*) 120 N/m
13. (*a*) 1.0 mm; (*b*) 0.75 m/s; (*c*) 570 m/s^2
14. (*a*) 2800 rad/s; (*b*) 2.1 m/s; (*c*) 5.7 km/s^2
15. (*a*) 1.29×10^5 N/m; (*b*) 2.68 Hz
16. (*a*) 3.0 m; (*b*) −49 m/s; (*c*) −270 m/s^2; (*d*) 20 rad; (*e*) 1.5 Hz; (*f*) 0.67 s
17. (*a*) 4.0 s; (*b*) $\pi/2$ rad/s; (*c*) 0.37 cm; (*d*) (0.37 cm) $\cos \frac{\pi}{2}t$; (*e*) (−0.58 cm/s) $\sin \frac{\pi}{2}t$; (*f*) 0.58 cm/s; (*g*) 0.91 cm/s^2; (*h*) 0; (*i*) 0.58 cm/s
18. 7.2 m/s
19. (*b*) 12.47 kg; (*c*) 54.43 kg
20. (*a*) 147 N/m; (*b*) 0.733 s
21. 1.6 kg
22. (*a*) 1.6×10^4 m/s^2; (*b*) 2.5 m/s; (*c*) 7.9×10^3 m/s^2; (*d*) 2.2 m/s
23. (*a*) 1.6 Hz; (*b*) 1.0 m/s, 0; (*c*) 10 m/s^2, ±10 cm; (*d*) $(-10\text{ N/m})x$
24. 2.08 h
25. 22 cm
26. 3.1 cm
27. (*a*) 25 cm; (*b*) 2.2 Hz
28. (*a*) 5.58 Hz; (*b*) 0.325 kg; (*c*) 0.400 m
29. (*a*) 0.500 m; (*b*) −0.251 m; (*c*) 3.06 m/s
30. (*a*) 2.2 Hz; (*b*) 56 cm/s; (*c*) 0.10 kg; (*d*) 20.0 cm below y_i
31. (*a*) $0.183A$; (*b*) same direction
32. $2\pi/3$ rad
33. (*a*) 0.525 m; (*b*) 0.686 s
37. (*a*) $k_1 = (n+1)k/n$, $k_2 = (n+1)k$;

(b) $f_1 = \sqrt{(n+1)/n}f$, $f_2 = \sqrt{n+1}f$
38. (a) 1.1 Hz; (b) 5.0 cm
39. 3.7×10^{-2} J
40. (a) 200 N/m; (b) 1.39 kg; (c) 1.91 Hz
41. (a) 130 N/m; (b) 0.62 s; (c) 1.6 Hz; (d) 5.0 cm; (e) -0.44 J, relative to the unstretched position; (f) 0.51 m/s
42. (a) 2.25 Hz; (b) 125 J; (c) 250 J; (d) 86.6 cm
43. (a) 7.25×10^6 N/m; (b) 49,400
44. (a) 3/4; (b) 1/4; (c) $x_m/\sqrt{2}$
45. (a) $mv/(m+M)$; (b) $mv/\sqrt{k(m+M)}$
46. (a) 3.5 m; (b) 0.75 s
47. (a) $-(80\text{ N})\cos[(2000\text{ rad/s})t - \pi/3\text{ rad}]$; ($b$) 3.1 ms; ($c$) 4.0 m/s; ($d$) 0.080 J
48. (a) 0.21 m; (b) 1.6 Hz; (c) 0.10 m
49. (a) 16.7 cm; (b) 1.23%
50. (a) 0.0625 J; (b) 0.03125 J
51. (a) 0.735 kg·m^2; (b) 0.024 N·m; (c) 0.181 rad/s
52. 12 s
53. 0.079 kg·m^2
54. (a) 39.5 rad/s; (b) 34.2 rad/s; (c) 124 rad/s^2
55. 9.79 in.
56. (a) 8.3 s; (b) no
57. 99 cm
58. 9.47 m/s^2
60. 8.77 s
61. $T/\sqrt{2}$
62. (a) $2\pi\sqrt{\dfrac{L^2+12d^2}{12gd}}$; ($b$) increases for $d < L/\sqrt{12}$, decreases for $d > L/\sqrt{12}$; (c) increases; (d) no change
63. 5.6 cm
64. $2\pi\sqrt{\dfrac{R^2+2d^2}{2gd}}$
65. (a) 0.869 s; (b) $r = R/2$
66. (a) 0.205 kg·m^2; (b) 47.7 cm; (c) 1.50 s
68. $\sqrt{2}f_0$
69. (a) $2\pi\sqrt{\dfrac{L^2+12x^2}{12gx}}$; ($b$) 0.289
70. (a) F/m; (b) $2F/mL_0$; (c) F/m
71. (a) 0.35 Hz; (b) 0.39 Hz; (c) 0
72. $\dfrac{1}{2\pi}\left(\dfrac{\sqrt{g^2+f^4/R^2}}{L}\right)^{1/2}$
73. 14.0°
74. (b) smaller
75. $2\pi\sqrt{m/3k}$
76. (a) $(r/R)\sqrt{k/m}$; (b) $\sqrt{k/m}$; (c) no oscillation
77. (a) 2.0 s; (b) 19.8 N·m/rad
79. $0.29L$
80. 6.0%
81. 0.39
82. (a) 14.3 s; (b) 5.27
83. (a) 0.102 kg/s; (b) 0.137 J
84. (a) 0.015; (b) no
85. $k = 490$ N/cm, $b = 1100$ kg/s
86. (a) $F_m/b\omega_0$; (b) F_m/b
87. 1.9 in.
89. (a) 0.20 s; (b) 0.20 kg; (c) -0.20 m; (d) -200 m/s^2; (e) 4.0 J
90. (a) 0.30 m; (b) 0.28 s; (c) 150 m/s^2; (d) 11.3 J
91. (a) 10 N, up; (b) 0.10 m; (c) 0.90 s; (d) 0.50 J

Chapter 15

1. 19 m
2. (a) 3×10^{-8} N; (b) 2×10^{-6} N; (c) 1×10^{-6} N; (d) no, the force due to the planet is 30 – 80 times greater
3. 2.16
4. 29 pN
5. 1/2
6. 2.60×10^5 km
7. 3.4×10^5 km
8. 1.1%
9. (a) 3.7×10^{-5} N, increasing y
10. 0.017 N, toward the 300-kg sphere
11. $M = m$
12. 4.4×10^{-6} N, perpendicular and toward the line between m_1 and m_2
13. 3.2×10^{-7} N
14. (a) $2GMm/\pi R^2$; (b) 0
15. $\dfrac{GmM}{d^2}\left(1 - \dfrac{1}{8(1-R/2d)^2}\right)$
16. 0.997 s
17. (a) 1.62 m/s^2; (b) 4.9 s
18. 2.6×10^6 m
19. 0.016 lb
20. (a) 1.3×10^{12} m/s; (b) 1.6×10^6 m/s
21. (a) 3.4×10^{-3} g; (b) 6.1×10^{-4} g; (c) 1.4×10^{-11} g
22. (a) 17 N; (b) 2.5
23. 9.78 m/s^2

24. (*a*) 0.05%; (*b*) 1.3 s
25. (*b*) 1.9 h
26. 4.7×10^{24} kg
27. (*b*) 3.2 m
28. 6.6×10^{-5}, due to ship's slightly different centripetal accelerations
30. (*b*) 250 m, 47 m; (*c*) 290 m, 7.0 m
31. (*a*) $G(M_1 + M_2)m/a^2$; (*b*) GM_1m/b^2; (*c*) 0
32. 7.91 km/s
34. (*a*) 9.83 m/s²; (*b*) 9.84 m/s²; (*c*) 9.79 m/s²
35. (*b*) 2.0×10^8 N/m²; (*c*) 360 km
36. (*b*) 42 min
37. (*a*) -4.4×10^{-11} J; (*b*) -2.9×10^{-11} J; (*c*) 2.9×10^{-11} J
38. (*a*) -1.4×10^{-4} J; (*b*) less; (*c*) positive; (*d*) negative
39. 1/2
40. (*a*) 0.74; (*b*) 3.8 m/s²; (*c*) 5.1 km/s
41. 220 km/s
42. (*a*) 0.0451; (*b*) 28.5
44. $-Gm(M_e/R + M_m/r)$
46. (*a*) 5.0×10^{-11} J; (*b*) -5.0×10^{-11} J
48. (*a*) 1700 m/s; (*b*) 250 km; (*c*) 1400 m/s
49. 2.6×10^4 km
50. (*a*) 2.2×10^{-7} rad/s; (*b*) 89.5 km/s
51. (*a*) 82 km/s; (*b*) 1.8×10^4 km/s
53. (*a*) $\dfrac{GMmx}{(x^2 + R^2)^{3/2}}$; (*b*) $v^2 = 2GM\left(\dfrac{1}{R} - \dfrac{1}{\sqrt{R^2 + x^2}}\right)$
55. 1.87 y
56. 6.5×10^{23} kg
57. 5.93×10^{24} kg
58. 5×10^{10}
59. 0.35 lunar months
60. (*a*) 7.82 km/s; (*b*) 87.5 min
61. 3.9 y
62. (*a*) 6640 km; (*b*) 0.0136
63. 5.01×10^9 m or 7.20 solar radii
64. (*a*) 39.5 AU³/M_S·y²; (*b*) $T^2 = r^3$
65. 3.58×10^4 km
66. every 12 hours
67. 81°
68. south, at 35.4° above the horizon
70. 0.71 y
72. $2\pi r^{3/2}/\sqrt{G(M + m/4)}$
73. (*a*) 42.1 km/s; (*b*) 12.3 km/s; (*c*) 16.6 km/s
74. $\sqrt{GM/L}$
75. 1.6 cm/s, to the west along the equator
76. (*a*) 2.8 y; (*b*) 1.0×10^{-4}
77. (*a*) $-GM_em/r$; (*b*) $-2GM_em/r$; (*c*) it falls radially to earth
78. (*a*) 1/2; (*b*) 1/2; (*c*) B, by 8.3×10^7 ft·lb
80. (*a*) 54 km/s; (*b*) 960 m/s; (*c*) $R_p/R_a = v_a/v_p$
81. (*a*) no; (*b*) same; (*c*) yes
82. (*a*) 4.6×10^5 J; (*b*) 260
83. (*a*) $T \propto r^{3/2}$; (*b*) $K \propto r^{-1}$; (*c*) $L \propto r^{1/2}$; (*d*) $v \propto r^{-1/2}$
84. (*a*) 7.5 km/s; (*b*) 97 min; (*c*) 410 km; (*d*) 7.7 km/s; (*e*) 92 min; (*f*) 3.2×10^{-3} N; (*g*) if the satellite-Earth system is considered isolated, its **L** is conserved
85. (*a*) 0; (*b*) 1.8×10^{32} J; (*c*) 1.8×10^{32} J; (*d*) 0.99 km/s
86. $2R$
87. (*a*) 2.2×10^7 J; (*b*) 6.9×10^7 J
88. (*a*) 8.0×10^8 J; (*b*) 36 N
89. 2.1×10^{-8} N
90. (*a*) -1.67×10^{-8} J; (*b*) 0.56×10^{-8} J
91. (*a*) $(3.0 \times 10^{-7}m)$ N; (*b*) $(3.3 \times 10^{-7}m)$ N; (*c*) $(6.7 \times 10^{-7}mr)$ N

Chapter 16

1. 1000 kg/m³
2. 18 N
3. 1.1×10^5 Pa or 1.1 atm
4. (*a*) 190 kPa; (*b*) 159/106
5. 2.9×10^4 N
6. 0.074
7. 6.0 lb/in.²
8. (*b*) 6000 lb
9. 1.90×10^4 Pa (143 mm Hg)
10. 1.62×10^6 Pa
11. 5.4×10^4 Pa
12. 130 km
13. 0.52 m
14. (*a*) 600, 30, 80 tons
15. (*a*) 6.06×10^9 N; (*b*) 20 atm
16. 7.2×10^5 N
17. 0.412 cm
18. $\frac{1}{4}\rho gA(h_2 - h_1)^2$
19. 2.0
21. 44 km
22. 1.7 km
23. (*a*) $\rho gWD^2/2$; (*b*) $\rho gWD^3/6$; (*c*) $D/3$
24. (*a*) 2.2; (*b*) 3.6

26. 10.3 m
27. -3.9×10^{-3} atm
28. (*a*) 8.0 km; (*b*) 16 km
29. (*a*) fA/a; (*b*) 20 lb
30. 0.21 in.
31. 1070 g
32. (*a*) 8000 lb; (*b*) decreases by 11.6 ft^3
33. 1.5 g/cm^3
34. (*a*) 2.04×10^{-2} m^3; (*b*) 1570 N
35. 600 kg/m^3
36. (*a*) 8720 lb; (*b*) 9240 lb; (*c*) 480 lb; (*d*) 512 lb
37. (*a*) 670 kg/m^3; (*b*) 740 kg/m^3
38. 4.9×10^6 N
39. 390 kg
40. 1800 m^3
41. (*a*) 1.2 kg; (*b*) 1300 kg/m^3
42. 57.3 cm
43. 0.126 m^3
44. (*a*) 45 m^2; (*b*) car should be over center of slab if slab is to be level
45. five
46. $0.12\left(\frac{1}{\rho} - \frac{1}{8}\right)$, with ρ in g/cm^3
47. (*a*) 1.80 m^3; (*b*) 4.75 m^3
48. (*a*) 1.8 kg; (*b*) 2.0 kg
49. 2.79 g/cm^3
50. 3.82 m/s^2
51. (*b*) 3.17 s
53. 4.0 m
54. (*a*) 15 gal/min; (*b*) 95%
55. 28 ft/s
56. 66 W
57. 43 cm/s
58. (*a*) 2.5 m/s; (*b*) 2.6×10^5 Pa
59. (*a*) 2.40 m/s; (*b*) 245 Pa
60. 254 $lb/in.^2$
61. (*a*) 12 ft/s; (*b*) 13 $lb/in.^2$
62. 1.4×10^5 N·m
63. 0.72 $ft\cdot lb/ft^3$
64. (*a*) 5.5×10^{-2} ft^3/s; (*b*) 3.0 ft
65. (*a*) 2; (*b*) $R_1/R_2 = 1/2$; (*c*) drain it until $h_2 = h_1/4$
67. 116 m/s
68. 862 kg; lift is (*a*) same as, (*b*) more than, and (*c*) less than the weight
70. 63.3 m/s
71. 110 m/s
72. (*a*) 560 Pa; (*b*) 5.0×10^4 N
73. 1.11×10^4 N
74. 40 m/s
75. $\frac{1}{2}\rho v^2 A$, where ρ is the density of air
76. 31 m/s
77. 1.4 cm
78. (*b*) $H - h$; (*c*) $H/2$
79. (*a*) 74 N; (*b*) 150 m^3
80. (*a*) $\sqrt{2g(h_2 + d)}$; (*b*) $p_{atm} - \rho g(h_2 + d + h_1)$; (*c*) 10.3 m
82. 2.4 ft^3/s
83. (*a*) $v = 4.1$ m/s, $v' = 21$ m/s; (*b*) 8.0×10^{-3} m^3/s
84. 5 min 42 s
85. (*a*) 830 W; (*b*) 1100 W
86. (*a*) 6.4 m^3; (*b*) 5.4 m/s; (*c*) 9.8×10^4 Pa
87. 1.5 cm
88. (*a*) 9.4 N; (*b*) 1.6 N
89. (*a*) 5.0×10^6 N; (*b*) 5.6×10^6 N
90. (*a*) 3.1 m/s; (*b*) 9.5 m/s

Chapter 17

1. (*a*) 75 Hz; (*b*) 13 ms
2. (*a*) 3.49 m^{-1}; (*b*) 31.5 m/s
3. (*a*) 1.7 s; (*b*) 2.0 m/s; (*c*) 3.3 m; (*d*) 15 cm
4. (*a*) 400 THz to 800 THz; (*b*) 1.0 m to 200 m; (*c*) 6.0×10^4 THz to 3.0×10^7 THz
5. (*a*) 0.68 s; (*b*) 1.47 Hz; (*c*) 2.06 m/s
6. $y = 0.010 \sin \pi(3.33x + 1100t)$, with x and y in meters and t in seconds
7. (*c*) 200 cm/s, negative x direction
9. 5 cm/s
11. (*a*) 2.0 mm, 96 Hz, 30 m/s, 31 cm; (*b*) 1.2 m/s
12. (*a*) $z = 3.0 \sin(60y - 10\pi t)$, with z in mm, y in cm, and t in s; (*b*) 9.4 cm/s
13. (*a*) 6.0 cm; (*b*) 100 cm; (*c*) 2.0 Hz; (*d*) 200 cm/s; (*e*) $-x$ direction; (*f*) 75 cm/s; (*g*) -2.0 cm
14. (*a*) $y = 2.0 \sin 2\pi(0.10x - 400t)$, with x and y in cm and t in s; (*b*) 50 m/s; (*c*) 40 m/s
16. (*a*) 11.7 cm; (*b*) $\pi/2$ rad
17. 129 m/s
18. 3.2
19. 135 N
20. $\sqrt{2}$
22. (*a*) 30 m/s; (*b*) 17 g/m
23. (*a*) 15 m/s; (*b*) 0.036 N
24. 300 m/s

25. $y = 0.12\sin(141x - 628t)$, with y in mm, x in m, and t in s
26. $2\pi y_m/\lambda$
27. (*a*) 5.0 cm ; (*b*) 40 cm; (*c*) 12 m/s; (*d*) 0.033 s; (*e*) 9.4 m/s; (*f*) $5.0\sin(16x + 190t + 0.79)$, with x in m, y in cm, and t in s
28. (*a*) 0.64 Hz; (*b*) 2π cm; (*c*) $y = 5\sin(0.1x - 4.0t)$, with x and y in cm and t in s; (*d*) 6400 dyne
29. (*a*) $v_1 = 28.6$ m/s, $v_2 = 22.1$ m/s; (*b*) $M_1 = 188$ g, $M_2 = 313$ g
30. 2.63 m from the end of the wire from which the later pulse originates
31. (*a*) $\sqrt{k(\Delta\ell)(\ell + \Delta\ell)/m}$
33. (*a*) $P_2 = 2P_1$; (*b*) $P_2 = P_1/4$
34. 45 Hz
35. (*a*) 3.77 m/s; (*b*) 12.3 N; (*c*) zero; (*d*) 44.3 W; (*e*) zero; (*f*) zero; (*g*) 0.50 cm
36. $1.4y_m$
37. 82.8°, 1.45 rad
38. $\lambda = 2\sqrt{d^2 + 4(H + h)^2} - 2\sqrt{d^2 + 4H^2}$
41. 5.0 cm
42. (*a*) $2f_3$; (*b*) λ_3
43. 10 cm
44. 880 Hz and 1320 Hz
45. (*a*) 82.0 m/s; (*b*) 16.8 m; (*c*) 4.88 Hz
46. (*a*) 140 m/s; (*b*) 60 cm; (*c*) 240 Hz
47. (*a*) −0.0390 m; (*b*) $y = 0.15\sin(0.79x + 13t)$; (*c*) −0.14 m
48. 240 cm, 120 cm, 80 cm
49. (*a*) 66.1 m/s; (*b*) 26.4 Hz
50. 7.91 Hz, 15.8 Hz, 23.7 Hz
51. (*a*) 144 m/s; (*b*) 3.00 m, 1.50 m; (*c*) 48.0 Hz, 96.0 Hz
52. 480 cm, 160 cm, 96 cm
53. first and second harmonics of A match fourth and eighth harmonics of B, respectively
54. (*a*) 2.0 Hz, 200 cm, 400 cm/s; (*b*) $x = 50$ cm, 150 cm, 250 cm, etc.; (*c*) $x = 0$, 100 cm, 200 cm, etc.
55. (*a*) 0.25 cm, 120 cm/s; (*b*) 3.0 cm; (*c*) zero
56. (*a*) 105 Hz; (*b*) 158 m/s
58. (*b*) the energy is entirely kinetic energy, of the transversely moving sections of the flat string
60. (*a*) 50 Hz; (*b*) $y = 0.50\sin[\pi(x \pm 100t)]$, with x in m, y in cm, and t in s
61. (*a*) 1.3 m; (*b*) $y = 0.002\sin(9.4x)\cos(3800t)$, with x and y in m and t in s
62. (*a*) 36 N
64. (*a*) 323 Hz; (*b*) eight
65. (*a*) 4.0 m; (*b*) 24 m/s; (*c*) 1.4 kg; (*d*) 0.11 s
66. (*b*) 2.0 cm/s; (*c*) $y = (4.0\text{ cm})\sin(\pi x/10 - \pi t/5 + \pi)$, where x is in cm and t is in s; (*d*) −2.5 cm/s
67. (*a*) 0.50 m; (*b*) 0, 0.25, 0.50 s

Chapter 18

1. (*a*) $\approx 6\%$
2. 170 m
3. The radio listener by about 0.85 s
4. (*a*) 79 m, 41 m; (*b*) 89 m
5. 7.9×10^{10} Pa
6. 0.144 MPa
7. If only the length is uncertain, it must be known to within 10^{-4} cm. If only the time is imprecise, the uncertainty must be no more than one part in 10^5.
8. (*a*) $L(V - v)/Vv$; (*b*) 364 m
9. 43.5 m
10. 1900 km
11. 40.7 m
12. 17 m and 1.7 cm, respectively
13. 100 kHz
14. (*a*) 0.0762 mm; (*b*) 0.333 mm
15. (*a*) 2.29, 0.229, 22.9 kHz; (*b*) 1.14, 0.114, 11.4 kHz
16. (*a*) 57 nm; (*b*) ≈ 35
17. (*a*) 6.0 m/s; (*b*) $y = 0.30\sin(\pi x/12 + 50\pi t)$, with x and y in cm and t in s
18. (*a*) 1.5 Pa; (*b*) 165 Hz; (*c*) 2.00 m; (*d*) 330 m/s
19. 4.12 rad
20. $2.00p_m$, $1.41p_m$, $1.73p_m$, $1.85p_m$
21. (*a*) $343(1+2m)$ Hz, with m being an integer from 0 to 28; (*b*) $686m$ Hz, with m being an integer from 1 to 29
22. (*a*) 141 Hz, 422 Hz, 703 Hz; (*b*) 281 Hz, 562 Hz, 844 Hz
23. (*a*) eight; (*b*) eight
24. 17.5 cm
25. 64.7 Hz, 129 Hz
26. at 0, ±0.572 m, ±1.14 m, ±1.72 m, ±2.29 m from the midpoint

27. (*a*) 0.080 W/m^2; (*b*) 0.013 W/m^2
28. 15.0 mW
29. 36.8 nm
30. 1.26
31. (*a*) 1000; (*b*) 32
32. yes: assuming a point source, the power is 18 W
33. (*a*) 39.7 μW/m^2; (*b*) 171 nm; (*c*) 0.893 Pa
34. (*a*) 8.84 nW/m^2; (*b*) 39.5 dB
35. (*a*) 59.7; (*b*) 2.81×10^{-4}
36. (*b*) 5.76×10^{-17} J/m^3
37. (*a*) $I \propto r^{-1}$; (*b*) $s_m \propto r^{-1/2}$
38. (*b*) length^2
39. (*a*) 5000; (*b*) 71; (*c*) 71
40. 1 cm
41. 171 m
42. (*a*) 77.8 dB; (*b*) 4.48 nW
43. 3.16 km
44. (*a*) 88 mW/m^2; (*b*) $A_4 = 0.75A_3$
45. (*a*) 5200 Hz; (*b*) $\text{Amplitude}_{\text{SAD}}/\text{Amplitude}_{\text{SBD}} = 2$
46. (*a*) phases differ by 30°; (*b*) 39 μW/m^2; (*c*) 16 μW/m^2; (*d*) 6.0 μW/m^2
47. 350 m
48. (*a*) 34.3 cm; (*b*) 3.60 μm; (*c*) 2.26 cm/s; (*d*) 17.2 cm
50. 20 kHz
51. (*a*) 833 Hz; (*b*) 0.418 m
52. (*a*) 405 m/s; (*b*) 596 N; (*c*) 44.0 cm; (*d*) 37.3 cm
53. by a factor of four
54. (*a*) 57.2 cm; (*b*) 42.9 cm
55. water filled to a height of 7/8, 5/8, 3/8, 1/8 meter
56. (*a*) 5.0 cm from one end; (*b*) 1.2; (*c*) 1.2
57. (*a*) $L(1 - 1/r)$; (*b*) 13 cm; (*c*) 5/6
58. (*a*) 1130, 1500, and 1880 Hz
59. 12.4 m
60. 230 Hz
61. (*a*) 71.5 Hz; (*b*) 64.8 N
62. (*a*) node; (*c*) 22 s
63. 45.3 N
64. 387 Hz
65. 2.25 ms
66. 505, 507, 508 Hz or 501, 503, 508 Hz
67. 0.02
68. (*a*) ten; (*b*) four
69. 3.8 Hz
70. zero
71. (*a*) 380 mi/h, away from owner; (*b*) 77 mi/h, away from owner
72. 17.5 kHz
73. 15.1 ft/s
74. (*a*) 526 Hz; (*b*) 555 Hz
75. 3.1 m/s
76. 500 m/s
77. 2.6×10^8 m/s
78. 30°
79. (*a*) 77.6 Hz; (*b*) 77.0 Hz
80. 33.0 km
81. (*a*) 42°; (*b*) 11 s
83. 0.189 MHz
84. 155 Hz
85. (*a*) 970 Hz; (*b*) 1030 Hz; (*c*) 60 Hz, no
86. (*a*) 467 Hz; (*b*) 494 Hz
87. (*a*) 1.02 kHz; (*b*) 1.04 kHz
88. (*a*) 0.818 ft; (*b*) 1580; (*c*) 1080 ft/s; (*d*) 0.683 ft; (*e*) 2160
89. 1540 m/s
90. 41 kHz
91. (*a*) 8.29 Hz; (*b*) 13.9 Hz
92. (*a*) 2.0 kHz; (*b*) 2.0 kHz
93. (*a*) 598 Hz; (*b*) 608 Hz; (*c*) 589 Hz
94. (*a*) 485.8 Hz; (*b*) 500.0 Hz; (*c*) 486.2 Hz; (*d*) 500.0 Hz
95. 0.073
96. 1×10^6 m/s, receding
97. (*a*) 1.94×10^4 km/s, (*b*) receding
98. 0.13*c*
99. ±3.78 pm
101. 49.5 m/s

Chapter 19

1. 2.71 K
2. 186° C
3. 31.5
4. 1.366
5. 291.1 K
6. 0.05 mm Hg; nitrogen
7. 348 K
8. 373 K
9. 7 Celsius
10. (*a*) 320° F; (*b*) −12.3° F
11. No; 310 K = 98.6° F
12. (*a*) −96° F; (*b*) 56.7° C
13. (*a*) 10,000° F; (*b*) 37.0° C; (*c*) −57° C; (*d*) −297° F; (*e*) 25° C = 77° F, for example

14. (*a*) −40°; (*b*) 575°; (*c*) Celsius and Kelvin cannot give the same reading
15. −91.9° X
16. (*a*) 10,000 Ω; $4.124 \times 10^{-3}/°C$; $-1.779 \times 10^{-6}/°C^2$
17. (*a*) Dimensions are inverse time
18. 1/2
20. 4.4×10^{-3} cm
21. 1.1 cm
22. 0.038 in.
23. 2.731 cm
24. (*a*) 9.996 cm; (*b*) 68° C
25. (*a*) $13 \times 10^{-6}/°F$; (*b*) 0.17 in.
26. 170 km
27. $23 \times 10^{-6}/°C$
28. 79.85 mm
29. 0.32 cm^2
30. 11 cm^2
31. 29 cm^3
32. 49.87 cm^3
33. 0.432 cm^3
34. 0.26 cm^3
35. −157° C
36. 360° C
40. (*a*) −0.69 %; (*b*) aluminum
42. (*a*) 0.36 %; (*b*) 0.18 %; (*c*) 0.54 %; (*d*) 0.00 %; (*e*) $1.8 \times 10^{-5}/°C$
43. 9.1 s; the clock running slow
44. +0.68 s/h
45. (*a*) $+9.0 \times 10^{-6}$; (*b*) -1.3×10^{-5}
46. (*b*) steel, 71 cm; brass, 41 cm
47. 7.5 cm
48. 0.217° C/s
49. increases by 0.1 mm
50. (*b*) use 39.3 cm of steel and 13.1 cm of brass
51. 2.64×10^8 Pa
52. ceiling clearance: 2.13 m, floor clearance: 1.5 m
53. 66° C
54. (*b*) $\left(\dfrac{E_1E_2}{E_1+E_2}\right)(\alpha_1+\alpha_2)\Delta T$
55. 0.266 mm

Chapter 20

1. 333 J
2. (*a*) 523 J/kg·K; (*b*) 0.600; (*c*) 26.2 J/mol·K
3. 35.7 m^3
4. 94.6 liters
5. 6.7×10^{12} J
6. 109 g
7. 42.7 kJ
8. 1.30 MJ
9. 250 g
10. 1.9 times as great
11. 220 m/s
12. (*a*) 33.9 Btu; (*b*) 172° F
13. 1.17° C
14. (*a*) 0.12° C
15. 160 s
16. (*a*) 411 g; (*b*) 3.1¢
17. (*a*) 20,300 cal; (*b*) 1110 cal; (*c*) 873° C
18. 0.54 kJ/kg · K
19. 45.4° C
20. 3.0 min
21. (*a*) 18,700; (*b*) 10.4 h
22. 73 kW
23. 2.8 days
24. 2.17 g
25. 82 cal
26. 33 m^2
27. 13.5° C
28. 33 g
29. (*a*) 5.3° C, no ice remaining; (*b*) 0° C, 60 g of ice left
30. (*a*) 0° C; (*b*) 2.5° C
31. 8.72 g
32. 2500 J/kg · K
33. $C \propto T^{-2/3}$
34. 25.3 kg
35. A: 120 J, B: 75 J, C: 30 J
36. +45 J along path *BC* and −45 J along path *BA*
37. (*a*) −200 J; (*b*) −293 J; (*c*) −93 J
38. (*a*) $A \to B$: + + +, $B \to C$: + 0 +, $C \to A$: − − −; (*b*) −20 J
39. −5.0 J
40. −30 J
41. 33.3 kJ
42. (*a*) 6.0 cal; (*b*) −43 cal; (*c*) 40 cal; (*d*) 18 cal, 18 cal
43. (*a*) 0.36 mg/s; (*b*) 0.81 J/s; (*c*) −0.69 J/s
44. 766° C
45. (*a*) 1.2 W/m · K, 0.70 Btu/ft · °F · h; (*b*) 0.030 ft^2 · °F · h/Btu
46. (*a*) 270 J/s; (*b*) heat flows out about 15 times as fast
47. 1660 J/s

48. (*a*) 16 J/s; (*b*) 0.048 g/s
51. arrangement (*b*)
52. 0.50 min
53. (*a*) 2.0 MW; (*b*) 220 W
54. (*a*) 1.8 W; (*b*) 0.025° C
55. 2.0×10^7 J
56. (*a*) 17 kW/m^2; (*b*) 18 W/m^2
57. 0.40 cm/h
58. 1.1 m
59. Cu-Al, 84.3° C; Al-Brass, 57.6° C
60. (*a*) 400 Btu/h; (*b*) 18; (*c*) 12%; (*d*) 33%
61. (*a*) $11p_iV_i$; (*b*) $6p_iV_i$
62. 42
63. 23 J

Chapter 21

1. (*a*) 0.0127; (*b*) 7.65×10^{21}
2. 0.933 kg
3. 6560
4. tablespoon
5. number of molecules in the ink $\approx 3 \times 10^{16}$; number of people $\approx 5 \times 10^{20}$; statement is wrong, by a factor of about 20,000
6. (*a*) 22.4 L
7. (*a*) 5.47×10^{-8} mol; (*b*) 3.29×10^{16}
8. 25
9. (*a*) 106; (*b*) 0.892 m^3
10. (*a*) 0.0388 mol; (*b*) 220° C
11. 27.0 lb/in.^{2}
12. 653 J
13. (*a*) 2.5×10^{25}; (*b*) 1.2 kg
14. $A(T_2 - T_1) - B(T_2^2 - T_1^2)$
15. 5600 N · m
17. 1/5
18. 3.4 m^3
19. 100 cm^3
20. 22.8 m
21. 198° F
22. (*a*) 38 L; (*b*) 71 g
23. 2.0×10^5 Pa
24. 2.50 km/s
25. 180 m/s
26. 442 m/s
27. 9.53×10^6 m/s
28. (*a*) 511 m/s; (*b*) −200° C, 899° C
29. 307° C
30. (*a*) 494 m/s; (*b*) 28 g/mol, N_2
31. 1.9×10^4 dyne/cm^2
32. (*a*) 3.3×10^{-20} J; (*b*) 0.21 eV
33. (*a*) 0.0353 eV, 0.0483 eV; (*b*) 3400 J, 4650 J
34. 7730 K
35. 9.1×10^{-6}
37. (*a*) 6.75×10^{-20} J; (*b*) 10.7
39. 0.32 nm
40. (*a*) 6×10^9 km
41. 15 cm
42. $\sqrt{2}\pi d^2 \overline{v} N/V$
43. (*a*) 3.27×10^{10}; (*b*) 172 m
44. 3.7 GHz
45. (*a*) 22.5 L; (*b*) 2.25; (*c*) 8.4×10^{-5} cm; (*d*) same as (*c*)
46. (*a*) 1.7; (*b*) 5.0×10^{-5} cm; (*c*) 7.9×10^{-6} cm
48. (*a*) 6.5 km/s; (*b*) 7.1 km/s
49. (*a*) 3.2 cm/s; (*b*) 3.4 cm/s; (*c*) 4.0 cm/s
50. (*a*) 420 m/s, 458 m/s
51. (*a*) $\overline{v}$, v_{rms}, v_P; (*b*) v_P, $\overline{v}$, v_{rms}
52. 1.50
53. (*a*) 1.0×10^4 K, 1.6×10^5 K; (*b*) 440 K, 7000 K
54. (*a*) 7.0 km/s; (*b*) 2.0×10^{-8} cm; (*c*) 3.5×10^{10} collisions/s
55. 4.7
56. (*a*) $3N/v_0^3$; (*b*) $0.750v_0$; (*c*) $0.775v_0$
57. (*a*) $2N/3v_0$; (*b*) $N/3$; (*c*) $1.22v_0$; (*d*) $1.31v_0$
58. (*a*) 3400 J; (*b*) no
59. $RT\ln(V_f/V_i)$
60. 3110 J/kg · K
61. (*a*) 15.9 J; (*b*) 34.4 J/mol · K; (*c*) 26.1 J/mol · K
62. (*a*) $1.5nRT_1$; (*b*) $4.5nRT_1$; (*c*) $6nRT_1$; (*d*) $2R$
63. $(n_1C_1 + n_2C_2 + n_3C_3)/(n_1 + n_2 + n_3)$
64. (*a*) 6.6×10^{-26} kg; (*b*) 40
65. constant volume
66. 8000 J
67. (*a*) 0.375 mol; (*b*) 1090 J; (*c*) 0.714
68. (*a*) 6980 J; (*b*) 4990 J; (*c*) 1990 J; (*d*) 2990 J
69. diatomic
70. (*a*) 14 atm; (*b*) 620 K
71. (*a*) 2.5 atm, 340 K; (*b*) 0.40 L
73. 1500 N · m$^{2.2}$
75. 1.40
78. 0.63
79. (*a*) $p_0/3$; (*b*) polyatomic (ideal); (*c*) $K_f/K_i = 1.44$
80. (*a*) monatomic; (*b*) 2.7×10^4 K; (*c*) 4.5×10^4 mol; (*d*) 3.4 kJ, 340 kJ; (*e*) 0.01

81. Q, W, and ΔE_{int} are all greatest for (*a*) and least for (*c*)
82. Constant volume: 900, 0, 900, 450 cal; Constant pressure: 1200, 300, 900, 450 cal; Adiabatic: 0, −900, 900, 450 cal
83. (*a*) In joules, in the order $Q, \Delta E_{\text{int}}, W$:
    $1 \to 2$: 3740, 3740, 0;
    $2 \to 3$: 0, −1810, 1810;
    $3 \to 1$: −3220, −1930, −1290;
    Cycle: 520, 0, 520;
    (*b*) $V_2 = 0.0246\,\text{m}^3$, $p_2 = 2.00\,\text{atm}$, $V_3 = 0.0373\,\text{m}^3$, $p_3 = 1.00\,\text{atm}$
84. 12 kW, 16 hp
85. (*a*) 1.5 mol; (*b*) 1800 K; (*c*) 600 K; (*d*) 5.0 kJ
86. (*a*) −45 J; (*b*) 180 K
87. (*a*) −5.0 kJ; (*b*) 2.0 kJ; (*c*) 5.0 kJ
88. (*a*) 8.0 atm, 300 K, 4400 J; (*b*) 3.2 atm, 120 K, 2900 J; (*c*) 4.6 atm, 170 K, 3400 J

Chapter 22

1. (*a*) 31%; (*b*) 16 kcal
2. (*a*) 33 kJ, 25 kJ; (*b*) 27 kJ, 18 kg
3. 25%
4. (*a*) 1.47×10^3 J; (*b*) 5.54×10^2 J; (*c*) 9.18×10^2 J; (*d*) 0.624
5. (*a*) 7200 J; (*b*) 960 J; (*c*) 13%
6. (*a*) 3.73; (*b*) 710 J
7. (*a*) 49 kcal; (*b*) 31 kJ
8. (*a*) 23.6%; (*b*) 62.2 kJ
9. (*a*) 0.071 J; (*b*) 0.50 J; (*c*) 2.0 J; (*d*) 5.0 J
10. (*a*) 2090 J; (*b*) 1570 J; (*c*) 1570 J
11. 99.999947%
12. 68° C, −7.0° C
13. 75
14. 20 J
16. (*b*) 6.7
17. (*a*) 94 J; (*b*) 230 J
18. 58%
19. 13 J
20. 440 W
21. (*a*) 1.11 kcal/s; (*b*) 0.995 kcal/s
22. 1.08 MJ
23. $e = 1/(K+1)$
24. (*b*) 75 cal; (*c*) 310 J
26. (*c*) 1150 J
27. $[1 - (T_2/T_1)]/[1 - (T_4/T_3)]$
28. 0.25 hp
29. (*a*) 78%; (*b*) 81 kg/s
30. (*a*) 2270 J; (*b*) 14,800 J; (*c*) 15.4%; (*d*) 75.0%
31. 0.139
32. (*a*) $T_2 = 3T_1$, $T_3 = 3T_1/4^{\gamma-1}$, $T_4 = T_1/4^{\gamma-1}$, $p_2 = 3p_1$, $p_3 = 3p_1/4^{\gamma}$, $p_4 = p_1/4^{\gamma}$; (*b*) $1 - 4^{1-\gamma}$
33. (*a*) +0.602 cal/K; (*b*) −0.602 cal/K
34. (*a*) 6900 cal; (*b*) 21 cal/K
35. −0.30 cal/g · K
36. (*a*) first; (*b*) first and second; (*c*) second; (*d*) neither
37. 8.79×10^{-3} cal/K
38. 2.75 mol
39. 4450 cal
40. 5.76 J/K
41. (*a*) 1.95 J/K; (*b*) 0.650 J/K; (*c*) 0.217 J/K; (*d*) 0.072 J/K
42. (*a*) +1.06 cal/K; (*b*) no
43. (*a*) 57° C; (*b*) −5.27 cal/K; (*c*) +5.95 cal/K; (*d*) +0.68 cal/K
44. 0.19 cal/K
45. +0.15 cal/K
46. (*a*) 9220 J; (*b*) 23.1 J/K; (*c*) 0
47. (*a*) 320 K; (*b*) 0; (*c*) +0.41 cal/K
48. (*a*) process I: $Q_T = pV \ln 2$, $Q_V = (9/2)pV$ process II: $Q_T = -pV \ln 2$, $Q_p = (15/2)pV$; (*b*) process I: $W_T = pV \ln 2$, $W_V = 0$ process II: $W_T = -pV \ln 2$, $W_p = 3pV$; (*c*) $(9/2)pV$ for either case; (*d*) $4R \ln 2$ for either case
49. (*a*) $p_1/3$, $p_1/3^{1.4}$, $T_1/3^{0.4}$;
    (*b*) in the order $W, Q, \Delta E_{\text{int}}, \Delta S$
    $1 \to 2$: $1.10RT_1$, $1.10RT_1$, 0, $1.10R$
    $2 \to 3$: 0, $-0.889RT_1$, $-0.889RT_1$, $-1.10R$
    $3 \to 1$: $-0.889RT_1$, 0, $0.889RT_1$, 0
51. (*a*) −225 cal/K; (*b*) +225 cal/K
52. (*a*) -6.2×10^{-3} cal/K · s; (*b*) 8.4×10^{-4} cal/K · s
53. (*a*) $3p_0V$; (*b*) $6RT_0$, $(3/2)R \ln 2$; (*c*) both are 0
54. (*a*) *AE*; (*b*) *AC*; (*c*) *AF*; (*d*) none
55. 97 K
56. (*a*) monatomic; (*b*) 0.75
57. (*a*) 1.84 kN/m²; (*b*) 441 K; (*c*) 3160 J; (*d*) 1.92 J/K
58. (*a*) 4500 J; (*b*) −5000 J; (*c*) 9500 J

Chapter 23

1. (a) 8.99×10^9 N; (b) 8990 N
2. 2.81 N on each
3. 1.38 m
4. 0.50 C
5. (a) 4.9×10^{-7} kg; (b) 7.1×10^{-11} C
6. (a) 1.60 N; (b) 2.77 N
7. $3F/8$
8. $q_1 = -4q_2$
9. (a) $q_1 = 9q_2$; (b) $q_1 = -25q_2$
10. $F_{hor} = 0.17$ N, $F_{vert} = -0.046$ N
11. 1.2×10^{-5} C and 3.8×10^{-5} C
12. either $-1.00\,\mu$C and $+3.00\,\mu$C or $+1.00\,\mu$C and $-3.00\,\mu$C
13. 14 cm from the positive charge, 24 cm from the negative charge
14. (a) 36 N, $-10°$ from the x axis; (b) $x = -8.3$ m, $y = +2.7$ cm
15. (a) A charge $-4q/9$ must be located on the line joining the two positive charges, a distance $L/3$ from the $+q$ charge.
16. (a) 5.7×10^{13} C, no; (b) 600
17. (a) $Q = -2\sqrt{2}q$; (b) no
18. $q = Q/2$
19. (b) $\pm 2.4 \times 10^{-8}$ C
20. 3.1 cm
21. (a) $\frac{L}{2}\left(1 + \frac{1}{4\pi\epsilon_0}\frac{qQ}{Wh^2}\right)$; (b) $\sqrt{3qQ/4\pi\epsilon_0 W}$
22. 2.89×10^{-9} N
23. 3.8 N
24. 1.32×10^{13} C
25. 1.9 MC
26. (a) 3.2×10^{-19} C; (b) two
27. (a) 8.99×10^{-19} N; (b) 625
28. (a) 6.3×10^{11}; (b) 7.3×10^{-15}
29. 11.9 cm
30. 5.1 m below the electron
31. 1.3 days
32. 122 mA
33. 1.3×10^7 C
34. (a) 0; (b) 1.9×10^{-9} N
35. 1.7×10^8 N
36. 10^{18} N
37. (a) positron; (b) electron
38. (a) ^{9}B; (b) ^{13}N; (c) ^{12}C
39. (a) 510 N; (b) 7.7×10^{28} m/s^2
41. (a) $(Gh/2\pi c^3)^{1/2}$; (b) 1.61×10^{-35} m
42. (a) $(hc/G)^{1/2}$; (b) 5.46×10^{-8} kg

Chapter 24

1. (a) 6.4×10^{-18} N; (b) 20 N/C
3. to the right in the figure
7. 0.111 nC
8. $+1.0\,\mu$C
9. 56 pC
10. (a) Q_2 produces a field of 1.3×10^5 N/C at Q_1; Q_1 produces a field of 5.3×10^4 N/C at Q_2; (b) 1.1×10^{-2} N for both
11. (a) 6.4×10^5 N/C, toward the negative charge; (b) 1.0×10^{-13} N, toward the positive charge
12. 3.07×10^{21} N/C, radially outward
13. $(1/4\pi\epsilon_0)(3q/d^2)$, directed toward $-2q$
15. (a) $1.7a$, to the right of the $+2q$ charge
16. (a) $q/8\pi\epsilon_0 d^2$, to the left; $3q/\pi\epsilon_0 d^2$, to the right; $7q/16\pi\epsilon_0 d^2$, to the left
17. 50 cm from q_1 and 100 cm from q_2
18. 0
19. 9:30
20. (a) 4.8×10^{-8} N/C; (b) 7.7×10^{-27} N
21. $E = q/\pi\epsilon_0 a^2$, along bisector, away from triangle
22. 1.02×10^5 N/C, upward
23. 6.88×10^{-28} C $\cdot$ m
24. 6.6×10^{-15} N
26. $(1/4\pi\epsilon_0)(p/r^3)$, antiparallel to **p**
29. $R/\sqrt{2}$
31. $(1/4\pi\epsilon_0)(4q/\pi R^2)$, toward increasing y
32. $Q/\pi^2\epsilon_0 r^2$, vertically downward
34. (a) $-q/L$; (b) $q/4\pi\epsilon_0 a(L+a)$
37. (a) $0.10\,\mu$C; (b) 1.3×10^{17}; (c) 5.0×10^{-6}
38. $R/\sqrt{3}$
39. 3.51×10^{15} m/s^2
40. 1.02×10^{-2} N/C, westward
41. (a) 4.8×10^{-13} N; (b) 4.8×10^{-13} N
42. 2.03×10^{-7} N/C, up
43. (a) 1.5×10^3 N/C; (b) 2.4×10^{-16} N, up; (c) 1.6×10^{-26} N; (d) 1.5×10^{10}
44. (a) -0.029 C; (b) repulsive forces would explode the sphere
45. (a) 2.46×10^{17} m/s^2; (b) 0.122 ns; (c) 1.83 mm
46. (a) 1.92×10^{12} m/s^2; (b) 1.96×10^5 m/s
47. (a) 7.12 cm; (b) 28.5 ns; (c) 11.2%
48. (a) 8.87×10^{-15} N; (b) 120
49. $-5e$
50. 1.64×10^{-19} C ($\approx$ 3% high)

51. (*a*) 0.245 N, 11.3° clockwise from the $+x$ axis; (*b*) $x = 108$ m, $y = -21.6$ m
52. (*a*) 2.7×10^6 m/s; (*b*) 1000 N/C
53. (*a*) $(-2.1 \times 10^{13}\,\mathbf{j})\,\text{m/s}^2$; (*b*) $(1.5 \times 10^5\,\mathbf{i} - 2.8 \times 10^6\,\mathbf{j})\,\text{m/s}$
54. 27 μm
55. (*a*) $2\pi\sqrt{\dfrac{\ell}{|g - qE/m|}}$; (*b*) $2\pi\sqrt{\dfrac{\ell}{g + qE/m}}$
56. (*a*) yes; (*b*) upper plate, 2.73 cm
57. (*a*) 9.30×10^{-15} C · m; (*b*) 2.05×10^{-11} J
58. (*a*) 0; (*b*) 8.5×10^{-22} N · m; (*c*) 0
59. $2pE\cos\theta_0$
60. $(1/2\pi)\sqrt{pE/I}$

Chapter 25

1. (*a*) 693 kg/s; (*b*) 693 kg/s; (*c*) 347 kg/s; (*d*) 347 kg/s; (*e*) 575 kg/s
2. $-0.015\,\text{N}\cdot\text{m}^2/\text{C}$
3. (*a*) 0; (*b*) $-3.92\,\text{N}\cdot\text{m}^2/\text{C}$; (*c*) 0; (*d*) 0 for each field
4. (*a*) $-\pi R^2 E$; (*b*) $\pi R^2 E$
5. (*a*) enclose $2q$ and $-2q$, or enclose all four charges; (*b*) enclose $2q$ and q; (*c*) not possible
6. $\Phi_1 = q/\epsilon_0$, $\Phi_2 = -q/\epsilon_0$, $\Phi_3 = q/\epsilon_0$, $\Phi_4 = 0$, $\Phi_5 = q/\epsilon_0$
7. $2.0 \times 10^5\,\text{N}\cdot\text{m}^2/\text{C}$
8. 26.2 fC
9. $q/6\epsilon_0$
10. $\pi a^2 E$
11. 3.54 μC
12. (*a*) $8.23\,\text{N}\cdot\text{m}^2/\text{C}$; (*b*) $8.23\,\text{N}\cdot\text{m}^2/\text{C}$; (*c*) 72.8 pC in each case
13. 0 through each of the three faces meeting at q, $q/24\epsilon_0$ through each of the other faces
15. $2.0\,\mu\text{C/m}^2$
16. (*a*) 37 μC; (*b*) $4.1 \times 10^6\,\text{N}\cdot\text{m}^2/\text{C}$
17. (*a*) $4.5 \times 10^{-7}\,\text{C/m}^2$; (*b*) 5.1×10^4 N/C
18. (*a*) $-Q$; (*b*) $-Q$; (*c*) $-(Q + q)$; (*d*) yes
19. (*a*) -3.0×10^{-6} C; (*b*) $+1.3 \times 10^{-5}$ C
21. 5.0 μC/m
22. (*a*) 0.32 μC; (*b*) 0.14 μC
23. (*a*) $E = \lambda/2\pi\epsilon_0 r$; (*b*) 0
25. $3.8 \times 10^{-8}\,\text{C/m}^2$
26. 3.6 nC
27. (*a*) $E = q/2\pi\epsilon_0 Lr$, radially inward; (*b*) $-q$ on both inner and outer surfaces; (*c*) $E = q/2\pi\epsilon_0 Lr$, radially outward
28. (*a*) 2.3×10^6 N/C, radially out; (*b*) 4.5×10^5 N/C, radially in
29. 270 eV
30. (*b*) $\rho R^2/2\epsilon_0 r$
31. (*a*) $E = \sigma/\epsilon_0$, to the left; (*b*) $E = 0$; (*c*) $E = \sigma/\epsilon_0$, to the right
32. (*a*) 5.3×10^7 N/C; (*b*) 60 N/C
33. $E = \dfrac{\sigma}{2\epsilon_0\sqrt{x^2 + R^2}}$
34. $5.0\,\text{nC/m}^2$
35. 0.44 mm
36. (*a*) 0; (*b*) $E = \sigma/\epsilon_0$, to the left; (*c*) 0
37. $\pm 4.9 \times 10^{-10}$ C
38. (*a*) $4.9 \times 10^{-22}\,\text{C/m}^2$; (*b*) downward
39. (*a*) $\rho x/\epsilon_0$; (*b*) $\rho d/2\epsilon_0$
40. -7.5 nC
41. (*a*) $-750\,\text{N}\cdot\text{m}^2/\text{C}$; (*b*) -6.64 nC
42. (*a*) 0; (*b*) 2.9×10^4 N/C; (*c*) 200 N/C
43. (*a*) 2.50×10^4 N/C; (*b*) 1.35×10^4 N/C
44. (*a*) 0; (*b*) $q_a/4\pi\epsilon_0 r^2$; (*c*) $(q_a + q_b)/4\pi\epsilon_0 r^2$
47. (*a*) $E = q/4\pi\epsilon_0 r^2$, radially outward; (*b*) same as (*a*); (*c*) no; (*d*) yes, charges are induced on the surface; (*e*) yes; (*f*) no; (*g*) no
48. (*a*) $E = (q/4\pi\epsilon_0 a^3)r$; (*b*) $E = q/4\pi\epsilon_0 r^2$; (*c*) 0; (*d*) 0; (*e*) inner, $-q$; outer, 0
50. (*a*) 4.0×10^6 N/C; (*b*) 0
51. -1.04 nC
53. $q/2\pi a^2$
56. $6Kr^3$
57. (*a*) $-e/\pi a_0^3$; (*b*) $5e\exp(-2)/4\pi\epsilon_0 a_0^2$, radially outward
58. (*a*) $e^2 r/4\pi\epsilon_0 a^3$; (*b*) $e/\sqrt{4\pi\epsilon_0 m a_0^3}$
60. (*a*) σ/ϵ_0; (*b*) $-q$
61. (*a*) $q^2/16\pi\epsilon_0 a^2$, attractive; (*b*) $-q$

Chapter 26

1. 1.2 GeV
2. (*a*) 3.0×10^5 C; (*b*) 3.6×10^6 J
3. (*a*) 3.0×10^{10} J; (*b*) 7.7 km/s; (*c*) 9.0×10^4 kg
5. (*a*) -2.46 V; (*b*) -2.46 V; (*c*) 0
6. (*a*) $q_0\sigma z/2\epsilon_0$
7. 2.90 kV
8. (*a*) 2.4×10^4 V/m; (*b*) 2900 V
9. 8.8 mm
11. (*a*) $-qr^2/(8\pi\epsilon_0 R^3)$; (*b*) $q/(8\pi\epsilon_0 R)$; (*c*) center

12. (*a*) 136 MV/m; (*b*) 8.82 kV/m
13. (*b*) because $V = 0$ point is chosen differently; (*c*) $q/(8\pi\epsilon_0 R)$; (*d*) potential differences are independent of the choice for the $V = 0$ point
14. (*a*) $Q/4\pi\epsilon_0 r$; (*b*) $\dfrac{\rho}{3\epsilon_0}\left(\dfrac{3}{2}r_2^2 - \dfrac{1}{2}r^2 - \dfrac{r_1^3}{r}\right)$, $\rho = Q\Big/\left(\dfrac{4\pi}{3}(r_2^3 - r_1^3)\right)$; (*c*) $\dfrac{\rho}{2\epsilon_0}(r_2^2 - r_1^2)$, with ρ as in (*b*); (*d*) yes
15. (*a*) −4500 V; (*b*) −4500 V
16. (*a*) 4.5 m; (*b*) no
17. 843 V
18. −1.1 nC
19. 2.8×10^5
22. (*a*) 4.0×10^5 V; (*b*) no
23. (*a*) 3.3 nC; (*b*) 12 nC/m^2
24. 640 mV
25. 200 mV
26. (*a*) 0.54 mm; (*b*) 790 V
27. (*a*) 38 s; (*b*) 280 d
28. $x = d/4$
29. none
30. 16.3 μV
31. (*a*) none; (*b*) 41 cm from $+q$, between the charges
32. (*a*) −4.8 nm; (*b*) 8.1 nm; (*c*) no
34. $-2.5q/4\pi\epsilon_0 d$
35. $-8q/4\pi\epsilon_0 d$
36. (*a*) $\dfrac{2\lambda}{4\pi\epsilon_0}\ln\left[\dfrac{L/2 + (L^2/4 + d^2)^{1/2}}{d}\right]$; (*b*) 0
37. $-Q/4\pi\epsilon_0 R$
38. (*a*), (*b*), and (*c*): $Q/4\pi\epsilon_0 R$; (*d*) rank of *a*, *b*, *c*.
39. (*a*) $-5Q/4\pi\epsilon_0 R$; (*b*) $-5Q/4\pi\epsilon_0(z^2 + R^2)^{1/2}$
40. $(\sigma/4\epsilon_0)\left[(z^2 + R^2)^{1/2} - z\right]$
41. $\dfrac{-Q/L}{4\pi\epsilon_0}\ln\left(\dfrac{L}{d} + 1\right)$
42. 670 V/m
43. *ab*: −6.0 V/m, *bc*: 0, *ce*: 3.0 V/m, *ef*: 15 V/m, *fg*: 0, *gh*: −3.0 V/m
44. $p\cos\theta/2\pi\epsilon_0 r^3$
46. 39 V/m, toward $x = 0$
49. (*a*) $\dfrac{\lambda}{4\pi\epsilon_0}\ln\left(\dfrac{L + x}{x}\right)$; (*b*) $\dfrac{\lambda}{4\pi\epsilon_0}\dfrac{L}{x(L + x)}$; (*c*) 0
50. (*a*) $(k/4\pi\epsilon_0)\left[\sqrt{L^2 + y^2} - y\right]$; (*b*) $\dfrac{k}{4\pi\epsilon_0}\left[1 - \dfrac{y}{\sqrt{L^2 + y^2}}\right]$
51. (*a*) $qd/2\pi\epsilon_0 a(a + d)$
52. (*a*) 2.5 MV; (*b*) 5.1 J; (*c*) 6.9 J
53. −1.9 J
54. (*a*) 2.72×10^{-14} J; (*b*) 3.02×10^{-31} kg, in error by a factor of about 3
55. (*a*) 0.484 MeV; (*b*) 0
56. $0.21q^2/\epsilon_0 a$
57. -1.2×10^{-6} J
58. 2.1 d
59. (*a*) $+6.0 \times 10^4$ V; (*b*) -7.8×10^5 V; (*c*) 2.5 J; (*d*) increase; (*e*) same; (*f*) same
60. 0
61. $W = \dfrac{qQ}{8\pi\epsilon_0}\left(\dfrac{1}{r_1} - \dfrac{1}{r_2}\right)$
62. (*a*) 27.2 V; (*b*) −27.2 eV; (*c*) 13.6 eV; (*d*) 13.6 eV
63. 1.8×10^{-10} J
64. 2.5 km/s
65. (*a*) 0.225 J; (*b*) 22.5 m/s^2; (*c*) *A*: 7.75 m/s, *B*: 3.87 m/s
66. 1.48×10^7 m/s
67. (*a*) 25 fm; (*b*) twice as much
68. $qQ/4\pi\epsilon_0 K$
69. $\sqrt{2eV/m}$
70. 0.32 km/s
71. 23 km/s
72. 1.6×10^{-9} m
73. 400 V
75. 2.5×10^{-8} C
76. (*a*) $V_1 = V_2$; (*b*) $q_1 = q/3$, $q_2 = 2q/3$; (*c*) 2
78. (*a*) −0.12 V; (*b*) 1.8×10^{-8} N/C, radially inward
79. (*a*) −180 V; (*b*) 3000 V, −9000 V
80. (*a*) 12,000 N/C; (*b*) 1800 V; (*c*) 5.8 cm
81. $r < R_1$: $E = 0$, $V = \dfrac{1}{4\pi\epsilon_0}\left(\dfrac{q_1}{R_1} + \dfrac{q_2}{R_2}\right)$; $R_1 < r < R_2$: $E = q_1/4\pi\epsilon_0 r^2$, $V = \dfrac{1}{4\pi\epsilon_0}\left(\dfrac{q_1}{r} + \dfrac{q_2}{R_2}\right)$; $r > R_2$: $E = (q_1 + q_2)/4\pi\epsilon_0 r^2$, $V = (q_1 + q_2)/4\pi\epsilon_0 r$
82. (*a*) 0.11 mC; (*b*) 1.1 μC; (*c*) larger field of smaller sphere causes electrical breakdown in the air
83. 9.52 kW
84. (*a*) 2.0 MeV ($= 3.2 \times 10^{-13}$ J); (*b*) 1.0 MeV ($= 1.6 \times 10^{-13}$ J; (*c*) the proton

86. (*a*) $r > 9.0$ cm; (*b*) 2.7 kW; (*c*) 20 μC/m^2
87. (*b*) 100 eV; (*c*) It would proceed to the right, speeding up between the left pair of screens, slowing down between the right pair, and leaving with its original velocity.
89. (*b*) $V = \dfrac{q}{4\pi\epsilon_0}\left(\dfrac{1}{a-x} + \dfrac{1}{a+x}\right)$; (*c*) $F = qQx/\pi\epsilon_0 a^3$; (*d*) $\omega = (qQ/\pi\epsilon_0 ma^3)^{1/2}$
90. (*b*) $q^2/16\pi\epsilon_0 d^2$
91. (*a*) $Q_1 = QR_1/(R_1+R_2)$, $Q_2 = QR_2/(R_1+R_2)$; (*b*) $(1/4\pi\epsilon_0)R_1R_2Q^2/(R_1+R_2)^2L^2$

Chapter 27

1. 7.5 pC
2. (*a*) 3.5 pF; (*b*) 3.5 pF; (*c*) 57 V
3. 3.0 mC
5. (*a*) 140 pF; (*b*) 17 nC
6. 8.85×10^{-12} m
7. 0.55 pF
8. (*a*) 84.5 pF; (*b*) 191 cm^2
9. 4×10^{-7} C
10. (*a*) 11 cm^2; (*b*) 11 pF; (*c*) 1.2 V
11. $5.05\pi\epsilon_0 R$
14. (*b*) 4.6×10^{-5} /°C
15. 9090
16. 3.16 μF
17. 7.33 μF
18. 315 mC
19. (*a*) 2.40 μF; (*b*) 0.480 mC on both; (*c*) $V_2 = 120$ V, $V_1 = 80$ V
20. (*a*) 10.0 μF; (*b*) $q_2 = 0.800$ mC, $q_1 = 1.20$ mC; (*c*) 200 V for both
21. (*a*) $d/3$; (*b*) $3d$
22. (*a*) $q_1 = q_2 = 0.48$ mC, $V_1 = 240$ V, $V_2 = 60$ V; (*b*) $q_1 = 0.19$ mC, $q_2 = 0.77$ mC, $V_1 = V_2 = 96$ V; (*c*) $q_1 = q_2 = 0$, $V_1 = V_2 = 0$
24. (*a*) 7.9×10^{-4} C; (*b*) 79 V
25. (*a*) five in a series; (*b*) three arrays as in (*a*) in parallel (and other possibilities)
27. 43 pF
28. (*a*) 50 V; (*b*) 0.50×10^{-4} C; (*c*) 1.5×10^{-4} C
29. $q_1 = \dfrac{C_1C_2 + C_1C_3}{C_1C_2 + C_1C_3 + C_2C_3} C_1V_0$, $q_2 = q_3 = \dfrac{C_2C_3}{C_1C_2 + C_1C_3 + C_2C_3} C_1V_0$
30. (*a*) $q_1 = 9.0\,\mu$C, $q_2 = 16\,\mu$C, $q_3 = 9.0\,\mu$C, $q_4 = 16\,\mu$C; (*b*) $q_1 = 8.4\,\mu$C, $q_2 = 17\,\mu$C, $q_3 = 11\,\mu$C, $q_4 = 14\,\mu$C
31. first case: 50.0 V, second case: 0
32. 99.6 nJ
33. (*a*) 3.05 MJ; (*b*) 0.847 kW · h
34. 72 F
35. (*a*) 0.204 μJ; (*b*) no
36. (*a*) 35 pF; (*b*) 21 nC; (*c*) 6.3 μJ; (*d*) 0.60 MV/m; (*e*) 1.6 J/m^3
37. 0.27 J
38. (*a*) $e^2/32\pi^2\epsilon_0 r^4$; (*b*) ∞
39. 4.9%
40. 0.11 J/m^3
41. 10.4 cents
42. (*a*) $U_2 = 58$ mJ, $U_8 = 14$ mJ; (*b*) $U_2 = 9.2$ mJ, $U_8 = 37$ mJ; (*c*) $U_2 = U_8 = 0$
43. (*a*) 2.0 J
44. (*a*) $q_1 = 0.21$ mC, $q_2 = 0.11$ mC, $q_3 = 0.32$ mC; (*b*) $V_1 = V_2 = 21$ V, $V_3 = 79$ V; (*c*) $U_1 = 2.2$ mJ, $U_2 = 1.1$ mJ, $U_3 = 13$ mJ
45. (*a*) $q_1 = q_2 = 0.33$ mC, $q_3 = 0.40$ mC; (*b*) $V_1 = 33$ V, $V_2 = 67$ V, $V_3 = 100$ V; (*c*) $U_1 = 5.4$ mJ, $U_2 = 11$ mJ, $U_3 = 20$ mJ
46. (*a*) $2V$; (*b*) $U_i = \epsilon_0 AV^2/2d$, $U_f = 2U_i$; (*c*) $\epsilon_0 AV^2/2d$
48. (*a*) $e^2/8\pi\epsilon_0 R$; (*b*) 1.41 fm
52. 4.0
53. Pyrex
54. the mica sheet
55. (*a*) 6.2 cm; (*b*) 280 pF
56. 81 pF/m
57. 0.63 m^2
58. (*a*) 0.73 nF; (*b*) 28 kV
59. (*a*) 2.85 m^3; (*b*) 1.01×10^4
60. (*a*) $2\kappa\epsilon_0 A/d$; (*b*) $\frac{1}{2}Q(1 + \Delta\kappa/\kappa)$
61. (*a*) $\epsilon_0 A/(d-b)$; (*b*) $d/(d-b)$; (*c*) $-q^2b/2\epsilon_0 A$, sucked in
62. (*a*) $\epsilon_0 A/(d-b)$; (*b*) $(d-b)/d$; (*c*) $\frac{1}{2}CV^2b/(d-b)$, pushed in
65. $\dfrac{\epsilon_0 A}{4d}\left(\kappa_1 + \dfrac{2\kappa_2\kappa_3}{\kappa_2 + \kappa_3}\right)$
66. (*a*) 10 kV/m; (*b*) 5.0 nC; (*c*) 4.1 nC
67. (*a*) 13.4 pF; (*b*) 1.15 nC; (*c*) 1.13×10^4 N/C; (*d*) 4.33×10^3 N/C
68. (*a*) 7.1; (*b*) 0.77 μC
69. (*a*) 89 pF; (*b*) 120 pF; (*c*) 11 nC, 11 nC; (*d*) 10 kV/m; (*e*) 2.1 kV/m; (*f*) 88 V; (*g*) 0.17 μJ

70. (*a*) 0.606; (*b*) 0.394

72. (*a*) $C = \frac{A}{2\Delta} \ln \frac{d+\Delta}{d-\Delta}$

73. (*a*) $C = 4\pi\epsilon_0\kappa\left(\frac{ab}{b-a}\right)$; (*b*) $q = 4\pi\epsilon_0\kappa V\left(\frac{ab}{b-a}\right)$; (*c*) $q' = q(1 - 1/\kappa)$

74. (*a*) 0; (*b*) $Q/4\pi\epsilon_0 r^2$; (*c*) $(Q^2/8\pi\epsilon_0 r^2)\,dr$; (*d*) $Q^2/8\pi\epsilon_0 R$; (*e*) $-Q^2\,\Delta R/8\pi\epsilon_0 R^2$

Chapter 28

1. (*a*) 1200 C; (*b*) 7.5×10^{21}
2. 1.25×10^{15}
3. 5.6 ms
4. 6.7 μC/m^2
5. (*a*) 6.4 A/m^2, north
6. (*a*) 2.4×10^{-5} A/m^2; (*b*) 1.8×10^{-15} m/s
7. 0.38 mm
8. 14–gauge
9. 0.67 A, toward the negative terminal
10. (*a*) 0.920 mA; (*b*) 1.08×10^4 A/m^2
11. (*a*) 0.654 μA/m^2; (*b*) 83.4 MA
12. 10 A/cm^2, to the east
13. 13 min
14. (*a*) 2×10^{12}; (*b*) 5000; (*c*) 10 MV
15. $J_0A/3$; (*b*) $2J_0A/3$
16. 0.536 Ω
17. 2.0×10^{-8} Ω · m
18. 100 V
19. 2.4 Ω
20. (*a*) 1.53 kA; (*b*) 54.1 MA/m^2; (*c*) 10.6×10^{-8} Ω · m, platinum
21. 2.0×10^6 $(\Omega \cdot \text{m})^{-1}$
22. (*a*) 253 °C; (*b*) yes
23. 57 °C
25. (*a*) 0.38 mV; (*b*) negative; (*c*) 3 min 58 s
26. (*a*) $A/2$; (*b*) $4R$
27. 54 Ω
28. $2R$
29. 2.9 mm
30. 3.0
31. (*a*) 2.39, iron being larger; (*b*) no
32. (*a*) 1.3 mΩ; (*b*) 4.6 mm
33. (*a*) silver; (*b*) 51.6 nΩ
34. (*a*) 6.00 mA; (*b*) 1.59×10^{-8} V; (*c*) 21.2 nΩ
35. 2000 K
36. 8.2×10^{-4} Ω · m
37. (*a*) 38.3 mA; (*b*) 109 A/m^2; (*c*) 1.28 cm/s; (*d*) 227 V/m
38. (*a*) copper: 5.32×10^5 A/m^2, aluminum: 3.27×10^5 A/m^2; (*b*) copper: 1.01 kg/m, aluminum: 0.495 kg/m
39. (*a*) 1.73 cm/s; (*b*) 3.24 pA/m^2
40. 0.40 Ω
41. (*a*) 0.43%, 0.0017%, 0.0034%
42. (*a*) $R = \rho L/\pi ab$
44. 14 kC
45. 560 W
46. 11.1 Ω
47. 0.20 hp
48. (*a*) 1.0 kW; (*b*) 25 cents
49. 0.135 W
50. (*a*) 28.8 Ω; (*b*) 2.60×10^{19} s^{-1}
51. (*a*) 4.9 MA/m^2; (*b*) 83 mV/m; (*c*) 25 V; (*d*) 640 W
52. (*a*) 1.74 A; (*b*) 2.15 MA/m^2; (*c*) 36.3 mV/m; (*d*) 2.09 W
53. new length = $1.369L$, new area = $0.730A$
54. (*a*) 1.3×10^5 A/m^2; (*b*) 94 mV
55. (*a*) 5.85 m; (*b*) 10.4 m
56. (*a*) 10.9 A; (*b*) 10.6 Ω; (*c*) 4.5 MJ
57. (*a*) $4.46 for a 31-day month; (*b*) 144 Ω; (*c*) 0.833 A
58. 660 W
59. (*a*) 9.4×10^{13} s^{-1}; (*b*) 240 W
60. (*a*) 3.1×10^{11}; (*b*) 25 μA; (*c*) 1300 W, 25 MW
61. 710 cal/g
62. 27 cm/s
63. (*a*) 8.6%; (*b*) smaller

Chapter 29

1. (*a*) 1.9×10^{-18} J (12 eV); (*b*) 6.5 W
2. 11 kJ
3. (*a*) $320; (*b*) 9.6 cents
4. 14 h 24 min
5. (*a*) counterclockwise; (*b*) battery 1; (*c*) B
6. (*a*) 80 J; (*b*) 67 J; (*c*) 13 J converted to thermal energy within battery
7. (*c*) third plot gives rate of energy dissipation by R
8. (*a*) 0.50 A; (*b*) $P_1 = 1.0$ W, $P_2 = 2.0$ W; (*c*) $P_1 = 6.0$ W supplied, $P_2 = 3.0$ W absorbed

9. (*a*) 14 V; (*b*) 100 W; (*c*) 600 W; (*d*) 10 V, 100 W
10. −10 V
11. (*a*) 50 V; (*b*) 48 V; (*c*) B is the negative terminal
12. 4.0 V
13. 2.5 V
14. (*a*) 80 mA; (*b*) 130 mA; (*c*) 400 mA
15. (*a*) 990 Ω; (*b*) 9.9×10^{-4} W
16. (*a*) 1.0 V; (*b*) 50 mΩ
17. 8.0 Ω
18. 10^{-6}
19. the cable
20. (*a*) $r_1 - r_2$; (*b*) r_1
21. (*a*) 1000 Ω; (*b*) 300 mV; (*c*) 2.3×10^{-3}
23. (*a*) 1.32×10^7 A/m^2 in each; (*b*) $V_A = 8.90$ V, $V_B = 51.1$ V; (*c*) A: copper, B: iron
24. (*a*) 3.41 A or 0.586 A; (*b*) 0.293 V or 1.71 V
25. silicon: 85.0 Ω, iron: 915 Ω
26. 5.56 A
27. 4.00 Ω, in parallel
28. 4.0 Ω and 12 Ω
29. $i_1 = 50$ mA, $i_2 = 60$ mA, $V_{ab} = 9.0$ V
30. 0.00, 2.00, 2.40, 2.86, 3.00, 3.60, 3.75, 3.94 A
31. (*a*) 6.67 Ω; (*b*) 6.67 Ω; (*c*) 0
32. 4.50 Ω
33. (*a*) R_2; (*b*) R_1
34. $V_d - V_c = +0.25$ V, by all paths
35. 3d
36. three
37. 7.5 V
38. 38 Ω or 260 Ω
39. nine
40. (*a*) $R = r/2$; (*b*) $P_{\max} = \mathcal{E}^2/2r$
41. (*a*) series: $2\mathcal{E}/(2r + R)$, parallel: $2\mathcal{E}/(r + 2R)$; (*b*) series; (*c*) parallel
43. (*a*) left branch: 0.67 A down, center branch: 0.33 A up, right branch: 0.33 A up; (*b*) 3.3 V
44. (*a*) low position connects larger resistance, middle position connects smaller resistance; high position connects filaments in parallel; (*b*) 72 Ω, 144 Ω
45. (*a*) 120 Ω; (*b*) $i_1 = 50$ mA, $i_2 = i_3 = 20$ mA, $i_4 = 10$ mA
46. (*a*) 0.346 W, 0.050 W, 0.709 W; (*b*) 1.26 W, −0.158 W
47. (*a*) 19.5 Ω; (*b*) 0; (*c*) ∞; (*d*) 82.3 W, 57.6 W
48. 1.43 Ω
49. (*a*) 2.50 Ω; (*b*) 3.13 Ω
50. (*a*) Cu: 1.11 A, Al: 0.893 A; (*b*) 126 m
51. $\dfrac{100R(\mathcal{E}x/R_0)^2}{(100R/R_0 + 10x - x^2)^2}$, x in cm
52. (*a*) $3R/4$; (*b*) $5R/6$
53. (*a*) 13.5 kΩ; (*b*) 1500 Ω; (*c*) 167 Ω; (*d*) 1480 Ω
54. (*a*) put r roughly in the middle of its range; adjust current roughly with B; make fine adjustment with A; (*b*) relatively large percentage changes in A cause only small percentage changes in the resistance of the parallel combination, thus permitting fine adjustment; any change in A causes 1/2 as much change in this combination.
55. 0.45 A
56. $\mathcal{E}/7R$
57. (*a*) 12.5 V; (*b*) 50 A
58. −3.0%
59. 0.9%
62. (*a*) top: 70.9 mA, 4.70 V; bottom: 55.2 mA, 4.86 V; (*b*) top: 68.2 Ω; bottom: 88.0 Ω
65. (*a*) 2.52 s; (*b*) 21.6 μC; (*c*) 3.40 s
66. 4.6
67. (*a*) 0.41τ; (*b*) 1.1τ
68. (*a*) 2.41 μs; (*b*) 161 pF
69. (*a*) 2.17 s; (*b*) 39.6 mV
70. 2.35 MΩ
71. (*a*) 1.0×10^{-3} C; (*b*) 10^{-3} A; (*c*) $V_C = 10^3 e^{-t}$, $V_R = -10^3 e^{-2t}$, volts
72. (*a*) 0.955 μC/s; (*b*) 1.08 μW; (*c*) 2.74 μW; (*d*) 3.82 μW
73. 0.72 MΩ
76. 24.8 Ω to 14.9 kΩ
77. decreases by 13 μC
78. (*a*) at $t = 0$, $i_1 = 1.1$ mA, $i_2 = i_3 = 0.55$ mA; at $t = \infty$, $i_1 = i_2 = 0.82$ mA, $i_3 = 0$; (*c*) at $t = 0$, $V_2 = 400$ V; at $t = \infty$, $V_2 = 600$ V; (*d*) after several time constants ($\tau = 7.1$ s) have elapsed
79. (*a*) 6.9 km; (*b*) 20 Ω

Chapter 30

1. M/QT
2. particle 1 is positive, particles 2 and 4 are negative, particle 3 is neutral
3. (*a*) 9.56×10^{-14} N, 0; (*b*) 0.267°
4. (*a*) 400 km/s; (*b*) 835 eV

5. (*a*) $(6.2\times10^{-14}\,\mathbf{k})$ N; (*b*) $(-6.2\times10^{-14}\,\mathbf{k})$ N
6. $(0.75\,\mathbf{k})$ T
7. (*a*) east; (*b*) $6.28\times10^{14}\ \text{m/s}^2$; (*c*) 2.98 mm
8. $(-11.4\,\mathbf{i}-6.00\,\mathbf{j}+4.80\,\mathbf{k})$ V/m
9. 2
10. (*a*) 3.4×10^{-4} T, horizontal and to the left as viewed along $\mathbf{v}_0$; (*b*) yes, if velocity is the same as the electron's velocity
11. (*a*) 3.75 km/s
12. 0.27 mT
13. 680 kV/m
14. (*a*) *a* and *c*, *a* at higher potential; (*b*) *b* and *d*, *b* at higher potential; (*c*) no Hall voltage
16. (*a*) 0.67 mm/s; (*b*) $2.8\times10^{29}\ \text{m}^{-3}$
17. (*b*) 2.84×10^{-3}
18. 38.2 cm/s
19. 21 μT
20. (*a*) 0.34 mm; (*b*) 2.6 keV
21. 1.6×10^{-8} T
22. (*a*) 2.05×10^{7} m/s; (*b*) 467 μT; (*c*) 13.1 MHz; (*d*) 76.3 ns
23. (*a*) 1.11×10^{7} m/s; (*b*) 0.316 mm
24. 127 u
25. (*a*) 2.60×10^{6} m/s; (*b*) 0.109 μs; (*c*) 0.140 MeV; (*d*) 70 kV
26. (*a*) 0.978 MHz; (*b*) 96.4 cm
28. (*a*) 1.0 MeV; (*b*) 0.5 MeV
29. (*a*) $K_p = K_d = \frac{1}{2}K_\alpha$; (*b*) $R_d = R_\alpha = 14$ cm
30. $R_d = \sqrt{2}R_p$; $R_\alpha = R_p$.
32. (*a*) $B\sqrt{mg/2V}\,\Delta x$; (*b*) 8.2 mm
33. (*a*) 495 mT; (*b*) 22.7 mA; (*c*) 8.17 mJ
35. (*a*) 0.36 ns; (*b*) 0.17 mm; (*c*) 1.5 mm
36. (*a*) $-q$; (*b*) $\pi m/qB$
37. (*a*) 2.9998×10^{8} m/s
38. (*a*) 18 MHz; (*b*) 17 MeV
39. (*a*) 22 cm; (*b*) 21 MHz
40. (*a*) 8.5 MeV; (*b*) 0.80 T; (*c*) 34 MeV; (*d*) 24 MHz; (*e*) 34 MeV, 1.6 T, 34 MeV, 12 MHz
41. neutron moves tangent to original path, proton moves in a circular orbit of radius 25 cm
42. 240 m
43. (*b*)
44. 28.2 N, horizontally west
45. 20.1 N
46. 467 mA, from left to right
47. $(-2.5\times10^{-3}\,\mathbf{j}+0.75\times10^{-3}\,\mathbf{k})$ N
48. $Bitd/m$, away from generator
50. $(-0.35\,\mathbf{k})$ N
51. (*a*) 3.3×10^{8} A; (*b*) 1.0×10^{17} W; (*c*) totally unrealistic
52. 0.10 T, at 31° from the vertical
53. (*a*) 0, 1.38 mN, 1.38 mN
54. 4.3×10^{-3} N·m, the torque vector is parallel to the long side of the coil and points down
55. (*a*) 20 min; (*b*) 5.9×10^{-2} N·m
59. $2\pi aiB\sin\theta$, normal to the plane of the loop (up)
60. (*a*) 540 Ω, connected in series; (*b*) 2.52 Ω, connected in parallel
61. 2.45 A
62. (*a*) 12.7 A; (*b*) 0.0805 N·m
63. 2.08 GA
64. (*a*) 0.184 A·m²; (*b*) 1.45 N·m
65. (*a*) 0.30 J/T; (*b*) 0.024 N·m
66. (*a*) 2.86 A·m²; (*b*) 1.10 A·m²
67. (*a*) $(8.0\times10^{-4}\ \text{N}\cdot\text{m})(-1.2\,\mathbf{i}-0.90\,\mathbf{j}+1.0\,\mathbf{k})$; (*b*) -6.0×10^{-4} J
68. 0.335 A·m², 297° counterclockwise from the positive *y* direction, in the *yz* plane
69. 0.53 m
70. $B_{\text{min}} = \sqrt{2mV/ed^2}$
71. $qvaB/2$
72. $\mathbf{v} = v_{0x}\,\mathbf{i} + v_{0y}\cos(\omega t)\,\mathbf{j} - v_{0y}\sin(\omega t)\,\mathbf{k}$, where $\omega = eB/m$

Chapter 31

1. 7.7 mT
2. 32.1 A
3. (*a*) 3.3×10^{-6} T; (*b*) yes
4. 12 nT
5. (*a*) $(0.24\,\mathbf{i})$ nT; (*b*) 0; (*c*) $(43\,\mathbf{k})$ pT; (*d*) $(0.14\,\mathbf{k})$ nT
6. along a line parallel to the wire and 4.0 mm from it
7. (*a*) 16 A; (*b*) west to east
8. (*a*) $\mu_0 qvi/2\pi d$, antiparallel to *i*; (*b*) same magnitude, parallel to *i*
9. (*a*) 3.2×10^{-16} N, parallel to the current; (*b*) 3.2×10^{-16} N, radially outward if **v** is parallel to the current; (*c*) 0
10. 0
11. (*a*) 0; (*b*) $\mu_0 i/4R$, into the page; (*c*) same as (*b*)
13. $\dfrac{\mu_0 i}{4}\left(\dfrac{1}{R_1}-\dfrac{1}{R_2}\right)$, into page

14. (a) 1.0 mT, out of the figure; (b) 0.80 mT, out of the figure
15. 2 rad
16. $\frac{\mu_0 i\theta}{4\pi}\left(\frac{1}{b}-\frac{1}{a}\right)$, out of page
24. $\sqrt{2}\mu_0 i/8\pi a$, into page
25. 200 μT, into page
26. $(\mu_0 i/2\pi w)\ln(1+w/d)$, up
27. (a) it is impossible to have other than $B=0$ midway between them; (b) 30 A
28. 4.3 A, out of page
29. at all points between the wires, on a line parallel to them, at a distance d/a from the wire carrying current i
30. $(46.9\mathbf{j})\,\mu$T, $(18.8\mathbf{j})\,\mu$T, 0, $(-18.8\mathbf{j})\,\mu$T, $-46.9\,\mathbf{j}\,\mu$T
32. (a) 400 μT, to the left; (b) 400 μT, up
34. 80 μT, up
35. $0.338\mu_0 i^2/a$, toward the center of the square
36. $0.791\mu_0 i^2/\pi a$, 162° counterclockwise from the horizontal
37. (b) to the right
38. 3.2 mN, toward the wire
39. (b) 2.3 km/s
40. (a) $-2.0\mu_0$; (b) 0
41. $+5\mu_0 i_0$
43. 4.5×10^{-6} T · m
44. 1: $-2.0\mu_0$; 2: $-13\mu_0$
47. (a) $\mu_0 ir/2\pi c^2$; (b) $\mu_0 i/2\pi r$; (c) $\frac{\mu_0 i}{2\pi(a^2-b^2)}\left(\frac{a^2-r^2}{r}\right)$; ($d$) 0
48. $\mu_0 J_0 r^2/3a$
49. $3i_0/8$, into page
52. (a) $8.0\times10^{-3}d$, in T, where d is in meters; (b) 3200 A; (c) **k**
53. 5.71 mT
54. 0.30 mT
55. 108 m
56. (a) 533 μT; (b) 400 μT
60. (a) 4.77 cm; (b) 35.5 μT
61. 0.272 A
62. (a) negative; (b) 9.68 cm
63. 0.47 A · m^2
64. (a) 4; (b) 1/2
65. $8\mu_0 Ni/5\sqrt{5}R$
66. (a) 2.4 A · m^2; (b) 46 cm
67. (b) ia^2
68. (a) $\frac{\mu_0 i}{4}\left(\frac{1}{a}+\frac{1}{b}\right)$, into page; ($b$) $\frac{1}{2}i\pi(a^2+b^2)$, into page
71. (a) 79 μT; (b) 1.1×10^{-6} N · m
72. (b) $(0.060\,\mathbf{j})$ A · m^2; (c) $(9.6\times10^{-11})\mathbf{j}$ T, $(4.8\times10^{-11}\,\mathbf{i})$ T
73. (a) $(\mu_0 i/2R)(1+1/\pi)$, out of page; (b) $(\mu_0 i/2\pi R)(\sqrt{1+\pi^2}$, 18° out of page

Chapter 32

1. 57 μWb
2. $-\mu_0 nAi_0\omega\cos\omega T$
3. 1.5 mV
4. $B_0(\pi/\tau)r^2 e^{-t/\tau}$
5. (a) 31 mV; (b) right to left
6. (a) -11 mV; (b) 0; (c) 11 mV
7. $A^2B^2/R\Delta t$
8. (a) $1.1\times10^{-3}\,\Omega$; (b) 1.4 T/s
9. (b) 58 mA
10. (a) 30 mA
11. 1.2 mV
12. $(\mu_0 Nih/2\pi)\ln(b/a)$
13. 1.15 μWb
14. 3.66 μW
15. 51 mV, clockwise when viewed along the direction of **B**
16. (a) $\mu_0 iR^2\pi r^2/2x^3$; (b) $3\mu_0 i\pi R^2 r^2 v/2x^4$; ($c$) in the same direction as the current in the large loop
17. (b) no
18. 29.5 mC
19. (a) 21.7 V; (b) counterclockwise
20. (a) 24 μV; (b) from c to b
21. (a) 13 μWb/m; (b) 17%; (c) zero flux
22. 0
23. (a) 48.1 mV; (b) 2.67 mA
24. (a) 600 mV, up; (b) 1.5 A, clockwise; (c) 0.90 W; (d) 0.18 N; (e) same as (c)
25. $BiLt/m$, away from G
26. 0.452 V
27. (a) 85.2 T · m^2; (b) 56.8 V; (c) linearly
28. (a) f; (b) π^2a^2fB
29. (b) design it so that $Nab=(5/2\pi)\,\mathrm{m}^2$
30. 5.50 kV
31. 268 W
32. (a) $\mathcal{E}/BL$ to the left; (b) 0
33. 15.5 μC
34. (a) 240 μV; (b) 0.600 mA; (c) 0.144 μW;

(*d*) 2.88×10^{-8} N; (*e*) same as (*c*)
35. (*a*) 0.598 μV; (*b*) counterclockwise
36. 80 μV, clockwise
37. (*a*) $\frac{\mu_0 ia}{2\pi}\left(\frac{2r+b}{2r-b}\right)$;
(*b*) $2\mu_0 iabv/\pi R(4r^2 - b^2)$
39. (*a*) $3.4(2+\theta)$ mΩ, θ in rad; (*b*) 4.3θ mWb, θ in rad; (*c*) 2.0 rad; (*d*) 2.2 A
40. (*a*) 71.5 μV/m; (*b*) 143 μV/m
41. 1: −1.07 mV, 2: −2.40 mV, 3: 1.33 mV
42. 0.15 V/m
43. at *a*: 4.4×10^7 m/s^2, to the right; at *b*: 0; at *c*: 4.4×10^7 m/s^2, to the left
45. (*a*) 1st, 2nd, 5th, 6th; (*b*) 1st, 4th, 5th, 8th; (*c*) 1st, 5th
46. (*a*) 33.8 V/m; (*b*) 5.94×10^{12} m/s^2
48. $v_t = mgR/B^2L^2$
49. (*a*) 1.26×10^{-4} T, 0, -1.26×10^{-4} T; (*b*) 5.04×10^{-8} V
50. $\omega I_0(\mu_0 W/2\pi)(\ln \frac{\sqrt{y^2+W^2}}{y}) \cos\omega t$
51. (*a*) 0.40 V; (*b*) 20 A

Chapter 33

1. 0.10 μWb
2. (*a*) 2.45 mWb; (*b*) 0.645 mH
3. (*a*) 800; (*b*) 2.5×10^{-4} H
5. (*b*) so that the changing magnetic field of one does not induce current in the other;
(*c*) $L_{eq} = \sum_{j=1}^{N} L_j$
6. (*b*) see answer to 5*b*; (*c*) $\frac{1}{L_{eq}} = \sum_{j=1}^{N} \frac{1}{L_j}$
7. (*a*) $\mu_0 i/W$; (*b*) $\pi\mu_0 R^2/W$
9. (*a*) decreasing; (*b*) 0.68 mH
10. let the current change at 5.0 A/s
11. (*a*) 0.10 H; (*b*) 1.3 V
12. (*a*) 0.60 mH; (*b*) 120
13. (*a*) 16 kV; (*b*) 3.1 kV; (*c*) 23 kV
14. 12.3 s
15. 6.91
16. 46 Ω
17. 1.54 s
18. (*a*) $\mathcal{E}$; (*b*) $0.135\mathcal{E}$; (*c*) $0.693\tau_i$
19. (*a*) 8.45 ns; (*b*) 7.37 mA
20. (*a*) 4.7 mH; (*b*) 2.4 ms
21. $(42 + 20t)$ V
22. (*a*) 24 V; (*b*) 3.6 ms
23. 12.0 A/s
24. (*a*) 0.29 mH; (*b*) 0.27 ms
25. (*a*) $i_1 = i_2 = 3.33$ A; (*b*) $i_1 = 4.55$ A, $i_2 = 2.73$ A; (*c*) $i_1 = 0$, $i_2 = 1.82$ A;
(*d*) $i_1 = i_2 = 0$
26. I. (*a*) 2.0 A; (*b*) 0; (*c*) 2.0 A; (*d*) 0; (*e*) 10 V; (*f*) 2.0 A/s
II. (*a*) 2.0 A; (*b*) 1.0 A; (*c*) 3.0 A; (*d*) 10 V; (*e*) 0; (*f*) 0
27. (*a*) 1.5 s
28. (*a*) $i(1 - e^{-Rt/L})$
29. (*a*) 13.9 H; (*b*) 120 mA
30. $1.23\tau_L$
31. (*a*) 10 A; (*b*) 100 J
32. (*a*) 240 W; (*b*) 150 W; (*c*) 390 W
33. 25.6 ms
34. (*a*) 97.9 H; (*b*) 0.196 mJ
35. (*a*) 18.7 J; (*b*) 5.10 J; (*c*) 13.6 J
36. (*a*) 10.5 mJ; (*b*) 14.1 mJ
38. (*a*) 34.2 J/m^3; (*b*) 49.4 mJ
39. 5.58 A
40. 1.5×10^8 V/m
41. 3×10^{36} J
42. $(\mu_0 \ell/2\pi)\ln(b/a)$
43. (*a*) 1.3 mT; (*b*) 0.63 J/m^3
44. (*a*) $\mu_0 i^2N^2/8\pi^2r^2$; (*b*) 0.306 mJ; (*c*) 0.306 mJ
45. (*a*) 1.0 J/m^3; (*b*) 4.8×10^{-15} J/m^3
46. (*a*) 1.0×10^{-3} J/m^3; (*b*) 8.4×10^{15} J
47. (*a*) 1.67 mH; (*b*) 6.00 mWb
48. (*a*) 1.5 μWb, 100 mV; (*b*) 90 nWb, 12 mV
49. (*b*) have the turns of the two solenoids wrapped in opposite directions
51. magnetic field exists only within the cross section of the solenoid
52. magnetic field exists only within the cross section of solenoid 1
53. (*a*) $\frac{\mu_0 Ni}{2\pi}\ln\left(1+\frac{a}{b}\right)$; (*b*) 13 μH
54. QR/i_f
55. 1.25 H
56. $\frac{\mu_0}{2\pi}\left[S - \frac{2d}{\sqrt{3}}\ln\frac{\sqrt{3}S+2d}{2d}\right]$
57. $\frac{L_1}{L_1+L_2}\frac{\mathcal{E}}{R}$

Chapter 34

2. (a) stable; (b) unstable; (c) stable; (d) unstable
4. (a) 5.1×10^{11} V/m; (b) 19 mT; (c) 660
5. (b) in the direction of the angular momentum vector
6. (b) sign is minus; (c) no, compensating positive flux through open end near magnet
7. +3 Wb
8. 47 μWb, inward
9. $(\mu_0 iL/\pi)\ln 3$
10. 55 μT
11. 13 MWb, outward
12. (a) 600 MA; (b) yes; (c) no
14. (a) 31.0 μT, 0°; (b) 55.9 μT, 73.9°; (c) 62.0 μT, 90°
15. 1660 km
16. 383 μT
17. 61.1 μT, 84.2°
18. 0.48 K
19. 20.8 mJ/T
21. yes
22. (a) 160 T; (b) 600 T
23. (a) 3.7 K; (b) 1.3 K
24. (b) K_i/B, opposite to the field; (c) 310 A/m
27. $\Delta\mu = e^2 r^2 B/4m$
28. 25 km
29. (a) 3.0 μT; (b) 5.6×10^{-10} eV
30. 5.15×10^{-24} A $\cdot$ m^2
31. (a) 8.9 A $\cdot$ m^2; (b) 13 N $\cdot$ m
32. (a) 180 km; (b) 2.3×10^{-5}
34. (a) 0.14 A; (b) 79 μC
35. 840 J/T
36. (b) $\Delta\mu = r^2 q^2 B/4m$
38. $F_z = -\mu(\mathrm{d}B_z/\mathrm{d}z)$

Chapter 35

1. 9.14 nF
2. 0.115 A
3. 45.2 mA
4. (a) 1.17 μJ; (b) 5.58 mA
5. (a) 6.00 μs; (b) 167 kHz; (c) 3.00 μs
6. with n an integer: (a) $t = n(5.00\,\mu\text{s})$; (b) $t = (2n-1)(2.50\,\mu\text{s})$; ($c$) $t = (2n-1)(1.25\,\mu\text{s})$
7. (a) 89 rad/s; (b) 70 ms; (c) 25 μF
8. (a) 1.25 kg; (b) 3.72×10^4 N/m; (c) 1.75×10^{-4} m; (d) 3.02 cm/s
9. 38 μH
10. 1.59 μF
11. 7.0×10^{-4} s
15. (a) 3.0 nC; (b) 1.7 mA; (c) 4.5 nJ
16. (a) 5770 rad/s; (b) 1.09 ms
17. (a) 3.60 mH; (b) 1.33 kHz; (c) 0.188 ms
18. (a) 275 Hz; (b) 364 mA
19. 600, 710, 1100, 1300 Hz
20. (a) 0.689 μH; (b) 17.9 pJ; (c) 0.110 μC
21. (a) $Q/\sqrt{3}$; (b) 0.152
22. (a) $Q/2$; (b) $0.866I$
24. (a) 6.0:1; (b) 36 pF, 0.22 mH
25. (a) 1.98 μJ; (b) 5.56 μC; (c) 12.6 mA; (d) −46.9°; (e) +46.9°
26. (a) 0.180 mC; (b) $T/8$; (c) 66.7 W
27. (a) 356 μs; (b) 2.50 mH; (c) 3.20 mJ
28. ω
29. (a) 0; (b) $2i(t)$
30. Let T_2 (= 0.596 s) be the period of the inductor plus the 900 μF capacitor and let T_1 (= 0.199 s) be the period of the inductor plus the 100 μF capacitor. Close S_2, wait $T_2/4$; quickly close S_1, then open S_2; wait $T_1/4$ and then open S_1.
31. 8.66 mΩ
33. $(L/R)\ln 2$
34. 5.85 μC; 5.52 μC; 1.93 μC
35. (b) 2.10×10^{-3}
37. 1.84 kHz
38. (a) 2.35 mH; (b) they move away from 1.40 kHz
39. 1.13 kHz, 1.45 kHz, 1.78 kHz, 2.30 kHz
40. (a) 796 Hz; (b) no change; (c) decreased; (d) increased
41. (a) $\pi/2$ rad; (b) $q = (I/\omega')e^{-Rt/2L}\sin\omega' t$
42. $L = 100\pi R/\omega p$, $C = p/100\pi R\omega$
43. $L = R^2 C$

Chapter 36

1. 377 rad/s
2. (a) 0.283 A; (b) 2.26 A
3. (a) 0.955 A; (b) 0.119 A
4. (a) 0.600 A; (b) 0.600 A
5. (a) 4.60 kHz; (b) 26.6 nF; (c) $X_L = 2.60$ kΩ, $X_C = 0.650$ kΩ
6. (a) 8.84 kHz; (b) 6.00 Ω
7. (a) 0.65 kHz; (b) 24 Ω

8. (a) 5.22 mA; (b) 0; (c) 4.51 mA; (d) taking energy
9. (a) 39.1 mA; (b) 0; (c) 33.8 mA; (d) supplying energy
10. (a) 6.37 ms; (b) 11.2 ms; (c) inductor; (d) 138 mH
11. (a) 6.73 ms; (b) 2.24 ms; (c) capacitor; (d) 59.0 μF
13. (a) $X_C = 0$, $X_L = 86.7\,\Omega$, $Z = 182\,\Omega$, $I = 198\,\text{mA}$, $\phi = 28.5°$
14. (a) $X_C = 177\,\Omega$, $X_L = 0$, $Z = 239\,\Omega$, $I = 151\,\text{mA}$, $\phi = -47.9°$
15. (a) $X_C = 37.9\,\Omega$, $X_L = 86.7\,\Omega$, $Z = 167\,\Omega$, $I = 216\,\text{mA}$, $\phi = 17.1°$
16. (b) resistive
18. 1000 V
19. 89 Ω
20. (a) 36.0 V; (b) 27.3 V; (c) 17.0 V; (d) −8.34 V
21. (a) 224 rad/s; (b) 6.00 A; (c) 228 rad/s, 219 rad/s; (d) 0.039
22. (a) 39.1 Ω; (b) 21.7 Ω; (c) capacitive
23. (a) 45.0°; (b) 70.7 Ω
24. (a) resonance at $f = 1/2\pi\sqrt{LC} = 85.7$ Hz; (b) 15.6 μF; (c) 225 mA
26. 100 V
28. 165 Ω, 313 mH, 14.9 μF
29. 141 V
30. 1.84 A
31. 0, 9.00 W, 3.14 W, 1.82 W
33. 177 Ω
34. (a) 12.1 Ω; (b) 1.19 kW
35. 7.61 A
38. (a) 41.4 μW; (b) −17.0 μW; (c) 44.1 μW; (d) 14.1 μW
40. (a) 0.743; (b) it lags; (c) capacitive; (d) no; (e) yes, no, yes; (f) 33.4 W
41. (a) 117 μF; (b) 0; (c) 90.0 W, 0; (d) 0°, 90°; (e) 1, 0
42. (a) 23.6 mH; (b) 17.8 Ω
43. (a) 2.59 A; (b) 38.8 V, 159 V, 224 V, 64.2 V, 75.0 V; (c) 100 W for R, 0 for L and C.
44. 1000 V
45. (a) 2.4 V; (b) 3.2 mA, 0.16 A
46. step up: 5.00, 4.00, 1.25; step down: 0.800, 0.250, 0.200
47. (a) 1.9 V, 5.8 W; (b) 19 V, 0.58 W; (c) 0.19 kV, 58 kW
48. 10
49. (a) 16.6 Ω; (b) 422 Ω; (c) 0.521 A; (d) increases; (e) decreases; (f) increases
50. (a) 707 Ω; (b) 14.3 mH; (c) 1.95 μF

Chapter 37

3. At $r = 27.5$ mm and $r = 110$ mm
4. 1.9 pT
7. Change the potential difference across the plates at the rate of 1.0 MV/s
11. (a) 0.63 μT; (b) 2.3×10^{12} V/m · s
12. (a) 710 mA; (b) 0; (c) 1.1 A
13. (a) 2.0 A; (b) 2.3×10^{11} V/m · s; (c) 0.50 A; (d) 0.63 μT · m
14. (a) 5.5 mA
15. (a) 7.60 μA; (b) 859 kV · m/s; (c) 3.39 mm; (d) 5.16 pT
20. (a) $\frac{1}{2}\alpha t^2$; (b) $\alpha t^2/2\pi\epsilon_0 R^2$; ($d$) $\mu_0 \alpha r t/2\pi R^2$; ($e$) same results
21. (a) 0.324 V/m; (b) 2.87×10^{-16} A; (c) 2.87×10^{-18}
22. 0

Chapter 38

1. (a) 4.7×10^{-3} Hz; (b) 3 min 32 s
2. (a) 0.50 ms; (b) 8.4 min; (c) 5500 B.C.
3. (a) 4.5×10^{24} Hz; (b) 1.0×10^4 km or 1.6 earth radii
4. (a) 515 nm, 610 nm; (b) 555 nm, 5.41×10^{14} Hz, 1.85×10^{-15} s
6. 7.49 GHz
7. it would steadily increase; (b) the summed discrepancies between the apparent time of eclipse and those observed from x; the radius of the earth's orbit
8. 15 m
9. 5.0×10^{-21} H
11. 1.07 pT
12. $B_x = B_y = -0.67 \times 10^{-8} \cos[\pi \times 10^{15}(t - x/c)]$, $B_z = 0$ in SI units
16. 0.10 MJ
17. 4.8×10^{-29} W/m^2
18. 0.33 μT, in the $-x$ direction
19. 4.51×10^{-10}
20. 8.78×10^{-2} km^2
21. 89 cm
23. 1.2 MW/m^2
24. (a) 16.7 nT; (b) 33.1 mW/m^2

25. 820 m
27. (*a*) 1.03 kV/m; 3.43 μT
28. (*a*) 6.7 nT; (*b*) 5.3 mW/m^2; (*c*) 6.7 W
29. (*a*) $\pm EBa^2/\mu_0$ for faces parallel to the xy plane, zero through each of the other four faces; (*b*) 0
30. (*a*) 1.4×10^{-22} W; (*b*) 1.1×10^{15} W
31. (*a*) 83 W/m^2; (*b*) 1.7 MW
32. (*a*) 87 mV/m; (*b*) 0.30 nT; (*c*) 13 kW
33. (*a*) 3.5 μW/m^2; (*b*) 0.78 μW; (*c*) 1.5×10^{-17} W/m^2; (*d*) 110 nV/m; (*e*) 0.25 T
34. 3.3×10^{-8} N/m^2
35. 1.0×10^7 N/m^2
36. (*a*) 4.7×10^{-6} N/m^2; (*b*) 2.1×10^{10} times smaller
37. (*a*) 6.0×10^8 N; (*b*) $F_{\text{grav}} = 3.6 \times 10^{22}$ N
38. 5.9×10^{-8} N/m^2
39. (*a*) 100 MHz; (*b*) 1.0 μT along the z axis; (*c*) 2.1 m^{-1}, 6.3×10^8 rad/s; (*d*) 120 W/m^2; (*e*) 8.0×10^{-7} N, 4.0×10^{-7} N/m^2
40. (*a*) 3.97 GW/m^2; (*b*) 13.2 N/m^2; (*c*) 1.67×10^{-11} N; (*d*) 3.14×10^3 m/s^2
41. 491 nm
42. $I(2 - \mathit{frac})/c$
45. 1.9 mm/s
46. 0.48 km^2
47. (*b*) 580 nm
48. (*a*) $-y$ direction; (*b*) $E_z = -cB\sin(ky+\omega t)$, $E_x = E_y = 0$; (*c*) plane polarized with **E** along the z axis
49. (*a*) 1.9 V/m; (*b*) 1.7×10^{-11} N/m^2
50. 35°
51. 1/8
52. 0
53. 47%
54. 0.21
55. 20° or 70°
56. 4.4 W/m^2
57. 19 W/m^2
58. 2/3
59. (*a*) 2 sheets; (*b*) 5 sheets
60. (*a*) 0.16; (*b*) 0.84
61. $E_m^2 A/2\mu_0 cmc_s$
62. $\theta = \cos^{-1}(\sqrt{p/50})$
63. $p_r(\theta) = p_{r\perp}\cos^2\theta$
64. $4RI/c$

Chapter 39

1. (*a*) 38.0°; (*b*) 52.9°
2. 1.48
3. 1.26
4. (*a*) yes; (*b*) 1.3
5. 1.07 m
9. (*a*) 1.8 m; (*b*) 1.2 m
10. 42 mm
11. 401 cm beneath the mirror surface
13. (*b*) 0.00605°
15. 1.41
16. 34°
17. 1.22
18. (*a*) 49°; (*b*) 29°
19. 41.2°
20. 182 cm
21. (*a*) cover the center of each face with an opaque disk of radius 4.5 mm; (*b*) about 0.63
22. (*a*) no; (*b*) yes; (*c*) about 43°
23. (*a*) $\sqrt{1+\sin^2\theta}$; (*b*) $\sqrt{2}$
24. (*a*) 35.6°; (*b*) 53.1°
25. (*b*) 0.170
26. (*b*) 23.2°
27. (*b*) 51.6 ns
28. (*a*) 53°; (*b*) yes
29. 49.0°
30. (*a*) 55.5°; (*b*) 55.8°
31. (*a*) 1.60; (*b*) 58.0°
32. (*a*) $2v$; (*b*) v
33. 180°
34. 40 cm
35. 9.10 m
36. $d/3$, $5d/3$, $7d/3$, $11d/3$
37. a and c
38. 3
39. (*a*) 7; (*b*) 5; (*c*) 1 to 3, depending on the position of O and your perspective
40. 1.5 m
42. 2.2 mm^2
43. new illumination is 10/9 of the old
44. (*a*) virtual; (*b*) upright; (*c*) $m = +1$; (*d*) $D + L$
46. 10.5 cm
48. (*a*) +, +40, −20, +2.0, no, yes; (*b*) plane, ∞, ∞, −10, yes; (*c*) concave, +40, +60, −2.0, yes, no; (*d*) concave, +20, +40, +30, yes, no; (*e*) convex, −20, +20, +0.50, no,

yes; (*f*) convex, −, −40, −18, +180, no, yes; (*g*) −20, −, −, +50, +0.8, no, yes; (*h*) concave, +8, +16, +12, −, yes

50. (*b*) 0.56 cm/s; (*c*) 11 m/s; (*d*) 5.1 cm/s
51. (*a*) 2.00; (*b*) none
52. (*a*) −18, no; (*b*) −33, no; (*c*) +71, yes; (*d*) any n_2 possible, no; (*e*) +30, no; (*f*) +10, no; (*g*) −26, no; (*h*) 1.1, yes
53. $i = \dfrac{(2-n)r}{2(n-1)}$, to the right of the right side of the sphere
54. $i = -12$ cm
55. (*b*) separate the lenses by a distance $f_2 - f_1$, where f_2 is the focal length of the converging lens
56. f_1^2/f_2^2
57. 45 mm, 90 mm
58. 1.85 mm
59. (*a*) +40 cm; (*b*) at infinity
60. (*a*) converging; (*b*) diverging; (*c*) converging; (*d*) diverging
62. 5.1 mm
63. (*a*) 40 cm, real; (*b*) 80 cm, real; (*c*) 240 cm, real; (*d*) −40 cm, virtual; (*e*) −80 cm, virtual; (*f*) −240 cm, virtual
65. An X means that the quantity cannot be found from the given data: (*a*) +, X, X, +20, X, −1.0, yes, no; (*b*) converging, X, X, −10, X, +2.0, no, yes; (*c*) converging, +, X, X, −10, X, no, yes; (*d*) diverging, −, X, X, −3.3, X, no, yes; (*e*) converging, +30, −15, +1.5, no, yes; (*f*) diverging, −30, −7.5, +0.75, no, yes; (*g*) diverging, −120, −9.2, +0.92, no, yes; (*h*) diverging, −10, X, X, −5, X, +, no; (*i*) converging, +3.3, X, X, +5, X, no
66. upright, virtual, 30 cm to the left of the second lens
67. (*a*) 0.60 m on the side of the lens away from the mirror; (*b*) real; (*c*) upright; (*d*) +0.20
68. (*a*) final image coincides in location with the object; it is real inverted, and $m = -1.0$
69. (*a*) converging; (*b*) 26.7 cm; (*c*) 8.89 cm
70. (*a*) coincides in location with the original object and is enlarged 5.0 times; (*c*) virtual; (*d*) yes
73. 22 cm
76. (*a*) 13.0 cm; (*b*) 1.23 cm; (*c*) −3.25; (*d*) 3.13; (*e*) −10.2
77. 2.1 mm
78. (*b*) when image is at near point
79. (*a*) 2.35 cm; (*b*) decrease
80. (*b*) farsighted
81. (*a*) 5.3 cm; (*b*) 3.0 mm
82. (*b*) 8.4 mm; (*c*) 2.5 cm
83. −75
84. (*a*) −0.50 m; (*b*) diverging; (*c*) −2.0 diopters
86. (*a*) $m = 1 + (25\text{ cm})/f$; (*b*) $m = (25\text{ cm})/f$; (*c*) 3.5, 2.5
87. 1.14
88. 100 cm

Chapter 40

1. (*a*) 5.09×10^{14} Hz; (*b*) 388 nm; (*c*) 1.97×10^8 m/s
2. 4.55×10^7 m/s
4. 1.56
5. 2.1×10^8 m/s
6. 1.9×10^8 m/s
7. the time is longer for the pipeline containing air, by about 1.55 ns
9. 22°, refraction reduces θ
10. (*a*) pulse 2; (*b*) $0.03L/c$
11. (*a*) 3.60 μm; (*b*) intermediate, close to fully constructive interference
12. (*a*) 1.70 (or 0.70); (*b*) 1.70 (or 0.70); (*c*) 1.30 (or 0.30); (*d*) brightness is identical, close to fully destructive interference
13. (*a*) 1.55 μm; (*b*) 4.65 μm
14. (*a*) 0.833; (*b*) intermediate, closer to fully destructive interference
15. (*a*) 0.216 rad; (*b*) 12.4°
16. $(2m+1)\pi$
17. the distance D must also be doubled
18. 2.25 mm
19. 33 μm
20. 648 nm
21. (*a*) 0.010 rad; (*b*) 5.0 mm
22. 1.6 mm
23. 0.15°
24. 0.072 mm
25. 23.1 Hz
26. 600 nm
27. 8.75λ
28. 6.64 μm
30. 16
31. 24 μm

32. (*a*) 0.253 mm; (*b*) there is a minimum in place of the central maximum
33. 8.0 μm
34. 0.03%
36. $y = 27\sin(\omega t + 8.5°)$
37. 0
39. 2.65
40. (*a*) 1.17 m, 3.00 m, 7.50 m; (*b*) no
41. $y = 17\sin(\omega t + 13°)$
43. $I = \frac{1}{9}I_m[1+8\cos^2(\pi d\sin\theta/\lambda)]$, I_m = intensity of central maximum
44. $\lambda/4$
45. $L = (m+\frac{1}{2})\lambda/2$, for m = 0, 1, 2, ...
46. all
47. fully constructive interference
48. 0.117 μm, 0.352 μm
49. 131 nm
50. $\lambda/5$
51. 492 nm
52. 70.0 nm
53. 120 nm
54. none
55. (*a*) and (*c*)
56. (*a*) 552 nm; (*b*) 442 nm
57. 673 nm
58. 338 nm
59. (*a*) 169 nm; (*c*) blue-violet will be sharply reduced
61. 840 nm
62. (*a*) bright; (*b*) 590 nm
63. (*a*) $\lambda/2n_2$; (*b*) $\lambda/4n_2$; (*c*) $\lambda/2n_2$
64. intensity is diminished by 88% at 450 nm and by 94% at 650 nm
65. 141
66. (*b*) violet; (*c*) yellow-red end of visible spectrum
67. (*a*) 1800 nm; (*b*) 8
68. spacing decreases by a factor of 3/4
69. 1.89 μm
70. 2.4 μm
71. 1.00025
72. $\sqrt{(m+\frac{1}{2})\lambda R}$
73. (*a*) 34; (*b*) 46
74. 1.00 m
77. (*a*) π rad; (*b*) dark; (*c*) $2h\sin\theta = m\lambda$ (minima), $2h\sin\theta = (m+\frac{1}{2})\lambda$ (maxima)
78. 588 nm
79. 5.2 μm
80. 1.003
81. $DI = I_m\cos^2(2\pi x/\lambda)$
82. $2n_2L\cos\theta_r = (m+\frac{1}{2})\lambda$, for m = 0, 1, 2, ..., where $\theta_r = \sin^{-1}\left(\frac{\sin\theta_i}{n_2}\right)$
84. $x = (D/2a)(m+\frac{1}{2})\lambda$, for m = 0, 1, 2, ...
85. 0.354 m

Chapter 41

1. 690 nm
2. (*a*) 0.430°; (*b*) 0.118 mm
3. 60.4 μm
4. (*a*) $\lambda_a = 2\lambda_b$; (*b*) coincidences occur when $m_b = 2m_a$
5. (*a*) 2.5 mm; (*b*) 2.2×10^{-4} rad
6. 1.41
7. (*a*) 70 cm; (*b*) 1.0 mm
8. 1.77 mm
9. 41.2 m from the perpendicular to the speaker
10. 24.0 mm
11. 160°
12. (*a*) 0.18°; (*b*) 0.46 rad; (*c*) 0.93
15. (*d*) 53°, 10°, 5.1°
16. (*b*) 0, 4.493 rad, etc.; (*c*) −0.50, 0.93, etc.
18. (*a*) 1.3×10^{-4} rad; (*b*) 10 km
19. (*a*) 1.3×10^{-4} rad; (*b*) 21 m
20. 51 m
21. 30 m
22. 30.5 μm
23. (*a*) 1.1×10^4 km; (*b*) 11 km
24. 1600 km
25. 53 m
26. (*a*) 17.1 m; (*b*) 1.37×10^{-10}
27. (*a*) 19 cm; (*b*) larger
28. (*a*) 6.7°; (*b*) since $1.22\lambda > d$, there is no answer for 1.0 kHz
29. 4.7 cm
30. 36 cm
31. (*a*) 0.347°; (*b*) 0.97°
32. (*a*) 0.18″; (*b*) 8.4×10^7 km; (*c*) 0.025 mm
33. (*a*) red; (*b*) 100 μm
34. about 10^{-12}
35. five
36. three
38. (*a*) set $d = 4a$; (*b*) every fourth fringe is missing
39. $\lambda D/d$
40. (*a*) 9; (*b*) 0.255

41. (*a*) 5.05 μm; (*b*) 20.2 μm
43. (*a*) 3330 nm; (*b*) 0, $\pm 10.2°$, $\pm 20.7°$, $\pm 32.0°$, $\pm 45.0°$, $\pm 62.2°$
44. (*a*) 625 nm, 500 nm, 416 nm; (*b*) orange, blue-green, violet
45. all wavelengths shorter than 635 nm
46. three
47. 13,600
48. 500 nm
49. (*a*) 6.0 μm; (*b*) 1.5 μm; (*c*) $m = 0$, 1, 2, 3, 5, 6, 7, 9
50. (*a*) three; (*b*) 0.051°
51. 1100
52. 523 nm
58. 470 nm to 560 nm
60. 491
61. (*a*) 56 pm; (*b*) none
62. 3650
63. (*a*) 23,100; (*b*) 28.7°
64. (*a*) 1.0×10^4 nm; (*b*) 3.3 mm
66. (*a*) 0.032° /nm, 0.076° /nm, 0.24° /nm; (*b*) 40,000, 80,000, 120,000
67. (*a*) 2400 nm; (*b*) 800 nm; (*c*) $m = 0$, 1, 2
68. 1.79
70. 0.26 nm
71. 2.9°
72. 6.8°
73. 26 pm, 39 pm
74. (*a*) 170 pm; (*b*) 130 pm
75. 39.8 pm
77. yes, $m = 3$ for $\lambda = 130$ pm, $m = 4$ for $\lambda = 97.2$ pm
78. 0.570 nm
79. (*a*) $a_0/\sqrt{2}$, $a_0/\sqrt{5}$, $a_0/\sqrt{10}$, $a_0/\sqrt{13}$, $a_0/\sqrt{17}$
80. 30.6°, 15.3° (clockwise); 3.08°, 37.8° (counterclockwise)

Chapter 42

1. (*a*) 3×10^{-18}; (*b*) 2×10^{-12}; (*c*) 8.2×10^{-8}; (*d*) 6.4×10^{-6}; (*e*) 1.1×10^{-6}; (*f*) 3.7×10^{-5}; (*g*) 9.9×10^{-5}; (*h*) 0.10
2. (*a*) 6.7×10^{-10} s; (*b*) 2.2×10^{-18} m
3. $0.75c$
4. (*a*) 0.140; (*b*) 0.9550; (*c*) 0.999950; (*d*) 0.99999950
5. $0.99c$
6. 0.445 ps
7. 55 m
8. $0.99999950c$
9. 1.32 m
10. (*a*) $0.866c$; (*b*) 2.00
11. 1.53 cm
12. 6.4 cm
13. (*a*) 87.4 m; (*b*) 394 ns
14. (*a*) 26 y; (*b*) 52 y; (*c*) 3.7 y
15. (*a*) 2.21×10^{-12}; (*b*) 5.25 d
16. (*b*) $0.99999915c$
17. $x' = 138$ km, $t' = -374\ \mu$s
18. (*a*) $x' = 0$, $t' = 2.29$ s; (*b*) $x' = 6.55 \times 10^8$ m, $t' = 3.16$ s
19. $t'_1 = 0$, $t'_2 = -2.5\ \mu$s
20. (*a*) 25.8 μs; (*b*) 'red' flash, Doppler shifted
22. (*a*) 1.25; (*b*) 0.800 μs
23. (*a*) S' must move toward S, along their common axis, at a speed of $0.480c$; (*b*) 'red' flash, Doppler shifted; (*c*) 4.39 μs
24. 2.40 μs
25. $0.81c$
26. (*a*) $0.84c$, in the direction of increasing x; (*b*) $0.21c$, in the direction of increasing x; the classical predictions are $1.1c$ and $0.15c$
27. $0.95c$
28. (*a*) $0.35c$; (*b*) $0.62c$
29. $0.588c$, recession
30. 1.2 μs
31. (*a*) 34,000 mi/h; (*b*) 6.4×10^{-10}
32. seven
33. 22.9 MHz
34. (*b*) $0.8c$
35. +2.97 nm
36. yellow (550 nm)
37. (*a*) $\tau_0/\sqrt{1 - v^2/c^2}$
38. (*a*) 79 keV; (*b*) 3.11 MeV; (*c*) 10.9 MeV
39. (*a*) $0.134c$; (*b*) 4.65 keV; (*c*) 1.1%
40. (*a*) 0.0625, 1.00196; (*b*) 0.941, 2.96; (*c*) 0.99999987, 1960
41. (*a*) 0.9988, 20.6; (*b*) 0.145, 1.01; (*c*) 0.073, 1.0027
43. (*a*) 5.71 GeV, 6.65 GeV, 6.58 GeV/c; (*b*) 3.11 MeV, 3.62 MeV, 3.59 MeV/c
44. 88 kg
45. 18 smu/y
46. (*a*) 1.0 keV; (*b*) 1.1 MeV
47. (*a*) $0.943c$; (*b*) $0.866c$
48. (*a*) 256 kV; (*b*) $0.746c$
49. (*a*) 1.41; (*b*) $0.707c$; (*c*) $0.414mc^2$

50. $\sqrt{8}mc$
51. (*a*) the photon; (*b*) the proton; (*c*) the proton; (*d*) the photon
52. 6.65×10^6 mi, or 270 earth circumferences
53. (*c*) $207m_e$, the particle is a muon
54. 110 km
55. (*a*) $0.948c$; (*b*) 226 MeV; (*c*) 314 MeV/c
56. (*a*) 2.7×10^{14} J; (*b*) 1.8×10^7 kg; (*c*) 6.0×10^6
57. (*a*) 0.776 mm; (*b*) 16.0 mm; (*c*) 0.335 ns, no
58. 4.00 u, probably a helium nucleus
59. 630 km
60. 330 mT
61. (*a*) 534; (*b*) 0.99999825; (*c*) 2.23 T
62. (*a*) 65.4 per minute; (*b*) 1570 m
63. (*a*) 2.6 y; (*b*) 2.8 y; (*c*) 6.0 y
64. 0.57 m
65. $0.27c$

Chapter 43

2. 2.11 eV
3. 2.1 μm, infrared
4. (*a*) 5.9 μeV; (*b*) 2.047 eV
5. (*a*) 35.4 keV; (*b*) 8.57×10^{18} Hz; (*c*) 35.4 keV/c = 1.89×10^{-23} kg · m/s
6. 3.6×10^{-17} W
7. (*a*) 1.24×10^{20} Hz; (*b*) 2.43 pm; (*c*) 2.73×10^{-22} kg · m/s = 0.511 MeV/c
9. (*a*) the infrared bulb; (*b*) 6.0×10^{20}
10. (*a*) 3.61 kW; (*b*) $1.00 \times 10^{22}\ s^{-1}$; (*c*) 60.3
11. 4.7×10^{26}
12. (*a*) 5.5×10^{23} photons/m^2 · s; (*b*) 3.8×10^{30} photons/m^2 · s
13. (*a*) 2.96×10^{20} photons/s; (*b*) 48,600 km; (*c*) 280 m; (*d*) $5.89 \times 10^{18}\ m^{-2} \cdot s^{-1}$, $1.96 \times 10^{10}\ m^{-3}$
14. barium and lithium
15. 233 nm
16. (*a*) no; (*b*) 544 nm, green
17. 10 eV
18. 170 nm
19. 676 km/s
20. (*a*) 2.00 eV; (*b*) 0; (*c*) 2.00 V; (*d*) 295 nm
21. (*a*) 1.3 V; (*b*) 680 km/s
22. 1.07 eV
23. (*a*) 382 nm; (*b*) 1.82 eV
24. (*a*) 6.60×10^{-34} J · s; (*b*) 2.27 eV; (*c*) 545 nm
26. 9.68×10^{-20} A
27. (*a*) 3.1 keV; (*b*) 14 keV
29. (*a*) 2.7 pm; (*b*) 6.05 pm
30. (*a*) 2.43 pm; (*b*) 4.86 pm; (*c*) 0.255 MeV
31. (*a*) +4.8 pm; (*b*) −41 keV; (*c*) 41 keV
32. (*a*) 2.43 pm; (*b*) 1.32 fm; (*c*) 0.511 MeV; (*d*) 938 MeV
33. (*a*) 8.1×10^{-9}%; (*b*) 4.9×10^{-4}%; (*c*) 8.8%; (*d*) 66%
34. 300%
35. 2.65 fm
36. (*a*) 7.4°; (*b*) 0.62 eV
38. 44°
40. 1.12 keV
42. 500 nm, blue-green; the sensitivity of the eye peaks at 550 nm (Fig. 32–2), giving the sun its yellow hue
43. 9.99 μm
44. 5270 K
45. 91 K
46. (*a*) 1.45 m; (*b*) radio region
47. (*a*) 0.97 mm, microwave; (*b*) 9.9 μm, infrared; (*c*) 1.6 μm, infrared; (*d*) 0.26 nm, x ray; (*e*) 2.9×10^{-41} m, very high energy gamma ray
50. 10 kW
52. 192 mW
53. $4\,\Delta T/T$, 0.0130
54. 1.5 W
55. 2.6 eV
56. 1.17 eV
57. −80.7 keV
59. (*a*) 121.5 nm; (*b*) 91.2 nm
60. (*a*) 12 eV; (*b*) 6.5×10^{-27} kg·m/s; (*c*) 103 nm
61. 656 nm, 486 nm, 434 nm
62. four
63. (*a*) 12.7 eV; (*b*) 12.7 eV ($4 \to 1$), 2.55 eV ($4 \to 2$), 0.66 eV ($4 \to 3$), 12.1 eV ($3 \to 1$), 1.89 eV ($3 \to 2$), 10.2 eV ($2 \to 1$)
64. (*a*) 13.6 eV; (*b*) 3.40 eV
65. (*a*) 30.5 nm, 291 nm, 1050 nm; (*b*) 8.25×10^{14} Hz, 3.65×10^{14} Hz, 2.06×10^{14} Hz
66. (*a*) $n = 4$ to $n = 2$; (*b*) Balmer series
68. (*a*) 2.6 eV; (*b*) $n = 4$ to $n = 2$
69. 4.1 m/s
72. (*b*) 166
74. four
75. (*a*) 3×10^{74}; (*b*) no
76. (*a*) 1; (*b*) 0.0529 nm; (*c*) $h/2\pi$; (*d*) 1.99×10^{-24} kg · m/s;

(e) 4.14×10^{16} rad/s; (f) 2.19×10^6 m/s; (g) 8.26×10^{-8} N; (h) 9.07×10^{22} m/s^2; (i) 13.6 eV; (j) −27.2 eV; (k) −13.6 eV

77. (b) n^2; (c) n; (d) $1/n$; (e) $1/n^3$; (f) $1/n$; (g) $1/n^4$; (h) $1/n^4$; (i) $1/n^2$; (j) $1/n^2$; (k) $1/n^2$
78. (a) $nh/\pi md^2$; (b) $n^2h^2/4\pi^2md^2$
79. 4.0×10^{15} Hz
80. 7.8×10^{-21} J
81. 0.20 nm
82. 2×10^{-22} J

Chapter 44

1. (a) 1.7×10^{-35} m
3. 7.75 rm
4. (a) 38.7 pm; (b) 1.24 nm; (c) 904 fm
5. (a) 3.3×10^{-24} kg · m/s for each; (b) 38 eV for the electron, 6.2 keV for the photon
6. 4.3 μeV
7. (a) 38.8 meV; (b) 146 pm
8. (a) 3.96×10^6 m/s; (b) 81.9 kV
9. (a) 73 pm, 3.4 nm; (b) yes
10. (a) 1.24 μm, 1.22 nm; (b) 1.24 fm, 1.24 fm
11. (a) 1.24 keV, 1.50 eV; (b) 1.24 GeV, 1.24 GeV
12. (a) 1.9×10^{-21} kg · m/s; (b) 346 fm
13. 0.062 fm
14. (a) 5.2 fm
15. a neutron
16. (a) 15 keV; (b) 120 keV
17. 9.70 kV (relativistic calculation), 9.79 kV (classical calculation)
18. (a) 2190 km/s; (b) 3.32 nm; (c) $\lambda = 2\pi r$
19. 11.5°, 23.6°, 36.9°, 53.1°
20. (a) the beams are not present; (b) 46°
21. (a) 20.5 meV; (b) 37.7 eV
22. 850 pm
23. (a) 1900 MeV; (b) no
24. 650 meV
25. 90.2 eV
26. (a) 3.9×10^{-22} eV; (b) the thermal energy is about 10^{20} times as great; (b) 3.0×10^{-18} K
27. 18.1, 36.2, 54.3, 66.3, 72.4 μeV
29. (a) 0.196; (b) 0.608; (c) 0.196
31. 0.323
32. (a) 2150 nm^{-3}, 0; (b) 291 nm^{-3}, 10.2 nm^{-1}
33. 0.439
34. (b) $1.34r_B$
36. 1.2×10^{-14}
37. proton: 9.2×10^{-6}, deuteron: 7.6×10^{-8}
38. 5.1 eV
39. 10^{104} y
40. (a) −20.0%; (b) −10.0%; (c) +15.0%
41. 7×10^{-23} kg · m/s
42. 1.33×10^{-23} kg · m/s
43. 1.2 m
44. (a) 8.46 keV; (b) 9.94×10^{-23} kg · m/s; (c) 100 pm
45. 0.41 μeV, $E_2 = -3.4$ eV
47. (a) 124 keV; (b) 40.5 keV
48. $A = 1/\sqrt{L}$
49. $1.5r_B$
50. (a) 0.612; (b) 28°
51. (a) $E = U$; (b) $E = U + h/8mL^2$

Chapter 45

3. 3.64×10^{-34} J · s
4. (a) 3; (b) 3
5. 24.1°
6. $n = 4$; $\ell = 3$; $m_\ell = +3, +2, +1, 0, -1, -2, -3$; $m_s = \pm\frac{1}{2}$
7. $n > 3$; $m_\ell = +3, +2, +1, 0, -1, -2, -3$; $m_s = +\frac{1}{2}, -\frac{1}{2}$
8. $\ell = 4$; $n \geq 5$; $m_s \pm \frac{1}{2}$
9. 50
12. (a) 3×10^{74}; (b) 6×10^{74}; (c) 6×10^{-38} rad
14. (c) 0.18°, 0.057°, 0.018°
16. 19 h
17. 5.54 nm^{-1}
18. 1.8×10^{-4}
19. 0.0054
21. (b) 16.4 nm$^{-3/2}$
22. (a) −2.02 nm$^{-3/2}$; (b) 4.07 nm^{-3}; (c) 3.58 nm^{-1}
24. $0.764r_B$, $5.236r_B$
25. 0.981 nm^{-1}, 3.61 nm^{-1}
26. 1.5×10^{-15}
27. 0.0019
28. 0.045
29. 54.7°, 125°
30. 73 km/s
31. (a) 58 μeV; (b) 14 GHz; (c) 2.1 cm, short radio wave region
32. (a) 1.5×10^{-21} N; (b) 2.0×10^{-5} m
33. 5.35 cm
34. 51 mT

35. 19 mT
36. (a) 2.13 meV; (b) 18 T
37. all statements are true
38. $n = 1, \ell = 0, m_\ell = 0, m_s = \pm\frac{1}{2}$
39. (a) $(2, 0, 0, \pm\frac{1}{2})$; (b) $n = 2, \ell = 1, m_\ell = 1, 0, -1, m_s = \pm\frac{1}{2}$
40. (a) 18 (36 if indistinguishability is not taken into account); (b) 6, the states where both electrons share the quantum numbers $(n, \ell, m_\ell, m_s) = (2, 1, 1, \frac{1}{2}), (2, 1, 1, -\frac{1}{2}), (2, 1, 0, \frac{1}{2})$,
$(2, 1, 0, -\frac{1}{2}), (2, 1, -1, \frac{1}{2}), (2, 1, -1, -\frac{1}{2})$
42. only argon would remain a noble gas
45. 12.4 kV
46. (a) 5.7 keV; (b) 87 pm, 14 keV, 220 pm, 5.7 keV
47. 49.6 pm, 99.2 pm
49. (a) 24.8 pm; (b) and (c) remain unchanged
50. (a) 35.4 pm, as for molybdenum; (b) 57 pm; (c) 50 pm
51. 6.44 keV, 10.2 eV
52. 2.2 keV
53. 9/16
55. (a) 69.5 kV; (b) 17.9 pm; (c) 21.4 pm
56. (a) 19.7 keV, 17.5 keV; (b) Zr or Nb, Zr better
57. 282 pm
58. (b) 57.5; (c) 2070
59. (b) 24%, 15%, 11%, 7.9%, 6.5%, 4.7%, 3.5%, 2.5%, 2.0%, 1.5%
60. 9.1×10^{-7}
61. 10,000 K
62. -2.75×10^5 K
63. 4.4×10^{17}
64. (a) 3.60 mm; (b) 5.25×10^{17}
65. $2.0 \times 10^{16}\ \mathrm{s^{-1}}$
66. 2×10^7
67. 4.8 km
68. (a) 3.03×10^5; (b) 1430 MHz; (c) 3.30×10^{-6}
69. 1.8 pm
70. (a) none; (b) 51 J
71. (a) 7.33 μm; (b) 707 kW/m^2; (c) 24.9 GW/m^2
72. (a) no; (b) 140 nm

Chapter 46

1. 3460 atm
2. $5.90 \times 10^{28}\ \mathrm{m^{-3}}$
3. (a) $2.7 \times 10^{25}\ \mathrm{m^{-3}}$; (b) $8.38 \times 10^{28}\ \mathrm{m^{-3}}$; (c) 3100; (d) 3.3 nm for oxygen, 0.228 nm for the electrons
4. (a) 0.900; (b) 0.125; (c) sodium
7. $1.92 \times 10^{28}\ \mathrm{m^{-3} \cdot eV^{-1}}$
8. (a) 0; (b) 0.0955
9. (a) 6.81 eV; (b) $1.77 \times 10^{28}\ \mathrm{m^{-3} \cdot eV^{-1}}$; (c) $1.59 \times 10^{28}\ \mathrm{m^{-3}} \cdot eV^{-1}$
11. 5.33 eV
12. 1.36, 1.67, 0.90, 0.10, 0.00 $\times 10^{28}\ \mathrm{m^{-3} \cdot eV^{-1}}$
13. $T \gg 10^5$ K
14. (a) 0, 0.986, 0.500, 0.014, 0; (b) 700 K
15. 3
18. (a) $1.31 \times 10^{29}\ \mathrm{m^{-3}}$; (b) 9.43 eV; (c) 1820 km/s; (d) 0.40 nm
19. (a) $5.86 \times 10^{28}\ \mathrm{m^{-3}}$; (b) 5.52 eV; (c) 1390 km/s; (d) 0.522 nm
20. 234 keV
21. 137 MeV
24. (a) 0; (b) 0.00554; (c) 0.0185
25. 200° C
26. 0.029
28. 57.1 kJ
29. (a) 19.8 kJ; (b) 3 min 18 s
30. (a) 2.37×10^{-6}; (b) 2.37×10^{-6}
31. (a) n-type; (b) $5 \times 10^{21}\ \mathrm{m^{-3}}$; (c) 2.5×10^5
32. 0.22 μg
33. (a) pure: 4.78×10^{-10}; doped: 0.0141; (b) 0.824
34. (a) 0.744 eV; (b) 7.29×10^{-7}
36. 6.02×10^5
37. (b) 2.42×10^8
38. (a) 177 nm; (b) ultraviolet
39. 4.20 eV
40. opaque

Chapter 47

1. 15.8 fm
2. 28.3 MeV
3. (a) 0.39 MeV; (b) 4.61 MeV
4. 12 km
5. (a) six; (b) eight
6. 27
8. (a) ^{142}Nd, ^{143}Nd, ^{146}Nd, ^{148}Nd, ^{150}Nd; (b) ^{97}Rb, ^{98}Sr, ^{99}Y, ^{100}Sr, ^{101}Nb, ^{102}Mo, ^{103}Tc, ^{105}Rh, ^{109}In, ^{110}Sn, ^{111}Sb, ^{112}Te; (c) ^{60}Zn, ^{60}Cu, ^{60}Ni, ^{60}Co, ^{60}Fe
9. (a) 1150 MeV; (b) 4.8 MeV/nucleon,

12 MeV/proton

10. ^{127}I, ^{89}Y, 19
12. (a) 2.3×10^{17} kg/m^3 for each; (b) 1.0×10^{25} C/m^3 for ^{55}Mn, 8.8×10^{24} C/m^3 for ^{209}Bi
14. (b) 0.50%, 0.81%, 0.83%, 0.81%, 0.78%, 0.74%, 0.72%, 0.71%
15. 4×10^{-22} s
16. (a) 6.2 fm; (b) yes
17. $K \approx 30$ MeV
18. (a) 1.000000 u, 11.90683 u, 236.2025 u
20. ^{25}Mg: 9.30%, ^{26}Mg: 11.71%
21. (a) 19.8 MeV, 6.26 MeV, 2.22 MeV; (b) 28.3 MeV; (c) 7.07 MeV
22. 1.0087 u
23. 1.6×10^{25} MeV
24. (a) +7.29 MeV; (b) +8.07 MeV; (c) −91.0 MeV
25. 7.92 MeV
26. 3.0×10^{19}
27. 280 d
28. (a) 1/4; (b) 1/8
29. (a) $7.6 \times 10^{16}\,s^{-1}$; ($b$) $4.9 \times 10^{16}\,s^{-1}$
30. (a) 64.2 h; (b) 0.125; (c) 0.0749
31. (a) $4.8 \times 10^{-18}\,s^{-1}$; ($b$) 4.6×10^{9} y
32. (a) 5.04×10^{18}; (b) $4.60 \times 10^{6}\,s^{-1}$
33. 5.3×10^{22}
34. (a) 2.0×10^{20}; (b) $2.8 \times 10^{9}\,s^{-1} = 75$ mCi
35. 265 mg
36. (a) 59.5 d; (b) 1.18
37. 209 d
38. 1.13×10^{11} y
39. 87.8 mg
40. 660 mg
41. 730 cm^2
44. (a) $8.88 \times 10^{10}\,s^{-1}$; ($b$) $(8.88 \times 10^{10}\,s^{-1}\left[1 - e^{-(7.46\times 10^{-5})t}\right]$, t in seconds; (c) 1.19×10^{15}; (d) 0.11 μg
45. (a) $3.66 \times 10^{7}\,s^{-1}$; ($b$) $t \gg 3.82$ d; (c) $3.66 \times 10^{7}\,s^{-1}$; ($d$) 6.42 ng
47. Pu: 5.5×10^{-9}, Cm, zero
48. 4.269 MeV
49. (a) 4.25 MeV; (b) −24.1 MeV; (c) 28.3 MeV
50. (a) 5.98 MeV, 31.8 MeV; (b) 78 MeV
51. $Q_3 = -9.50$ MeV, $Q_4 = 4.66$ MeV, $Q_5 = -1.30$ MeV
52. ^{7}Li
53. 1.40 MeV
54. (b) $4n+3$, $4n$, $4n+2$, $4n+3$, $4n$, $4n+1$, $4n+2$, $4n+1$, $4n+1$
55. 0.782 MeV
56. (a) 900 fm; (b) 5.7 fm; (c) no; (d) yes
58. 600 keV
59. 0.961 MeV
60. (b) 2.6×10^{13} W
61. 78.4 eV
62. (a) U: 1.06×10^{19}, Pb: 0.624×10^{19}; (b) 1.69×10^{19}; (c) 2.98×10^{9} y
63. 1600 y
64. 132 μg
65. 1.7 mg
66. 3.92 nCi
67. 1.02 mg
68. 0.73 rem
69. (a) 18 mJ; (b) 0.29 rem
70. 13 mJ
71. (a) 6.3×10^{18}; (b) 2.5×10^{11}; (c) 0.20 J; (d) 0.23 rad; (e) 3.0 rem
72. (a) 6.6 MeV; (b) no
73. 3.87×10^{10} K
74. (a) ^{18}O, ^{60}Ni, ^{92}Mo, ^{144}Sm, ^{207}Pb; (b) ^{40}K, ^{91}Zr, ^{121}Sb, ^{143}Nd; (c) ^{13}C, ^{40}K, ^{49}Ti, ^{205}Tl, ^{207}Pb
75. (a) 25.35 MeV; (b) 12.8 MeV; (c) 25.0 MeV
76. (a) 3.55 MeV; (b) 7.72 MeV; (c) 3.26 MeV
77. (a) 3.85 MeV, 7.95 MeV; (b) 3.98 MeV, 7.33 MeV
78. (a) 7.21 MeV; (b) 12.0 MeV; (c) 8.69 MeV
80. hmc^2/E^2
81. (a) 1.93×10^{-12} m; (b) 4.7×10^{-15} m; (c) no

Chapter 48

1. (a) 2.6×10^{24}; (b) 8.2×10^{13} J; (c) 2.6×10^{4} y
2. 4.54×10^{26} MeV
3. $3.1 \times 10^{10}\,s^{-1}$
4. by rows: ^{95}Sr, ^{95}Y, ^{134}Te; 3
6. −23.0 MeV
7. +5.00 MeV
8. 181 MeV
9. (a) 16 fissions/day; (b) 4.3×10^{8}
11. (a) 10; (b) 226 MeV
12. (a) ^{153}Nd; (b) 110 MeV to ^{83}Ge, 60 MeV to ^{153}Nd; (c) 1.6×10^{7} m/s for ^{83}Ge, 8.7×10^{6} m/s for ^{153}Nd
13. (a) 252 MeV; (b) typical fission energy is 200 MeV

14. (a) +25%; (b) zero; (c) −82%
15. 463 kg
16. 617 kg
17. yes
18. 557 W
19. (a) 1.15 MeV; (b) 3.2 kg
20. $^{238}\mathrm{U} + n \rightarrow {}^{239}\mathrm{U} \rightarrow {}^{239}\mathrm{Np} + e$, $^{239}\mathrm{Np} \rightarrow {}^{239}\mathrm{Pu} + e$
21. (a) 44 kton
22. (a) 84 kg; (b) 1.7×10^{25}; (c) 1.3×10^{25}
24. 0.99938
25. 1.7×10^{16}
26. 8030 MW
27. (b) 1.0, 0.89, 0.28, 0.019; (c) 8
28. 3.6×10^9 y
29. (a) 75 kW; (b) 5800 kg
30. 1.7×10^9 y
32. 450 keV
33. (a) 30 MeV; (b) 6 MeV; (c) 170 keV
35. (a) 170 kV
36. 0.151
37. 1.41 MeV
38. (b) 500 km/s
41. (a) 3.1×10^{31} photons/m^3; (b) 1.2×10^6 times
43. (a) 4.0×10^{27} MeV; (b) 5.1×10^{26} MeV
44. (a) 4.3×10^9 kg/s; (b) 3.1×10^{-4}
45. (a) $1.83 \times 10^{38}\ \mathrm{s}^{-1}$; ($b$) $8.25 \times 10^{28}\ \mathrm{s}^{-1}$
46. (a) 4.1 eV/atom; (b) 9.0 MJ/kg; (c) 1500 y
48. 5×10^9 y
49. (a) 6.3×10^{14} J/kg; (b) 6.2×10^{11} kg/s; (c) 4.3×10^9 kg/s; (d) 15×10^9 y
50. 7.2×10^7 y
51. (a) 24.9 MeV; (b) 8.65 Mton
53. $K_\alpha = 3.5$ MeV, $K_n = 14.1$ MeV
54. 14.4 kW
56. (a) 900 km/s; (b) 200 nm
57. (a) 230 kJ; (b) 0.11 lb; (c) 23 MW
58. 3.6×10^{-14} J
59. $(1 - 2^{-2/3}) \dfrac{3}{5} \dfrac{Z^2 e^2}{4\pi\epsilon_0 R_0 A^{1/3}}$
60. 1.2×10^{23} J/s

Chapter 49

1. 6.03×10^{-29} kg
2. 9.19 fm
3. 2.4×10^{-43}
4. $\pi^- \rightarrow \mu^- + \overline{\nu}$
5. 1.08×10^{42} J
6. 31 nm
7. 27 cm/s
8. one
9. (a) 1.90×10^{-18} kg · m/s; (b) 9.90 m
10. 769 MeV
13. (a) strangeness; (b) charge; (c) energy
14. (b) boson, meson, $B = 0$
15. $Q = 0$, $B = -1$, $S = 0$
16. (b), (d)
17. (a) energy; (b) angular momentum; (c) charge
18. (a) −181 MeV; (b) 605 MeV
19. 338 MeV
20. (a) K^+; (b) $\overline{n}$; (c) π^0
22. (a) 37.7 MeV; (b) 5.35 MeV; (c) 32.4 MeV
23. (a) $\overline{u}\overline{u}\overline{d}$; ($b$) $\overline{u}\overline{d}\overline{d}$
24. (a) n; (b) Σ^+; (c) Ξ^-
25. (a) sud; (b) uss
26. (a) not possible; (b) uuu
29. Σ^0, 7530 km/s
30. 1.76×10^{10} ly
31. 665 nm
32. 3.59×10^8 ly
33. (b) 0.934; (c) 1.65×10^{10} ly
34. (b) 3.46 H-atoms/cm^3
35. (a) 256 μeV; (b) 4.84 mm
36. $102M$
37. (a) 122 m/s; (b) 246 y
38. (b) $2\pi r^{3/2}/\sqrt{GM}$
39. (a) 2.6 K; (b) 29 nm
40. (b) 2.38×10^9 K

SECTION SEVEN
COMPUTER PROJECTS

Computational physics has become a part of many introductory courses. Its inclusion serves two purposes: students use their programming skills and they learn to apply the fundamental principles they are studying to situations that are more complicated than those presented in their physics text. The act of writing and debugging a computer program often helps students understand more precisely the equations they are working with. Perhaps more importantly from a pedagogical viewpoint, a computer can be programmed to solve the same problem over and over with different input data. With proper instruction, a student can use this technique to develop insight into the physics being studied.

To help in these endeavors some computer projects are suggested on the following pages. In addition, outlines of programs are supplied to help with the work on the projects. The projects and programs are keyed to the chapter of *Fundamentals of Physics* in which the physical principles used are discussed. You or the students must flesh out the programs to make them compatible with the programming language or spreadsheet that is used and to make the input and output routines work on the specific computer that is used. You may want to revise some projects to bring them more in line with your goals for the course.

Many of the projects require graphs to be drawn. Beginners may want to use a spreadsheet with graphing capability or else a simple BASIC or Pascal program to generate the required data, then draw the graphs manually. More advanced programmers will want to include programming steps to draw the plots on a monitor screen. Either technique is satisfactory.

Projects may be assigned as homework or carried out in groups during the laboratory portion of the course. The latter is especially beneficial since the projects can then be used to generate class discussions of the physics.

Chapter 2

A computer can be used to calculate the position and velocity of an object when its acceleration is a known function of time. Divide the time axis into a large number of small intervals, each of duration Δt. If Δt is sufficiently small the acceleration can be approximated by a constant in each interval, perhaps with a different value in different intervals. If v_b is the velocity at the beginning of an interval then $v_e = v_b + a\Delta t$ can be used to calculate the velocity at the end of the interval. If a is the average acceleration for the interval then this expression is exact. In most cases, however, the average acceleration is not known and so must be approximated. For the projects in this section it is approximated by $[a(t_b) + a(t_e)]/2$, where $a(t_b)$ is the acceleration at the beginning of the interval and $a(t_e)$ is the acceleration at the end. For each of the projects below the function $a(t)$ must be supplied.

If x_b is the coordinate of the object at the beginning of the interval, then the coordinate at the end of the interval is given by $x_e = x_b + v\Delta t$, where v is the average velocity in the interval. Approximate v by $(v_b + v_e)/2$. Smaller values for Δt make the approximations for v_e and x_e better. Δt, however, cannot be taken to be so small that significance is lost when the computer sums the terms in these equations. Note that the coordinate and velocity at the end of any interval are the coordinate and velocity at the beginning of the next interval. A skeleton program might look like this:

```
input initial values: t_0, x_0, v_0
input final time and interval width: t_f, Δt
set t_b = t_0, x_b = x_0, and v_b = v_0
calculate acceleration at beginning of first interval: a_b = a(t_b)
begin loop over intervals
    calculate time at end of interval: t_e = t_b + Δt
    calculate acceleration at end of interval: a_e = a(t_e)
    calculate "average" acceleration in interval: a = (a_b + a_e)/2
    calculate velocity at end of interval: v_e = v_b + aΔt
    calculate "average" velocity in interval: v = (v_b + v_e)/2
    calculate coordinate at end of interval: x_e = x_b + vΔt
  * if t_e ≥ t_f then
        print or display t_b, x_b, v_b
        print or display t_e, x_e, v_e
        exit loop
    end of if statement
    set t_b = t_e, x_b = x_e, v_b = v_e, a_b = a_e in preparation for next interval
end loop over intervals
stop
```

The line marked with an asterisk will be different for different applications. You will change this line, for example, if you want to display results and exit the loop when the velocity or coordinate have certain values.

Because Δt is arbitrary, the end of the last interval may not correspond exactly to t_f. The first line of output corresponds to a time just before t_f and the second line corresponds to a time just after. You can force $t = t_f$ at the end of the last interval by asking the computer to recalculate Δt near the beginning of the program. First use $(t_f - t_0)/\Delta t$ to estimate the number of intervals, then round the result to the nearest integer N. Finally take $\Delta t = (t_f - t_0)/N$.

Most programming languages allow you to write a separate section of the program to define the function $a(t)$. Then the lines implementing $a_b = a(t_b)$ and $a_e = a(t_e)$ simply refer to the function definition. This is more efficient than writing the instructions for $a(t)$ twice, once for $t = t_b$ and once for $t = t_e$.

PROJECT 1. Test the program on a problem you can analyze analytically. Suppose the acceleration in m/s^2 is given by $12t$, for t in seconds. Take the initial coordinate (at $t = 0$) to be 5.0 m and the initial velocity to be $-120\,\text{m/s}$. Find the coordinate and velocity at $t = 4.5\,\text{s}$.

Start with $\Delta t = 0.5\,\text{s}$ and carry out the computation. Repeat several times, each time halving Δt, until you get the same answers to 3 significant figures on successive trials. Work the problem analytically to obtain an exact solution and carefully compare the answers with those obtained by the program. Check your program code if they differ.

PROJECT 2. Now use the program to determine when and where the particle is instantaneously stopped. You want the program to stop when the velocity changes sign in some interval. Use $v_b v_e \leq 0$ as the condition for displaying results and exiting the loop. The correct answer is between the two results displayed. If they are the same to within the number of significant figures you desire you are finished. If they are not, run the program again with a smaller value of Δt. Try to obtain 3 significant figure accuracy.

Now find when the particle is at $x = -100\,\text{m}$ and its velocity when it is there. The condition for displaying results and exiting the loop is now $(x_b + 100)(x_e + 100) \leq 0$. Obtain 3 significant figure accuracy.

PROJECT 3. Try a more complicated example. Suppose the particle starts at $x = 0$ with a velocity of $-120\,\text{m/s}$ and has an acceleration that is given in m/s^2 by $a(t) = 30e^{-t/8}$, where t is the time in seconds. Find the time at which it is instantaneously at rest and its position at that time. [ans: 5.55 s; −295 m]

You can use the program to generate a list of values ready to be plotted by hand or to plot values on the monitor screen. Suppose the time interval desired between displayed points is Δt_d. Input the value of Δt_d just after the values for t_f and Δt are read, then set $t_d = t_0 + \Delta t_d$. Replace the last statements of the program, from the if statement, with

> if $t_e \geq t_d$ then
>   print, display, or plot x_e or v_e
>   increment t_d by Δt_d
> end of if statement
> set $t_b = t_e$, $x_b = x_e$, $v_b = v_e$, $a_b = a_e$ in preparation for next interval
> if $t_e \geq t_f$ then exit loop
> **end loop** over intervals
> stop

Because the computer may produce values of t_e that are in error in the last place carried, the set of points you get may be somewhat different from what you anticipate. If this is intolerable, round the values of t_e so the position of the last non-zero digit is a few less than the number of figures carried by the machine.

PROJECT 4. A particle starts from rest at the origin and has an acceleration that is given in m/s^2 by $a(t) = 30te^{-t}$, where t is the time in seconds. Draw graphs of its coordinate and velocity as functions of time from $t = 0$ to $t = 5\,\text{s}$. Plot points every 0.5 s. You should obtain 2 significant figure accuracy with $N = 100$.

The acceleration is quite small near $t = 0$ and increases as t increases. This is the effect of the factor t. It reaches a maximum at $t = 1\,\text{s}$, then decreases. This is the effect of the exponential factor. Your graph of $v(t)$ should have its greatest slope in the vicinity of $t = 1\,\text{s}$, then approach a line with a slope of zero. The velocity has become nearly constant. What do you think the limiting value of the velocity is?

The graph of $x(t)$ should have a slope of zero at $t = 0$ (the initial velocity is zero) but it soon curves upward and eventually approaches a straight line, the slope of which is the limiting value of the velocity.

Chapter 4

The program outlined for Chapter 2 projects can be revised to investigate two-dimensional motion. You must now supply two components of the acceleration as functions of t and have the computer carry out calculations for each of the two components of the velocity and position vectors. The outline of a sample program might be:

> input initial values: t_0, x_0, y_0, v_{0x}, v_{0y}
> input final time and interval width: t_f, Δt
> set $t_b = t_0$, $x_b = x_0$, $y_b = y_0$, $v_{xb} = v_{0x}$, $v_{yb} = v_{0y}$
> calculate acceleration at beginning of first interval: $a_{xb} = a_x(t_b)$, $a_{yb} = a_y(t_b)$
> **begin loop** over intervals
>   calculate time at end of interval: $t_e = t_b + \Delta t$
>   calculate acceleration at end of interval: $a_{xe} = a_x(t_e)$, $a_{ye} = a_y(t_e)$

```
        calculate "average" acceleration: a_x = (a_xb + a_xe)/2, a_y = (a_yb + a_ye)/2
        calculate velocity at end of interval: v_xe = v_xb + a_x Δt, v_ye = v_yb + a_y Δt
        calculate "average" velocity : v_x = (v_xb + v_xe)/2, v_y = (v_yb + v_ye)/2
        calculate coordinates at end of interval: x_e = x_b + v_x Δt, y_e = y_b + v_y Δt
      * if t_e ≥ t_f then
            print or display t_b, x_b, y_b, v_xb, v_yb
            print or display t_e, x_e, y_e, v_xe, v_ye
          exit loop
        end of if statement
        set t_b = t_e, x_b = x_e, y_b = y_e, v_xb = v_xe, v_yb = v_ye, a_xb = a_xe, a_yb = a_ye
    end loop over intervals
    stop
```

As before, the line with the asterisk may be changed for other applications.

Instead of v_{0x} and v_{0y} you may wish to input the initial speed v_0 and angle ϕ_0 between the velocity and the x axis, then have the computer calculate v_{0x} and v_{0y} using $v_{0x} = v_0 \cos\phi_0$ and $v_{0y} = v_0 \sin\phi_0$. You may also wish to define $a_x(t)$ and $a_y(t)$ in a separate section of the program.

PROJECT 1. Start with a projectile motion problem that can be solved analytically. Suppose a projectile is fired over level ground at 350 m/s, at an angle of 25° above the horizontal. When does it reach the highest point? How high and how far down range is the highest point? When does it hit the ground and what is the range? Compare your answers with the analytic solutions.

If the x axis is horizontal and the y axis is positive in the upward direction then the components of the acceleration are given by $a_x = 0$ and $a_y = -9.8\,\text{m/s}^2$. To find the highest point exit the loop when $v_{ye} \le 0$. To find the range exit the loop when $y_e \le 0$. You must experiment a little to find an appropriate value for Δt. Start with 0.1 s and reduce it by a factor of 5 in successive calculations. Stop when you get the same results to 3 significant figures.

PROJECT 2. Now suppose the acceleration of the projectile has a horizontal component that varies with time, as a rocket might have. Take $a_x = 3.0t\,\text{m/s}^2$ and $a_y = -9.8\,\text{m/s}^2$, where t is in seconds. Use the same initial conditions (an initial velocity of 350 m/s, 25° above the horizontal) and find the time the projectile reaches its highest point and the coordinates of the highest point, assuming it is fired from the origin. Find the range over level ground and the velocity of the object just before it lands. [ans: highest point: $t = 15.1\,\text{s}$, $x = 6.51 \times 10^3\,\text{m}$, $y = 1.12 \times 10^3\,\text{m}$; range: $t = 30.2\,\text{s}$, $x = 2.33 \times 10^4\,\text{m}$, $v_x = 1.68 \times 10^3\,\text{m/s}$, $v_y = -148\,\text{m/s}$]

PROJECT 3. Now suppose the horizontal component of the acceleration is given in m/s² by $a_x(t) = 30te^{-t}$ while the y component is still $a_y = -9.8\,\text{m/s}^2$. The x axis is horizontal and the y axis is positive in the upward direction. The initial velocity is still 350 m/s, 25° above the horizontal. At what time does the object reach the highest point on its trajectory and what are the coordinates of that point? At what time does it return to the level of the firing point and what is the range? What are the components of its velocity just before landing? [ans: highest point: $t = 15.1\,\text{s}$, $x = 5.18 \times 10^3\,\text{m}$, $y = 1.12 \times 10^3\,\text{m}$; range: $t = 30.2\,\text{s}$, $x = 1.04 \times 10^4\,\text{m}$; velocity: $v_x = 347\,\text{m/s}$, $v_y = -148\,\text{m/s}$]

Chapter 6

The computer program outlined for the projects of Chapter 2 must be revised if the acceleration is a function of position or velocity. The average acceleration in an interval cannot be approximated by $a = (a_b + a_e)/2$ because a_e cannot be computed until v_e or x_e are known. You can, however, use the acceleration at the beginning of the interval and write $v_e = v_b + a_b\Delta t$. This is usually a poor approximation to the average acceleration, so compared to the program of Chapter 2, much smaller intervals must be used to obtain the same accuracy.

Errors arise when a large number of intervals are used because the computer normally carries only a small number of significant figures (eight or ten) and the last is often in error. These so-called truncation errors accumulate. You can decrease the effect significantly by carrying out the calculation in double precision.

The outline of a possible program for one-dimensional motion is:

input initial values: t_0, x_0, v_0
input final time and interval width: t_f, Δt
set $t_b = t_0$, $x_b = x_0$, $v_b = v_0$
begin loop over intervals
  calculate acceleration at beginning of interval: $a_b = a(t_b)$
  calculate velocity at end of interval: $v_e = v_b + a_b\Delta t$
  calculate "average" velocity: $v = (v_b + v_e)/2$
  calculate coordinate at end of interval: $x_e = x_b + v\Delta t$
  calculate time at end of interval: $t_e = t_b + \Delta t$
 * if $t_e \geq t_f$ then
    print or display t_b, x_b, v_b
    print or display t_e, x_e, v_e
    exit loop
  end of if statement
  set $t_b = t_e$, $x_b = x_e$, $v_b = v_e$
end loop over intervals
stop

You may want to revise the program so it displays values for a sequence of times. See the Computer Projects section for Chapter 2.

Sometimes air resistance must be taken into account when an object moves in the air. Consider an object that is moving vertically and take its acceleration to be given by $a = -g - (b/m)v$, where the positive direction is upward, m is the mass of the object, and b is a constant that depends on the interaction of the object with the air.

PROJECT 1. Suppose an object with $(b/m) = 0.100\,\text{s}^{-1}$ is fired upward from ground level with an initial speed of $100\,\text{m/s}$. On separate graphs plot the coordinate and velocity every $0.5\,\text{s}$ from $t = 0$ to $t = 17\,\text{s}$. First test the program to see what interval width it needs to obtain 2 significant figure accuracy over that range of time.

How much time does the projectile take to get to the highest point on its trajectory and how high is that point? How long does it take to get back to the ground and what is its velocity just before it reaches the ground? You might use your graph to obtain approximate answers, then use the program to refine them. [ans: highest point: 7.03 s, 311 m; ground: 16.2 s, 58.9 m/s]

Notice that the projectile takes longer to fall from the highest point than it does to reach that point and that it returns to ground level with a speed that is less than the firing speed. Compare your answers with those you would obtain if the object were in free fall. The time to reach the highest point is ______ and the highest point is ______ with air resistance than without. The total flight time is ______ and the speed on impact is ______ with air resistance than without.

PROJECT 2. Here's a problem for which the acceleration depends on position. Starting from rest at $x = 0$, a 3.5 kg box is dragged along the ground in the positive x direction by a constant force of 12 N acting horizontally. The ground is rougher toward larger x and the coefficient of kinetic friction increases according to $\mu_k = 0.070\sqrt{x}$. Use the program to find its position and velocity every second from $t = 0$ to the time it comes to rest again. At what time does the box come to rest? How far has it been dragged? [ans: 11.4 s; 56.2 m]

What is the maximum speed of the box? When does it have this speed? Where is it when it has this speed? [ans: 7.56 m/s; 4.84 s; 25.0 m from the starting point]

For the special case of an acceleration that is linear in the velocity $a = (a_b + a_e)/2$ *can* be used to calculate the velocity at the end of the interval. Here's how. The acceleration at the beginning of the interval is given by $a_b = -g - (b/m)v_b$ and the acceleration at the end of the interval is given by $a_e = -g - (b/m)v_e$, so $a = (a_b + a_e)/2 = -g - (b/2m)v_b - (b/2m)v_e$. The velocity at the end of the interval is given by $v_e = v_b + a\Delta t = v_b - g\Delta t - (b\Delta t/2m)v_b - (b\Delta t/2m)v_e$. Solve this expression for v_e. The result is $v_e = [v_b(1 - b\Delta t/2m) - g\Delta t]/(1 + b\Delta t/2m)$. If $h = b\Delta t/2m$ this becomes $v_e = [v_b(1 - h) - g\Delta t]/(1 + h)$. An outline of a program is:

```
input initial values: t0, x0, v0
input final time and interval width: tf, Δt
calculate parameter: h = bΔt/2m
set tb = t0, xb = x0, vb = v0
begin loop over intervals
    calculate velocity at end of interval: ve = [vb(1 − h) − gΔt]/(1 + h)
    calculate "average" velocity: v = (vb + ve)/2
    calculate coordinate at end of interval: xe = xb + vΔt
    calculate time at end of interval: te = tb + Δt
  * if te ≥ tf then
        print or display tb, xb, vb
        print or display te, xe, ve
        exit loop
    end of if statement
    set tb = te, xb = xe, vb = ve
end loop over intervals
stop
```

PROJECT 3. Use this program to find the position and velocity of the projectile of the first project at the end of 17 s. Notice that far fewer intervals are required to obtain 2 significant figure accuracy.

Now use the program to investigate the influence of the air on a projectile fired straight upward. Make a table of the time to reach the highest point, the coordinate of the highest point, the time to reach the ground, and the velocity just before reaching ground, all as functions of the resistance parameter b/m. Try $b/m = 1.00\,s^{-1}$, $0.100\,s^{-1}$, and $0.0100\,s^{-1}$.

This exercise is designed to give you some idea of the effect of air resistance. A javelin has a small value of b/m, a basketball has a larger value, and a ping-pong ball has a still larger value. The larger b/m the greater the effects.

You might try to find a value for the resistance coefficient b/m of a ping-pong ball or other object. Shoot it into the air with a spring gun, measure the initial speed with a photogate timer, and time its return to the firing level. Then try various values of b in your computer program until you find the value that reproduces the experimentally determined time of flight.

PROJECT 4. You can use the program to investigate the approach to terminal velocity. Suppose an object with $b/m = 0.100\,\mathrm{s}^{-1}$ is dropped from a high cliff. Use the program to plot its velocity every 2 s over the first minute of its fall. The speed should approach the value $mg/b = 98\,\mathrm{m/s}$. Repeat for objects with $b/m = 0.0500\,\mathrm{s}^{-1}$ and $0.500\,\mathrm{s}^{-1}$.

The program can be modified to deal with a projectile moving in two dimensions, and subjected to a drag force that is proportional to its velocity. The acceleration components are now given by $a_x = -(b/m)v_x$ and $a_y = -g - (b/m)v_y$. The outline of a program is:

```
input initial values: t0, x0, y0, v0x, v0y
input final time and interval width: tf, Δt
calculate parameter: h = bΔt/2m
set tb = t0, xb = x0, yb = y0, vxb = v0x, vyb = v0y
begin loop over intervals
     calculate velocity at end of interval:
          vxe = vxb(1 − h)/(1 + h)
          vye = [vyb(1 − h) − gΔt]/(1 + h)
     calculate "average" velocity: vx = (vxb + vxe)/2, vy = (vyb + vye)/2
     calculate coordinates at end of interval: xe = xb + vxΔt, ye = yb + vyΔt
     calculate time at end of interval: te = tb + Δt
   * if te ≥ tf then
          print or display tb, xb, yb, vxb, vyb
          print or display te, xe, ye, vxe, vye
          exit loop
     end of if statement
     set tb = te, xb = xe, yb = ye, vxb = vxe, vyb = vye
end loop over intervals
stop
```

PROJECT 5. A projectile with $b/m = 0.100\,\mathrm{s}^{-1}$ is fired over level ground with an initial velocity of 100 m/s, 45° above the horizontal. Find its coordinates every 0.5 s from $t = 0$ (the time of firing) to $t = 12.5\,\mathrm{s}$. Plot its trajectory. First find the value of Δt required to obtain 3 significant figure accuracy over the entire time interval.

Find the time the projectile reaches the highest point and the coordinates of the highest point. Take the x axis to be horizontal and the y axis to be positive in the upward direction. [ans: $t = 5.43\,\mathrm{s}$, $x = 296\,\mathrm{m}$, $y = 175\,\mathrm{m}$]

Find the time of flight and range. What are its velocity components just before it lands? [ans: $t = 12.1\,\mathrm{s}$, $x = 495\,\mathrm{m}$, $v_x = 21.2\,\mathrm{m/s}$, $v_y = -47.5\,\mathrm{m/s}$]

PROJECT 6. If air resistance is significant, maximum range is obtained for a firing angle that is different from 45°. Consider a projectile with $b/m = 0.100\,\mathrm{s}^{-1}$, fired with an initial speed of 100 m/s over level ground. Find the range to 3 significant figures for firing angles of 30°, 35°, and 40°. You already know the range for 45°. The firing angle for maximum range is between ______ and ______. You may wish to refine the interval to obtain the firing angle to 2 significant figures. [ans: 510 m; 519 m; 514 m]

PROJECT 7. Suppose the projectile of the last project ($b/m = 0.100$) is fired from the edge of a high cliff instead of over level ground. Assume the same initial conditions (fired from the origin with a speed of 100 m/s, 45° above the horizontal) and plot its trajectory for the first minute of its flight.

Notice that the trajectory is not symmetric about the highest point but is blunted in the forward direction. Near the end of the time interval the projectile is falling nearly straight down. Both the horizontal and vertical components of the velocity approach limiting values. The limiting value of the horizontal component is ______ and the limiting value of the vertical component is ______.

Chapter 7

A force with x component $F(x)$ does work given by $W = \int_{x_i}^{x_f} F(x)\,dx$ as the object on which it acts moves from x_i to x_f along the x axis. A computer can be used to evaluate integrals of this form. One of the simplest techniques to use is Simpson's rule. An interval of the x axis, from x_b to x_e, is divided into two equal parts of width Δx. Let F_b be the value of the force at the beginning of the interval ($x = x_b$), F_m be the value at the middle ($x = x_b + \Delta x$), and F_e be the value of the end ($x = x_e$). Then according to Simpson's rule the integral over the interval can be approximated by

$$\int_{x_b}^{x_e} F(x)\,dx = \frac{\Delta x}{3}(F_b + 4F_m + F_e)$$

This expression can be derived easily by fitting the function $F(x)$ to a quadratic of the form $F(x) = A_0 + A_1(x - x_b) + A_2(x - x_b)^2$. The coefficients are chosen so the quadratic yields the correct values of the function at $x = x_b$, $x = x_m$, and $x = x_e$. They are $A_0 = F_b$, $A_1 = -(F_e - 4F_m + 3F_b)/2\Delta x$, and $A_2 = (F_e - 2F_m + F_b)/2(\Delta x)^2$. The quadratic can be integrated easily. The approximation becomes better as the interval becomes narrower.

To evaluate an integral for an extended portion of the x axis divide the region from x_i to x_f into N intervals, where N is an even number. The interval width is given by $\Delta x = (x_f - x_i)/N$. Label the points x_0 $(= x_i)$, x_1, x_2, ..., x_N $(= x_f)$ and the corresponding values of the force F_0, F_1, F_2, ..., F_N. Apply the formula given above to each pair of intervals; for example, x_0 is the first point, x_1 is the second point, and x_2 is the third point in the first application; x_2 is the first point, x_3 is the second point, and x_4 is the third point in the next application. Except for x_0 and x_N, each point with an even label enters the final formula twice: once as the first point in an integration and once as a third point. Each point with an odd label enters as a midpoint of an integration. Thus

$$\int_{x_i}^{x_f} F(x)\,dx = \frac{\Delta x}{3}\Big[(F_N - F_0) + 2(F_0 + F_2 + F_4 + \dots + F_{N-2}) + 4(F_1 + F_3 + \dots + F_{N-1})\Big]$$

F_0 is subtracted at the beginning of the equation because it is also included in the sum over values with even labels. There it is multiplied by 2 but it should be included only once. F_N is not included in any of the sums but it must be included once in the final equation, so it is added at the beginning.

Write a computer program to evaluate a work integral for one-dimensional motion. Input the lower and upper limits of the integral and the number of intervals. You can force N to be an even integer by dividing the value read by 2, rounding the result to the nearest integer, then multiplying by 2. Calculate Δx. Now write a loop to sum all the force values with even labels and sum all the force values with odd labels. Instructions in the loop will be executed $N/2$ times. An outline might be:

input limits of integral: x_i, x_f
input number of intervals: N
replace N with nearest even integer
calculate interval width: $\Delta x = (x_f - x_i)/N$
initialize quantity to hold sum of values with even labels: $S_e = 0$
initialize quantity to hold sum of values with odd labels: $S_o = 0$
set $x = x_i$
begin loop over intervals: counter runs from 1 to $N/2$
    calculate $F(x)$ and add it to the sum of values with even labels:
        replace S_e with $S_e + F(x)$
    increment x by Δx
    calculate $F(x)$ and add it to the sum of values with odd labels:
        replace S_o with $S_o + F(x)$
    increment x by Δx
end loop
calculate force for upper and lower limits: $F_0 = F(x_i)$, $F_N = F(x_f)$
evaluate integral: $W = (\Delta x/3)(F_N - F_0 + 2S_e + 4S_o)$
display result
stop

You may want to define the force function $F(x)$ in a separate section of the program.

PROJECT 1. This problem can be solved analytically. Use it to check your program. A 2.00 kg block moves along the x axis, subjected to the force $\mathbf{F} = (6 - 4x)\,\mathbf{i}$, where $\mathbf{F}$ is in newtons and x is in meters. What work is done by this force as the block moves from $x = 0$ to $x = 1.00$ m? from $x = 0$ to $x = 5.00$ m? Try $N = 2$ and double its value on successive runs until you get the same first three significant figures. Check the answers by working the problem analytically. [ans: 4.00 J; −20.0 J]

If $\mathbf{F}$ is the total force acting on the block then the work that it does is equal to the change in the kinetic energy of the block. Suppose the block has a speed of 6.00 m/s when it is at $x = 0$. What is its speed when it is at $x = 1.00$ m? at 5.00 m? [ans: 6.32 m/s; 4.00 m/s]

PROJECT 2. A certain non-ideal spring exerts a force that is given by $F(x) = -250x - 125xe^{-x}$, where x is its extension (if positive) or its compression (if negative). A 2.3 kg block is attached to one end and the other end is held fixed. The block is pulled out so the spring is extended by 25 cm, then it is released from rest. How much work does the spring do on the block as the block moves from its initial position to the position for which the spring is neither extended or compressed? If the force of the spring is the only force acting on the block what is its speed as it passes this point? [ans: 11.1 J; 3.11 m/s]

The block continues to move, compressing the spring. By how much is it compressed when the block comes to rest instantaneously? Since the block is at rest both initially and finally the total work done by the spring over the entire trip is 0. Use trial and error to find the final value of x. Alternatively you might imbed the program in a loop in which x_f is incremented each time around and search for the values of x_f that straddle $W = 0$. [ans: −0.237 m]

If the force is given as a function of time and you are asked for the work it does over a given time interval you must change your strategy somewhat. If the force is the total force you can use one of the programs of Chapters 2 or 4 to find the velocity. The total work done, of course, is just the change in kinetic energy. If the force is not the total force you can use the power equation $dW/dt = Fv$ to write $W = \int Fv\,dt$. The program is the same as before except the integrand is now Fv. You must know the velocity as a function of time to use this method.

PROJECT 3. The position as a function of time for a 25-kg crate sliding across the floor (along the x axis) is given by $x(t) = 3te^{-0.60t}$, where x is in meters and t is in seconds. Numerically evaluate the integral for the work done by the total force during the first 2.0 s. You must find $v(t)$ and $F(t)$ by differentiating $x(t)$ and using $F = ma$. Select an integration interval for 3 significant figure accuracy. [ans: -112 J]

Use the expression you found for $v(t)$ to calculate the change in the kinetic energy during the first 2.0 s. Your result should agree with the result of the numerical integration.

PROJECT 4. A 2.0-kg block starts from rest at the origin and moves along the x axis. The total force acting on it is given by $F(t) = (8.0 + 12t^2)$, where F is in newtons and t is in seconds. Use analytical means to show that $v(t) = 4.0t + 2.0t^3$ and $x(t) = 2.0t^2 + 0.50t^4$.

One of the forces acting on the block is given by $F_1(t) = 3.0t^2$. Find the work done by F_1 during the first 3.0 s of the motion. Do this by analytical evaluation of $\int Fv\,dt$ and by numerical integration. [ans: 972 J]

PROJECT 5. The velocity of a body with mass m dropped from rest near the surface of the earth is given by $v(t) = (mg/b)(1 - e^{-bt/m})$, where b is the drag coefficient and down is taken to be the positive direction. Suppose an object with mass $m = 5.0$ kg and drag coefficient $b = 75$ kg/s is dropped from the edge of a high cliff. Use numerical integration to find the work done by gravity during the first 5.0 s. [ans: 5280 J]

The work done by gravity, of course, is given by mgs, where s is the distance fallen. The expression for $v(t)$ can be integrated with respect to time, with the result $s(t) = (mg/b)[t - (m/b) + (m/b)e^{-bt/m}]$. Use this expression to calculate the distance fallen in 5.0 s, then use $W = mgs$ to calculate the work done by gravity. You should obtain agreement with your numerical calculation.

If the object moves in two dimensions the integral for the work becomes $W = \int \mathbf{F} \cdot d\mathbf{s} = \int (F_x\,dx + F_y\,dy)$, where $d\mathbf{s} = \mathbf{i}\,dx + \mathbf{j}\,dy$ is an infinitesimal segment of the path. It is a vector tangent to the path. Because the object is on a specified path in the xy plane the infinitesimals dx and dy are related to each other. You must know the path to determine the relationship.

There are several ways of specifying a path. One way is to give the functional relationship between the coordinates of points on the path: $y = g(x)$, where $g(x)$ is a specified function. Then $dy = (dg/dx)\,dx$ and the work integral becomes $W = \int [F_x + (dg/dx)F_y]\,dx$. For example, a straight line with slope S and intercept A is given by $y = A + Sx$ and the work is given by $W = \int (F_x + SF_y)\,dx$. The limits of integration are the x coordinates of the points at the beginning and end of the path.

You must use the functional relationship between x and y in another way. The components of the force will be given as functions of both x and y but there can be only a single variable of integration. You must substitute the function $g(x)$ wherever y occurs in the expressions for the force components. The following is an example.

PROJECT 6. A particle moves along a straight line from $x = 1.5$ m, $y = 2.0$ m to $x = 3.0$ m, $y = 5.0$ m. One of the forces acting on it is given by $\mathbf{F} = 3x^2y^2\,\mathbf{i} + 2x^3y\,\mathbf{j}$. What work does this force do?

The path is given by $y = -1.0 + 2.0x$, so $dy = 2.0\,dx$. The force components are given by $F_x = 3x^2y^2 = 3x^2(-1 + 2x)^2$ and $F_y = 2x^3y = 2x^3(-1 + 2x)$. The work integral is $\int_{1.5}^{3.0}[3x^2(-1+2x)^2 + 4x^3(-1+2x)]\,dx$. The integration program can now be used to calculate the work done. [ans: 662 J]

The method does not work if the path is parallel to the y axis and it requires a large number of intervals if the path is nearly parallel to the y axis. Then you should use y as the variable of integration. The expression for work becomes $\int[(dx/dy)F_x + F_y]\,dy$.

A path can also be specified by giving both x and y as functions of some parameter. The parameter need not have any physical significance. For the straight line path of the previous project you might write $x = 1.5 + 1.5s$ and $y = 2.0 + 3.0s$. When the parameter $s = 0$ then $x = 1.5$ and $y = 2.0$. When $s = 1$ then $x = 3.0$ and $y = 5.0$. Now $dx = 1.5\,ds$ and $dy = 3.0\,ds$. The integral for the work becomes $W = \int_0^1[(dx/ds)F_x + (dy/ds)F_y]\,ds$. You must now substitute for both x and y in terms of s in the expressions for F_x and F_y.

If the path is an arc of a circle you might use the angle between the x axis and the radial line to the point as a parameter. Thus $x = R\cos\theta$ and $y = R\sin\theta$, where R is radius of the path. The integral for the work becomes $W = \int[-F_x\sin\theta + F_y\cos\theta]R\,d\theta$. Notice that $-F_x\sin\theta + F_y\cos\theta$ is the component of the force along a line that is tangent to the path and $R\,d\theta$ is the length of a path segment that subtends the angle $d\theta$ at the center of the circle.

PROJECT 7. A particle travels counterclockwise around a circular path with a radius of 2.0 m, centered at the origin. What work is done by the force $\mathbf{F} = 3x^2y^2\,\mathbf{i} + 2x^3y\,\mathbf{j}$ as the particle goes from $x = 2.0$ m, $y = 0$ to $x = -\sqrt{2.0}$ m, $y = \sqrt{2.0}$ m? This is an arc of $3\pi/4$ radians or 135°. [ans: −5.66 J]

Chapter 8

The force $\mathbf{F} = 3x^2y^2\,\mathbf{i} + 2x^3y\,\mathbf{j}$ is a conservative force. You can see what this means by computing the work it does as the particle on which it acts goes via different paths from the same initial point to the same final point.

PROJECT 1. Use the program of the previous chapter to calculate the work done by the force $\mathbf{F} = 3x^2y^2\,\mathbf{i} + 2x^3y\,\mathbf{j}$ as the particle goes from $x = 2.0$ m, $y = 0$ to $x = 4.0$ m, $y = 2.0$ m along each of the following paths.

a. Along the x axis from $x = 2.0$ m to $x = 4.0$ m, then along the line $x = 4.0$ m from $y = 0$ to $y = 2.0$ m. Carry out the integration in two parts, one for each segment.
b. Along the line $x = 2.0$ m from $y = 0$ to $y = 2.0$ m, then along the line $y = 2.0$ m from $x = 2.0$ m to $x = 4.0$ m.
c. Along the straight line that joins the to points. You might take $x = 2.0 + 2.0s$ and $y = 2.0s$, where s is a parameter that varies between 0 and 1.
d. Along the perimeter of a 2.0-m radius circle centered at $x = 4.0$ m, $y = 0$. For this path $x = 4.0 - 2.0\cos\theta$ and $y = 2.0\sin\theta$, where θ varies from 0 to $\pi/2$ radians (90°).

Notice that in every case the work done is 256 J.

PROJECT 2. Since the work done by the force of the previous project is independent of the path a potential energy function is associated with it. The value of this function at any point is the negative of work done by the force as the particle moves from some reference point to the point in question. The table below is actually a grid of points in the xy plane. The x coordinate is given at the top and the y coordinate is given at the left side. Take the origin to be the reference point ($U = 0$ there) and calculate the potential energy associated with each of the other points. You may want to automate the process by reading in the coordinates x_f and y_f of the point. Use the path given by $x = x_f s$ and $y = y_f s$, where s is a parameter than varies from 0 to 1. Since $dx/ds = x_f$ and $dy/ds = y_f$ the potential energy is given by $U(x_f, y_f) = -\int_0^1 (F_x x_f + F_y y_f)\, ds$. You must write the force components in terms of s. Fill in the table below with values of U.

$y(m)$ \ $x(m)$	0	1	2	3	4
4					
3					
2					
1					
0	0				

The analytic form for the potential energy function is $U(x, y) = -x^3y^2$. You can easily check this since the derivative of U with respect to x must give the negative of the x component of the force and the derivative of U with respect to y must give the negative of the y component. Use the exact analytic expression to check the results of your numerical calculations. You might construct a program consisting of two loops, one over values of the x coordinate and the other over values of the y coordinate, to compute and display values of U.

Use the table to compute the following quantities:

a. The change in the potential energy when the particle moves from $x = 2\,\text{m}$, $y = 1\,\text{m}$ to $x = 3\,\text{m}$, $y = 3\,\text{m}$. [ans: $-235\,\text{J}$]

b. The work done by the force when the particle moves from $x = 3\,\text{m}$, $y = 4\,\text{m}$ to $x = 1\,\text{m}$, $y = 1\,\text{m}$. [ans: $-431\,\text{J}$]

c. The change in the kinetic energy of the particle when it moves from $x = 3\,\text{m}$, $y = 1\,\text{m}$ to $x = 2\,\text{m}$, $y = 4\,\text{m}$. Assume only one force acts on it. [ans: $+101\,\text{J}$]

PROJECT 3. By way of contrast, the force $\mathbf{F} = (3x^2 - 6)y^2\,\mathbf{i} + 2x^3y\,\mathbf{j}$ is not conservative. Use the program to calculate the work done by the force as the particle goes from $x = 2.0\,\text{m}$, $y = 0$ to $x = 4.0\,\text{m}$, $y = 2.0\,\text{m}$ along each of the following paths.

a. Along the x axis from $x = 2.0\,\text{m}$ to $x = 4.0\,\text{m}$, then along the line $x = 4.0\,\text{m}$ from $y = 0$ to $y = 2.0\,\text{m}$.

b. Along the line $x = 2.0\,\text{m}$ from $y = 0$ to $y = 2.0\,\text{m}$, then along the line $y = 2.0\,\text{m}$ from $x = 2.0\,\text{m}$ to $x = 4.0\,\text{m}$.

c. Along the straight line that joins the to points.

d. Along the perimeter of a 2.0-m radius circle centered at $x = 4.0\,\text{m}$, $y = 0$.

Notice that the work done is different for different paths. You can easily see that it is impossible to assign a potential energy to each point in space so that the work done by the force equals the difference in the values assigned to the end points of the path. [ans: 256 J; 208 J; 240 J; 224 J]

Chapter 9

A computer can be used to follow individual objects in a system as they interact with each other and to follow the center of mass of the system. As an example consider two carts connected by an ideal spring on a horizontal frictionless air track. If the spring has a natural length ℓ_0 (when it is neither compressed or extended), then when its length is ℓ it exerts a force of magnitude $k|\ell - \ell_0|$ on each of the carts. Here k is the force constant for the spring. If the spring is extended ($\ell > \ell_0$) then both forces are toward the center of the spring. If the spring is compressed ($\ell < \ell_0$) then both forces are away from the center of the spring.

Let x_1 be the coordinate of one cart and x_2 be the coordinate of the other and take $x_2 > x_1$. Then $\ell = x_2 - x_1$. The force on cart 1 is given by $F_1 = k(x_2 - x_1 - \ell_0)$ and the force on cart 2 is given by $F_2 = -k(x_2 - x_1 - \ell_0)$.

A program that will calculate the coordinate and velocity of each cart as functions of time can be modeled after the first program of Chapter 6. In addition, the program calculates the coordinate and velocity of the center of mass. Here's an outline.

input initial values: t_0, x_{10}, v_{10}, x_{20}, v_{20}
input final time and interval width: t_f, Δt
input display interval: Δt_d
set $t_b = t_0$, $t_d = t_0 + \Delta t_d$, $x_{1b} = x_{10}$, $v_{1b} = v_{10}$, $x_{2b} = x_{20}$, $v_{2b} = v_{20}$
calculate coordinate and velocity of center of mass:
    $x_{\text{cm}} = (m_1 x_{1b} + m_2 x_{2b})/(m_1 + m_2)$
    $v_{\text{cm}} = (m_1 v_{1b} + m_2 v_{2b})/(m_1 + m_2)$
display x_{1b}, x_{2b}, x_{cm}
begin loop over intervals
    calculate force on 1 at beginning of interval: $F_1 = k(x_{2b} - x_{1b} - \ell_0)$
    calculate accelerations at beginning of interval: $a_1 = F_1/m_1$, $a_2 = -F_1/m_2$
    calculate velocities at end of interval: $v_{1e} = v_{1b} + a_1 \Delta t$, $v_{2e} = v_{2b} + a_2 \Delta t$
    calculate "average" velocities: $v_1 = (v_{1b} + v_{1e})/2$, $v_2 = (v_{2b} + v_{2e})/2$
    calculate coordinates at end of interval: $x_{1e} = x_{1b} + v_1 \Delta t$, $x_{2e} = x_{2b} + v_2 \Delta t$
    calculate time at end of interval: $t_e = t_b + \Delta t$
    if $t_e \geq t_d$ then
        calculate coordinate and velocity of center of mass:
            $x_{\text{cm}} = (m_1 x_{1e} + m_2 x_{2e})/(m_1 + m_2)$
            $v_{\text{cm}} = (m_1 v_{1e} + m_2 v_{2e})/(m_1 + m_2)$
      * display x_{1e}, x_{2e}, x_{cm}
        increment t_d by Δt_d
    end of if statement
    if $t_e \geq t_f$ then exit loop
    set $t_b = t_f$, $x_{1b} = x_{1e}$, $v_{1b} = v_{1e}$, $x_{2b} = x_{2e}$, $v_{2b} = v_{2e}$
end loop over intervals
stop

For some applications the line marked with an asterisk is replaced by: display v_{1e}, v_{2e}, v_{cm}. You will need to use a large number of extremely small intervals, so truncation errors may be significant. If possible, use double precision variables.

PROJECT 1. Take $m_1 = 250\,\text{g}$, $m_2 = 600\,\text{g}$, $\ell_0 = 0.50\,\text{m}$, and $k = 5.00\,\text{N/m}$. Initially ($t = 0$) cart 1 is at rest at the origin and cart 2 is at rest at $x_2 = 0.70\,\text{m}$. Plot the coordinates of the carts and the coordinate of the center of mass every 0.10 s from $t = 0$ to $t = 2.0\,\text{s}$. You will need to take $\Delta t = 0.02\,\text{s}$ to obtain 2 significant figure accuracy.

Notice that the center of mass is initially at rest and remains at rest even though the carts oscillate back and forth. Now repeat the calculation with $v_{20} = 0.30\,\text{m/s}$. The carts again oscillate but as they do the center of mass moves with constant velocity in the positive x direction. Check that the calculation gives the same result as $v_{cm} = (m_1 v_{10} + m_2 v_{20})/(m_1 + m_2)$ for the velocity of the center of mass.

For both of the situations considered the total momentum of the system, given by $P = (m_1 + m_2)v_{cm}$, is conserved. It is. Now suppose an external force of 0.50 N is applied to the first cart in the positive x direction. The acceleration of that cart is computed using $a_1 = (F_1 + 0.50)/m_1$, where F_1 is still the force of the spring on cart 1. The acceleration of cart 2 is $a_2 = -F_1/m_2$, as before. Now plot the coordinates of the carts and the coordinate of the center of mass every 0.10 s from $t = 0$ to $t = 2.0\,\text{s}$. The curve representing the coordinate of the center of mass should be parabolic. In fact, $x_{cm} = \frac{1}{2}a_{cm}t^2$, where $a_{cm} = F_e/(m_1 + m_2)$. Here F_e is the external force, 0.50 N. Plot v_{cm} as a function of time and verify that it is a straight line with the correct slope.

A computer program can also be used to investigate the motion of a rocket. If the rocket and fuel have mass $M(t)$ and the fuel is ejected with velocity **u**, measured relative to the rocket, then the acceleration of the rocket is given by

$$M\frac{d\mathbf{v}}{dt} = \mathbf{F} + \frac{dM}{dt}\mathbf{u}\,,$$

where **F** is the external force acting on the rocket (the force of gravity, for example).

First consider a rocket that is fired straight upward and assume the acceleration due to gravity is uniform. If the positive y axis is upward then $\mathbf{u} = -u\mathbf{j}$, $\mathbf{v} = v\mathbf{j}$, and $\mathbf{F} = -Mg\mathbf{j}$. If the fuel is expended uniformly you may write $M(t) = M_0 - Kt$, where K is a constant. The differential equation becomes

$$(M_0 - Kt)\frac{dv}{dt} = uK - (M_0 - Kt)g\,.$$

This expression gives the acceleration as a function of time and can be used in conjunction with the computer program of Chapter 2 to calculate the position and velocity of the rocket at any time until the rocket runs out of fuel.

The differential equation above can also be solved analytically to obtain

$$v(t) = v_0 - gt + u\ln\frac{M_0}{M_0 - Kt}\,,$$

where v_0 is the velocity at $t = 0$. You will use this expression in the next computer project to check your program.

PROJECT 2. Consider a rocket that carries 80% of its original mass as fuel and ejects it uniformly for 5.0 s, at which time the fuel is gone and the engine shuts off. The rocket starts from rest and the speed of the fuel relative to the rocket is $u = 5000\,\text{m/s}$. Use the program of Chapter 2 to find the speed and altitude of the rocket at burnout. First you need to find a value for K. At the end of 5.0 s M is the mass of the rocket alone and this is $0.20M_0$. Thus $M_0 - 5.0K = 0.20M_0$, so $K = 0.16M_0$. While fuel is being expended the acceleration of the rocket is given by $a(t) = -g + 0.16u/(1 - 0.16t)$. Use this expression in the program. Check your answer by using the analytic expression above to calculate the speed. [ans: 8.00×10^3 m/s; 1.48×10^4 m]

The next project deals with a rocket that moves in two dimensions. If the acceleration due to gravity is uniform then

$$\frac{d\mathbf{v}}{dt} = -\frac{K\mathbf{u}}{M_0 - Kt} - g\mathbf{j},$$

where the y axis is positive in the upward direction. Since $\mathbf{u}$ is always opposite to $\mathbf{v}$ we may write $\mathbf{u} = -u\mathbf{v}/v$, where the unit vector $\mathbf{v}/v$ is in the direction of $\mathbf{v}$. Thus

$$\frac{d\mathbf{v}}{dt} = \left[\frac{Ku}{M_0 - Kt}\right]\frac{\mathbf{v}}{v} - g\mathbf{j}.$$

This equation holds while fuel is being ejected. After burnout the rocket becomes an ordinary projectile and $d\mathbf{v}/dt = -g\mathbf{j}$.

Write a program to investigate the motion of a rocket fired from rest over level ground. Since each component of the acceleration depends on both components of the velocity you cannot use $\mathbf{s} = (\mathbf{a}_b + \mathbf{a}_e)/2$ to approximate the average acceleration in an interval. Use instead the acceleration at the beginning of the interval. For the first interval the program will have trouble with the unit vector $\mathbf{v}/v$ since it involves division by 0. For that interval replace v_x/v by $\cos\phi_0$ and v_y/v by $\sin\phi_0$, where ϕ_0 is the firing angle. At the end of the interval, after the velocity components have been computed, calculate $v = \sqrt{v_{xe}^2 + v_{ye}^2}$ and the ratios v_{xe}/v and v_{ye}/v, in preparation for the next interval. This can be handled using the unit vector components α_x and α_y as you will see in the outline below.

The program should provide for the possibility of different interval widths before and after burnout. Before burnout the acceleration may be great and both the speed and direction of travel may change rapidly. You will want to use a small interval. After burnout you will want to use a larger interval to save time. Here's an outline of a program.

input burnout time: t_{bo}
input firing angle: ϕ_0
input interval widths: $(\Delta t)_b$ for $t < t_{bo}$, $(\Delta t)_a$ for $t > t_{bo}$
set interval width: $\Delta t = (\Delta t)_b$
calculate unit vector components: $\alpha_x = \cos\phi_0$, $\alpha_y = \sin\phi_0$
set initial values: $t_b = 0$, $x_b = 0$, $v_{xb} = 0$, $y_b = 0$, $v_{yb} = 0$
begin loop over intervals
    calculate x component of acceleration at beginning of interval:
        if $t_b < t_{bo}$ then $a_x = \alpha_x Ku/(M_0 - Kt_b)$
        if $t_b \geq t_{bo}$ then $a_x = 0$
    calculate y component of acceleration at beginning of interval:
        if $t_b < t_{bo}$ then $a_y = -g + \alpha_y Ku/(M_0 - Kt_b)$
        if $t_b \geq t_{bo}$ then $a_y = -9.8$
    calculate velocity at end of interval: $v_{xe} = v_{xb} + a_x\Delta t$, $v_{ye} = v_{yb} + a_y\Delta t$
    calculate "average" velocity: $v_x = (v_{xb} + v_{xe})/2$, $v_y = (v_{yb} + v_{ye})/2$
    calculate x coordinates at end of interval: $x_e = x_b + v_x\Delta t$, $y_e = y_b + v_y\Delta t$
    calculate time at end of interval: $t_e = t_b + \Delta t$
  * if $t_e \geq t_f$ then
        display or print t_b, x_b, y_b, v_{xb}, v_{yb}
        display or print t_e, x_e, y_e, v_{xe}, v_{ye}
        exit loop
    end of if statement
    set $t_b = t_e$, $x_b = x_e$, $v_{xb} = v_{xe}$, $y_b = y_e$, $v_{yb} = v_{ye}$
    if $t_b < t_{bo}$ set $\Delta t = (\Delta t)_b$, if $t_b \geq t_{bo}$ set $\Delta t = (\Delta t)_a$

calculate unit vector components:
$v = \sqrt{v_{xe}^2 + v_{ye}^2}$, $\alpha_x = v_{xe}/v$, $\alpha_y = v_{ye}/v$
end loop over intervals when $y_e < 0$
stop

The line marked with an asterisk will be different for different applications. To calculate the coordinates and velocity for the highest point on the trajectory, end the loop when $v_{ye} \le 0$. To calculate the range over level ground, end the loop when $y_e \le 0$.

PROJECT 3. A toy rocket has a mass (without fuel) of 250 g. It carries 10 g of fuel and is fired from rest at 45° above the horizontal. The fuel leaves the rocket at 4000 m/s and burns out after 0.200 s of flight. Use the program to find the coordinates and velocity components at burnout ($t = 0.20$ s). For 2 significant figure accuracy use $\Delta t = 0.001$ s before burnout and $\Delta t = 0.01$ s after burnout. [ans: $x = 12.0$ m, $y = 10.7$ m, $v_x = 121$ m/s, $v_y = 108$ m/s]

Use the program to find the time the rocket reaches the highest point on its trajectory. Find its coordinates and speed when it is there. [ans: $t = 11.2$ s, $x = 1.35 \times 10^3$ m, $y = 605$ m, $v = 121$ m/s]

Use the program to find the time the rocket lands if it is fired over level ground. Find its coordinates and speed just before it lands. [ans: $t = 22.3$ s, $x = 2.70 \times 10^3$ m, $v_x = 121$ m/s, $v_y = -109$ m/s]

Chapter 10

Many details of a two-body collision are controlled by the impulse exerted by each object on the other. Consider the situation in which an object of mass m_1, moving with velocity v_{1b}, impinges on an object of mass m_2, initially at rest. Suppose that after the collision both objects move along the line of incidence. Let J be the impulse object 1 exerts on object 2. Then the velocity of object 2 after the collision is given by $v_{2a} = J/m_2$. Object 2 exerts an impulse $-J$ on object 1 and its velocity after the collision is given by $v_{1a} = v_{1b} - J/m_1$. Both final velocities can be computed if the initial velocity of object 1 and the impulse J are known (along with the masses, of course).

Once the velocities are known the change in the kinetic energy of the system can be found. Before the collision the kinetic energy is $K_b = \frac{1}{2}m_1 v_{1b}^2$ and after the collision it is $K_a = \frac{1}{2}m_1 v_{1a}^2 + \frac{1}{2}m_2 v_{2a}^2$. The kinetic energy lost during the collision is given by $Q = K_b - K_a$. A computer can be used to investigate the outcome of a one-dimensional collision as a function of the impulse.

Write a program that calculates the final velocities and the change in kinetic energy for a given range of impulses. Input the masses and the initial velocity of object 1. Also input the limits of the range of impulses to be considered. The outline of a program is:

input masses, initial velocity: m_1, m_2, v_{1b}
calculate initial kinetic energy: $K_b = \frac{1}{2}m_1 v_{1b}^2$
input limits of range of impulse: J_i, J_f
number of values of J: $N = 21$
calculate impulse increment: $\Delta J = (J_f - J_i)/(N-1)$
set $J = J_i$
begin loop over impulse values; counter runs from 1 to N
calculate final velocities: $v_{1a} = v_{1b} - J/m_1$, $v_{2a} = J/m_2$
calculate final kinetic energy: $K_a = \frac{1}{2}m_1 v_{1a}^2 + \frac{1}{2}m_2 v_{2a}^2$
calculate kinetic energy loss: $Q = K_b - K_a$
print or display J, v_{1a}, v_{2a}, Q

increment J: replace J with $J + \Delta J$
end loop
stop

The number N was chosen to be 21 because 21 lines nicely fill the usual monitor screen. You may want to use a different value. Both J_i and J_f are included in the list of values considered.

PROJECT 1. An air-track cart of mass $m_1 = 250\,\text{g}$ impinges at $3.0\,\text{m/s}$ on a cart of mass $m_2 = 600\,\text{g}$, initially at rest. Use the computer program to list the final velocities and kinetic energy loss for values of the impulse J from 0 to $2\,\text{kg·m/s}$.

From the table you generated pick out the two values of the impulse that straddle the condition $v_{1a} = v_{2a}$. This is the completely inelastic collision. Note that Q is a maximum in this region of the table. Rerun the program with a narrower range of impulse values. Repeat if necessary and find J, v_{1a}, and Q for a completely inelastic collision to 3 significant figure accuracy. [ans: $0.529\,\text{kg·m/s}$; $0.882\,\text{m/s}$; $0.794\,\text{J}$]

From the original table ($J = 0$ to $2\,\text{kg·m/s}$) pick out the two values that straddle the condition $v_{1a} = 0$. Find to 3 significant figures the impulse, the final velocity of cart 2, and the kinetic energy loss. [ans: $0.750\,\text{kg·m/s}$; $1.250\,\text{m/s}$; $0.656\,\text{J}$]

From the original table pick out the two values that straddle the condition $Q = 0$. This is the elastic collision. Find to 3 significant figures the impulse and final velocities. [ans: $1.06\,\text{kg·m/s}$; $-1.24\,\text{m/s}$; $1.77\,\text{m/s}$]

PROJECT 2. Now repeat the calculations for two carts with the same mass. Take $m_1 = m_2 = 250\,\text{g}$ and $v_{1b} = 3.0\,\text{m/s}$. Notice that for this special case the impulse that stops the incident cart is the same as the impulse that produces an elastic collision. [ans: completely inelastic: $J = 0.375\,\text{kg·m/s}$, $v_{1a} = v_{2a} = 1.50\,\text{m/s}$, $Q = 0.563\,\text{J}$; elastic: $J = 0.750\,\text{kg·m/s}$, $v_{1a} = 0$, $v_{2a} = 3.00\,\text{m/s}$, $Q = 0$]

Two-dimensional collisions are somewhat more difficult to analyze. Consider a collision between two pucks on an air table. One puck with mass m_1 and speed v_{1b} is incident along the x axis on a puck with mass m_2, initially at rest. Suppose the impulse exerted by the moving puck on the puck at rest has magnitude J and makes the angle α with the x axis. Puck 2 leaves the collision along the line of the impulse; its velocity makes the angle $\theta_2 = \alpha$ with the x axis. Its speed after the collision is given by $v_{2a} = J/m_2$. The impulse-momentum equations for the incident puck are $-J\cos\alpha = m_1 v_{1a}\cos\theta_1 - m_1 v_{1b}$ and $-J\sin\alpha = m_1 v_{1a}\sin\theta_1$. These can be solved for v_{1a} and θ_1, with the result $v_{1a} = \sqrt{(J^2 + m_1^2 v_{1b}^2 - 2m_1 v_{1b} J\cos\alpha)}\Big/ m_1$ and $\theta_1 = -\arctan[J\sin\alpha/(m_1 v_{1b} - J\cos\alpha)]$. To conserve momentum the sign of θ_1 must be opposite that of θ_2; that is, the two pucks must leave the collision on opposite sides of the line of incidence. If the sign of θ_1 produced by the computer is not correct, add 180° to it or subtract 180° from it.

Write a program that accepts values for the masses, initial velocity of the incident puck, and angle of the impulse with the x axis, then calculates the magnitudes and directions of the velocities after the collision and the loss in kinetic energy, all as functions of the magnitude of the impulse. An outline might be:

input masses, initial velocity: m_1, m_2, v_{1b}
calculate initial kinetic energy: $K_a = \frac{1}{2}m_1 v_{1a}^2$
input angle of impulse: α
input limits of range of impulse: J_i, J_f
number of values of J: $N = 21$
calculate impulse increment: $\Delta J = (J_f - J_i)/(N-1)$
set $J = J_i$

begin loop over impulse values; counter runs from 1 to $N+1$
calculate final speeds:
$v_{1a} = \sqrt{(J^2 + m_1^2 v_{1b}^2 - 2m_1 v_{1b} J \cos\alpha)} \Big/ m_1$
$v_{2a} = J/m_2$
calculate angles:
$\theta_1 = -\arctan[J \sin\alpha/(m_1 v_{1b} - J\cos\alpha)]$
$\theta_2 = \alpha$
check sign of θ_1 and correct if necessary
calculate final kinetic energy: $K_a = \frac{1}{2}m_1 v_{1a}^2 + \frac{1}{2}m_2 v_{2a}^2$
calculate kinetic energy loss: $Q = K_b - K_a$
print or display J, v_{1a}, v_{2a}, θ_1, θ_2, Q
increment J: replace J with $J + \Delta J$
end loop
stop

PROJECT 3. Puck 1 has a mass of 250 g and is incident at 3.0 m/s on puck 2, which has a mass of 600 g and is initially at rest. The impulse of puck 1 on puck 2 makes an angle of 20° with the axis of incidence (the x axis). Use the program to construct a table of the final speeds, angles of motion, and kinetic energy losses for impulses with magnitudes in the range from 0 to 2 kg·m/s.

Use the table to find two values of the impulse magnitude that straddle the value for an elastic collision. Rerun the program to find the impulse for an elastic collision to 3 significant figures. What are the final speeds and angle of motion then? [ans: $J =$ 0.995 kg·m/s; $\mathbf{v}_{1a} = 1.55$ m/s at −119°; $\mathbf{v}_{2a} = 1.66$ m/s at 20°]

Now find the magnitude of the impulse for which puck 2 is scattered through 90°. What are the velocities of the pucks after the collision and what kinetic energy is lost? [ans: $J = 0.798$ kg·m/s; $\mathbf{v}_{1a} = 1.09$ m/s at −90°; $\mathbf{v}_{2a} = 1.33$ m/s at 20°; $Q = 1.33$ J]

Now suppose the impulse makes an angle of 60° with the line of incidence. What magnitude of impulse produces an elastic collision? What are the final velocities? [ans: $J = 0.529$ kg·m/s; $\mathbf{v}_{1a} = 2.67$ m/s at −43.4°; $\mathbf{v}_{2a} = 0.882$ m/s at 60°]

For what magnitude of impulse is puck 1 scattered through 90°? What are the final velocities and the kinetic energy loss? [ans: $J = 1.50$ kg·m/s; $\mathbf{v}_{1a} = 5.20$ m/s at −90°; $\mathbf{v}_{2a} = 2.50$ m/s at 60°; $Q = -4.13$ J]

The negative value of Q in the last case means kinetic energy must be *added* during the collision (in an explosion, for example).

PROJECT 4. What happens if the incident puck has greater mass than the target puck? Take $m_1 = 600$ g, $v_{1b} = 3.0$ m/s, and $m_2 = 250$ g. For $\alpha = 20°$ and 60° calculate the magnitude of the impulse that will produce an elastic collision and find the velocities of the pucks after such a collision. [ans: 20°: $J = 0.995$ kg·m/s, $\mathbf{v}_{1a} = 1.55$ m/s at −21.5°, $\mathbf{v}_{2a} = 3.78$ m/s at 20°; 60°: $J = 0.529$ kg·m/s, $\mathbf{v}_{1a} = 2.67$ m/s at −16.6°, $\mathbf{v}_{2a} = 2.12$ m/s at 60°]

PROJECT 5. If the masses are equal the pucks always leave an elastic collision with their lines of motion making an angle of 90° with each other. Use the program to verify this for $m_1 = m_2 = 250$ g, $v_{1b} = 3.0$ m/s, and $\alpha = 20°$. Also verify it for $\alpha = 60°$. [ans: 20°: $J = 0.705$ kg·m/s, $\mathbf{v}_{1a} = 1.03$ m/s at −70.0°, $\mathbf{v}_{2a} = 2.82$ m/s at 20°; 60°: $J = 0.375$ kg·m/s, $\mathbf{v}_{1a} = 2.60$ m/s at −30.0°, $\mathbf{v}_{2a} = 1.50$ m/s at 60°]

Chapter 12

Conservation of angular momentum problems usually involve a system of two or more parts that exert torques on each other. The net external torque, however, vanishes. The angular momentum of each part changes, but the sum of the changes vanishes. You can use a computer to examine details of the motion.

Consider two wheels that are free to rotate on the same axle and that exert torques on each other. If the torque on wheel 1 is τ, then the angular velocity ω_1 of that wheel obeys $\tau = I_1\, d\omega_1/dt$, where I_1 is the rotational inertia of the wheel. The torque on wheel 2 is $-\tau$ and the angular velocity of that wheel obeys $-\tau = I_2\, d\omega_2/dt$. If τ is given as a function of time a computer can be used to solve for θ_1, ω_1, θ_2, and ω_2, all as functions of time. Here's an outline, modeled after the program given in Chapter 2. The program also contains instructions to compute the angular momenta of the wheels and the total angular momentum of the system so you can check on angular momentum conservation.

input initial angular velocities: ω_{10}, ω_{20}
input final time, interval width: t_f, Δt
input display interval: Δt_d
set $t_b = 0$, $t_d = \Delta t_d$, $\theta_{1b} = 0$, $\omega_{1b} = \omega_{10}$, $\omega_{2b} = \omega_{20}$
calculate torque on 1: $\tau_b = \tau(t_b)$
begin loop over intervals
    calculate time at end of interval: $t_e = t_b + \Delta t$
    calculate torque on 1 at end of interval: $\tau_e = \tau(t_e)$
    calculate "average" torque: $\tau = (\tau_b + \tau_e)/2$
    calculate angular velocities at end of interval:
      $\omega_{1e} = \omega_{1b} + \tau\Delta t/I_1$
      $\omega_{2e} = \omega_{2b} - \tau\Delta t/I_2$
    calculate "average" angular velocities:
      $\omega_1 = (\omega_{1b} + \omega_{1e})/2$
      $\omega_2 = (\omega_{2b} + \omega_{2e})/2$
    calculate angular positions at end of interval:
      $\theta_{1e} = \theta_{1b} + \omega_1\Delta t$
      $\theta_{2e} = \theta_{2b} + \omega_2\Delta t$
    if $t_e \geq t_d$ then
        calculate angular momenta: $L_1 = I_1\omega_{1e}$, $L_2 = I_2\omega_{2e}$, $L = L_1 + L_2$
        print or display t_e, θ_{1e}, ω_{1e}, θ_{2e}, ω_{2e}, L_1, L_2, L
        increment t_d by Δt_d
    end of if statement
    if $t_e \geq t_f$ then exit loop
    set $t_b = t_e$, $\theta_{1b} = \theta_{1e}$, $\omega_{1b} = \omega_{1e}$, $\theta_{2b} = \theta_{2e}$, $\omega_{2b} = \omega_{2e}$, $\tau_b = \tau_e$
end loop over intervals
stop

PROJECT 1. Suppose wheel 1 has a rotational inertia of $2.5\,\text{kg}\cdot\text{m}^2$ and is initially rotating at 100 rad/s. Wheel 2 has a rotational inertia of $1.5\,\text{kg}\cdot\text{m}^2$ and is initially at rest. The wheels come in contact and exert torques on each other, the torque of wheel 2 on wheel 1 being $\tau(t) = -20te^{-t/2}$, in N·m for t in seconds. Use the program to plot the angular velocities of the wheels from $t = 0$ to $t = 15\,\text{s}$. Use an integration interval of 0.005 s and a display interval of 0.5 s.

What are the initial values of the angular momenta of the wheels and the total angular momentum of the system? What are the values after 15 s? [ans: $L_1 = 250\,\text{kg·m}^2/\text{s}$, $L_2 = 0$, $L = 250\,\text{kg·m}^2/\text{s}$; $L_1 = 170\,\text{kg·m}^2/\text{s}$, $L_2 = 80\,\text{kg·m}^2/\text{s}$, $L = 250\,\text{kg·m}^2/\text{s}$]

Suppose that in addition to the torque that wheel 2 exerts on wheel 1 an external agent exerts a torque of 8.0 N·m on wheel 1, in the direction of its angular velocity. What now are the angular momenta of the wheels and the total angular momentum of the system after 15 s? The torque on wheel 1 is $8.0 - 20te^{-t/2}$ and the torque on wheel 2 is $20te^{-t/2}$. [ans: $L_1 = 290\,\text{kg·m}^2/\text{s}$, $L_2 = 80\,\text{kg·m}^2/\text{s}$, $L = 370\,\text{kg·m}^2/\text{s}$]

Notice that the angular momenta of the individual wheels change with time in both cases. The total angular momentum, however, is conserved in the first situation but not in the second. Since $\mathrm{d}L/dt = \tau_{\text{ext}}$ and in the second situation $\tau_{\text{ext}} = 8.0\,\text{N·m}$ the total angular momentum at the end of 15 s should be $L_f = 8.0 \times 15 = 120\,\text{kg·m}^2/\text{s}$. Is this result produced by the program?

PROJECT 2. Conservation of angular momentum is sometimes demonstrated by considering the inelastic "collision" between two wheels on the same axle. Suppose wheel 1 of the previous project starts with an angular velocity of 100 rad/s and exerts a constant torque of 8.0 N·m on wheel 2, which starts from rest. Wheel 2, of course, exerts a torque of −8.0 N·m on wheel 1. Use the program to find the time for which $\omega_1 = \omega_2$ and the value of the angular velocity then. Through what angle has each wheel rotated by the time they reach the same angular velocity? Compare the value obtained for the final angular velocity with that predicted by $I_1\omega_{10} = I_1\omega_1 + I_2\omega_2$. [ans: 11.7 s, 62.5 rad/s, 952 rad, 366 rad]

The time for the wheels to reach the same angular velocity and the angle through which they turn depend on the torque they exert on each other but the final value of the angular velocity does not. To verify this statement repeat the calculation for the same initial conditions but for a torque on wheel 1 that is given by $\tau = -20te^{-t/3}$, in N·m for t in seconds. [ans: 5.24 s, 62.5 rad/s, 438 rad, 144 rad]

PROJECT 3. The kinetic energy of the wheels is not conserved during the inelastic "collision" of the previous project. You can see this by calculating the initial and final kinetic energies, but you can also use a computer to calculate the work done by each torque. If a wheel turns through the small angle $\Delta\theta$ while the torque τ acts on it, the work done by the torque is given by $\Delta W = \tau\Delta\theta$. Modify the program so it computes the work done on each wheel. Just before the loop over intervals set $W_1 = 0$ and $W_2 = 0$. These variables will be used to sum the work done in the intervals. Just before displaying results have the computer increment W_1 by $\tau(\theta_{1e} - \theta_{1b})$ and W_2 by $-\tau(\theta_{2e} - \theta_{2b})$. Here τ is the average torque exerted by wheel 2 on wheel 1. Also have the computer calculate the total work $W = W_1 + W_2$ and display the result.

Take the initial conditions to be as before ($\omega_{10} = 100\,\text{rad/s}$, $\omega_{20} = 0$). Take the torque of wheel 1 on wheel 2 to be a constant 8.0 N·m. Find the work done on each wheel and the total work done from $t = 0$ to the time when the wheels have the same angular velocity. Compare the total work done to the change in the total kinetic energy: $\Delta K = \frac{1}{2}(I_1 + I_2)\omega_f^2 - \frac{1}{2}I_1\omega_{10}^2$, where ω_f is the final angular velocity. [ans: -7.62×10^3 J, 2.93×10^3 J, -4.69×10^3 J]

Now repeat the calculation for a torque on wheel 1 that is given by $\tau = -20te^{-t/3}$, in N·m for t in seconds. You should get the same answers.

Can the "collision" be elastic? Take the torque on wheel 1 to be −8.0 N·m and calculate the total work done as a function of time. Search for a time such that the total work is zero. You might use the condition $W \geq 0$ to exit the loop over intervals. If the wheels no longer interact after this time the collision is elastic. What are the final angular velocities of the wheels? [ans: 23.4 s, 25.0 rad/s, 125 rad/s]

The text showed you how to analyze the change in angular velocity of a spinning ice skater as she brings her arms toward the center of her torso from an outstretched position. Before she brings them in her rotational inertia is I_i and her angular velocity is ω_i. After she brings them in her rotational inertia is I_f and her angular velocity is ω_f. Conservation of angular momentum leads to $I_i\omega_i = I_f\omega_f$, so $\omega_f = (I_i/I_f)\omega_i$.

Here you will use a computer program to investigate some of the details. In particular, you will be able to follow her angular velocity as she brings her arms in. Let I_T be the rotational inertia of her torso alone and let I_A be the rotational inertia of her arms, including any weights she might hold to heighten the effect. I_A is a function of time, determined by the rate with which she moves her arms. Angular momentum is conserved at each stage of the process so

$$\omega(t) = \frac{I_{Ai} + I_T}{I_A(t) + I_T}\omega_i\,,$$

where I_{Ai} is the initial value of the rotational inertia of her arms.

If $\omega(t)$ is known a computer program can be used to calculate the angle $\theta(t)$ through which the skater has rotated since $t = 0$. The angle θ_e at the end of any interval of width Δt is related to the angle θ_b at the beginning by $\theta_e = \theta_b + \omega\Delta t$, where ω is the average angular velocity in the interval. As usual, we approximate the average angular velocity by $(\omega_b + \omega_e)/2$. Here is an outline of a program.

```
input final time and interval width: t_f, Δt
input display interval: Δt_d
set t_b = 0, t_d = Δt_d, θ_b = 0
calculate angular velocity at beginning of first interval: ω_b = ω(t_b)
begin loop over intervals
     calculate time at end of interval: t_e = t_b + Δt
     calculate angular velocity at end of interval: ω_e = ω(t_e)
     calculate "average" angular velocity: ω = (ω_b + ω_e)/2
     calculate angle at end of interval: θ_e = θ_b + ωΔt
     if t_e ≥ t_d then
          calculate torque at end of interval: τ_e
          print or display t_e, θ_e, ω_e
          increment t_d by Δt_d
     end of if statement
     if t_e ≥ t_f then exit loop
     set t_b = t_e, θ_b = θ_e, ω_b = ω_e in preparation for next interval
end loop over intervals
stop
```

Both the angular momentum of the skater's torso and the angular momentum of her arms change. Her torso exerts an torque on her arms and her arms exert a torque of equal magnitude but opposite direction on her torso. If τ represents the torque on her torso then $\tau = I_T\,\mathrm{d}\omega/\mathrm{d}t$. Differentiation of the expression above for $\omega(t)$ produces

$$\tau = -\frac{I_T\omega_i(I_{Ai} + I_T)}{(I_A + I_T)^2}\frac{\mathrm{d}I_A}{\mathrm{d}t}$$

Use this expression in the program to calculate the torque just before t_e, θ_e, and ω_e are displayed, then display it along with the other quantities. A line for the instruction has already been indicated in the outline above.

PROJECT 4. A skater is initially turning at 12 rad/s. Her torso has a rotational inertia of $0.65\,\text{kg}\cdot\text{m}^2$. She carries weights in her hands so that as she pulls her arms in their rotational inertia changes from $0.50\,\text{kg}\cdot\text{m}^2$ to $0.04\,\text{kg}\cdot\text{m}^2$. Suppose she moves her arms in such a way that their rotational inertia changes uniformly over 2.0 s. That is, $I_A(t) = I_{Ai} - Kt$, where $I_{Ai} = 0.50\,\text{kg}\cdot\text{m}^2$ and K is a constant chosen so $I_A = 0.04$ kg·m2 when $t = 2.0$ s.

Use the program to plot her angular velocity and the torque acting on her torso as functions of time from $t = 0$ to $t = 2.0$ s. What is her final angular velocity? Through what angle did she rotate during the process? At what time did she exert the maximum torque on her arms and what was the magnitude of that torque? [ans: 20.0 rad/s; 30.7 rad (4.88 rev); 1.73 N·m at 2.0 s]

Suppose now that all of the rotational inertia of her arms is actually due to the weights she is carrying and that she pulls them in at a constant rate. If m is the mass of one weight and r is it distance from the axis of rotation then the rotational inertia of both weights together is $I_A = 2mr^2$. Since the weights are brought in at a constant rate, we may write $r = r_i + Kt$, where r_i is the initial distance and K is a constant. Take $m = 1.0$ kg and pick values of r_i and K so $I_{Ai} = 0.50\,\text{kg}\cdot\text{m}^2$ and $I_A = 0.05\,\text{kg}\cdot\text{m}^2$ at the end of 2.0 s. Then use the program to plot her angular velocity and the torque acting on her torso as functions of time from $t = 0$ to $t = 2.0$ s. What is her final angular velocity? Through what angle did she rotate during the process? At what time did she exert the maximum torque on her arms and what was the magnitude of that torque? [ans: 20.0 rad/s; 32.2 rad (5.12 rev); 2.82 N·m at 0.95 s]

Chapter 14

Here numerical integration is used to investigate the oscillation of a mass m on an ideal spring with spring constant k. The acceleration of the mass is given by $a = -(k/m)x$, where x is the coordinate of the mass, measured from the equilibrium point. The motion is periodic with an angular frequency that is given by $\omega = \sqrt{k/m}$, a period that is given by $2\pi\sqrt{m/k}$, and an amplitude that is given by $x_m = \sqrt{x_0^2 + (v_0/\omega)^2}$, where x_0 is the initial coordinate and v_0 is the initial speed.

The first program discussed in the Computer Projects section of Chapter 6 can be used but the interval width Δt must be quite small and the number of intervals correspondingly large to avoid accumulating errors. The accuracy can be improved considerably by using more terms in the equations for the velocity and coordinate at the end of an interval. Including terms that are proportional to $(\Delta t)^2$, the velocity is given by $v_e = v_b + a_b\Delta t + \frac{1}{2}(da/dt)(\Delta t)^2$, where the derivative da/dt is evaluated for the beginning of the interval. Since $a = -(k/m)x$, $da/dt = -(k/m)v_b$. The derivative of the acceleration with respect to time is sometimes called the "jerk". Define $j_e = -(k/m)v_e$ and use $v_e = (\frac{1}{2}j_e\Delta t + a_e)\Delta t + v_b$ to calculate the velocity at the end of an interval. Some factoring has been carried out to save computational time and to reduce truncation errors. Similarly $x_e = x_b + v_b\Delta t + \frac{1}{2}a_b(\Delta t)^2$, which is best written $x_e = (\frac{1}{2}a_b\Delta t + v_b)\Delta t + x_b$. You must change appropriate lines of the program.

PROJECT 1. Consider a 2.0-kg mass attached to an ideal spring with spring constant $k = 350$ N/m. It is released from rest at $x = 0.070$ m. Take $\Delta t = 0.001$ s and plot the coordinate for every 0.05 s from $t = 0$ (the time of release) to $t = 1$ s. For use later have the computer calculate the potential energy ($U = \frac{1}{2}kx^2$), the kinetic energy ($K = \frac{1}{2}mv^2$), and the total mechanical energy ($E = K + U$) for each point plotted.

Notice that the graph predicts an oscillatory motion. Use the graph to estimate the amplitude and compare the value with 0.070 m. Use the graph to estimate the period and compare the value with $2\pi\sqrt{m/k}$.

Also notice that at times when the spring has its greatest extension (0.070 m) the speed is 0, the kinetic energy is 0, the potential energy is maximum, and the magnitude of the acceleration is maximum. Also notice that when the spring is neither extended or compressed ($x = 0$) the speed is maximum, the kinetic energy is maximum, the potential energy is 0, and the acceleration is 0. Finally, notice that the total mechanical energy is constant. Your results may show some fluctuation after the third or fourth significant figure but this is due to computational errors.

The coordinate as a function of time can be written $x(t) = x_m \cos(\omega t + \phi)$, where ϕ is the phase constant. For any value of ϕ the initial coordinate is given by $x_0 = x_m \cos\phi$ and the initial velocity is given by $v_0 = -\omega x_m \sin\phi$. For the initial conditions you used above (x_0 positive and $v_0 = 0$) the phase constant is zero. Other initial conditions result in different values for the phase constant. If the amplitude is the same, a different phase constant produces a function $x(t)$ that is shifted along the time axis relative to the function for $\phi = 0$.

Suppose the phase constant is $\pi/6$ rad (30°). Take the amplitude to be $x_m = 0.070$ m and calculate the initial coordinate and velocity. Now use the program to plot $x(t)$ for the first second of the motion and find the time at which the first maximum ($x = 0.070$ m) occurs. Since the first maximum occurs at $t = 0$ for $\phi = 0$, the result you obtain is the amount by which the plot is shifted.

A phase constant of 2π radians (360°) corresponds to a shift equal to one period of the motion. A phase constant of $\pi/6$ radians corresponds to one twelfth of a period. Does this agree with your result?

PROJECT 2. Suppose the mass is also subjected to a resistive force, proportional to its velocity. Then the acceleration is given by $a = -(k/m)x - (b/m)v$, where b is the drag coefficient. Take $m = 2.0$ kg, $k = 350$ N/m, $b = 2.8$ kg/s, $x_0 = 0.070$ m, and $v_0 = 0$. Use the computer program, with $\Delta t = 0.001$ s, to plot the coordinate at intervals of 0.050 s from $t = 0$ (when the mass is released) to $t = 1.0$ s. Also calculate the potential energy, the kinetic energy, and the total mechanical energy for these times. The rate of change of the acceleration is given by $j = -(k/m)v - (b/m)a$.

Notice that the amplitude decreases as time goes on. This decrease is related, of course, to the decrease in total mechanical energy. The resistive force does negative work on the mass. Plot the total mechanical energy as a function of time.

Does the drag force also change the period of the motion? Measure the period as the time between successive maxima and compare the result with $2\pi\sqrt{m/k}$.

PROJECT 3. The oscillator of the previous project is said to execute damped harmonic motion. The motion is said to be *underdamped* because the mass continues to oscillate. If the value of b is increased sufficiently no oscillations occur and the motion is said to be *overdamped.*

Take $m = 2.0$ kg, $k = 350$ N/m, $b = 90$ kg/s, $x_0 = 0.070$ m, and $v_0 = 0$. Plot $x(t)$ every 0.050 s from $t = 0$ (when the mass is released) to $t = 1.0$ s.

The motion of the mass is changed by an external force. Consider a force given by $F_m \cos(\omega'' t)$, where F_m and ω'' are constants. The angular frequency of the impressed force should not be confused with the natural angular frequency $\omega = \sqrt{k/m}$ of the mass and spring alone. The acceleration of the mass is now given by $a = -(k/m)x + (F_m/m)\cos(\omega'' t)$ and its derivative is given by $j = -(k/m)v - (F_m\omega''/m)\sin(\omega''t)$. Change the program accordingly. The following project is designed to help you investigate forced oscillations without damping.

PROJECT 4. Take $m = 2.0\,\text{kg}$, $k = 350\,\text{N/m}$, $b = 0$, $x_0 = 0.070\,\text{m}$, and $v_0 = 0$. The natural angular frequency is $\omega = \sqrt{350/2} = 13.2\,\text{rad/s}$, corresponding to a period of 0.475 s. For each of the following impressed forces of the form $F_m\cos(\omega''t)$ plot $x(t)$ from $t = 0$ to $t = 1.0\,\text{s}$. Use an integration interval of $\Delta t = 0.001\,\text{s}$ and a display interval of $\Delta t_d = 0.050\,\text{s}$. (a.) $F_m = 18\,\text{N}$, $\omega'' = 35\,\text{rad/s}$ (b.) $F_m = 18\,\text{N}$, $\omega'' = 15\,\text{rad/s}$.

For the conditions of part a, the motion is nearly sinusoidal with a period of about 0.47 s, the natural period. The influence of the impressed force is seen in deviations from a sinusoidal shape. When the impressed frequency is closer to the natural frequency, as in part b, the influence of the impressed force is more pronounced. The amplitude grows with time, then levels off. If the impressed frequency is made exactly equal to the natural frequency the amplitude grows without bound.

The increase in amplitude can easily be accounted for in terms of the work done by the impressed force. If ω'' is different from ω the impressed force is in the same direction as the velocity of the mass over some portions of the motion and in the opposite direction over other portions. If the frequencies are very different, as in part a, the net work done over a time period that is long compared with the period is almost zero. On the other hand, when the two frequencies are nearly the same, as in part b, the impressed force and velocity are in the same direction over a large portion of the motion and the amplitude grows with time. As time goes on the impressed force and velocity are in the same direction over smaller portions of the motion and in opposite directions over larger portions. Eventually the net work vanishes over a long time interval and the amplitude becomes constant.

You can easily verify these assertions by plotting the total mechanical energy as a function of time. Modify the program so it calculates and lists or plots the total mechanical energy every 0.05 s from $t = 0$ to $t = 2.0\,\text{s}$. Note intervals during which the impressed force does positive work (the energy increases) and intervals during which it does negative work (the energy decreases).

PROJECT 5. Now investigate the motion when a damping force is present. Consider the same oscillator ($m = 2.0\,\text{kg}$, $k = 350\,\text{N/m}$) subjected to an impressed force with $F_m = 18\,\text{N}$ and $\omega'' = 35\,\text{rad/s}$, but with the drag coefficient $b = 15\,\text{kg/s}$. Use the same initial conditions ($x_0 = 0.070\,\text{m}$, $v_0 = 0$) and integration interval ($\Delta t = 0.001\,\text{s}$). Plot $x(t)$ every 0.050 s from $t = 0$ to $t = 2.0\,\text{s}$.

Notice that motion starts out very much the same as when there is no damping. Its period is nearly the natural period. Now, however, this motion is quickly damped and what remains is a sinusoidal motion at the frequency of the impressed force. Verify that the period is about 0.18 s, corresponding to an angular frequency of 35 rad/s. The amplitude eventually becomes constant. The energy supplied by the impressed force is dissipated by damping.

Repeat the calculation for an impressed angular frequency of 15 rad/s, near the natural frequency. Note that the final amplitude is much larger than when the impressed frequency was far from the natural frequency.

A simple pendulum consists of a mass m at the end of a light rod of length ℓ, free to swing in a uniform gravitational field. The angular position is given by the angle $\theta(t)$, measured from the vertical. If g is the acceleration due to gravity then θ obeys the differential equation

$$\frac{d^2\theta}{dt^2} = -\frac{g}{\ell}\sin\theta\,.$$

You have learned that if θ is always small and is measured in radians then $\sin\theta$ may be replaced by θ itself and the solution to the resulting differential equation is $\theta(t) = \theta_m \sin(\omega t + \phi)$, where the angular frequency is given by $\omega = \sqrt{g/\ell}$.

Use a computer to investigate the motion when the amplitude is not small. You may want to change the program so it is written in terms of θ and its derivative Ω ($= d\theta/dt$): simply replace x with θ and v with Ω. The angular acceleration, which replaces a, is $\alpha = -(g/\ell)\sin\theta$ and its derivative is $j = -(g\Omega/\ell)\cos\theta$.

PROJECT 6. First use the program to check the small angle approximation. A simple pendulum with a length of 1.2 m is pulled aside 10° (0.175 rad) and released from rest. What is its period? Use an integration interval of 0.001 s and search for the second time after starting that Ω is zero. Compare the result with $2\pi\sqrt{\ell/g}$. [ans: 2.20 s]

Repeat the calculation for an initial angular displacement of 45° (0.785 rad). [ans: 2.29 s]

Repeat the calculation for an initial angular displacement of 75° (1.31 rad). [ans: 2.46 s]

Now check to see how close the motion is to simple harmonic. Take the initial angular displacement to be 75° and use the program to generate a table of $\theta(t)$ for the first 2.5 s of the motion. You might obtain values at intervals of 0.1 s. At the same time have the program generate values of $\theta_m \cos(\omega t)$, where the value of ω is calculated from the value you found for the period.

The conversion of kinetic energy to potential energy and back again to kinetic energy is quite similar to the conversion that takes place for a spring-mass system, although the mechanism is the work done by the force of gravity, not the force of a spring. Suppose the pendulum is released from rest with an initial angular displacement of 75°. For each of the displayed points have the computer calculate the kinetic energy per unit mass ($\frac{1}{2}\ell^2\Omega^2$), the potential energy per unit mass [$g\ell(1 - \cos\theta)$], and their sum.

Chapter 15

In this section a computer is used to investigate satellite motion. A massive central body (the sun or earth, for example) is at the origin and is assumed to remain motionless. If the coordinates of the satellite are x and y then the force exerted by the central body on the satellite has components that are given by $F_x = -GMmx/r^3$ and $F_y = -GMmy/r^3$, where G is the universal gravitational constant ($6.67 \times 10^{-11}\,\text{m}^3/\text{s}^3\cdot\text{kg}$), M is the mass of the central body, m is the mass of the satellite, and $r = \sqrt{x^2 + y^2}$ is the center-to-center distance of the satellite from the central body. These expressions assume both the central body and satellite have spherically symmetric mass distributions. Notice that the force is directed radially from the center of the satellite toward the center of the central body.

The acceleration of the satellite has components that are given by $a_x = -GMx/r^3$ and $a_y = -GMy/r^3$. Since the force depends on the coordinates of the satellite, you cannot use $\mathbf{a} = (\mathbf{a}_b + \mathbf{a}_e)$ to approximate the average acceleration in an interval. The acceleration at the beginning of an interval is an extremely poor approximation to the average acceleration and if it is used the intervals must be quite narrow and the running time must be quite long. The Chapter 14 projects introduced you to one trick, using the derivative of the acceleration. Here you will learn another.

The program outlined below calculates the coordinates and acceleration at the beginning of each interval and the velocity at the midpoint of each interval. The acceleration at the beginning of an interval and the velocity at the midpoint of the previous interval are used to compute the velocity at the midpoint of the interval, then that velocity and the coordinates at the beginning of the interval are used to compute the coordinates at the end of the interval.

To start, the velocity half an interval before t_0 must be computed using the acceleration and velocity at t_0. When the computer is asked to display results, in the loop over intervals, x_b and y_b are the coordinates at the beginning of the interval while v_{xm} and v_{ym} are the velocity components for the midpoint of the interval. We wish to display the velocity components for the beginning and end of the interval, so the acceleration for the beginning is used to calculate these quantities before printing. Here's the outline.

> input initial conditions: x_0, y_0, v_{x0}, v_{y0}
> input final time and interval width: t_f, Δt
> input display interval: Δt_d
> calculate acceleration at t_0: a_x, a_y
> calculate velocity at $t_0 - \Delta t/2$:
> $v_{xm} = v_{x0} - a_x\Delta t/2$
> $v_{ym} = v_{y0} - a_y\Delta t/2$
> set $t_b = t_0$, $t_d = t_0 + \Delta t_d$, $x_b = x_0$, $y_b = y_0$
> **begin loop** over intervals
>   calculate acceleration at beginning of interval: a_x, a_y
>   calculate velocity at midpoint of interval:
>   replace v_{xm} with $v_{xm} + a_x\Delta t$
>   replace v_{ym} with $v_{ym} + a_y\Delta t$
>   calculate coordinates at end of interval:
> $x_e = x_b + v_{xm}\Delta t$
> $y_e = y_b + v_{ym}\Delta t$
>   calculate time at end of interval: $t_e = t_b + \Delta t$
>   if $t \geq t_d$ then
>   calculate velocity at beginning of interval:
> $v_{xb} = v_{xm} - a_x\Delta t/2$
> $v_{yb} = v_{ym} - a_y\Delta t/2$
>   print or display t_b, x_b, y_b, v_{xb}, v_{yb}
>   calculate velocity at end of interval:
> $v_{xe} = v_{xm} + a_x\Delta t/2$
> $v_{ye} = v_{ym} + a_y\Delta t/2$
>   print or display t_e, x_e, y_e, v_{xe}, v_{ye}
>   set $t_d = t_d + \Delta t_d$
>   end of if statement
>   if $t_e \geq t_f$ then exit loop
>   set $t_b = t_e$, $x_b = x_e$, $y_b = y_e$
> **end of loop** over intervals
> stop

To write an efficient program define the constant C by $C = -GM\Delta t$, then use $a_x = Cx_b/r^3$ and $a_y = Cy_b/r^3$ to compute the acceleration components.

In the following project you will use the program to verify Kepler's laws of planetary motion. First you will verify the conservation of energy and angular momentum.

PROJECT 1. Consider a satellite in orbit around the earth ($M = 5.98 \times 10^{24}$ kg). At $t = 0$ it is at $x = 7.2 \times 10^6$ m, $y = 0$ and has velocity components $v_{x0} = 0$, $v_{y0} = 9.0 \times 10^3$ m/s. Use the program with $\Delta t = 10$ s to plot the position for every 500 s from $t = 0$ to $t = 1.55 \times 10^4$ s.

Modify the program so it computes the total mechanical energy and angular momentum, both per unit satellite mass, for each point displayed. The total mechanical energy per unit mass is given by $E = \frac{1}{2}(v_{xe}^2 + v_{ye}^2) - GM/r$ and the angular momentum per unit mass is given by $\ell = x_e v_{ye} - y_e v_{xe}$. The results should be constant to 2 or 3 significant figures.

Now use the program to prove the orbit is an ellipse. An ellipse is a geometric figure such that the sum of the distances from any point on the figure to two fixed points, called foci, is the same as for any other point. For the orbit you generated above, one focus is at the origin and the other is on the x axis the same distance from the geometric center as the first. First locate the geometric center of the orbit. Use the coordinates generated by the program to find the x coordinates of the points where the orbit crosses the x axis, then calculate half their sum. Let this coordinate be $-x_c$. The second focus is at $x_F = -2x_c$.

Modify the program so it computes the sum of the distances from the point displayed to the foci. It is given by $\sqrt{x_e^2 + y_e^2} + \sqrt{(x_e - x_F)^2 + y_e^2}$. Check that this sum is the same for all displayed points to 3 or more significant figures.

Kepler's second law says that the vector from the central body to the satellite sweeps out equal areas in equal times. Let $\mathbf{r}_b$ be the position vector at the beginning of an interval and let $\mathbf{r}_e$ be the position vector at the end of the interval. The triangle defined by these two vectors has an area that is given by $A = \frac{1}{2}|\mathbf{r}_b \times \mathbf{r}_e| = \frac{1}{2}|x_b y_e - x_e y_b|$. This is the area swept out by the position vector in time Δt. Have the computer calculate A just before it displays results and ask it to display A along with the other quantities. Check to see if it is constant to a reasonable number of significant figures.

You should recognize that this calculation essentially repeats the calculation above to check on the constancy of the angular momentum. Since $\mathbf{r}_e = \mathbf{r}_b + \mathbf{v}\Delta t$, $A = \frac{1}{2}|\mathbf{r}_b \times \mathbf{v}|\Delta t$. Since the angular momentum is given by $\boldsymbol{\ell} = m\mathbf{r} \times \mathbf{v}$, this becomes $A = \frac{1}{2}\ell\Delta t$. A is constant because ℓ is constant.

Kepler's third law tells us that the period T is related to the length a of the semimajor axis by $T^2 = (4\pi^2/GM)a^3$. For the orbit generated above the length of the semimajor axis is half the distance between the two points where the orbit crosses the x axis. Calculate this distance and compute the right side of the third law equation. Use the program to find the period. It is, for example, twice the time between two successive crossings of the x axis. Compute the left side of the third law equation and compare the value with the value you obtained for the right side.

PROJECT 2. The length of the semimajor axis of an elliptical orbit depends only on the total mechanical energy. Suppose the satellite of the previous project is started from $x = 5.5 \times 10^6$ m, $y = 0$ with a velocity in the positive y direction. Select the initial speed so its total mechanical energy is the same as before. Plot the trajectory and find the length of the semimajor axis. Compare with the length of the semimajor axis for the previous project.

The length a of the semimajor axis is related to the total mechanical energy E by

$$E = -\frac{GMm}{2a}.$$

Suppose the satellite is started at $x = 7.2 \times 10^6$ m, $y = 0$ with a velocity in the positive y direction. What initial speed should it have if its orbit is to be circular? For a circular orbit the semimajor axis is the same as the radius and for an initial position on the x axis it must be the same as x_0. Equate $-GMm/2x_0$ to $\frac{1}{2}mv_{y0}^2 - GMm/x_0$ and solve for v_{y0}. Use the program to plot the orbit. You might have the computer calculate $r^2 = x_e^2 + y_e^2$

for every displayed point and see if this quantity remains constant. Compared to the orbit of the first project is the total mechanical energy greater or less for the circular orbit?

If the total mechanical energy is negative the satellite is bound and its orbit is an ellipse. This is so for the satellites studied above. If the total mechanical energy is zero the orbit is a parabola and if the total mechanical energy is positive the orbit is a hyperbola. In either of these cases the satellite is not bound and it eventually escapes from the gravitational pull of the central body. Here's an example.

PROJECT 3. Consider a spacecraft in the gravitational field of the earth, with initial coordinates $x_0 = 7.2 \times 10^6\,\mathrm{m}$, $y_0 = 0$. Take its initial velocity to be $v_{x0} = 0$, $v_{y0} = 1.2 \times 10^4\,\mathrm{m/s}$. Use the program to plot its position every 300 s from $t = 0$ to $t = 4500\,\mathrm{s}$. Also calculate the total mechanical energy for these times and note it is constant (within the limits of the calculation, of course) and positive. Notice that the trajectory tends to become a straight line as the spacecraft recedes from the central body. Find the angle this line makes with the x axis. [ans: 51°]

What happens to the line if the initial speed is increased? Use the program to verify your conjecture.

Chapter 17

Start with a rather simple program to plot the displacement $y(x,t) = y_m \sin(kx - \omega t)$ for a sinusoidal traveling wave at a given time. Input the amplitude y_m, wavelength λ, frequency f, and time t. Have the computer calculate values of $y(x,t)$ for x from 0 to some final value x_f, at intervals of width Δx. Here's an outline.

```
input amplitude, wavelength, frequency: y_m, λ, f
input final value of x, interval width: x_f, Δx
input time: t
calculate angular wave number, angular frequency: k = 2π/λ, ω = 2πf
set x = 0
begin loop over intervals
      calculate y(x,t): y = y_m sin(kx − ωt)
      display or plot x, y
      increment x by Δx
if x > x_f then end loop over intervals
stop
```

A reasonably good graph can be obtained if Δx is chosen to be about $\lambda/20$ and x_f is chosen to be about 2λ. If you program the computer to plot the wave on your monitor you might have the program calculate values of Δx and x_f rather than read them. If you are plotting by hand you will want values that are 1, 2, or 5 times a power of ten.

PROJECT 1. Take $y_m = 2.0\,\mathrm{mm}$, $\lambda = 5.0\,\mathrm{cm}$, and $f = 10\,\mathrm{Hz}$. Make a graph of the wave at $t = 0$ and note the position of the maximum nearest the origin.

Now make a graph of the wave at $t = 4.0 \times 10^{-2}\,\mathrm{s}$ and again note the position of the maximum nearest the origin. Measure the distance this maximum has moved in $4.0 \times 10^{-2}\,\mathrm{s}$ and use the value you obtain to calculate the wave speed. Compare your answer with $v = \lambda f$. The time was chosen so that no other maximum appears between the origin and the maximum you are following.

Two sinusoidal waves with the same frequency and wavelength, traveling in the same direction, sum to form another sinusoidal wave. You can use a computer program to plot the sum of the waves, then read the resultant amplitude and phase from the graph. Suppose $y_1(x,t) = y_{1m}\sin(kx - \omega t)$ and $y_2(x,t) = y_{2m}\sin(kx - \omega t - \phi)$. Then the resultant wave is given by $y(x,t) = y_{1m}\sin(kx - \omega t) + y_{2m}\sin(kx - \omega t - \phi)$. Here's the outline of a program.

> input amplitudes, wavelength, frequency, phase constant: y_{1m}, y_{2m}, λ, f, ϕ
> input final value of x, interval width: x_f, Δx
> input time: t
> calculate angular wave number, angular frequency: $k = 2\pi/\lambda$, $\omega = 2\pi f$
> set $x = 0$
> **begin loop** over intervals
>   calculate $y(x,t)$: $y = y_{1m}\sin(kx - \omega t) + y_{2m}\sin(kx - \omega t - \phi)$
>   display or plot x, y
>   increment x by Δx
> if $x > x_f$ then **end loop** over intervals
> stop

PROJECT 2. No matter what the amplitudes and the phase difference, the resultant wave is sinusoidal and moves with same speed as the constituent waves. Take $y_{1m} = 2.0\,\text{mm}$, $y_{2m} = 4.0\,\text{mm}$, $\lambda = 5.0\,\text{cm}$, $f = 10\,\text{Hz}$, and $\phi = 65°$. Plot the resultant wave from $x = 0$ to $x = 10\,\text{cm}$ for $t = 0$. Then plot the resultant wave for $t = 4.0 \times 10^{-2}\,\text{s}$ and calculate the wave speed. You should get the same answer as you obtained for the first project.

PROJECT 3. The amplitude of the resultant wave depends on the phase difference ϕ. First consider two waves with the same amplitude. Then the resultant amplitude is given by $y_m = 2y_{1m}\cos(\phi/2)$. Take $y_{1m} = y_{2m} = 2.0\,\text{mm}$, $\lambda = 5.0\,\text{cm}$, $f = 10\,\text{Hz}$, and $\phi = 65°$. Plot the resultant wave from $x = 0$ to $x = 10\,\text{cm}$ for $t = 0$. Measure the amplitude and compare the result with the calculated value.

The phase constant α of the resultant wave also depends on the phases of the constituent waves. For equal amplitudes $\alpha = \phi/2$. Use your graph to find α. At $t = 0$ the first constituent wave y_1 is zero and has positive slope at the origin. Use your graph to find the coordinate where the resultant wave is zero and has positive slope. The ratio of this coordinate to a wavelength is the same as the ratio of the phase constant to 360° or 2π rad. The coordinate should be $(32.5/360)\lambda = 0.45\,\text{cm}$. Is it?

What happens when the amplitudes are not equal? Take $y_{1m} = 2.0\,\text{mm}$, $y_{2m} = 4.0\,\text{mm}$, $\lambda = 5.0\,\text{cm}$, and $f = 10\,\text{Hz}$. For each of the following values of ϕ find the amplitude and phase constant of the resultant wave: (a.) 0; (b.) 30°; (c.) 45°; (d.) 90°; (e.) 180°. [ans: 6.0 mm, 0; 5.8 mm, 20°; 5.6 mm, 30°; 4.5 mm, 63°; 2.0 mm, 180°]

Your program can also be used to investigate standing waves. A standing wave is composed of two traveling waves with the same amplitude, frequency, and wavelength, but traveling in opposite directions. Revise the program so only one amplitude is read and so $y = y_m\sin(kx - \omega t) + y_m\sin(kx + \omega t - \phi)$ is used to calculate the resultant wave.

PROJECT 4. Take $y_m = 2.0\,\text{mm}$, $\lambda = 5.0\,\text{cm}$, and $f = 10\,\text{Hz}$. Make a graph of the wave at $t = 0$ and note the position of the maximum nearest the origin. Now make a graph of the wave at $t = 0.020\,\text{s}$ and again note the position of the maximum nearest the origin. During this time each of the traveling waves moved 1.0 cm, one fifth of a wavelength. But the standing wave maximum did not move.

Notice that the maximum displacement is less than at $t = 0$. In fact, all parts of the string move together toward $y = 0$. At $t = 0.025\,\text{s}$ the displacement is everywhere 0. Use the program to verify this. Also plot the wave for $t = 0.05$ s, one half period after the start.

PROJECT 5. Identify the nodes on the graph you made as part of the last project for $t = 0$ and verify that they are half a wavelength apart. How do the phases of the traveling waves affect the positions of the nodes? For the same wavelength and frequency as the last project and for each of the following values of ϕ find the coordinate of the node nearest the origin and verify that the node separation is the same: (a.) 0; (b.) 45°; (c.) 90°. [ans: 0; 0.31 cm; 0.63 cm]

Chapter 18

Modify the program of the last chapter to investigate beats. Two waves with slightly different frequencies are summed: $y(x,t) = y_{1m}\sin(k_1x - \omega_1 t) + y_{2m}\sin(k_2x - \omega_2 t - \phi)$. Once a value is chosen for x this can be written $y = y_{1m}\sin(\omega_1 t - \phi_1) + y_{2m}\sin(\omega_2 t - \phi_2)$, where $\phi_1 = k_1x - \pi$ and $\phi_2 = k_2x + \phi - \pi$. For the first few projects choose x and ϕ so both ϕ_1 and ϕ_2 vanish. Then $y = y_{1m}\sin(\omega_1 t) + y_{2m}\sin(\omega_2 t)$. Input the amplitudes and frequencies, then have the computer generate a plot of y for times from $t = 0$ to $t = t_f$, with an interval of Δt. Here's an outline.

> input amplitudes, frequencies: y_{1m}, y_{2m}, f_1, f_2
> input final value of t, interval width: t_f, Δt
> calculate angular frequencies: $\omega_1 = 2\pi f_1$, $\omega_2 = 2\pi f_2$
> set $t = 0$
> **begin loop** over intervals
>   calculate y: $y = y_{1m}\sin(\omega_1 t) + y_{2m}\sin(\omega_2 t)$
>   plot y
>   increment t by Δt
> if $t > t_f$ then **end loop** over intervals
> stop

Reasonable graphs are produced if Δt is taken to be about $1/50(f_1 + f_2)$ and t_f is taken to be about $2/|f_1 - f_2|$. The displacement y oscillates with a frequency of $(f_1 + f_2)/2$ so this value of Δt produces 100 points per period of oscillation. The beat frequency is $|f_1 - f_2|$ so this value of t_f lets you see two periods of the beat. Since a large number of points are generated you should program the computer to produce the graph on a monitor screen rather than plot by hand.

PROJECT 1. First take $y_1 = y_2 = 2.0\,\text{mm}$, $f_1 = 100\,\text{Hz}$, and $f_2 = 110\,\text{Hz}$. The graph shows a rapidly varying oscillation inside a more slowly varying envelope. Measure the period of the rapid oscillation and verify that it is $2/(f_1 + f_2)$. Measure the period of the envelope and verify that it is $1/|f_1 - f_2|$.

Now try $y_{1m} = y_{2m} = 2.0\,\text{mm}$, $f_1 = 100\,\text{Hz}$, and $f_2 = 105\,\text{Hz}$. Verify that the period of the rapid oscillation is nearly the same and that the period of the envelope has doubled.

Finally try $y_{1m} = y_{2m} = 2.0\,\text{mm}$, $f_1 = 500\,\text{Hz}$, and $f_2 = 510\,\text{Hz}$. Verify that the period of the rapid oscillation has decreased by a factor of about 5 while the beat period is the same as for the first case above ($f_1 = 100\,\text{Hz}$, $f_2 = 110\,\text{Hz}$).

PROJECT 2. What happens if the amplitudes are different? Try $y_{1m} = 2.0\,\text{mm}$, $y_{2m} = 4.0\,\text{mm}$, $f_1 = 100\,\text{Hz}$, and $f_2 = 110\,\text{Hz}$. Pay special attention to the regions between beats and tell how these compare to the same regions when the amplitudes are the same: ____

Now increase the amplitude of the second wave to $y_{2m} = 6.0\,\text{mm}$. Does the result substantiate your statement?

PROJECT 3. Now investigate the influence of a phase difference. Rewrite the program instruction for the calculation of y so it reads $y = y_{1m}\sin(\omega_1 t) + y_{2m}\sin(\omega_2 t - \phi)$ and add an input statement for ϕ near the beginning of the program. Take $y_{1m} = y_{2m} = 2.0\,\text{mm}$, $f_1 = 100\,\text{Hz}$, and $f_2 = 110\,\text{Hz}$. Plot the displacement as a function of time for $\phi = 30°$, $60°$, $90°$, and $180°$. You already have a graph for $\phi = 0$. As you look at the graphs pay particular attention to the position of the beat maxima along the time axis. What is the influence of the phase? ____

Chapter 24

A computer can be programmed to calculate the electric field due to a line of charge if the charge distribution along the line is known. Suppose the charge is on the x axis from x_0 to x_f and the charge density is given by the known function $\lambda(x)$. Place the y axis so the field point is in the xy plane and has coordinates x and y. Then the electric field at that point is given by

$$E_x = \frac{1}{4\pi\epsilon_0}\int_{x_0}^{x_f} \frac{\lambda(x')(x-x')\,dx'}{[(x-x')^2+y^2]^{3/2}}\,,$$

$$E_y = \frac{y}{4\pi\epsilon_0}\int_{x_0}^{x_f} \frac{\lambda(x')\,dx'}{[(x-x')^2+y^2]^{3/2}}\,.$$

The z component is zero.

The Simpson's rule program discussed in Chapter 7 can be used to carry out the integrations. For most problems, however, this technique requires a large number of intervals because the functions $(x-x')/\left[(x-x')^2+y^2\right]^{3/2}$ and $1/\left[(x-x')^2+y^2\right]^{3/2}$ in the integrands vary rapidly with x'. On the other hand, λ usually varies much more slowly. In order to avoid this problem divide the x axis from x_0 to x_f into N segments, each of width $\Delta x'$, and approximate λ by a constant in each segment. The integral over each segment can be evaluated analytically. The indefinite integrals are $\int(x-x')\left[(x-x')^2+y^2\right]^{-3/2}dx' = \left[(x-x')^2+y^2\right]^{-1/2}$ and $\int\left[(x-x')^2+y^2\right]^{-3/2}dx' = -(1/y^2)(x-x')\left[(x-x')^2+y^2\right]^{-1/2}$. Let an interval start at x_b and evaluate λ at the midpoint of the interval: $\lambda_m = \lambda(x_b+\Delta x/2)$. Then the components of the electric field are given approximately by

$$E_x = \frac{1}{4\pi\epsilon_0}\sum_{i=1}^{N}\lambda_m\left\{\frac{1}{[(x-x_b-\Delta x)^2+y^2]^{1/2}} - \frac{1}{[(x-x_b)^2+y^2]^{1/2}}\right\},$$

$$E_y = -\frac{1}{4\pi\epsilon_0}\frac{1}{y}\sum_{i=1}^{N}\lambda_m\left\{\frac{(x-x_b-\Delta x)}{[(x-x_b-\Delta x)^2+y^2]^{1/2}} - \frac{(x-x_b)}{[(x-x_b)^2+y^2]^{-1/2}}\right\}.$$

Write a program to calculate the components of the electric field produced by a line of charge. Input the coordinates x and y and the limits of integration x_0 and x_f, then have the program evaluate the sums and multiply the results by $1/4\pi\epsilon_0$ or $(1/4\pi\epsilon_0)(1/y)$, as appropriate. You will need to supply a programming line to define the function $\lambda(x)$. Here's an outline.

input number of intervals, limits of integration: N, x_0, x_f
calculate segment length: $\Delta x = (x_f - x_0)/N$
input coordinates: x, y
set $S_x = 0$, $S_y = 0$
set $x_b = x_0$
calculate $f_{1x} = 1/\left[(x - x_b)^2 + y^2\right]^{1/2}$
calculate $f_{1y} = (x - x_b)/\left[(x - x_b)^2 + y^2\right]^{1/2}$
begin loop over intervals: counter runs from 1 to N
    set $x_e = x_b + \Delta x$
    calculate $f_{2x} = 1/\left[(x - x_e)^2 + y^2\right]^{1/2}$
    calculate $f_{2y} = (x - x_e)/\left[(x - x_e)^2 + y^2\right]^{1/2}$
    calculate linear charge density at center of segment: λ_m
    $S_x = S_x + \lambda_m(f_{2x} - f_{1x})$
    $S_y = S_y + \lambda_m(f_{2y} - f_{1y})$
    replace x_b with x_e, f_{1x} with f_{2x}, f_{1y} with f_{2y}
end loop
calculate field components: $E_x = (1/4\pi\epsilon_0)S_x$, $E_y = -(1/4\pi\epsilon_0)(1/y)S_y$
display E_x, E_y
go back to enter another set of coordinates or quit

Test the program by considering a line of uniform charge density.

PROJECT 1. A line of charge runs along the x axis from the origin to $x = 0.10\,\text{m}$. Suppose the line contains 5.5×10^{-9} C of charge, distributed uniformly (λ = constant). Use the program to calculate the electric field components at $x = 0$, $y = 0.05\,\text{m}$. Evaluate the analytic expressions and compare answers.

Now suppose the same total charge is distributed on the same line with a linear density that is given by $\lambda = Ax^2$, where A is a constant. First show that $\lambda = 1.65 \times 10^{-5}x^2$ C/m for x in meters. Then use the program to find the electric field components at points along the line $y = 0.050\,\text{m}$. Take points every 0.020 m from $x = -0.060\,\text{m}$ to $x = 0.100\,\text{m}$.

Estimate the value of x for which the electric field is in the y direction. Explain why it is not on the center line of the wire, $x = 0.050\,\text{m}$.

At points far away from the line of charge the field tends to become like that of a point charge. Use the program to calculate the field components along the line $x = 0$. Take $y = 0.10$, 1.0, 10, 100, and 1000 m. Modify the program so it also calculates the electric field at the same points for a point charge $q = 5.5 \times 10^{-9}$ C at the origin. Compare the fields of the line and point charge. Does the field of the line become more like that of the point charge at far away points?

Given a charge distribution that creates an electric field, a computer can be used to plot the electric field lines. We consider a distribution of point charges. It is usual to start at a point close to one of the charges, where the field line is along the line that joins the point and the charge. The electric field at the point is calculated and the result is used to locate a neighboring point on the same field line. It is a short distance away in the direction of the electric field vector. The process is then continued to locate other points on the same field line.

For simplicity we will deal with charges, fields, and field lines in the x, y plane. Suppose the electric field at a point with coordinates x and y has components E_x and E_y and we wish to find another point on the field line through x, y. $(E_x/E)\mathbf{i} + (E_y/E)\mathbf{j}$ is a unit vector tangent to the field line and $x + (E_x/E)\Delta s$, $y + (E_y/E)\Delta s$ are the coordinates of a point on the line Δs distant from

x, y. The following is the outline of a program that calculates a sequence of points on a single field line. You must supply the coordinates x_0, y_0 of the first point, the distance Δs between points, and the program instructions to calculate the electric field.

```
input distance between points, number of points: Δs, N
input the coordinates of the first point: x0, y0
set xb = x0, yb = y0
plot xb, yb
begin loop over points; counter runs from 1 to N
     calculate field components at xb, yb: Ex, Ey
     calculate magnitude of field at xb, yb: E = sqrt(Ex^2 + Ey^2)
     calculate coordinates of new point:
        xe = xb + (Ex/E)Δs
        ye = yb + (Ey/E)Δs
     plot xe, ye
     set xb = xe, yb = ye
end loop over points
go back to get another starting point or quit
```

You may want to provide instructions so the field lines are plotted on the monitor screen. Alternatively, you may have the computer display the coordinates of the points so you can plot the line by hand. You should realize that the lines generated are approximate but the approximation becomes more accurate as Δs is made smaller. Do not make Δs so small that significance is lost in the calculation. If you plot by hand you will not want to display and plot every point, particularly if Δs is small. Add program instructions so that only every 10 or 20 calculated points are displayed.

You might also want to stop plotting when the line gets close to a charge. One way to do this is to save the coordinates of the charges, then check x_e and y_e to see if they are near a corresponding charge coordinate. If they are, have the computer go to the last line of the program.

PROJECT 2. Check the program by considering a dipole. Charge $q_1 = 7.1 \times 10^{-9}$ C is located at the origin and charge $q_2 = -7.1 \times 10^{-9}$ C is located on the y axis at $y = -0.40$ m. Use the program to plot 2 field lines. Start one at $x = 1.0 \times 10^{-3}$ m, $y = 1.0 \times 10^{-3}$ m and the second at $x = 1.0 \times 10^{-3}$ m, $y = -1.0 \times 10^{-3}$ m. Take $\Delta s = 0.005$ m and plot about 100 points on each line. Do these lines look like the lines of a dipole?

PROJECT 3. Four identical charges, each with $q = 5.6 \times 10^{-9}$ C, are placed at the corners of a square with edge length $a = 0.36$ m. The square is centered at the origin and its sides are parallel to the coordinate axes. Take $\Delta s = 0.005$ m and use the program to plot 6 field lines, each starting on a circle of radius 0.10 m centered on the charge at $x = 0.18$ m, $y = 0.18$ m. One line starts parallel to the x axis and the others start at intervals of $\pi/3$ radians around the circle. When a line is within 0.10 m of any charge or more than 0.70 m away from all charges, do not continue it.

One of the lines goes toward the center of the square where, the electric field vanishes. The program does not properly evaluate E_x/E and E_y/E in the limit of vanishing field. Stop plotting when the field becomes less than 10^{-4} times the field at the initial points.

Plot 6 field lines emanating from each of the other charges. This can be done using the data generated for the first charge and a symmetry argument. It is not necessary to calculate new points.

Far away from all charges the distribution has the same electric field as that of a single charge equal to the net charge in the distribution and located at the origin. The field lines are then radially outward if the net charge does not vanish and is positive. Look at your plot and notice that the lines tend to become more nearly in the radial direction as the distance from the origin increases. Because of the choice of starting points for the lines, they also tend to be arranged with equal angles between adjacent lines. You should see 24 lines, all nearly in a radial direction, with adjacent lines separated by $\pi/12$ radians. Do you?

Now change the sign of the charges at $x = -0.18\,\text{m}$, $y = 0.18\,\text{m}$ and at $x = 0.18\,\text{m}$, $y = -0.18\,\text{m}$ and use the program to plot lines starting at the same points as before. Notice that now the net charge in the distribution is zero and that all lines start and stop at charges within the distribution.

Chapter 25

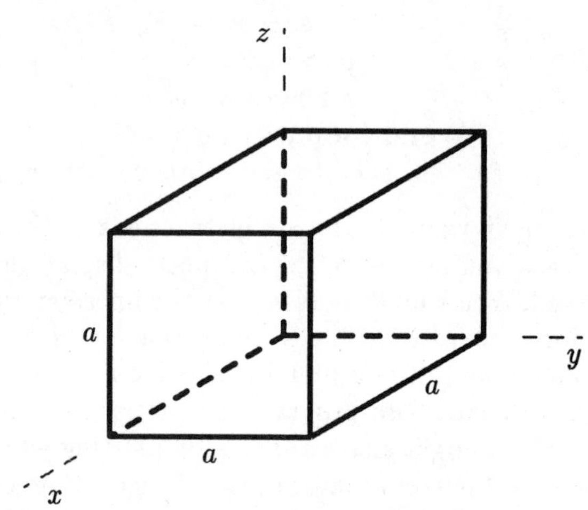

Gauss's law can be verified directly by using a computer to evaluate the integral $\oint \mathbf{E}\cdot d\mathbf{A}$ over a closed surface. Consider the special case of a cube bounded by the 6 planes $x = 0$, $x = a$, $y = 0$, $y = a$, $z = 0$, and $z = a$, as shown in the figure. Place a single charge q on the line $x = a/2$, $z = a/2$, through the cube center, and carry out the integration one face at a time.

First consider the top face. Divide it into small rectangles of width Δx and length Δy, evaluate the electric field at the center of each rectangle, and calculate $E_z \Delta x \Delta y$. Finally, sum the results. Then carry out similar calculations for the other faces.

If the y coordinate of the charge is y' then its electric field at x, y, z is given by

$$\mathbf{E} = \frac{q}{4\pi\epsilon_0}\,\frac{(x-a/2)\,\mathbf{i} + (y-y')\,\mathbf{j} + (z-a/2)\,\mathbf{k}}{[(x-a/2)^2 + (y-y')^2 + (z-a/2)^2]^{3/2}}\,.$$

By symmetry the flux through the top, bottom, front, and back faces are all the same. Calculate the flux through the top face, say, and multiply by 4. On the top face $z = a$ and the quantity to be summed is

$$E_z \Delta x \Delta y = \frac{q}{4\pi\epsilon_0}\,\frac{a/2}{[(x-a/2)^2 + (y-y')^2 + a^2/4]^{3/2}}\,.$$

For the left face $y = 0$ and the quantity to be summed is

$$-E_y \Delta x \Delta z = \frac{q}{4\pi\epsilon_0}\,\frac{y'}{[(x-a/2)^2 + (y')^2 + (z-a/2)^2]^{3/2}}\,.$$

For the right face $y = a$ and the quantity to be summed is

$$E_y \Delta x \Delta z = \frac{q}{4\pi\epsilon_0}\,\frac{a-y'}{[(x-a/2)^2 + (a-y')^2 + (z-a/2)^2]^{3/2}}\,.$$

Here is the outline of a program to calculate the flux through the upper face. The x axis and the y axis, both from 0 to a, are each divided into N segments and $\Delta x = \Delta y = a/N$ is computed.

In the program this quantity is called $\Delta\ell$. There are two loops: an outer loop over x and an inner loop over y. The first value of x is $\Delta\ell/2$, at the center of the first segment, and x is incremented by $\Delta\ell$ each time around the loop over x. For each value of x, y starts at $\Delta\ell/2$ and is incremented by $\Delta\ell$ each time around the loop over y. The sum is saved in S. The factor $q(a/2)(\Delta\ell)^2/4\pi\epsilon_0$ appears in every term so it is not included until after the sum is completed.

input charge and its coordinate: q, y'
input edge of cube, number of segments: a, N
calculate segment length: $\Delta\ell = a/N$
set $S = 0$
set $x = \Delta\ell/2$
begin loop over x
    set $A = (x - a/2)^2 + a/4$
    set $y = \Delta\ell/2$
    begin loop over y
        replace S with $S = S + \left[A + (y - y')^2\right]^{-3/2}$
        increment y by $\Delta\ell$
    end loop over y
    increment x by $\Delta\ell$
end loop over x
multiply S by $q(a/2)(\Delta\ell)^2/4\pi\epsilon_0$ and display result
stop

Program lines for other faces are similar. For the left face the loop lines are:

set $S = 0$
set $x = \Delta\ell/2$
begin loop over x
    set $A = (x - a/2)^2 + (y')^2$
    set $z = \Delta\ell/2$
    begin loop over z
        replace S with $S = S + \left[A + (z - a/2)^2\right]^{-3/2}$
        increment z by $\Delta\ell$
    end loop over z
    increment x by $\Delta\ell$
end loop over x
multiply S by $qy'(\Delta\ell)^2/4\pi\epsilon_0$ and display result

For the right face the loop lines are:

set $S = 0$
set $x = \Delta\ell/2$
begin loop over x
    set $A = (x - a/2)^2 + (a - y')^2$
    set $z = \Delta\ell/2$
    begin loop over z
        replace S with $S = S + \left[A + (z - a/2)^2\right]^{-3/2}$
        increment z by $\Delta\ell$
    end loop over z
    increment x by $\Delta\ell$

end loop over x
multiply S by $q(a - y')(\Delta\ell)^2/4\pi\epsilon_0$ and display result

PROJECT 1. Use the program to evaluate $\oint \mathbf{E} \cdot d\mathbf{A}$ for the surface of a cube with edge $a = 10$ m and with charge $q = 3.7 \times 10^{-9}$ C inside. Use $N = 15$ for 3 significant figure accuracy. First place the charge at the center of the cube: $x' = 5$ m, $y' = 5$ m, $z' = 5$ m. The flux is the same through each face so you need run the program for only one face, then multiply by 6. Compare the result with q/ϵ_0.

Now place the charge at $x' = 5$ m, $y' = 7.5$ m, $z' = 5$ m. Notice that the value of the flux through each face has changed from the previous situation. Also notice that the flux through the left, right, and top faces differ from each other. Nevertheless, the total flux is the same to within the accuracy of the calculation.

Finally, place the charge at $x' = 5$ m, $y' = 12.5$ m, $z' = 5$ m. It is outside the cube and the total flux through the cube should be zero. The result of the program may not be exactly zero because the flux through each face was computed to only about 3 significant figures. The total, however, should be significantly less than the flux through any individual face and should be still less if the calculation is done with a larger value of N.

Compare the calculations by filling out the following table with values of the flux:

y'	5 m	7.5 m	12.5 m
top	________	________	________
bottom	________	________	________
right	________	________	________
left	________	________	________
back	________	________	________
front	________	________	________
total	________	________	________

PROJECT 2. If the net charge inside the cube is zero then, according to Gauss' law, the total flux through the surface of the cube is zero. The flux through any particular face, however, does not necessarily vanish. Suppose $q_1 = -3.7 \times 10^{-9}$ C is at $y' = 2.5$ m on the line through the center of the cube and $q_2 = +3.7 \times 10^{-9}$ C is at $y' = 7.5$ m on the same line. Use the program to find the flux through each face of the cube due to each charge separately, then fill out the following table with values of the flux. If you completed the first project you already have values for q_1.

	q_1	q_2
top	________	________
bottom	________	________
right	________	________
left	________	________
back	________	________
front	________	________
total	________	________

The two total fluxes may not sum to zero because of errors in the calculation. However, it should be considerably less than the flux through any of the faces.

Chapter 26

If charge is distributed along the x axis with linear charge density $\lambda(x')$, then the electric potential at a point in the xy plane is given by

$$V(x,y) = \frac{1}{4\pi\epsilon_0} \int \frac{\lambda(x')}{[(x-x')^2+y^2]^{1/2}}\, dx',$$

where the integral extends over the charge distribution. You can use the Simpson's rule program of Chapter 7 to evaluate the integral. After evaluating the integral, multiply by $1/4\pi\epsilon_0$.

Here is an outline of the program, modified slightly to handle the calculation of an electric potential.

input limits of integral: x'_i, x'_f
input number of intervals: N
replace N with nearest even integer
calculate interval width: $\Delta x' = (x'_f - x'_i)/N$
input coordinates of field point: x, y
initialize quantity to hold sum of values with even labels: $S_e = 0$
initialize quantity to hold sum of values with odd labels: $S_o = 0$
set $x' = x'_i$
begin loop over intervals: counter runs from 1 to $N/2$
    calculate integrand: $I(x') = \lambda(x')\left[(x-x')^2+y^2\right]^{-1/2}$
        add it sum of values with even labels: replace S_e with $S_e + I(x')$
    increment x' by $\Delta x'$
    calculate integrand: $I(x') = \lambda(x')\left[(x-x')^2+y^2\right]^{-1/2}$
        add it to the sum of values with odd labels: replace S_o with $S_o + I(x')$
    increment x' by $\Delta x'$
end loop
calculate integrand at upper and lower limits:
    $I_0 = \lambda(x'_i)\left[(x-x'_i)^2+y^2\right]^{-1/2}$
    $I_N = \lambda(x'_f)\left[(x-x'_f)+y^2\right]^{-1/2}$
evaluate integral: $(\Delta x/3)(I_N - I_0 + 2S_e + 4S_o)$
multiply by $1/4\pi\epsilon_0$ and display result
go back for coordinates of another field point or stop

First use the program to investigate the relationship between the electric potential and the electric field: $E_x = -\partial V/\partial x$, $E_y = -\partial V/\partial y$.

PROJECT 1. Charge is distributed from $x' = 0$ to $x' = 0.10\,\text{m}$ along the x axis with a linear charge density given by $\lambda(x') = 1.83 \times 10^{-5}\sqrt{x'}\,\text{C/m}$, where x' is in meters. Use the Simpson's rule program to find values for the electric potential at $x = -0.005\,\text{m}$ and at $x = +0.005\,\text{m}$ on the line $y = 0.20\,\text{m}$. Start with $N = 20$ and repeat the calculation with N doubled each time until you get the same results to 3 significant figures.

Estimate the x component of the electric field at $x = 0$, $y = 0.2\,\text{m}$ by evaluating $-(V_2 - V_1)/\Delta x$, where V_1 is the potential at $x = -0.005\,\text{m}$, V_2 is the potential at $x =$

+0.005 m, and $\Delta x = 0.01$ m. Check your answer by using the program of Chapter 24 to compute the electric field directly. To how many figures do you obtain agreement? Some significance is lost when you subtract the two values of the potential.

Use the Simpson's rule program to calculate the electric potential at $y = 0.195$ m and at $y = 0.205$ m on the y axis. Use the results to estimate the y component of the electric field at $x = 0$, $y = 0.2$ m. Check your result by using the program of Chapter 24.

PROJECT 2. You can use the program to plot equipotential surfaces. In this project you will consider the line charge of the previous project and plot a line in the xy plane along which the electric potential has a given value V. Start with $x = -0.01$ m and use trial and error to find the y coordinate of the point for which the potential has the value V. You might start with $y = .01$ m and increment y by 0.1 m until you find two points with potentials that straddle V, then narrow the gap until the coordinates of the two points at its ends are the same to 2 significant figures. Increment x by 0.01 m and repeat. Continue until you reach $x = +0.11$ m. Once you have found the first few points a pattern should emerge and later points should be easier to find. Try potentials of 3, 5, and 10 V.

Chapter 29

Many circuit problems involve the solution of simultaneous linear equations. They can be solved on a computer. We describe what is known as the Gauss-Seidel iteration scheme, in which a solution is guessed and the given equations are used to improve the guess.

Suppose there are N equations and the unknowns are $i_1, i_2, \ldots, i_N$. For many problems they are the currents in the various branches of a circuit. Equation number j may be written

$$\sum_{k=1}^{N} A_{jk} i_k = B_j$$

where A_{jk} is the coefficient of unknown i_k in equation j and B_j is a term that contains no unknown. Solve the first equation for i_1 in terms of the other unknowns, the second equation for i_2 in terms of the other unknowns, etc. The result is

$$i_j = \frac{\left[B_j - \sum\limits_{\substack{k=1 \\ k \neq j}}^{N} A_{jk} i_k \right]}{A_{jj}} .$$

The first step is to guess values for $i_1, i_2, \ldots, i_N$. These guesses are used in the above equations to calculate new values. The process is then carried out again using the results of the first run. Iteration is continued until two successive runs yield the same results to within an acceptable error. The most current values are used on the right side of the equations as soon as they are calculated.

Care must be taken to arrange the equations so A_{jj} is not zero for any equation in the set. Even so, the results do not converge for some sets of equations. After many iterations successive results may differ greatly and may show no sign of getting closer in value. When this occurs, the original set of equations must be modified by adding (repeatedly, perhaps) some equations to others or subtracting some equations from others and using the resulting equation to replace one of the originals. Such manipulations do not change the solution.

We state without proof that the Gauss-Seidel iteration scheme converges toward the correct solution if, for every equation in the set, the so-called diagonal term (A_{jj} for equation j) is larger in magnitude than the sum of the magnitudes of the other coefficients in the equation. That is,

$$|A_{jj}| > \sum_{\substack{k=1 \\ k \neq j}}^{N} |A_{jk}| \,.$$

The goal of any modifications made to the original set is to obtain a new set for which this inequality holds.

The first step in the modification process is to put the equations in optimal order. Search for the largest coefficient and arrange the equations so this coefficient becomes a diagonal coefficient. Now search for the largest coefficient that is not in the same equation and does not multiply the same unknown as the first one found, then arrange the equations so this one is also diagonal. Continue until all equations have been considered. For example, suppose you wish to solve the set

$$3i_1 + 2i_2 + 2i_3 = 8$$
$$3i_1 + 4i_2 + 3i_3 = 5$$
$$7i_1 + 5i_2 + 3i_3 = 3\,.$$

The largest coefficient is 7. It multiplies i_1 in the third equation, so this equation should be the first. Of the coefficients that multiply other unknowns in other equations, the largest is 4, which multiplies i_2 in the second equation, so this equation should remain the second. The optimal order is

$$7i_1 + 5i_2 + 3i_3 = 3$$
$$3i_1 + 4i_2 + 3i_3 = 5$$
$$3i_1 + 2i_2 + 2i_3 = 8\,.$$

None of these equations obey the inequality. Subtract the second from the first and use the result to replace the first. The new set is

$$4i_1 + i_2 = -2$$
$$3i_1 + 4i_2 + 3i_3 = 5$$
$$3i_1 + 2i_2 + 2i_3 = 8\,.$$

Now the first equation obeys the inequality but the others do not. Subtract the third from the second and use the result to replace the second. The equations are now

$$4i_1 + i_2 = -2$$
$$2i_2 + i_3 = -3$$
$$3i_1 + 2i_2 + 2i_3 = 8\,.$$

To bring the third equation into line, multiply the first by 3, the third by 4, and subtract. Replace the third equation with the result. The new set is

$$4i_1 + i_2 = -2$$
$$2i_2 + i_3 = -3$$
$$5i_2 + 8i_3 = 38\,.$$

All three equations now satisfy the inequality and we expect the Gauss-Seidel iteration scheme to work. The scheme may work even if the inequality is not satisfied, so you may want to run through a few iterations before spending time in modifying the equations.

Write a program to solve a set of linear simultaneous equations. Store the coefficients in a subscripted variable $A(j,k)$, the constant terms in the subscripted variable $B(j)$, and the unknowns in the subscripted variable $I(j)$. Take the initial guesses to all be zero. Here's an outline.

input number of equations: N
begin loop over equations; counter j runs from 1 to N
  begin loop over variables; counter k runs from 1 to N
    input coefficients: $A(j,k)$
  end loop over variables
  input constant term: $B(j)$
  set $I(j) = 0$
end loop over equations
* **begin loop** over equations; counter j runs from 1 to N
  set $S = 0$ in preparation for computing sum
  begin loop over variables; counter k runs from 1 to N
    if $k \neq j$ replace S with $S + A(j,k)I(k)$
  end loop over variables
  replace $I(j)$ by $[B(j) - S]/A(j,j)$
end loop over equations
display solution
go back to starred instruction for another iteration or stop

When the solution is displayed you must judge whether convergence has been reached or not. If values of $I(j)$ have not changed much from the last iteration you will want to stop the program. If they have changed you will want the program to perform another iteration. You might arrange the display so that results of two or more iterations are on the screen simultaneously.

PROJECT 1. To test the program, use it to solve the following set of 4 simultaneous equations. Obtain 3 significant figure accuracy.

$$2i_1 - i_2 = 5$$

$$2i_2 - i_3 = 7$$

$$2i_3 - i_4 = 9$$

$$2i_4 - i_1 = 11\,.$$

[ans: $i_1 = 6.47$, $i_2 = 7.93$, $i_3 = 8.87$, $i_4 = 8.73$]

PROJECT 2. Consider the circuit shown below with current arrows and labels. Write down 2 junction and 3 loop equations. Modify the set of equations so the convergence conditions are met. Take $\mathcal{E}_1 = 10\,\mathrm{V}$, $R_1 = R_2 = 5\,\Omega$, $R_3 = R_4 = 8\,\Omega$, and $R_5 = 12\,\Omega$. Use the program to find values of the 5 currents for each of the following values of $\mathcal{E}_2$: 5, 7.5, 10, 12, and 15 V. Assume the given values are exact and find the solutions with an accuracy of at least 3 significant figures. For each value of $\mathcal{E}_2$ tell if the seats of emf are charging or discharging.

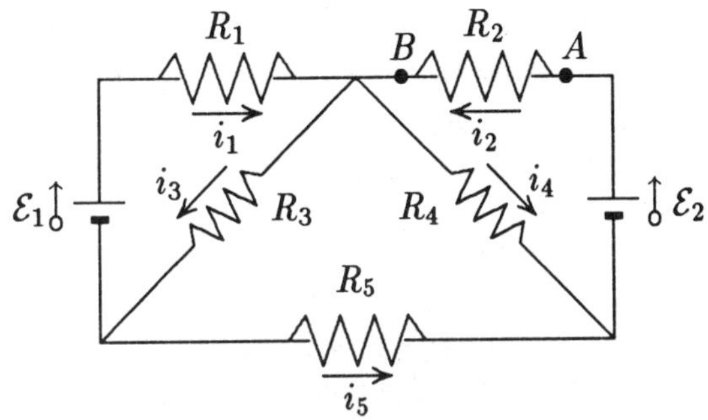

PROJECT 3. Suppose the ends of resistor R_2 in the circuit of the previous project are also connected to an external circuit. Current i_6 enters the external circuit at A and returns to the circuit of the diagram at B.

Take $\mathcal{E}_2 = 15\,\mathrm{V}$ and the other quantities as given in the previous project. For each of the following values of i_6 solve for the values of the other currents: 0, 2, 4, 6, 8, and 10 A. Obtain 3 significant figure accuracy.

For each of the cases considered calculate the potential difference ΔV across R_2. Plot ΔV vs. i_6. Notice that it is a straight line. As far as the external circuit is concerned the circuit shown in the diagram above can be replaced by a seat of emf and a resistor in series. The emf has the value of ΔV for $i_6 = 0$ and the value of the resistance is the slope of the line. These values are $\mathcal{E} =$ ______ V and $R =$ ______ Ω.

We now use the program to investigate the operation of a measuring instrument. In this case the instrument is a Wheatstone bridge and is used to measure resistance. It is typical of many different bridge circuits used for various electrical measurements.

PROJECT 4. The circuit for a Wheatstone bridge is shown below. The symbol G stands for a galvanometer, an instrument that can detect small currents. We suppose the unknown resistor is R_3 and the other resistors are variable. They are set so the galvanometer reads 0 and $i_6 = 0$. Then $R_3 = R_1 R_4 / R_2$. The resistances R_1, R_2, and R_4 are read and their values are used to calculate R_3.

First verify that $i_6 = 0$ when $R_3 = R_1 R_4 / R_2$. Take $R_1 = R_2 = 12\,\Omega$, $R_3 = R_4 = 18\,\Omega$, and $R_5 = 3.2\,\Omega$. The resistance of the galvanometer is $2.0\,\Omega$ and $\mathcal{E} = 10\,\mathrm{V}$. Note that the balance condition is met. Write 3 junction and 3 loop equations, then use the program to solve for the 6 currents. Obtain 3 significant figure accuracy. Don't forget to modify the equations so the convergence conditions are met. You should find that $i_6 = 0$ to within the accuracy of the calculation.

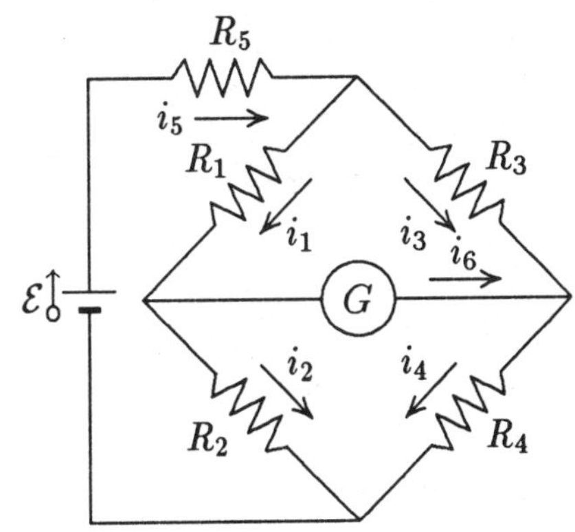

It is usually of some interest to know how sensitive a Wheatstone bridge is. You can test the sensitivity by making an error in the setting of one of the resistors, then seeing if the galvanometer can detect the resulting current i_6.

Take $R_1 = R_2 = 12\,\Omega$, $R_3 = 18\,\Omega$, $R_4 = 19\,\Omega$, and $R_5 = 3.2\,\Omega$. R_g is still $2.0\,\Omega$ and $\mathcal{E}$ is still 10 V. The balance condition is not met, the current in the galvanometer does not vanish, and the balance equation predicts that $R_3 = 19\,\Omega$ instead of the correct value, $18\,\Omega$. Use the program to solve for i_6. The galvanometer must be able to detect a current of this value or less if the bridge is to measure the unknown resistance with an error of less than $1\,\Omega$. Is this a reasonable current for a galvanometer to detect? [ans: -7.67×10^{-3} A; easily detected]

Chapter 31

A computer program can be used to integrate the Biot-Savart equation for the magnetic field of a current loop. In general the field is given by the integral around the current loop:

$$\mathbf{B}(\mathbf{r}) = \frac{\mu_0}{4\pi} i \int \frac{d\mathbf{r}' \times (\mathbf{r} - \mathbf{r}')}{|\mathbf{r} - \mathbf{r}'|^3},$$

where i is the current, $\mathbf{r}$ is the position vector of the point where the field is $\mathbf{B}$, and $\mathbf{r}'$ is the position vector of a point on the loop.

Consider circular a loop in the xy plane and use an angular variable of integration. Since the situation has cylindrical symmetry we can without loss of generality specialize to a point in the yz plane and write $\mathbf{r} = y\,\mathbf{j} + z\,\mathbf{k}$. Take $\mathbf{r}' = R(\cos\theta\,\mathbf{i} + \sin\theta\,\mathbf{j})$. Then $d\mathbf{r}' = R(-\sin\theta\,\mathbf{i} + \cos\theta\,\mathbf{j})\,d\theta$. When these substitutions are made the expressions for the components of the field are

$$B_x = 0,$$

$$B_y = \frac{\mu_0}{4\pi}\frac{iz}{R^2}\int_0^{2\pi} \frac{\sin\theta\,d\theta}{[1 + (y/R)^2 + (z/R)^2 - 2(y/R)\sin\theta]^{3/2}},$$

$$B_z = -\frac{\mu_0}{4\pi}\frac{i}{R}\int_0^{2\pi} \frac{[(y/R)\sin\theta - 1]\,d\theta}{[1 + (y/R)^2 + (z/R)^2 - 2(y/R)\sin\theta]^{3/2}},$$

Both integrals have the form

$$B_i = \frac{\mu_0 i}{4\pi R}\int_0^{2\pi} f(\theta)\,d\theta.$$

The Simpson's rule program of Chapter 7 can be used to carry out the integrations. Divide the interval in θ from 0 to 2π into N segments, each of length $\Delta\theta = 2\pi/N$. Then B_i is approximated by $(\Delta\theta/3)(2S_e + 4S_o)$, where S_e is the sum of the values of the integrand at the beginning of segments with even labels and S_o is the sum of the values at the beginning of segments with odd labels. The term $f_N - f_0$ does not appear because the integrands have the same value at $\theta = 0$ and $\theta = 2\pi$. N must be an even integer.

To evaluate both integrals (for B_y and B_z) within the same program, let S_{ye} and S_{yo} collect the sums for the y component and S_{ze} and S_{zo} collect the sums for the z component. Here's an outline.

input number of intervals: N
replace N with nearest even integer

calculate interval width: $\Delta\theta = 2\pi/N$
input coordinates of field point: y, z
initialize quantities to hold sum of values with even labels: $S_{ye} = 0$, $S_{ze} = 0$
initialize quantities to hold sum of values with odd labels: $S_{yo} = 0$, $S_{zo} = 0$
set $\theta = 0$
begin loop over intervals: counter runs from 1 to $N/2$
  calculate $A = \sin\theta$ for future use
  calculate $B = \left[1 + (y/R)^2 + (z/R)^2 - 2(y/R)\sin\theta\right]^{3/2}$ for future use
  update sums over even terms
    replace S_{ye} with $S_{ye} + A/B$
    replace S_{ze} with $S_{ze} + [(y/R)A - 1]/B$
  increment θ by $\Delta\theta$
  calculate $A = \sin\theta$ for future use
  calculate $B = \left[1 + (y/R)^2 + (z/R)^2 - 2(y/R)\sin\theta\right]^{3/2}$ for future use
  update sums over odd terms
    replace S_{yo} with $S_{yo} + A/B$
    replace S_{zo} with $S_{zo} + [(y/R)A - 1]/B$
  increment θ by $\Delta\theta$
end loop
evaluate integrals:
  $B_y = (\mu_0 iz/4\pi R^2)(\Delta\theta/3)(2S_{ye} + 4S_{yo})$
  $B_z = -(\mu_0 i/4\pi R(\Delta\theta/3)(2S_{ze} + 4S_{zo})$
display result
go back to input new field coordinates or stop

Running time can be reduced if you add instructions to calculate the new variables $y' = y/R$ and $z' = z/R$ immediately after y and z are read. Then use y' and z' in succeeding instructions. They equations have been written in a convenient form to do this.

The purpose of the first project is to test the program and to obtain a rough idea of the number of intervals that should be used.

PROJECT 1. If the field point is on the z axis the integrals can be evaluated analytically, with the result $B_x = 0$, $B_y = 0$, and

$$B_z = \frac{\mu_0 i R^2}{2(R^2 + z^2)^{3/2}}.$$

Consider a 1.0-m radius loop carrying a current of 1.0 A and use the program to calculate the field at $z = 0$, 0.50, 1.5, and 2.5 m on the z axis. Since the integrand is constant you should obtain the correct answers with $N = 2$. Check your answers by evaluating the analytic expression.

Now find the magnetic field at the point $x = 0$, $y = 0.50$ m, $z = 0$. Start with $N = 2$ and on the next trial double N. Continue to double N until the results of two successive trials agree to 3 significant figures.

Repeat the calculation for the following points along the $x = 0$, $y = 0.50$ m line: $z = 0.10$, 1.0, and 10 m. Note the value of N for which 3 significant figure accuracy is obtained.

A set of Helmholtz coils consists of two identical loops parallel to each other and carrying the same magnitude of current, in the same direction. In the region between the coils the two magnetic fields tend to be in roughly the same direction and when the distance between the coils is equal to the radius of one of the loops, the field in the region between is particularly uniform. For this reason Helmholtz coils are often used to produce magnetic fields in the laboratory. In the following project you use the integration program to investigate the uniformity of the magnetic field between two coils.

PROJECT 2. Two circular loops of wire, each with a radius of 1.0 m, are placed parallel to the xy plane, with their centers on the z axis. The center of one is at $z = 0$ while the center of the other is at $z = 2.0$ m. Each carries a current of 1.0 A in the counterclockwise direction when viewed from the positive z axis. Note that the separation is twice the radius.

Use the program to calculate the magnetic field with 3 significant figure accuracy for field points at $z = 0.60$, 0.80, and 1.0 m, all on the z axis. The last point is at the center of the region between the loops. You will need to run the program twice, once for each loop. The first field point ($z = 0.60$ m) is 0.60 m from the first coil and 1.4 m from the second. Run the program with $z = 0.60$ m, then with $z = -1.4$ m. Vectorially sum the two fields.

Use these calculated fields to test the uniformity of the field along the z axis between the loops. For each of the first two points, subtract the magnitude of the field at the center ($z = 1.0$ m) from the field at the point and divide by the magnitude of the field at the center. If we denote this measure of homogeneity by h, then

$$h(z) = \frac{|B(z) - B_c|}{|B_c|},$$

where B_c is the magnitude of the field at the center of the region.

If h is small the field does not change much with position and is said to be homogeneous. If h is large the field is said to be inhomogeneous. For good laboratory magnets h may be on the order of 10^{-6} or less over several centimeters. The distance between the loops, of course, is usually less than 2 m.

Now consider the same loops, but separated by 1.0 m, the radius of one of them. Use the program to calculate the magnetic field at $z = 0.10$, 0.30, and 0.50 m, all on the z axis. The last point is at the center of the region between the loops. Calculate h for the first two points. Has the field become more or less homogeneous?

PROJECT 3. You can also investigate homogeneity in a transverse direction. Now the field has two non-vanishing components and we define a measure of homogeneity for each:

$$h_z = \frac{|B_z(y) - B_z(y = 0)|}{|B_z(y = 0)|}$$

$$h_y = \frac{|B_y(y)|}{|B_z(y = 0)|}.$$

Both are zero for a perfectly homogeneous field.

Consider the loops of the previous project, with a separation of 2.0 m, and calculate h_y and h_z for $y = 0.20$ and 0.40 m on the line $x = 0$, $z = 1.0$ m.

Now take the separation to be 1.0 m and calculate h_y and h_z for $y = 0.20$ and 0.50 m on the line $x = 0$, $z = 0.50$ m. Has the homogeneity increased or decreased?

If the loops are moved still closer to each other so their separation is less than the radius, the field becomes less uniform. Consider a separation of 0.50 m and calculate h_y and h_z for $y = 0.20$ and 0.40 m along the line $x = 0$, $z = 0.25$ m.

In the following projects you will use an integration program to investigate Ampere's law. The square shown is in the xy plane, is centered at the origin, and has edges of length a. A wire carrying current i pierces the plane of the square at the point on the x axis with coordinate ℓ. It produces a magnetic field at each point on the perimeter of the square (as well as at other points). According to Ampere's law

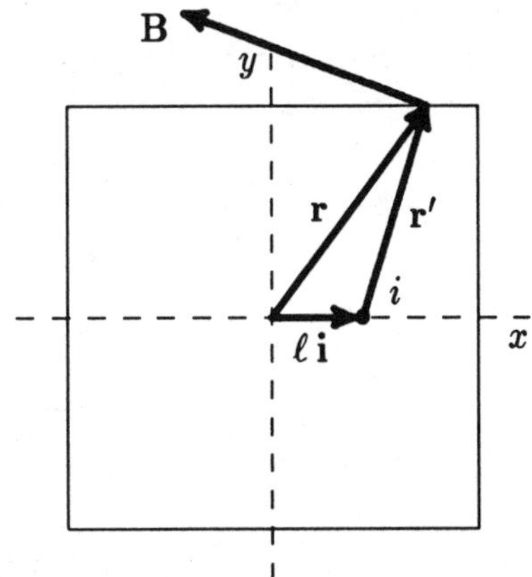

$$\oint \mathbf{B} \cdot d\mathbf{s} = \mu_0 i \,,$$

where the integral is around the perimeter and i is the current in the wire. Take the current to be positive if it is out of the page and carry out the integration in the counterclockwise direction around the square.

The magnitude of the magnetic field is given by $B = \mu_0 i/2\pi r'$, where r' is the distance from the wire. The field components are $B_x = -(\mu_0 i/2\pi)y'/(r')^2$ and $B_y = (\mu_0 i/2\pi)x'/(r')^2$. As the diagram shows, $y' = y$, $x' = x - \ell$, and $(r')^2 = (x-\ell)^2 + y^2$ so $B_x = -(\mu_0 i/2\pi)y/[(x-\ell)^2 + y^2]$ and $B_y = (\mu_0 i/2\pi)(x-\ell)/[(x-\ell)^2+y^2]$.

Consider each of the four sides of the square separately. Across the top $y = a/2$, $d\mathbf{s} = dx\,\mathbf{i}$ and

$$\int \mathbf{B} \cdot d\mathbf{s} = -\frac{\mu_0}{2\pi}\frac{a}{2}\int_{a/2}^{-a/2} \frac{dx}{(x-\ell)^2 + (a/2)^2} \,.$$

The contribution of the bottom is exactly the same. Down the left side $x = a/2$, $d\mathbf{s} = dy\,\mathbf{j}$, and

$$\int \mathbf{B} \cdot d\mathbf{s} = \frac{\mu_0 i}{2\pi}(\frac{a}{2} + \ell)\int_{a/2}^{-a/2} \frac{dy}{(\ell + a/2)^2 + y^2} \,.$$

Up the right side $x = a/2$, $d\mathbf{s} = dy\,\mathbf{j}$, and

$$\int \mathbf{B} \cdot d\mathbf{s} = \frac{\mu_0 i}{2\pi}(\frac{a}{2} - \ell)\int_{-a/2}^{a/2} \frac{dy}{(-\ell + a/2)^2 + y^2} \,.$$

Each of the integrals can be evaluated using the Simpson's rule program.

PROJECT 4. Suppose the wire carries a current of 1.0 A, out of the page, and the square has sides of length $a = 2.0$ m. Evaluate $\oint \mathbf{B} \cdot d\mathbf{s}$, one side at a time, for each of the following positions of the wire: $\ell = 0$, 0.40, 0.80, and 1.2 m. The first three points are inside the square and the fourth is outside. Fill in the table below with values of $\int \mathbf{B} \cdot d\mathbf{s}$.

ℓ	0	0.40 m	0.80 m	1.2 m
top				
bottom				
right				
left				
total				

For each of the situations compute $\mu_0 i$, where i is the net current through the square. Compare the result with the totals above.

PROJECT 5. For the square of the previous project evaluate $\oint \mathbf{B} \cdot d\mathbf{s}$ for a current of 2.0 A, out of the page at $\ell = 0$. Evaluate the integral for a current of 1.0 A, out of the page at $\ell = 0.50$ m. Finally, evaluate the integral for a current of 3.0 A, out of the page at $\ell = 0.75$ m. Compare the last answer to the sum of the first two.

Evaluate the integral for the a current of 3.0 A, out of the page at $\ell = 0$. Evaluate the integral for a current of 2.0 A, into the page at $\ell = 0.50$ m. Compare the difference in these results with the value of the integral for a current out of the page at $\ell = 0.40$ m (see the results of Project 4).

Chapter 41

The text deals with the diffraction of light by a single slit for the special case when the viewing screen is far away. Then the Huygen wavelets emanating from the slit are essentially plane waves. When the viewing screen is close to the slit or when the slit is wide you must take into account the true spherical nature of the wavelets. The diffraction pattern for such wavelets cannot be described analytically but you can use a computer to describe it numerically. As the slit is widened you will be able to see the diffraction pattern change into the geometrical image of the slit.

The diagram shows a plane wave impinging on a single slit from the left. A spherical wave emanates from each point in the slit. One of them is shown. The viewing screen is to the right, a distance x from the slit. Take the origin to be at the center of the slit and let y' be the coordinate of the point within the slit from which the spherical wave emanates. You will calculate the light intensity at the point on the screen a distance y from the center line. According to Huygen's principle the wavelet from y' has the form

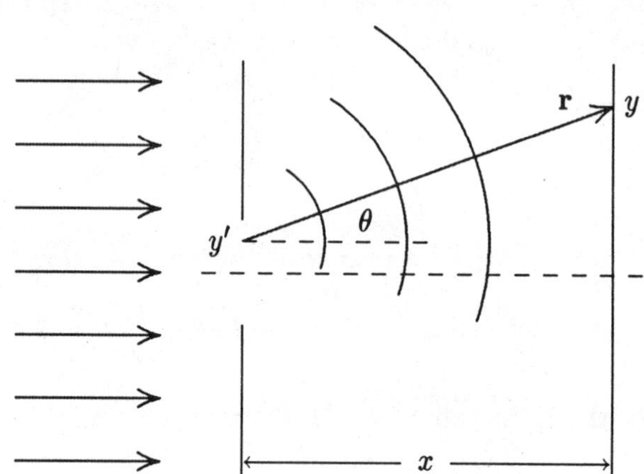

$$E = \frac{B}{r}(1 + \cos\theta)\sin(kr - \omega t),$$

where r is the distance from the wavelet source to the observation point and θ is the angle between the ray from the source to the observation point and the x axis. B is a constant and is chosen so that in the absence of an obstacle the wavelets sum to produce a plane wave with the same amplitude as the original plane wave. Since we are concerned only with changes in the intensity as the observation point moves we shall choose its value for computational convenience.

The amplitude of the spherical wavelet is not the same as the amplitude of the incident plane wave. Since there are an infinite number of wavelets with sources in the slit B must be infinitesimal. In addition, the amplitude contains the factor $(1+\cos\theta)$. The wavelet has a larger amplitude in the forward direction (θ near 0) than in the backward direction (θ near 180°). This factor is important if the sum of the wavelets is to reproduce a wave that is traveling in the same direction as the original wave. If the slit is narrow and the viewing screen is far away then $\cos\theta \approx 1$ for all wavelets that reach the observation point and this factor does not play an important role in determining the intensity pattern. If the slit is wide and the screen is nearby, however, it is important.

The wavelet is spherical and its amplitude decreases as $1/r$. Since the coordinates of the observation point are x and y, $r = [x^2 + (y - y')^2]^{1/2}$ and $\cos\theta = x/r$.

All of the wavelets that originate at points in the slit must now be summed to find the total disturbance at the observation point. Since there is a continuous distribution of wavelet sources in the slit, the sum takes the form of an integral. The amplitude of a wavelet must be infinitesimal and we take $B = dy'/a$, where a is the width of the slit. The total disturbance at the observation point is given by

$$E = \frac{1}{a}\int_{-a/2}^{a/2} \frac{(1+\cos\theta)\,\sin(kr-\omega t)}{r}\,dy'\,.$$

The resultant wave has the form $E = E_0 \sin(-\omega t + \alpha)$, where

$$E_0^2 = \frac{1}{a^2}\left[\int_{-a/2}^{a/2} \frac{(1+\cos\theta)\,\sin(kr)}{r}\,dy'\right]^2 + \frac{1}{a^2}\left[\int_{-a/2}^{a/2} \frac{(1+\cos\theta)\,\cos(kr)}{r}\,dy'\right]^2$$

The intensity at the observation point is proportional to E_0^2.

The Simpson's rule integration program of Chapter 7 can be modified to evaluate the integrals. S_{1e} and S_{1o} collect the sums of even and odd terms, respectively, for the first integral and S_{2e} and S_{2o} collect the even and odd terms, respectively, for the second integral. The program calculates the intensity at a series of observation points starting at $y = 0$ and ending at $y = y_f$, with an interval of Δy. The intensity pattern is symmetrical about $y = 0$ ($E_0^2(-y) = E_0^2(y)$), so only the upper half need be computed. Here's the outline of a program.

input slit width: a
input number of segments: N
replace N with the nearest even integer
calculate wavelet source interval: $\Delta y' = a/N$
input wavelength: λ
calculate $k = 2\pi/\lambda$
input observation interval and final observation coordinate: Δy, y_f
input distance from slit to viewing screen: x
set $y = 0$
begin loop over observation coordinate
  set $S_{1e} = 0$, $S_{1o} = 0$, $S_{2e} = 0$, $S_{2o} = 0$, $y' = 0$
  begin loop over wavelet sources; counter runs from 1 to $N/2$
    calculate r: $r = \left[x^2 + (y-y')^2\right]^{1/2}$
    calculate $(1+\cos\theta)/r$: $A = (1 + x/r)/r$
    update sums of even terms:
      replace S_{1e} with $S_{1e} + A\sin(kr)$
      replace S_{2e} with $S_{2e} + A\cos(kr)$
    increment y' by $\Delta y'$
    calculate r: $r = \left[x^2 + (y-y')^2\right]^{1/2}$
    calculate $(1+\cos\theta)/r$: $A = (1 + x/r)/r$
    update sums of odd terms:
      replace S_{1o} with $S_{1o} + A\sin(kr)$
      replace S_{2o} with $S_{2o} + A\cos(kr)$
    increment y' by $\Delta y'$
  end loop over wavelet sources
  calculate integrands at upper and lower limits: I_{1N}, I_{10}, I_{2N}, I_{20}

calculate intensity:

$$E_0^2 = (\Delta y'/3a)^2[(I_{1N} - I_{10} + 2S_{1e} + 4S_{1o})^2 + (I_{2N} - I_{20} + 2S_{2e} + 4S_{2o})^2]$$

display or plot intensity
if $y \geq y_f$ exit loop over observation coordinate
increment y by Δy
end loop over observation coordinate
stop

You might write instructions to plot the intensity pattern directly on the monitor screen or you might plot it by hand using data generated by the program.

First use the program to investigate the pattern when the viewing screen is far from the slits. This project also acts as a check on the program. The plot of the intensity as a function of viewing coordinate should agree with the diagram in the text.

PROJECT 1. A 1.0×10^{-4}-m wide slit is illuminated by a plane wave with a wavelength of 5.0×10^{-7} m and the intensity pattern is viewed on a screen 1.0 m away. Use the numerical integration program with $N = 16$ to plot E_0^2 every 1.0×10^{-3} m from $y = -15 \times 10^{-3}$ m to $y = +15 \times 10^{-3}$ m. The intensity for negative y is exactly the same as that for positive y, so you need to run the program for positive y only. Indicate on the plot the geometric image of the slit. It extends from $y = -5.0 \times 10^{-5}$ m to $y = +5.0 \times 10^{-5}$ m.

The pattern is dominated by a broad, intense, central region, with a maximum at $y = 0$. This is followed on either side by a series of less intense bright fringes, separated by regions where the intensity is low. The fringes, of course, come about as a result of the interference of the Huygen wavelets. At $y = 0$ the wavelets are all nearly in phase with each other and add constructively. As y increases the various wavelets travel different distances to the observation point and arrive with different phases. At a minimum of the intensity the phases are such that the sum of the wavelets is small.

Notice that the intensity pattern spreads well beyond the region of the geometric image. In fact, the central bright region alone is roughly 100 times as wide as the geometric image. The appearance of secondary maxima makes the pattern even wider. This is diffraction.

The pattern spreads as the slit width is narrowed. Suppose a 7.0×10^{-5}-m wide slit is illuminated by a plane wave with a wavelength of 5.0×10^{-7} m and the intensity pattern is viewed on a screen 1.0 m away. Use the program to plot the intensity as a function of the observation coordinate y. Use the graph to find the coordinates of the minima of intensity closest to the central bright area. Compare the values of these coordinates to those when the slit width is 1.0×10^{-4} m.

An increase in wavelength also produces a broadening of the intensity pattern. Suppose a 1.0×10^{-4}-m wide slit is illuminated by a plane wave with a wavelength of 6.5×10^{-7} m and the intensity pattern is viewed on a screen 1.0 m away. Use the program to plot the intensity as a function of the observation coordinate y. Find the coordinates of the minima of intensity closest to the central bright area. Compare with the coordinates of these minima when the wavelength is 5.0×10^{-7} m and the slit width is the same (1.0×10^{-4} m).

As the slit is widened or the viewing screen is brought closer to the slit, the fringe system narrows. Eventually the central maximum of intensity occupies approximately the region of the geometric image and very little light reaches regions of the geometric shadow. There is still some fringing near the edges of the geometric image, however. Something else happens as the slit is widened: fringes appear in the region of the geometric image. The intensities at minima within the image region are not zero so the fringes are not as noticeable as the fringes of a narrow slit.

The following project is designed to show you how the intensity pattern changes as the slit widens. To carry it out requires considerable running time. To shorten the time the number of integration intervals has been selected so the calculation is accurate to only 2 significant figures. If a fast machine is available and higher accuracy is desired, you might double the number of intervals in each case.

PROJECT 2. A plane wave with a wavelength of 5.0×10^{-7} m illuminates a 7.0×10^{-5}-m wide slit and the intensity pattern is viewed on a screen 0.020 m away. Use the program to plot the intensity as a function of the coordinate y of the observation point. Take $N = 16$ and plot points every 2.0×10^{-5} m from $y = -3.6 \times 10^{-4}$ m to $y = +3.6 \times 10^{-4}$ m. Locate the edges of the geometric image on the graph.

Qualitatively the pattern is quite similar to the patterns obtained in the previous project. The bright central region extends a considerable distance beyond the geometric image and this region is followed by a series of fringes. The central maximum is not quite as bright and the dark regions between secondary maxima are slightly broader than for greater slit-to-screen distances.

The slit is now widened to 1.4×10^{-6} m. The wavelength and slit-to-screen distance remain the same. Use the program, with $N = 32$, to plot the intensity every 1.0×10^{-5} m from $y = -2.2 \times 10^{-4}$ m to $y = +2.2 \times 10^{-4}$ m. Locate the edges of the geometric image on the graph.

Notice that the pattern is much more narrow than before. The central bright region is now within the geometric image and the pattern has a shoulder near the edge of the image. This shoulder is a remnant of the first minimum of the pattern for a narrow slit.

The slit is now widened to 2.8×10^{-4} m. Use the program, with $N = 50$, to plot the intensity every 1.0×10^{-5} m from $y = -2.0 \times 10^{-4}$ m to $y = +2.0 \times 10^{-4}$ m. Locate the edges of the geometric image on the graph.

The geometric image is now discernible in the intensity pattern. There are fringes deep within the image but the intensity does not become zero anywhere in that region. There is a gray central area where the intensity is about half that at the maximum but this region merges into a bright region and then, as the edge of geometric image is approached, the intensity falls off rapidly. There is fringing in the neighborhood of the image edge but the illuminated region does not extend very far beyond the edge.

Chapter 42

First write a program that carries out the Lorentz transformation for a succession of values of the relative velocity v of the reference frames. Input the coordinate x and time t of the event, as measured in frame S. Then input the first value of v to be considered, the last, and the interval Δv. Positive values of v mean S' is moving in the positive x direction; negative values mean it is moving in the negative x direction. Use $x' = \gamma(x - vt)$ and $t' = \gamma(t - vx/c^2)$, where $\gamma = 1/\sqrt{1 - v^2/c^2}$ and c is the speed of light, to calculate the coordinate and time in the primed frame. Here's an outline.

input coordinate and time in S: x, t
input first velocity, last velocity, and increment: v_i, v_f, Δv
set $v = v_i$
begin loop over intervals
    exit loop if $|c - v|/c < 10^{-5}$
    calculate γ: $\gamma = 1/\sqrt{1 - v^2/c^2}$
    calculate x': $x' = \gamma(x - vt)$
    calculate t': $t' = \gamma(t - vx/c^2)$

```
        display v, x, t, x', t'
        increment v by Δv
    end loop if v > v_f
    stop
```

The computer will overflow if it tries to calculate γ for v nearly equal to c so these values are rejected. If $v_f = v_i$ then results for only one velocity are produced.

First use the program to investigate simultaneity and length measurements.

PROJECT 1. Two events occur at time $t = 0$ in reference frame S, one at the origin and the other at $x = 5.0 \times 10^7$ m. Plot the distance $|\Delta x'|$ between the events as measured in frame S', as a function of the velocity v of that frame relative to frame S. Consider values from $-0.95c$ to $+0.95c$.

For the range considered which value of v results in the greatest value of $|\Delta x'|$? Which results in the smallest value? [ans: $\pm 0.95c$; 0]

Plot the time interval $|\Delta t'|$ between the events as measured in S', as a function of v. For the range considered which value of v results in the greatest value of $|\Delta t'|$? Which results in the smallest value? [ans: $\pm 0.95c$, 0]

For what range of v does the second event occur before the first ($\Delta t' < 0$)? For what range does it occur after the first ($\Delta t' > 0$)? [ans: $v > 0$; $v < 0$]

You may think of the values of $|\Delta x'|$ as the results of a series of length measurements, with the object being measured traveling at the various velocities considered. Its length is S is measured by simultaneously making marks at the front and back ends on the x axis and measuring the distance between them. Since the marks are made simultaneously $\Delta t = 0$ and $\Delta x' = \gamma \Delta x$, where $\Delta x'$ is the length in the rest frame of the object. Note that $|\Delta x'| > |\Delta x|$ and the discrepancy becomes greater as the speed becomes greater.

PROJECT 2. Two events at separated places that are simultaneous in one frame are not simultaneous in any other frame that is moving with respect to the first. The closer together the two events are in space, however, the smaller is the time interval between them in S'. Verify this statement by repeating the calculations of the last project, but take $x = 2.5 \times 10^7$ m. Compare the results you obtain with those of the last project for the same value of v.

If two events occur at the same coordinate then the time between the events is the proper time interval. The time between the events, as measured in another frame, is longer than the proper time by the factor γ. This is the phenomenon of time dilation.

PROJECT 3. Two events occur at $x = 0$ in reference frame S, one at $t = 0$ and the other at $t = 8.0 \times 10^{-5}$ s. Use the program to plot the time interval $|\Delta t'|$ between the events as measured in S', as a function of the velocity v of that frame relative to S. Consider values of v from $-0.95c$ to $+0.95c$.

For the range considered which value of v results in the greatest value of $|\Delta t'|$? Which results in the smallest value? [ans: $\pm 0.95c$, 0]

The proper time interval for these events is measured in frame S, where both events occur at the same place. Notice that the time interval, as measured in any other frame, is longer.

Plot the distance $|\Delta x'|$ in frame S' as a function of the velocity of that frame. For what range of v is $\Delta x'$ negative? For what range is it positive? [ans: $v < 0$; $v > 0$]

The two events considered may not occur at the same coordinate in either S or S'. Then neither Δt nor $\Delta t'$ is the proper time interval between them and these two quantities are not related to each other by a factor of γ. If $\Delta x/\Delta t < c$ then a frame, traveling at less than the speed of light, exists for which the two events occur at the same coordinate. The time interval, as measured in that frame, *is* the proper time interval between the events.

PROJECT 4. One event occurs at $x = 0$, $t = 0$ and another occurs at $x = 200\,\text{m}$, $t = 9.5 \times 10^{-7}\,\text{s}$. Use the program and a trial and error technique to find the velocity of a frame for which the events occur at the same place. Check your answer by a direction calculation: $\Delta x' = 0$ means $\gamma(\Delta x - v\Delta t) = 0$, or $v = \Delta x/\Delta t$. [ans: $2.11 \times 10^8\,\text{m/s}$ ($0.702c$)]

The next project gives a nice demonstration of time dilation. Each observer signals the other, giving the time read by his clocks. Each finds that the other's clocks run slowly. You may wish to revise the program so it will calculate x' and t' for various values of x and t, all for the same value of the relative velocity v. Here's an outline.

input velocity of S' relative to S: v
calculate γ: $\gamma = 1/\sqrt{1 - v^2/c^2}$
input coordinate and time in S: x, t
calculate x': $x' = \gamma(x - vt)$
calculate t': $t' = \gamma(t - vx/c^2)$
display v, x, t, x', t'
another calculation?
    if yes go back to third line
    if no stop

PROJECT 5. A rocket starts on earth and travels away at $0.97c$. Every hour for the first five hours an earth bound transmitter at the launch pad sends a radio signal to the rocket. These signals are electromagnetic and travel at the speed of light. Assume the rocket starts at time $t = 0$ and the first signal is sent at $t_{s1} = 3600\,\text{s}$. Signal n is sent at $t_{sn} = 3600n$, where $n = 1, 2, 3, 4, 5$. Take the launch pad to be at the origin of the earth's rest frame S and the rocket to be at the origin of its rest frame S'.

Fill in the first 3 columns of the following table (t_{sn} is the time signal n is sent, t_{rn} is the time signal n is received, x_{rn} is the coordinate of the rocket when signal n is received, all in the rest frame of the earth. You will need to show that signal n is received at time $t_{rn} = t_{sn}c/(c - v)$, where v is the speed of the rocket. This is easy since the coordinate of the rocket is given by $x = vt$ and the coordinate of signal n is given by $x = c(t - t_{sn})$. The signal is received when these are equal. The distance from the earth to the rocket when it receives signal n is given by $x_{rn} = vt_{rn}$.

n	t_{sn}	t_{rn}	x_{rn}	t'_{rn}	t'_{sn}
1	______	______	______	______	______
2	______	______	______	______	______
3	______	______	______	______	______
4	______	______	______	______	______
5	______	______	______	______	______

Now use the program to find the time t'_{rn} when signal n is received, as measured by a clock on board the rocket. Fill in the fourth column of the table.

An observer on the rocket can calculate the time, according to his clock, when each signal was sent. The signal starts from the coordinate of the launch pad at time t'_{sn}. This is $x' = -vt'_{sn}$, where the negative sign appears because in the frame of the rocket the earth is moving in the negative x direction. The coordinate of the signal is given by $x' = -vt'_{sn} + c(t' - t'_{sn})$. This must be zero when $t' = t'_{rn}$ (the rocket receives the signal at the origin of S'. So $t'_{sn} = t'_{rn}c/(c+v)$. Fill in the last column of the table. According to clock on the rocket is the earth clock slow or fast? __________

Now suppose the rocket emits a signal every hour, according to on-board clocks. Let t'_{sn} be the time signal n is sent from the rocket (at $x' = 0$), let t'_{rn} be the time it is received on earth, and let x'_{rn} be the coordinate of the earth when the signal is received. Fill out the following table for those signals.

n	t'_{sn}	t'_{rn}	x'_{rn}	t_{rn}	t_{sn}
1	__________	__________	__________	__________	__________
2	__________	__________	__________	__________	__________
3	__________	__________	__________	__________	__________
4	__________	__________	__________	__________	__________
5	__________	__________	__________	__________	__________

According to earth clocks is the clock on the rocket slow or fast? __________

The first program above can be used to investigate the velocity of a particle as measured by observers moving at various velocities. Suppose the particle has constant velocity u along the x axis of S. If reference frame S is placed so its origin is at the position of the particle at $t = 0$, then the coordinate of the particle is given by $x = ut$. Pick a value for t and calculate x. Now use the program to find x' and t', the coordinate and time in another frame. Finally use $u' = x'/t'$ to compute the velocity in the primed frame.

PROJECT 6. A particle moves along the x axis of reference frame S with a velocity of $0.10c$. Plot the velocity in frame S' as a function of the velocity v of that frame. Consider values from $-0.95c$ to $+0.95c$. Notice that for v close to zero the Galilean transformation is nearly correct. That is, u' is nearly $u - v$. But for v close to the speed of light u' is also close to the speed of light.

If the particle speed is c in frame S then it is c in every frame. Take $u = c$ and find the particle velocity for values of v in the range from $-0.95c$ to $+0.95c$. The answer should be c for every value of v.

Now consider a particle moving along the y axis of S. The y component of its velocity in S' is not the same as the y component in S because $\Delta t' \neq \Delta t$. In addition, the velocity in S' has an x component equal to $-v$. Test these assertions. Take the y component of the velocity in S to be $u_y = 0.10c$ and the x component to be $u_x = 0$. Modify the program to find the components of the velocity and the speed in S' as function of the velocity of that frame. Let $x = 0$ and $y = u_y t$. Use $u'_x = x'/t'$ and $u'_y = y'/t'$ to calculate the components of the particle velocity in S'. Consider values of v from $-0.95c$ to $+0.95c$.

Notice that the y' component becomes smaller as the speed of S' approaches the speed of light. This is because $\Delta t'$ becomes larger. In the limit as $v \to c$, $u'_y \to 0$ and $u'_x \to c$. The particle moves along the x' axis at the speed of light.

As the speed of a particle with mass increases toward the speed of light its kinetic energy increases and, for speeds near the speed of light, the increase is dramatic. According to the defining equations, $\mathbf{p} = m\mathbf{u}/(1 - u^2/c^2)^{1/2}$ and $E = mc^2/(1 - u^2/c^2)^{1/2}$, both the energy and momentum become infinite as u approaches c. The ratio of the magnitude of the momentum to the energy, however, does not blow up but approaches $1/c$ in the limit as the speed approaches the speed of light. When pc is much greater than mc^2 then $E \approx pc$.

The first program of this chapter can be modified to plot the energy and the magnitude of the momentum as functions of the particle velocity. Consider a particle that moves along the x axis. Here's an outline.

> input mass, first velocity, last velocity, and increment: m, u_i, u_f, Δu
> set $u = u_i$
> **begin loop** over intervals
>   exit loop if $|c - u|/c < 10^{-5}$
>   calculate constant α: $\alpha = \sqrt{1 - u^2/c^2}$
>   calculate p: $p = mu/\alpha$
>   calculate E: $E = mc^2/\alpha$
>   display u, p, E, p/E
>   increment u by Δu
> **end loop** if $u > u_f$
> stop

PROJECT 7. A 6.5×10^{-29}-kg particle travels along the x axis. Use the program to make separate plots of the energy, momentum, and ratio p/E as function of the particle speed. Plot points every $0.05c$ from $u = 0$ to $u = 0.95c$. Mark the value of the rest energy on the energy graph.

At slow speeds the energy is nearly the rest energy. Added to this is the kinetic energy, which increases as the square of the speed. Close to the speed of light the energy increases more rapidly with the speed of the particle and becomes infinite at $u = c$. The momentum starts at zero and increases linearly with the speed of the particle. At relativistic speeds it increases more rapidly and becomes infinite at $u = c$. The ratio p/E is small near $u = 0$ since p is small. It increases since E is nearly constant and p increases. In the relativistic region, p/E approaches $1/c$ as a limiting value.

Newton's second law is valid in the form $\mathbf{F} = d\mathbf{p}/dt$, but emphatically not in the form $\mathbf{F} = m\mathbf{a}$. If the force is given as a function of time then $\mathbf{p} = \mathbf{p}_0 + \int_0^t \mathbf{F}\,dt$ can be used to find the momentum at time t. The velocity $\mathbf{u}$ of the particle is given in terms of its momentum by $\mathbf{u} = \mathbf{p}c/(m^2c^2 + p^2)^{1/2}$ and the position vector is given by $\mathbf{r} = \mathbf{r}_0 + \int_0^t \mathbf{u}(t)\,dt = \mathbf{r}_0 + \int_0^t [\mathbf{p}c/(m^2c^2 + p^2)^{1/2}]\,dt$.

Consider motion along the x axis and suppose the momentum is given as a function of time. Write a program, essentially a modification of the Simpson's rule program of Chapter 7, to compute the velocity and position as functions of time. Take the initial position to be at the origin. Let Δt_p be the display interval, let N be the number of integration intervals used in each display interval, and let t_f be the final time. Here's an outline.

> input mass and initial velocity: m, u_0
> input display interval, number of integration intervals, final time: Δt_p, N, t_f
> replace N with nearest even integer
> calculate integration interval: $\Delta t = \Delta t_p/N$
> initialize quantity to hold sum of values with even labels: $S_e = 0$
> initialize quantity to hold sum of values with odd labels: $S_o = 0$
> set $t = 0$
> calculate momentum at time $t = 0$: p_0
> calculate velocity at time $t = 0$: $u_0 = p_0c/(m^2c^2 + p_0^2)^{1/2}$

begin loop over display intervals
begin loop over integration intervals: counter runs from 1 to $N/2$
calculate momentum at time t: p
calculate velocity at time t: $u = pc/(m^2c^2 + p^2)^{1/2}$
add velocity to sum of values with even labels: replace S_e with $S_e + u$
increment t by Δt
calculate momentum at time t: p
calculate velocity at time t: $u = pc/(m^2c^2 + p^2)^{1/2}$
add velocity to sum of values with odd labels: replace S_o with $S_o + u$
increment t by Δt
end loop over integration intervals
calculate momentum at time t: p_N
calculate velocity at time t: $u_N = p_Nc/(m^2c^2 + p_N^2)^{1/2}$
evaluate integral: $x = (\Delta t/3)(u_N - u_0 + 2S_e + 4S_o)$
display t, u_N, x
if $t \geq t_f$ **end loop** over display intervals
stop

PROJECT 8. A 7.6×10^{-25}-kg particle starts from rest and is acted on by a constant force of $F = 4.1 \times 10^{-18}$ N, in the positive x direction. Use the program to plot the velocity u as a function of time from $t = 0$ to $t = 200$ s. Use $\Delta t_p = 5$ s and $N = 30$. Use $p = Ft$ to calculate the momentum.

Although the momentum increases uniformly the velocity does not. At first, when $p \ll mc$, $u \approx p/m$ and the velocity does increase uniformly. But later, when p is a significant fraction of mc or larger, u changes much more slowly. In fact it approaches the speed of light as a limit. It cannot increase beyond that limit no matter how large the momentum becomes. To verify that the acceleration decreases as the speed increases, use your data to estimate the average acceleration in the first and last 5 s intervals. [ans: 5×10^6 m/s^2; 1×10^5 m/s^2]

On the same graph plot the velocity as a function of time using the classical kinematic relation $u(t) = (F/m)t$, where F is the force. Estimate the time for which the velocity deviates from the classical approximation by 10%. [ans: 25 s]

Now plot the coordinate x for the same time interval. Notice that at first the curve has the shape predicted by classical kinematics: $x = \frac{1}{2}(F/m)t^2$ but later it becomes nearly a straight line with slope equal to c.

No matter how large the force becomes the qualitative results are the same: it cannot accelerate the particle beyond the speed of light. Take the force to be $4.1 \times 10^{-20}t^2$ N, where the time t is in seconds. Use the program to plot the velocity u as a function of time from $t = 0$ to $t = 50$ s.